QUANTUM SCIENCE

METHODS AND STRUCTURE

A Tribute to Per-Olov Löwdin

PER-OLOV LÖWDIN

QUANTUM SCIENCE
METHODS AND STRUCTURE

A Tribute to Per-Olov Löwdin

Edited by

Jean-Louis Calais and Osvaldo Goscinski
University of Uppsala
Uppsala, Sweden

Jan Linderberg
Aarhus University
Aarhus, Denmark

and

Yngve Öhrn
University of Florida
Gainesville, Florida

SPRINGER SCIENCE+BUSINESS MEDIA, LLC

Library of Congress Cataloging in Publication Data

Main entry under title:

Quantum science.

Includes index.

1. Quantum theory—Addresses, essays, lectures. 2. Löwdin, Per-Olov, 1916- I. Löwdin, Per-Olov, 1916- II. Calais, Jean-Louis.

QCI74.125.Q37 530.1'2 76-21354

ISBN 978-1-4757-1661-0 ISBN 978-1-4757-1659-7 (eBook)
DOI 10.1007/978-1-4757-1659-7

Originally published by Plenum Press, New York in 1976
Softcover reprint of the hardcover 1st edition 1976

Preface

A "Festschrift" volume fulfils a more far-reaching purpose than the laudatory one. It shows how science develops as a result of the activities - scientific and organizational - of an individual person. Scientific achievement cannot be subjected to the very refined measurement techniques of science itself, but there is a continuous mutual evaluation among scientists which manifests itself through refereeing, literature citation and dedicatory volumes like the present one. Near and distant associates of Per-Olov Löwdin were enthusiastic about the idea of a tribute to him in the form of a collection of scientific papers on the occasion of his sixtieth birthday.

Monographs and journals have fairly well-defined readerships. This book is directed to a wider group of scientists. It presents reviews of areas where Löwdin's work has influenced the development as well as research papers with original results. We feel that it can serve as a source on the current status of the quantum theory of matter for scientists in neighbouring fields. It might also provide stimulus for renewed scientific efforts among scientists turned administrators and will certainly be relevant for teachers and students of quantum theory.

The organization of the contributions attempts to delineate the present frontiers in the research areas where students and associates of Löwdin are working to implement his ideas or to develop new concepts in the reference frame he has provided. The first contribution is an appraisal and personal account of Per-Olov Löwdin from a close associate, and it is followed by a list of publications, and an analysis of the impact in the scientific literature. We have also found it appropriate to present here some previously unpublished material from the Summer Institute in Vålådalen, where Löwdin and others made remarks of definite significance also some twenty years later. This is an example of the scientific climate which Löwdin creates twice a year in his unprecedented schools, workshops and symposia. There are some six contributions that are concerned with the overlap problem as it appears in molecules and solids.

Related to these are three contributions to the theory of cohesion in crystals. Löwdin's unique contributions in the development of numerical procedures is stressed by Professor Slater and other contributors, while several formal aspects have been treated by others. Perturbation theory has been a long time interest with Löwdin and is represented here by four papers, which in turn lead to the section on the electronic structure of atoms and molecules where both general and specific approaches to the correlation problem appear. The six contributions that end the volume are centered around biological applications of the theory and particularly hydrogen bonds.

The editors have been fortunate to be able to draw assistance from many interested parties and thanks are due to them. Anders Fröman and Kimio Ohno have offered invaluable help and advice to us. The preparation of the photoready copy was in the hands of the indefatigable Mrs. Hanne Kirkegaard. Her fast and competent typing secured the completion of the volume in a minimum of time. The publishers have also supported the project beyond the pure business aspects.

The emphasis on clear concepts, "decent" manipulations and the stubbornness not to have the easy way out are trademarks of Per-Olov Löwdin. This has been and will be a continuous source of inspiration to very many.

It is a privilege to express our warmest wishes to him on behalf of all his students, friends and colleagues.

The Editors

Contents

Per-Olov Löwdin - Conqueror of Scientific, Educational and Rocky Mountains
Kimio Ohno 1

Publications of Per-Olov Löwdin 1939-1976................. 13

Per-Olov Löwdin in the Scientific Literature 1965-1974
Rolf Manne 25

Acta Vålådalensia Revisited:
Per-Olov Löwdin in Scientific Discussion
Anders Fröman and Osvaldo Goscinski.................. 33

The Non-Orthogonality Problem and Orthogonalization Procedures
Giuseppe del Re...................................... 53

Biorthonormal Bases in Hilbert Space
Luc Lathouwers....................................... 77

Energy Weighted Maximum Overlap (EWMO)
Jan Linderberg, Yngve Öhrn and Poul W. Thulstrup...... 93

The Calculation of the Exchange Parameter $J = \frac{1}{2}(E_{singlet} - E_{triplet})$ for Two Equivalent Electrons using Canonical Molecular orbitals
R.D. Harcourt.. 105

Importance of Overlap in the Analysis of Weak Exchange Interactions by Perturbation Methods
R. Block and L. Jansen............................... 123

Inelastic Scattering of Photons from Ionic Crystals and Effects of Overlap
K.-F. Berggren....................................... 141

Some Comments on the Quantum-Mechanical Treatment of Defects in Ionic Crystals
Jean-Louis Calais.................................... 171

Properties of Compressed Atoms from a Spherical Cellular Model
Anders Fröman.......... 179

Static and Dynamic Correlations in Many-Electron Systems
Stig Lundqvist.......... 201

Power Series Method for Cellular Calculations on Atoms, Molecules and Solids
J.C. Slater.......... 215

Test of Conventional Quantum Chemistry Methods on the Hydrogen Atom
T.L. Bailey and J.L. Kinsey.......... 249

Numerical Aspects of Weyl's Theory
Michael Hehenberger.......... 265

Quantization and a Green's Function for Systems of Linear Ordinary Differential Equations
Harold V. McIntosh.......... 277

On Resonant Potential Scattering
M. Berrondo and G. García-Calderón.......... 295

Laguerre Polynomials, Reminiscences from Uppsala
L.B. Rédei.......... 305

Partitioning Technique for Determinantal Equations
Frank Weinhold.......... 307

Lower Bounds to Energy Eigenvalues
Charles E. Reid.......... 315

Bounds to the Sum-rule Function from Inner-Projections
Ragnar Ahlberg.......... 349

Investigations into the Properties of Projected Spin Functions
Ruben Pauncz.......... 357

Many-Body Theory of Molecular Collisions
David A. Micha.......... 367

On the Löwdin Bracketing Function
Erkki Brändas.......... 381

Numerical Infinite-Order Perturbation Theory
Rodney J. Bartlett and David M. Silver.......... 393

Calculation of the Bromine Nuclear Pseudoquadrupole Coupling Constant in the LiBr Molecule Using a Density-of-States Function Deduced from Overlap Integrals
H.B. Jansen and P. Pyykkö.......... 409

On Inversion Symmetry in Momentum Space
Per Kaijser and Vedene H. Smith Jr.................... 417

Bonding Character of Inner-Shell Orbitals in Diatomic Molecules
Osvaldo Goscinski.................................... 427

A New Formulation of the Correlation Problem
G.G. Hall.. 433

The Chemical Bond as a Many-Electron Problem
Raymond Daudel....................................... 445

Orbital Methods and Correlation Errors in Expectation Values
S. Larsson and R.E. Brown............................ 459

Long-range Interaction in some Two-Electron Systems
W. Kolos... 465

The Generator Coordinate Method Illustrated on the Hydrogen Molecule
Bernard Laskowski.................................... 479

Projected Hartree-Fock Calculations on the Ground and First Excited $^1\Sigma_g^+$ States of the Hydrogen Molecule
G. Howat and Sten Lunell............................. 491

Deformed Atoms and the Projected Hartree-Fock Method
P. van Leuven.. 499

MCSCF Studies of Chemical Reactions: Natural Reaction Orbitals and Localized Reaction Orbitals
Klaus Ruedenberg and Kenneth R. Sundberg............. 505

The Phosphate Group in Quantum Biochemistry
Alberte Pullman and Bernard Pullman.................. 517

Towards the Theoretical Determination of the Conformation of Biological Macromolecules
János J. Ladik....................................... 539

Energetics and Mechanism of 2-Aminopurine Induced Mutations
Robert Rein and Ramon Garduno........................ 549

External Electrical Field and Proton Transfer
H. Chojnacki... 561

Proton Tunnelling in DNA Base Pairs and Mutagenesis
Suheil F. Abdulnur................................... 567

On Proton Mobilities in Individual Hydrogen Bonds
Mark A. Ratner and J.R. Sabin........................ 577

Index.. 589

PER-OLOV LÖWDIN - CONQUEROR OF SCIENTIFIC, EDUCATIONAL AND ROCKY MOUNTAINS

Kimio Ohno

Department of Chemistry, Faculty of Science, Hokkaido University, Sapporo 060, Japan

FOREWORD

I happened to be the first Japanese scientist to meet Per Löwdin on Japanese soil. In September 1953, an international conference on theoretical physics was held in Japan. Professor Kotani gave me the responsibility of meeting some foreign guests at the Tokyo airport. Thus on one hot summer day, I met Per as well as Professor Waller and Professor Mulliken at the airport. At that time, I did not realize that this was to be the beginning of a long and happy association with Per and his groups.

Four years later, in the summer of 1957, I was working with Roy McWeeny at Keele. Professor Daudel organized an international colloquium on quantum chemistry in Paris in October of that year. I took advantage of attending this colloquium by extending the trip to Uppsala in late September. The quantum chemistry group was busy in preparation for its move from Kemikum to a new building at Rundelsgränd 2A, but Per took very good care of us. I had been able to meet Professor Kotani at the Stockholm station, and we spent five nights in the group's apartment at Luthagesplanaden. The weather was miserable and it even sleeted on one day. Nevertheless I enjoyed my first visit to Uppsala immensely. The group members I remember having met on this occasion include Anders Fröman, Klaus and Marianne Appel and Jean-Louis Calais. Although tickets were bought quite independently in England and Sweden, Per and I were in the same compartment in the continental train from Copenhagen to Paris. We had plenty of time for discussion and when we arrived at Paris, I must confess that I was quite exhausted from the brain labour to which I had been subjected in the train.

The next summer, on my way back to Japan, I visited the Uppsala Quantum Chemistry Group again, at Vålådalen this time. Natural surroundings were beautiful and although I stayed only a week I was greatly impressed by the vigorous scientific activities and warm international friendship of the first, now famous, Summer School.

With this background, it was no wonder that I gave a most enthusiastic "yes" when Professor Kotani passed on to me Per's invitation to Uppsala in May, 1961. The invitation was the opportunity for me to serve as the Assistant Director of the academic year 1961-1962. It was indeed a great honour for me to follow the excellent series of Harrison Shull, Ruben Pauncz and Roy McWeeny.

I stayed in Uppsala for a little more than two years and moved to Gainesville to spend a year with the Quantum Theory Project. Through the long association with Per, what I learned from him was certainly not limited to science. Mutual understanding of people of different nations is absolutely necessary for peaceful co-existence of different nations and different ideologies. The efforts of Per to this direction is clear - he never gets tired of organizing Summer Schools and Winter Institutes - thus he is not only spreding quantum chemistry among scientists but is also making a definite contribution towards true peace in the world.

To me Per Löwdin is an inspired, great teacher and a close, personal friend. However, in what follows, I try to be impersonal and to present a review of what he has achieved from a scientist's point of view.

I. A BRIEF BIOGRAPHY

Per-Olov Löwdin was born on October 28, 1916, at Uppsala where he also attended school and in 1935 started his studies in natural sciences at the university. He obtained the Swedish degree of Filosofie licentiat in 1942 and the same year he became a lecturer in Mechanics and Mathematical Physics at Uppsala University. During a six months' visit to Pauli's group in 1946 he studied some problems in quantum electrodynamics, a field appropriate to his keen interest in mathematics. The influence of Pauli's crystal clear thinking is perceivable in Löwdin's later work, writing and teaching. In 1948 he presented his doctoral thesis devoted to quite a different subject, namely a theoretical investigation of ionic crystals. He was appointed Docent at the Uppsala University and the following year he spent five months with N.F. Mott in England continuing his solid-state research.

In 1950-51 he worked in USA together with Hertha Sponer at Duke University, Robert S. Mulliken at Chicago and John C. Slater at MIT and he attended the 1951 Shelter Island Conference. This visit to USA seems to have greatly influenced Löwdin's research interest. With a vigour and capability seldom found in a single person he was instrumental in establishing quantum chemical research in Sweden. In 1955 he organized together with Inga Fischer-Hjalmars the Stockholm-Uppsala symposium and also launched the Quantum Chemistry Group with himself as a captain. During 1955 to 1960 he held a Laboratorship at the Swedish Natural Sciences Research Council and on May 13, 1960, he was appointed the first professor of quantum chemistry at the Uppsala University. This formal recognition of quantum chemistry in the Swedish University system represents a milestone in Löwdin's effort in strengthening this field of research. Not content with this achievement he found an outlet for his never-failing enthusiasm and energy in activities on the international level. He has been and still is the organizer and driving force in numerous Summer and Winter Institutes, he serves on the editorial board of several international journals, and by extensive travelling he has made the whole world his lecture room. In 1960 he started the Quantum Theory Project at the University of Florida where he is Graduate Research Professor of Physics and Chemistry. He became a member of the Royal Swedish Academy of Sciences in 1969 and now serves on the Academy's Nobel Committee for physics.

For an active member of the international science community a brief note like this one has to be incomplete. But after all, the important aspects of a scientist's work are his contributions with scientific ideas, methods and concepts leading to an increased knowledge of nature. This is the theme of the following sections.

II. CRYSTAL PHYSICS

The first major scientific papers by Löwdin contain an investigation of the ionic crystals (2, 3, 4). This topic was suggested by his teacher Waller, who apparently had in his mind an analysis of the failure of the Cauchy relations and a calculation of the lattice dynamics. The main emphasis of Löwdin's work, however, turned out to be on cohesion and high pressure properties. The calculation was big and difficult. A number of mathematical and numerical techniques, particularly in connection with the overlap and non-orthogonality problem, had to be developed. In addition, he stood in need of the atomic orbitals and a great number of molecular integrals - overlap, kinetic, Coulomb, exchange, hybrid, and many-center - defined over these atomic orbitals. The numerical work constituted a feat. It was carried out on primitive, electric

FACIT desk-calculators by a group of students. Some of them later became professors. When Mulliken visited Uppsala and was introduced to the students working in the basement at Trädgårdsgatan, his opening remark was "So this is the famous student computer". Learning to appreciate proper numerical techniques is a good training for scientists-to-be. Löwdin handled this part of his work with elegance and efficiency and he cultivated a lasting interest in practical calculations.

The calculation on ionic crystals, a landmark in ab-initio calculations, was carried out before the advent of electronic computers and it is, indeed, the first successful ab-initio calculation on crystals. The following quotations may suffice to show how the calculations were evaluated. Kittel wrote in his "Introduction to Solid State Physics" (1st ed., 1953), that "The most basic quantum-mechanical discussion of ionic crystals has been made by Löwdin". Similarly in "Perspectives in Material Research" (1963) the committee chaired by Brooks stated; "The most satisfactory cohesive energy calculation is probably that carried out for the alkali halides by Löwdin". When writing on the development of solid state physics (Intern. J. Quantum Chem. 1, 37 (1967)) Slater mentioned the stimulus resulting from the invention in 1948 of the transistor and also Löwdin's thesis work which "gave physicists a new hope that some of the problems that seemed too hard before the war might be capable of being handled". These three comments are certainly very unusual appraisals of a doctoral thesis.

After the work on ionic crystals,the general energy formula for the ground state of molecules and crystals in the MO-LCAO scheme was further developed (14). He then embarked on the adventure of applying this technique to metallic sodium (15). This required handling of much larger overlap than the one in ionic crystals. The calculation was made at the restricted Hartree-Fock level and it gave a good result for the cohesive energy. It was clear that there was some cancellation of errors since the correlation energies of the crystal and the free atoms would not cancel each other for the alkali metals. This interested him in the correlation problem in general which became one of his main research interests.These kinds of ab-initio calculations were certainly very time-consuming and tedious. He always insists on a clear connection with the Schrödinger equation, and he dislikes to neglect complicated terms by assuming,without check, that they are not important. These are among his outstanding scientific characteristics.

In 1956 Löwdin published an often quoted review paper on the quantum theory of cohesive properties of solids (33). Many of the ingenious techniques developed earlier then became available to a wider public. The readers were also confronted with projection operators and density matrices, two of the main fields of Löwdin's research interests at that time. During later years he published papers on cohesive properties (59, 89, 119) and on correlation and exchange phenomena in solids (56, 57, 116, 117), but these papers do not appear to have had such a great impact as the early work on ionic crystals.

III. EARLY WORK ON MOLECULES

Löwdin treated crystals as large molecules, therefore his method was directly applicable to molecules. He developed a method of fitting the numerical Hartree Fock atomic orbitals with analytic functions (20, 21, 37). This was a continuation of Slater's pioneering study and was followed by works of Watson, Roothaan, Clementi and others. Molecular integrals were evaluated by expanding the atomic orbitals at various centers in spherical harmonics around one and the same center (2,3,4,10,25,33). The same basic idea was applied to integrals defined over Slater-type orbitals by Barnett and Coulson.

Among his work related to molecules, the Löwdin orthogonalization (2, 3, 4, 9, 33) has probably the greatest influence on the present day molecular calculations. By using the orthogonal atomic orbitals obtained by his transformation, he showed that there is no "non-orthogonality catastrophe", although the overlap integrals are of essential importance (19). These orthogonal orbitals are important in the Pariser-Parr-Pople, CNDO, INDO, and many other NDO schemes. One of the basic assumptions in these schemes is the neglect of differential overlap. This seemingly crude procedure can be justified, to some extent at least, by assuming that the basic orbitals are Löwdin orthogonalized AO's. Löwdin pointed this out in a letter to Parr dated December 3rd, 1952 (cf also 30).

Correlation effects are important also in molecules. At the Nikko Symposium on Molecular Physics in 1953, he proposed a simple, intuitive but ingenious method to deal with the effect (24 see also 28, 29, 33, 44). This is the alternant-molecular-orbitals (AMO) method. In an alternant system, the electrons having different spins may accumulate on different subsystems and their separation may be regulated by one or more parameters to be determined by the variational principle. This is an example of using different orbitals for electrons with different spins. The AMO method has been successfully applied to conjugated systems by Löwdin himself (60, 61), Itoh, Yoshizumi, Pauncz, de Heer and others and to the alkali metals by Calais and collaborators. At the same time, the AMO method opened the door to a number of important theoretical questions. A single Slater determinant with different orbitals for different spins usually contains a set of different spin multiplets. The question

No 1 is what are the weights of these multiplets. The question No 2 is what we should do if we want to recover a pure spin multiplicity. These questions were duely solved mainly by Löwdin (28, 42), and his collaborators.

IV. UPPSALA QUANTUM CHEMISTRY GROUP AND QUANTUM THEORY PROJECT IN GAINESVILLE

In 1955, the Uppsala Quantum Chemistry Group was founded by Löwdin. In the beginning the Group had its office in Kemikum overlooking the English Park with the playing squirrels. In autumn of 1957 the Group moved to the present modern building at Rundelsgränd. The international character of the Group was distinct from the beginning. Among the first members, we find Roberto Fieschi and Hiroyuki Yoshizumi in addition to the Swedes; Anders Fröman, Klaus and Marianne Appel. Fröman recalls the following episode. One day at a formal lunch at Flustret a university official commented that he found it most remarkable that an Italian could work in Sweden as a Texas-Swedish Cultural Foundation Fellow.

By that time Löwdin had done extensive travelling abroad. He attended a large number of international conferences held in Europe, U.S.A. and Japan. He was bringing back ideas and news as well as distinguished scientists to Uppsala. When the first Summer School was held at Vålådalen in 1958, Pauling, Mulliken, Matsen, Preuss, Jansen, Shull, Hall, and McWeeny were among the visitors and guest lecturers.

The Summer Schools, which have been held almost every year since 1958, are now well-known by their heavy schedule - both intellectual and physical - and the warm international atmosphere. Mountain climbing has been an important ingredient of the school's physical activities in recent years.

The Quantum Theory Project was established in the winter of 1960 at the University of Florida. At the very beginning a large part of the staff consisted of visiting scientists from the Uppsala Quantum Chemistry Group. The Project has developed rapidly and it has now four professors, four associate professors and one visiting associate professor.

Corresponding to the Summer Schools, the Winter Institutes have been held each winter since 1961. International Symposia in Atomic, Molecular, Solid-State Theories have also been organized following the Institute at the Sanibel Island. The Symposium in 1963 was dedicated to E.A. Hyllerass for his pioneering work on the atomic and molecular theory. Since then R.S. Mulliken (1965), J.C. Slater (1967), H. Eyring (1969), J.H. Van Vleck (1971), E.U.

Condon (1973), and L.H. Thomas (1975) were honored in the odd year symposia.

Counting both Summer Schools, Winter Institutes and Sanibel Symposia, there must be between 4000 and 5000 scientists from 35-40 different countries all over the world, who have participated in these activities. There have been students of students of students and so on, participating so that academically fifth or sixth generation students are now being sent to these institutes. It is clear that these schools and symposia, which were organized by Per-Olov Löwdin, have had great impact on quantum chemistry and related fields. By bringing together people from all over the world, these meetings also encouraged the mutual understanding and generosity for the building of a better world which is a deep-rooted concern of Löwdin.

V. QUANTUM THEORY OF MANY PARTICLE SYSTEMS

Löwdin published a series of three papers on the subject in the Physical Review in 1955 (26, 27, 28). In these papers a number of important concepts and methods concerning the many body problem was proposed and discussed. First of all the use of reduced density matrices in analysing general wave functions was advocated. Although there had been several authors who had explicitly or implicitly used these concepts, he and McWeeny were the first quantum chemists who emphasized their use and potentialities in the quantum theory of many body problems at absolute zero temperature. Their observation that the energy can be determined from the reduced 2-matrix alone evoked a long series of sophisticated researches to obtain 2-matrices directly without going through wave functions. These include the N-representability problem and general mathematical structure of 1- and 2-matrices.

Natural spin orbitals are a genuine Löwdin concept. They are defined as the orbital bases in which the 1-matrix has a diagonal form. It is argued that they lead to a configuration interaction expansion of most rapid convergence. While the knowledge of the wave function is required, the general shape and occupation number characteristics of the natural orbitals derived from different approximate wave functions enable us to make a meaningful comparison which otherwise would elude us. The earliest example of this are found in Löwdin and Shull's papers on He and H_2 (32, 34, 35, 46). The natural spin orbitals thus perform a function very similar to that the charge and bond order matrix serves in a LCAO-MO calculation. An iterative scheme using the natural spin orbitals are also widely used in numerical computations to speed up the convergence of CI expansions. There is again a great number of papers concerning properties of natural spin orbitals and natural orbitals. Their symmetry properties and various modifications are discussed and utilized in a wide range of the theory of many electron systems.

In the series of Phys. Rev. papers, Löwdin proposed two extensions of the restricted Hartree-Fock (RHF) approximation. One is the extended Hartree-Fock method and the other is the projected Hartree-Fock method. The former is a precursor of the multi-configuration SCF method and he derived the equation, which determines the optimal orbitals, in terms of the density matrix. The latter started from an optimal single determinant, not RHF, which in general does not have symmetry. The symmetry properties were reinstated by applying appropriate projection operators. The projected HF method goes beyond the RHF limit but it preserves the physical simplicity and visuality of RHF. Löwdin insisted that we should not look at a projected determinant as a sum of determinants, but as an entity, just as we think of the determinant as an entity and not as a sum of products. The AMO method which was mentioned in section II is an example of the projected HF method.

Projection operators are one of the most used tools in his work. The use is made in two ways. One way is its use in actual calculations such as in the construction of eigenfunctions of spin and space angular momenta (42, 50, 78) and for the construction of symmetry functions for finite groups (62, 92). The other way is more abstract. That is an introduction of the concept "projection on a manifold", which was used extensively in the development of the perturbation theory and the theory of bounds (cf section VI).

It was already mentioned that one important factor which was missing in his Na metal calculation was correlation between electrons and that the correlation problem became a center of his research interests and is probably so even now. He wrote a treatise on the subject which was published in 1959 (44). In this article he proposed the definition of the correlation energy E_{corr}, that is

$$E_{corr} = E_{exact} - E_{RHF},$$

where E_{exact} is the non-relativistic exact energy of the system and E_{RHF} is the energy obtained by the RHF approximation. The correlation energy has now become one of the most important concepts in quantum chemistry. It has a tremendous practical value too. Being combined with Nesbet and Sinanoğlu's idea of the pair correlation energy, it can give a fair estimate of E_{exact} for most systems from the RHF calculation.

The Virial theorem is a unique relation in that it holds rigorously for the exact wave function and also for any scaled approximate wave function. He has emphasized its importance and has tried to make every use of it (45, 57, 59, 111, 116, 117).

V. PERTURBATION THEORY

There is another long series of papers by Löwdin entitled "Studies in Perturbation Theory" (65, 73, 74, 63, 64, 75, 76, 77, 80, 81, 82, 87, 99 and 115). The starting point of the whole approach is older and is found in a paper published in 1951 (13), in which the solution of the eigenvalue problem by means of "partitioning" was proposed. The technique has been reformulated in terms of two associated variables ε and ε_1, defining a "bracketing function" $\varepsilon_1 = f(\varepsilon)$. This has the property that, between the two real numbers ε and ε_1, there is at least one eigenvalue E. The bracketing theorem offers successful methods for solving eigenvalue problems and also constitutes the basis for bounding the eigenvalues. The very significant concept of a reduced resolvent was introduced. A meticulous study of the properties of the reduced resolvents was made and various approximations for the resolvents were put forward. The strongest tool was the inner projection. It has the form $A' = A^{\frac{1}{2}}OA^{\frac{1}{2}}$ where A is a positive definite operator and O is a projection operator. It can be shown that the expectation values of A' are smaller or equal to those of A. By appropriate choice of O, e.g. the projection on a linear manifold, a number of interesting applications to concrete problems were made. For example, a number of numerical calculations of lower bounds were carried out in the Florida Quantum Theory Project and the Uppsala Quantum Chemistry Group.

The perturbation theory has been extended in many directions. Simple and unified treatments of degenerate eigenvalues and linearly dependent reference functions were developed. Bounding of physical properties in terms of others and the reduction of the Dirac equation to the Pauli form were accomplished. Contact with Padé approximant theory and continued fractions was clarified. The exact self-consistent field theory was proposed and the symmetry-adapted perturbation theory was established. The idea of inner-projections has had an impact in the field of propagator theory when discussing decoupling schemes.

These developments are rather formal and may seem too abstract. However, there are singular merits in Löwdin's persistence in having an exact, closed form, as a basis for physical and also semi-empirical discussions. The elegant and powerful formalism makes it feasible to compare in detail most of the various, often bewildering, perturbation methods in quantum mechanics. Being written in a typically Löwdin style - stringent and transparent reasoning, logical sequence, natural language - this series on perturbation theory is the best guide for anyone who wishes to understand the essentials of modern theory in this field of study.

VI. PROTON TUNNELING IN DNA

Löwdin became interested in biophysics rather suddenly around 1961. Biophysics was an entirely new field for him. With his characteristic energy and concentration, he read through books and papers and listened to lectures and seminars. Two years later in 1963, he proposed the theory of proton tunneling in DNA and then discussed its possible consequences (66, 68, 70, 72, 79, 83, 86, 88, 94, 98, 101, 107). Based upon the Watson-Crick model of DNA, he pointed out that there is a certain intrinsic probability of a proton movement in DNA. Namely the proton in one or more of the hydrogen bonds between a base pair changes its position in time from the most favourable position to the next most favourable position. This spontaneous shift of the position, which is characteristic and inherent to a quantum mechanical particle, transforms a base to its tautomeric form. The tautomeric form can make a pair only with a base different from the normal partner. This would cause an error in the genetic code and accumulated errors of this kind could be responsible for mutation, aging and spontaneous tumors. At the present time, as far as the writer of this article is aware, this interesting hypothesis has been neither proved nor disproved. It remains to be seen. There is no doubt, however, that the proposal is a major philosophical contribution in that it brings together the concept of quantum mechanics and the question of mutation. It is also clear that this bold attempt provided a great stimulus to those scientists interested in quantum biology.

VII. THE SCIENTIST AND THE TEACHER

One of the roots of Löwdin's scientific strength is undoubtedly his mastery of the mathematical "craft". This, in addition to his background and experience, makes it possible to "see through" a paper or a seminar very quickly and to put his finger on a "sore point" if there is any. This also gives him the ability to translate different scientific languages to a unified one. His insistence of finding good notations is closely connected with this. Other roots of his scientific strength, which are perhaps more basic, are his enthusiasm and devotion to research and his enormous working capacity. In the winter of 1963, when he was working on the lower bound of the energy and got the idea of inner-projections, he continued to give two seminars per week for a few months on the same subject. Each seminar contained one ore more new steps towards his goal and it was fascinating to watch an evolution of the new theoretical approach to this difficult problem.

In all his scientific work, there is a very obvious desire to present the material in a pedagogical fashion. Per Löwdin is undoubtedly an outstanding scientist but as a teacher he is also uniquely successful. He has an unusual ability to enthusiastically lead and work with younger colleagues and never get tired of teaching his methods and mode of thinking. As as teacher, he is clear and deceivingly simple in his approach, which allows him to present even very sophisticated ideas to beginning students. He likes and always strives for "the one-line proof" and a short and elegant way to a result is made as important as the result itself. Thus comes his insistence on clear notations, simple algebra, and a complete bibliography.

Another activity of his, which should be mentioned, is his extensive editorial commitments. He is the chief editor of the International Journal of Quantum Chemistry and the Advances in Quantum Chemistry. He has also been serving as a member of the advisory Editorial Board of a large number of international journals. He has edited several books including "Molecular Orbitals in Chemistry, Physics, and Biology - Mulliken Dedicatory Volume - " and "Quantum Theory of Atoms, Molecules, and the Solids - Slater Dedicatory Volume - ".

Per Löwdin's name is closely associated with quantum chemistry. Quantum chemistry, however, is perhaps a too narrow concept to use in describing his scientific work, which has had a much wider impact. He is and will be remembered by his colleagues, friends and students as an excellent, original scientist, an inspiring and tireless teacher, and a warm and sympathetic person combined in one.

ACKNOWLEDGEMENT

The article would never have been written without the help from Anders Fröman, Jean-Louis Calais, Jan Linderberg, Yngve Öhrn and Osvaldo Goscinski to whom the author expresses his sincere thanks and great indebtedness. Nikolaj Stepanov has read a part of the manuscript and has offered a number of useful comments which are very much appreciated. He stayed with our little group in Sapporo recently and this is an example of many products produced by the catalytic activity of Per Löwdin in the international cooperation.

PUBLICATIONS OF PER-OLOV LÖWDIN

1939 - 1976

1. Lorentz-Transformationerna och den Kinematiska Relativitetsprincipen (Elementa 22, 161 - 169, 1939)

2. A Quantum Mechanical Calculation of the Cohesive Energy, the Interionic Distance, and the Elastic Constants of Some Ionic Crystals. I. (Ark. Mat., Astr., Fys. 35A. No 9, 1 - 10, 1947)

3. A Quantum Mechanical Calculation of the Cohesive Energy, the Interionic Distance, and the Elastic Constants of Some Ionic Crystals. II. The Elastic Constants c_{12} and c_{44}. (Ark. Mat., Astr., Fys. 35A, No 30. 1-18, 1948)

4. A Theoretical Investigation into Some Properties of Ionic Crystals (Thesis, Almqvist & Wiksells, Uppsala, 1948)

5. A Quantum Mechanical Calculation of the Cohesive Energy, the Interionic Distance, and the Elastic Constants of Some Ionic Crystals (Reports from the Conference of the Swedish National Committee for Physics in 1947. Ark. Mat., Astr., Fys. 34A, No 29, 18-19, 1948)

6. On the Occurrence of Many-Body-Forces in Molecules and in Crystals. (Reports from the conference of the Swedish National Committee for Physics in 1948. Ark. Fysik 1, 543, 1949)

7. The Band Theory of Metals and the Importance of the Overlap Integrals. (Reports from the Conference of the Swedish National Committee for Physics in 1949. Ark. Fysik 2, 220, 1950)

8. A Note on the Method of Steepest Descents with a Remark on T. Ljunggren's Paper "Contributions to the Theory of Diffraction of Electromagnetic Waves by Spherical Particles" (Ark. Fysik 2, 367-370, 1950)

9. On the Non-Orthogonality Problem Connected with the Use of Atomic Wave Functions in the Theory of Molecules and Crystals (J.Chem. Phys. 18, 365-375, 1950)

10. (with S.O. Lundqvist) On the Calculation of Certain Integrals Occurring in the Theory of Molecules, Especially Three-Centre and Four-Centre Integrals (Ark. Fysik 3, 147-154, 1951)

11. (with A Sjölander) A Note on the Numerical Calculation of Asymptotic Phases with a Numerical Study of Hulthén's Variational Principle (Ark. Fysik 3, 155-166, 1951)

12. Calculation of Electric Dipole Moments of Some Heterocyclics (J. Chem. Phys. 19, 1323-1324, 1951)

13. A Note on the Quantum-Mechanical Perturbation Theory. (J. Chem. Phys. 19, 1396-1401, 1951)

14. On the Quantum-Mechanical Calculation of the Cohesive Energy of Molecules and Crystals. I. A General Energy Formula for the Ground State. (J. Chem. Phys. 19, 1570-1578, 1951)

15. On the Quantum-Mechanical Calculation of the Cohesive Energy of Molecules and Crystals. II. Treatment of the Alkali Metals with Numerical Applications to Sodium. (J. Chem. Phys. 19, 1579-1591, 1951)

16. On the Methods of Numerical Integration Used in Determining Self-Consistent Fields. (NAS-ONR Report from the Shelter Island Conference in 1951, 187-194, 1951)

17. On The Numerical Integration of Ordinary Differential Equations of the First Order. (Quart. Appl. Math. 10, 97-111, 1952)

18. Approximate Formulas for Many-Center Integrals in the Theory of Molecules and Crystals (J. Chem. Phys. 21, 374-375, 1953)

19. On the Molecular-Orbital Theory of Conjugated Organic Compounds with Application to the Perturbed Benzene Ring (J. Chem. Phys. 21, 496-515, 1953)

20. Studies of Atomic Self-Consistent Fields. I. Calculation of Slater Functions. (Phys. Rev. 90, 120-125, 1953)

21. Studies of Atomic Self-Consistent Fields. II. Interpolation Problems. (Phys. Rev. 94, 1600-1609, 1954)

22. (with H. Sponer) Les Niveaux d'Énergie Électronique dans l'Éthylène (J. Phys. Rad. 15, 607-611, 1954)

23. Recent Simplifications in the Molecular Orbital Theory of Calculating Energy Levels. (Proceedings of the International Conference of the Theoretical Physics at Kyoto and Tokyo, Japan in 1953, 599-611, 1954)

24. A Method of Alternant Molecular Orbitals. (Symposium on Molecular Physics at Nikko, Japan in 1953, 13-16, 1954)

25. Calculations of Molecular Integrals in Uppsala. (Symposium on Molecular Physics at Nikko, Japan in 1953, 113-117, 1954)

26. Quantum Theory of Many-Particle Systems. I. Physical Interpretations by Means of Density Matrices, Natural Spin-Orbitals, and Convergence Problems in the Method of Configurational Interaction. (Phys. Rev. 97, 1474-1489, 1955)

27. Quantum Theory of Many-Particle Systems. II. Study of the Ordinary Hartree-Fock Approximation. (Phys. Rev. 97, 1490-1508, 1955)

28. Quantum Theory of Many-Particle Systems. III. Extension of the Hartree-Fock Scheme to Include Degenerate Systems and Correlation Effects. (Phys. Rev. 97, 1509-1520, 1955)

29. An Extension of the Hartree-Fock Method to Include Correlation Effects (Les Electrons dans les Métaux, Dixième Conférence Solvay, Bruxelles in 1954, 71-88, 1955)

30. (with I. Fischer-Hjalmars) Report From the Symposium on Quantum Theory of Molecules, Stockholm and Uppsala, 1955. (Sv.Kem.Tidskrift 67, 365-398, 1955; esp. 369,370,373,375, 379,380,383)

31. (with H. Shull) Role of the Continuum in Superposition of Configurations. (J. Chem. Phys. 23, 1362, 1955)

32. (with H. Shull) Natural Spin-Orbitals for Helium. (J. Chem. Phys. 23, 1565, 1955)

33. Quantum Theory of Cohesive Properties of Solids. (Adv. Phys. 5, 1-172, 1956

34. (with H. Shull) Natural Orbitals in the Quantum Theory of

Two-Electron Systems (Phys. Rev. 101, 1730-1739, 1956).

35. (with H. Shull) Correlation Splitting in Helium-Like Ions (J. Chem. Phys. 25, 1035-1040, 1956).

36. Electronic Correlation in the Theory of Molecular Energy Levels. (A Report from the Molecular Quantum Mechanics Conference in Austin, Texas in 1955, 30; Texas J. Sci. 8, 163, 1956).

37. (with K. Appel) Studies of Atomic Self-Consistent Fields. III. Analytic Wave Functions for the Argon-Like Ions and for the First Row of the Transition Metals (Phys. Rev. 103, 1746-1755, 1956).

38. Present Situation of Quantum Chemistry (J. Phys. Chem. 61, 55-68, 1957).

39. Den Kovalenta Kemiska Bindningen i Kvantmekanisk Belysning (Elementa 40, 9-24, 1957).

40. Generalizations of the Hartree-Fock Scheme (Ann. Acad. Reg. Sci. Upsaliensis, 2, 127-135, 1958).

41. (with H. Shull) Variation Theorem for Excited States (Phys. Rev. 110, 1466-1467, 1958).

42. Nature des Fonctions de la Mésomérie (Ed. du Centre Nat. Rech. Sci. LXXXII, 23-37, 1958).

43. (with A.J. Freeman) Quantum Mechanical Kinetic Energy Transformation (Phys. Rev. 111, 1212-1213, 1958).

44. Spin Degeneracy Problem (Coll. Int. Centre Nat. Rech. Sci. 82, 23, 1958).

45. Correlation Problem in Many-Electron Quantum Mechanics. I. Review of Different Approaches and Discussion of Some Current Ideas (Adv. Chem. Phys. 2, 207-322, 1959).

46. Scaling Problem, Virial Theorem and Connected Relations in Quantum Mechanics (J. Mol. Spect. 3, 46-66, 1959).

47. (with H. Shull) Superposition of Configurations and Natural Spin Orbitals. Applications to the He Problem (J. Chem. Phys. 30, 617-626, 1959).

48. (with J.O. Hirschfelder) The Long-Range Interaction of Two 1s Hydrogen Atoms Expressed in Terms of Natural Spin Orbitals (Mol. Phys. 2, 229-258, 1959).

49. (with L. Rédei) Combined Use of the Method of Superposition of Configurations and Correlation Factor on the Ground States of the Helium-Like Ions (Phys. Rev. 114, 752-757, 1959)

50. En Iterations-Variationsmetod För Att Lösa Egenvärdesproblem. (Nord-SAM, Karlskrona and Lund 1959, 199-209, 1959)

51. Some Aspects on the Recent Development of the Theory of the Electronic Structure of Atoms. (Proceedings of the Robert A. Welch Foundation Conferences on Chemical Research. II. Atomic Structure, 5-75, 1960)

52. Expansion Theorems of the Total Wave Function and Extended Hartree-Fock Schemes (Rev. Mod. Phys. 32, 328-334, 1960)

53. Quantum Theory of Electronic Structure of Molecules (Ann. Rev. Phys. Chem. 11, 107-132, 1960)

54. (with R. Pauncz and J. de Heer) On the Calculation of the Inverse of the Overlap Matrix in Cyclic Systems (J. Math. Phys. 1, 461-467, 1960)

55. The Principle of Causality, the Chemical Bond and Modern Quantum Chemistry (Ann. Acad. Reg. Sci. Upsaliensis, 5, 63-78, 1961)

56. Note on the Separability Theorem for Electron Pairs (J. Chem. Phys. 35, 78-81, 1961)

57. Band Theory, Valence Bond and Tight-Binding Calculations (J. Appl. Phys. 33, 251-280, 1962

58. Exchange, Correlation and Spin Effects in Molecular and Solid-State Theory (Rev. Mod. Phys. 34, 80-87, 1962)

59. (with J.-L. Calais) A Simple Method of Treating Atomic Integrals Containing Functions of r_{12} (J. Mol. Spect. 8, 203-211, 1962)

60. (with A. Fröman) Virial Theorem and Cohesive Energy of Solids, Particularly Ionic Crystals (J. Phys. Chem. Sol 23, 75-84, 1962)

61. (with R. Pauncz and J. de Heer) Studies on the Alternant Molecular Orbital Method. I. General Energy Expression for an Alternant System with Closed-Shell Structure. (J. Chem. Phys. 36, 2247-2256, 1962)

62. (with R. Pauncz and J. de Heer) Studies on the Alternant Molecular Orbital Method. II. Application to Cyclic Systems.

(J. Chem. Phys. 36, 2257-2265, 1962).

63. The Normal Constants of Motion in Quantum Mechanics Treated by Projection Technique. (Rev. Mod. Phys. 34, 520-530, 1962).

64. Studies in Perturbation Theory. IV. Solution of Eigenvalue Problem by Projection Operator Formalism. (J. Math. Phys. 3, 969-982, 1962).

65. Studies in Perturbation Theory V. Some Aspects on the Exact Self-Consistent Field Theory. (J. Math. Phys. 3, 1171-1184, 1962).

66. Studies in Perturbation Theory. I. An Elementary Iteration-Variation Procedure for Solving the Schrödinger Equation by Partitioning Technique (J. Mol. Spect. 10, 12-33, 1963).

67. Quantum Genetics (Int. Sci. Tech. 17, 64, 1963).

68. Wave and Reaction Operators in the Quantum Theory of Many-Particle Systems; (Rev. Mod. Phys. 35, 702-708, 1963).

69. Discussion on the Hartree-Fock Approximation. (Rev. Mod. Phys. 35, 496-498, 1963).

70. Discussion on Natural Expansions and Properties of the Chemical Bond (Rev. Mod. Phys. 35, 629-630, 1963).

71. Proton Tunnelling in DNA and its Biological Implications. (Rev. Mod. Phys. 35, 724-732, 1963).

72. Effect of Proton Tunnelling in DNA on Genetic Information and Problems of Mutations, Aging, and Tumors. (Bio. Symp. 1, 161-181, 1964).

73. Some Aspects of Quantum Biology. (Bio. Symp. 1, 293-311, 1964).

74. Molecular Orbitals in the Exact SCF Theory. (Molecular Orbitals in Chemistry, Physics, and Biology, Mulliken Dedicatory Volume, Academic Press, Inc., New York, 37-55, 1964).

75. Some Aspects on DNA Replication; Incorporation Errors and Proton Transfer. (Electronic Aspects of Biochemistry, Academic Press, Inc., New York, 167-201, 1964).

76. Studies in Perturbation Theory. II. Generalization of the Brillouin-Wigner Formalism. (J. Mol. Spect. 13, 326-331, 1964).

77. Studies in Perturbation Theory. III. Solution of the Schrödinger

Equation Under a Variation of a Parameter. (J. Mol. Spect. 13, 331-337, 1964).

78. Studies in Perturbation Theory. VI. Contraction of Secular Equations.(J. Mol. Spect. 14, 112-118, 1964).

79. Studies in Perturbation Theory. VII. Localized Perturbation (J. Mol. Spect. 14, 119-130, 1964).

80. Studies in Perturbation Theory. VIII. Separation of the Dirac Equation and Study of the Spin-Orbital Coupling and Fermi Contact Terms. (J. Mol. Spect. 14, 131-144, 1964).

81. Angular Momentum Wavefunctions Constructed by Projection Operators. (Rev. Mod. Phys. 36, 966-976, 1964).

82. Datamaskinupprustning på universitetsområdet. (ULF, organ för Universitetslärarförbundet, 1,1965).

83. Isotope Effect in Tunneling and Its influence on Mutation Rates. (Mutation Research 2, 218-221, 1965).

84. Studies in Perturbation Theory. IX. Connection Between Various Approaches in the Recent Development. Evaluation of Upper Bounds to Energy Eigenvalues in Schrödinger's Perturbation Theory. (J. Math. Phys. 6, 1341-1353, 1965).

85. Studies in Perturbation Theory. X. Lower Bounds to Energy Eigenvalues in Perturbation-Theory Ground State. (Phys. Rev. 139, A357-A372, 1965).

86. Studies in Perturbation Theory. XI. Lower Bounds to Energy Eigenvalues, Ground State, and Excited States. (J. Chem. Phys. 43, S175-S185, 1965).

87. Quantum Genetics and the Aperiodic Solid. Some Aspects on the Biological Problems of Heredity, Mutations, Ageing, and Tumors in View of the Quantum Theory of the DNA-Molecule (Adv. Quant. Chem. II, 1965).

88. Arvsanlagen och Deras Förändringar - Ur kvantgenetisk synpunkt. (Sv. Naturvetenskap, 1965).

89. (with J. O. Hirschfelder) Long-Range Interaction of Two 1s-Hydrogen Atoms Expressed in Terms of Natural Spin-Orbitals. (Mol. Phys. 9, 491-496, 1965).

90. Some Recent Developments in the Quantum Theory of Many-Electron Systems and the Correlation Problem. (Adv. Chem. Phys. 8, 3-4, 1965).

91. Some Aspects of the Biological Problems of Heredity, Mutations, Ageing and Tumours in View of the Quantum Theory of the DNA Molecule. (Adv. Chem. Phys. 8, 177-179, 1965)

92. The Calculation of Upper and Lower Bounds of Energy Eigenvalues in Perturbation Theory by Means of Partitioning Techniques. (Perturbation Theory and its Application in Quantum Mechanics, Ed., C.H. Wilcox, Proceedings of Madison Symposium, 255-294, John Wiley and Sons, Inc., 1966)

93. Quantum Genetics and the Aperiodic Solid. Some Aspects on the Biological Problems of Heredity, Mutations, Ageing and Tumours in View of the Quantum Theory of the DNA Molecule. (Adv. Quant. Chem. 2, 213, 1965)

94. Comments on Professor John C. Slater's Paper, "Cohesion in Monovalent Metals". (Quantum Theory of Atoms, Molecules, Solid State, 15, 1966)

95. The Projected Hartree-Fock Method. An Extension of the Independent-Particle Scheme. (Quantum Theory of Atoms, Molecules, and Solid State, 601, 1966)

96. Program. (Int. J. Quant. Chem. 1, 1-6, 1967)

97. Nature of Quantum Chemistry. (Int. J. Quant. Chem. 1, 7-12, 1967)

98. Group Algebra, Convolution Algebra, and Applications to Quantum Mechanics. (Rev. Mod. Phys. 39, 259-287, 1967)

99. Quantum Theory of Time-Dependent Phenomena Treated by the Evolution Operator Technique. (Adv. Quant. Chem. 3, 323-381, 1967)

100. Some Properties of the Hydrogen Bonds in Biochemistry with Particular Reference to the Stability of the Genetic Code. (Pontificiae Academiae Scientiarum Scripta Varia 31, "Semaine d'Etude sur les Forces Moléculaires", 637-708, 1967)

101. Eigenvalue Problem in a Linearly Dependent Basis and the Super-Secular-Equation. (Int. J. Quant. Chem. 1S, 811-827, 1967)

102. Molecular Associations in Biology - A Brief Summary. (Molecular Associations in Biology, Academic Press, Inc., New York, 539-549, 1968)

103. (with P. Lindner) Upper and Lower Bounds in Second-Order Perturbation Theory and the Unsöld Approximation. (Int. J. Quant. Chem. 2S, 161-173, 1968)

104. Some Aspects on the Possible Importance of the Reading Mechanism of DNA in Carcinogenesis. (Proceedings of the Israel Academy of Sciences Symposium on Physico-Chemical Mechanism of Carcinogenesis, Jerusalem in 1968)

105. Studies in Perturbation Theory. XIII. Treatment of Constants of Motion in Resolvent Method, Partitioning Technique, and Perturbation Theory. (Int. J. Quant. Chem. 2, 867-931, 1968)

106. Some Comments on the Treatment of Symmetry Properties in Perturbation Theory. (Int. J. Quant. Chem. 2S, 137-150, 1968)

107. (with W.M. MacIntyre) Electronic Energy of the DNA Replication Plane. (Int. J. Quant. Chem. 2S, 207-217, 1968)

108. Some Aspects on the Correlation Problem and Possible Extensions of the Independent-Particle Model. (Proceedings of Frascati Summer School on the Correlation Problem, 1967; Eds. R. Lefebvre and C. Moser, Interscience, 1968; Adv. Chem. Phys. 14, 283, 1969)

109. (with M. Berrondo) The Projection Operator for a Space Spanned by a Linearly Dependent Set. (Int. J. Quant. Chem. 3, 767-780, 1969)

110. Some Comments on the Periodic System of the Elements. (Int. J. Quant. Chem. 3S, 331-334, 1969)

111. (with O. Goscinski) The Exchange Phenomenon, the Symmetric Group, and the Spin Degeneracy Problem. (Int. J. Quant. Chem. 3S, 533-591, 1969)

112. (with T.K. Lim) Calculation of Lower Bounds to Energy Eigenvalues by Reduced Density Matrices and the Representability Problem. (Int. J. Quant. Chem. 3S, 697-702, 1969)

113. Some Aspects of the Hydrogen Bond in Molecular Biology. (Ann. N. Y. Acad. Sci. 158, 86-95, 1969)

114. Energia Sajátékek Alsóés Felsó Korlátjának Számítása Perturbációs - Zámításban Particionálási Technikával. (Magyar Fizikai Folyóirat XVIII, 515-540, 1969)

115. On the Non-Orthogonality Problem. (Adv. Quant. Chem. 5, 185-199, 1970).

116. Some Aspects on the Research Process in the Natural Sciences. (Scientists at Work. Festschrift in Honour of Herman Wold, Almqvist & Wiksells, Uppsala, 112-135, 1970)

117. (with J. Gruninger and Y. Öhrn) Comments on the Analysis of Atomic Correlation Energies). (J. Chem. Phys. 52, 5551-5554, 1970)

118. Recent Research in the Uppsala & Florida Quantum Theory Projects. (IBM Ludwigsburg Meeting Report, 47-83, 1970)

119. Theoretical Treatment of Impurities in Solid State Physics and Quantum Biology. (Proceedings of the Seminar on Impurity Centers in Crystals at Tallinn, USSR in 1970)

120. Some Properties of Inner Projections. (Int. J. Quant. Chem. 4S, 231-237, 1971)

121. (with O. Goscinski) Studies in Perturbation Theory. XIV. Treatment of Constants of Motion, Degeneracies and Symmetry Properties by Means of Multi-Dimensional Partitioning. (Int. J. Quant. Chem. 5S, 685-705, 1971)

122. Treatment of Exchange and Correlation Effects in Crystals. An Introduction. (Proc. Wildbad Symposium, 1971)

123. Quantum Chemistry and Molecular Physics of Solids. (Coll. Int. C. Nat. Res. Sci. 197, 207-266, 1971)

124. (with J.-L. Calais and M.R. Hayns) A Theoretical Study of the Behaviour of Solids under High Pressure and the Borelius' Law, with Applications to the Alkali Halides. (J. of Non-Metals 1, 63-78, 1972)

125. (with P.K. Mukherjee) Some Comments on the Time Dependent Variation Principle. (Chem. Phys. Lett. 14, 1-7, 1972)

126. (with B. Laskowski) Treatment of Constants of Motion in the Variation Principle. Symmetry Properties of Variational Wave Functions. (Chem.Phys. Lett. 16, 1-4, 1972)

127. Electronic Mobility and Transfer of Energy and Momentum as Time-Dependent Processes. (Proc. 3rd Conf. From Theoretical Physics to Biology, Versailles 1971, 145-146 (Karger, Basel 1973))

128. (with T. Ahlenius and J.-L. Calais) Some Comments on the Construction of an Orthonormal Set of LCAO Basis Functions for Crystals. (J. Phys. C: 6, 1896-1908, 1973)

129. (with B. Laskowski, J.-L. Calais and P.V. Leuven) Electron Gas Test for the Alternant Molecular Orbital Method. (J. Phys. C : 6, 2777-2787, 1973).

130. Josef - Maria Jauch, In Memoriam (Eur. Phys. News. 5, No. 12, 7, 1974)

131. Människan och hennes psyke i den moderna kvantteorins världsbild (Forskning och Praktik 6, 121-125, 1974)

132. Some Aspects on the American-Swedish Exchange in Quantum Sciences particularly the Uppsala-Florida Exchange Project. (Uppsala TN 470, to be published, 1976)

133. Internationella aktiviteter å Kvantkemiska Institutionen och Forskargruppen vid Uppsala Universitet. (Universen, Uppsala Universitet informerar 1/76, 1976)

134. Set Theory and Linear Algebra. (Uppsala TN 472, unpublished lecture notes, 1976)

PER-OLOV LÖWDIN IN THE SCIENTIFIC LITERATURE 1965-1974

Rolf Manne

Department of Chemistry, University of Bergen

Norway

Scientific writing differs from many other types of text by the abundance of references to sources of information, to procedures used, to ideas and hypotheses discussed, etc. Science Citation Index[1] provides a yearly compilation of the references in the majority of the scientific journals of the world. In fields like chemistry and physics with established bibliographical procedures and age-old traditions of international dissemination of scientific results Science Citation Index offers an almost complete coverage.

Under the name of the first author Science Citation Index lists chronologically the cited papers and references to the citing papers. One is thus able to find all papers in a given year which cite a given key publication. At the present occasion it might be of some interest not to use Science Citation Index as it was intended - for a search of relevant scientific material - but for a study of the impact of the work of one man - Per-Olov Löwdin. It is obvious that this impact is manifested also in other ways than in citations, and that the number of citations is not a measure of scientific importance. Other authors in this volume, however, discuss other aspects of Per-Olov Löwdin's scientific achievements, and in this little study I will concentrate on how his work has influenced others as shown in Science Citation Index. For this purpose I have gone through Science Citation Index for the years 1965-1974 counting the number of citations to each publication. Science Citation Index is full of errors, partly due to erroneous citations, partly due to mistakes in preparing the data. A third source of error is introduced here by the small print, of the order of 8 lines to the centimeter, which makes reading strenuous even with a magnifying glass. For these reasons,

TABLE I

MOST-CITED ARTICLES BY PER-OLOV LÖWDIN 1965-1974

References are to the list of publications on page 12-22.

	Name	Reference	Year	No. of Citations
1.	Quantum theory of many-particle systems I	26	1955	383
2.	On the non-orthogonality problem	9	1950	351
3.	Cohesive properties of solids	33	1956	333
4.	Correlation problem in quantum mechanics	45	1959	242
5.	Quantum theory of many-particle systems III	28	1955	238
6.	Scaling problem and virial theorem	46	1959	162
7.	Natural orbitals of two-electron systems (with H. Shull)	34	1956	109
8.	Quantum theory of many-particle systems II	27	1955	106
9.	Perturbation theory IV	64	1962	90
10.	A note on perturbation theory	13	1951	81
11.	Perturbation theory X	85	1965	71
12.	Superposition of configurations and natural spin orbitals (with H. Shull)	47	1959	68
13.	Band theory, valence bond and tight-binding calculations	57	1962	60
14.	Angular momentum wavefunctions	81	1964	57
15.	Long-range interaction of two hydrogen atoms (with J.O. Hirschfelder)	48	1959	56
16.	Proton tunneling in DNA	71	1963	54
17.	A quantum mechanical calculation of the cohesive energy I and II	2 3	1947 1948	53 (together)
18.	A theoretical investigation into some properties of ionic crystals	4	1948	53
19.	Studies on the alternant molecular orbital method I (with R. Pauncz and J. DeHeer)	61	1962	50
20.	Studies of atomic self-consistent fields I	20	1953	50

the numbers obtained are not without errors. Nevertheless, I believe they are accurate enough for the present purpose.

A list of the most-cited papers in the ten-year period is given in Table I. Out of a total of about 120 papers this table includes those 20 which have received 50 or more citations, i.e. 5 or more citations per year on the average.

The paper receiving most citations is the first of the three papers on the quantum theory of many-particle systems from 1955. The number of citations is 383 which is close to 40 citations per year. Most of these citations are to the definition of natural orbitals, but there are also some to matrix elements between determinants constructed from non-orthogonal orbitals. A check of some articles citing other papers by P.O.L. reveals a fair number discussing natural orbitals without a citation to the original definition. The authors of these papers apparently consider natural orbitals as such a well-established concept that citation is no longer necessary. One may assume this development to continue in the future in the same way as we leave out references to the Schrödinger equation or the definition of molecular orbitals. The peak of the number of citations, however, does not yet seem to have been reached, since the greatest number of citations for a given year was in 1974.

The second-most cited paper is the important 1950 paper on the non-orthogonality problem where the symmetric orthogonalization was introduced to a wider audience. It is not the original reference, though, since symmetric orthogonalization was used also in the thesis and in the first account of the calculations of cohesion in ionic crystals.

The high number of citations to the review article on cohesive properties of solids may at first be a little surprising. Unlike many other fields toucned by P.-O. L. this field has remained more or less an Uppsala speciality. A spot check shows that most citing papers describe self-consistent-field energy band calculations using the muffin-tin potential approximation. The citation is to the definition of the α-functions used for the construction of the potential inside the muffin-tin spheres. These functions were also used in the early papers on ionic crystals, but as mentioned by P.-O.L. himself, seem to have been used first by Coolidge[2] in a study of the water molecule.

The review article on correlation from 1959 is No. 4 on the list, and No. 5 is the third of the celebrated 1955 papers. For both these papers a major reason for citation is the extension of the single-particle model.

The remaining papers on the list cover a wide variety of fields, some also covered by the papers mentioned so far. For new fields we may give key words like perturbation theory, projection operators, and quantum biology. Of the many papers on perturbation theory by P.-O.L. one finds the highest citation counts for three. These are, in chronological order, the note from 1951 introducing the partitioning technique, and Perturbation Theory IV from 1962 giving a concise formulation of the same technique in terms of projection operators, resolvents, wave and reaction operators. The third paper is Perturbation Theory X describing methods for the calculation of lower bounds.

It has been stated that the life-time of a publication in physics is rather short, perhaps only five years. Consequently, the number of citations in Science Citation Index would decrease as the paper gets older. This is not the case for most of the heavily-cited papers by P.-O.L. that I have been able to check. There are several explanations possible for this. One is that the journal coverage by Science Citation Index has improved and that consequently the figures for the early years are too low. This is a contributing factor as is the fact that the number of quantum chemists has increased. However, there is also an explanation relating to the cited work itself. In the 1950's when several of the much-cited papers were written only few scientists had experience with large-scale numerical calculations - one of them was P.-O.L. himself. Almost all his work is related to problems requiring heavy numerical computations. It is thus only with the present technology and programming knowledge that a larger number of scientists have been able to use the methods developed by P.-O. L. in the pioneering years.

Table II gives the number of citations in 1973 in the most heavily citing journals. The full list is quite long, about 60 journals (some citations are dubious). The most heavily citing journal is - quite properly - the International Journal of Quantum Chemistry. Off the table are journals with fewer than 5 citations. Some of these cover fields removed from P.-O.L.'s own research interests such as nuclear and high-energy physics and in one case - agronomy.

Would Science Citation Index be of value for a literature search on a topic covered by the work of P.-O.L? For natural orbitals and the non-orthogonality problem the answer is no. There are too many references to make the search worthwhile, and there are also too many references left out for these well-established topics. In many cases one finds that popular citations have a rather small information content. They might even be made by copying reference lists of other papers in the same field as is likely to be the case of many of the citations of the α-functions

TABLE II

JOURNALS CONTAINING MOST CITATIONS TO THE WORK OF PER-OLOV LÖWDIN IN 1973

	Name	Number of citations
1.	International Journal of Quantum Chemistry	43
2.	Journal of Chemical Physics	31
3.	Physical Review B	16
4.	Molecular Physics	13
5.	Theoretica Chimica Acta	12
6.	Physical Review A	11
7.	Journal of Physics C	10
8.	Chemical Physics Letters	9
9.	Bulletin of the Chemical Society of Japan	7

by users of standardized computer programs. Citations which are made after a careful study of the paper on Cohesive Properties of Solids are thus drowned in the flood of references to a standard technique. This may hold for lesser-cited papers as well. On the other hand, negative citations of the type "Contrary to X we have found..." are not likely to be propagated very far in a citation index. Once a mistake is corrected it is likely to receive but few citations in the future.

One may thus state that the citation count is a measure of positive scientific influence. It is a crude measure, though, since high counts are given to concepts and methods which are easily applicable. Educational achievements in the form of university and summer-school teaching or in the more subtle forms of setting standards for good science and worthwhile scientific problems are not honored this way. All such factors have to be added before one gets a true picture of the influence of the work of an eminent scientist.

ACKNOWLEDGEMENTS

In my tribute to Per-Olov Löwdin I would like to thank him for continually stressing the importance of a proper literature search. Part of this work was done with the help of the library facilities of Chalmers University of Technology, Gothenburg, Sweden.

REFERENCES

1. Science Citation Index is published by the Institute for Scientific Information, Philadelphia, Pa., U.S.A.

2. A.S. Coolidge, Phys. Rev. 42, 189 (1932).

ACTA VÅLÅDALENSIA REVISITED:

PER-OLOV LÖWDIN IN SCIENTIFIC DISCUSSION

Anders Fröman and Osvaldo Goscinski

A characteristic trait of Per-Olov Löwdin is that every scientific question which interested him lead to a more general formulation. As an intrinsic structuralist he very soon goes beyond a specific problem towards the conceptual borderlines. Quantum Science as a whole is his problem - this justifies the title of this book. In order to illustrate this eagerness to draw a map of the mountain range at the same time as climbing a top we include a paper by Löwdin on the connection between chemistry and quantum chemistry. It was presented during the first summer school organized by him, in Vålådalen, Sweden, in 1958. It finished with a panel discussion which is a precursor of the many Symposia in Florida and Scandinavia. A rather complete version of the proceedings was circulated as Technical Note 16 (no longer available) from the Uppsala Quantum Chemistry Group. It had the unofficial name of Acta Vålådalensia.

Löwdin's article has the value of old maps. They tell us much about the world as it was and as it is now. The similarities astonish us and make us appreciate the drafsman. To publish the 1958 lively discussions is forbidden by space limitations. A glimpse of their depth, relevance and spontaneity is possible, however. An isolated sentence or discussion remark can tell about a school of thought, a philosophy, a controversy or a person. In this vein, the fragments included can be stimulating as the letters and charts preserved in the Codex Argenteus room at Carolina Rediviva. They tell us about their authors, their time and about ourselves. We apologize to the Vålådalen speakers for the arbitrary selection and to George Hall and Anders Fröman for abridging the result of their enormous personal effort.

FROM THE FOREWORD TO TECHNICAL NOTE 16

Per-Olov Löwdin

"During the last week of the international summer school in quantum chemistry in Vålådalen, Sweden, July 26 - August 30, 1958, there was arranged a special panel meeting on the subject "Correspondence between Concepts in Chemistry and in Quantum Chemistry". The purpose of this meeting was to state the basic problems in this field, to try to clarify what the chemists actually mean when they speak about "atoms" and "ions" in a compound, and ultimately to try to find the explicit connection between such concepts as atomic and ionic radii, electronegativity, covalent and ionic nature etc., and pure quantum-mechanical quantities.

Depending on the restricted accommodations available, invitations were issued only to a limited number of chemists and physicists. We were particularly happy to have among us two of the great pioneers within the field of quantum chemistry, namely Prof. Linus Pauling (California Institute of Technology) and Prof. Robert S. Mulliken (University of Chicago). The characteristic difference in their approaches to the subject created a most stimulating atmosphere for the discussions. I am also grateful to the other members of the panel for their valuable contributions. ...."

"SOME COMMENTS ON THE QUANTUM MECHANICAL METHODS FOR TREATING MANY-ELECTRON SYSTEMS AND THE CONNECTION BETWEEN CHEMISTRY AND QUANTUM CHEMISTRY."

Per-Olov Löwdin

In this session, we will discuss the relation between chemistry and quantum chemistry in greater detail. Today we know that the existence of the covalent chemical bond depends on a typical quantum-mechanical phenomenon, namely the identity principle and the general symmetry law for measurable quantities. The saturation of the bond follows further from the Pauli exclusion principle contained in the antisymmetry postulate. Usually classical mechanics provides at least a first approximation to a quantum-mechanical description, but, as regards symmetry properties, no such simple correspondence seems to exist.

In the development of modern chemistry, the influence of the quantum-mechanical ideas has so far been of a more qualitative than quantitative nature. In order to bridge the gap between chemistry and electron physics, four main approaches have been used:

1. Highly accurate solution of the Schrödinger equation.
2. Approximate solution of the Schrödinger equation.
3. Semi-empirical theories.
4. Chemistry in quantum-mechanical language.

A sketch of the sitation is given in the figure. Of course, the four methods may "interact" and influence each other, and this phenomenon will be discussed by Hall in the next talk. A few comments are perhaps worthwhile:

1. Heitler and London's treatment of the hydrogen molecule in 1927 gave only an approximate solution of the Schrödinger equation. However, by using the same technique as had previously

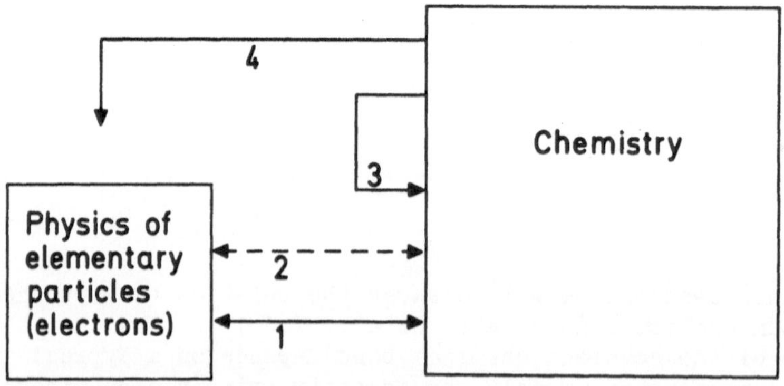

Fig. - Quantum Chemistry as a link between chemistry and the physics of the elementary particles.

1. Theory of covalent bond (H_2).

2. Theory of van der Waals forces, repulsive forces in ionic crystals, Hume-Rothery rules for alloys, etc.

3. Semi-empirical theories of dipole moments, spectra of molecules, conjugation and aromatic bond, directing power of substituents, stability of free radicals, etc.

4. Chemistry in quantum-mechanical language, theory of resonance etc.

been developed by Hylleraas for helium, James and Coolidge could derive the eigenfunction for the ground state of H_2 and show that the theoretical energy was in agreement with the experimental value. Since all previous attempts to treat two-electron systems by classical mechanics or classical quantum theory had failed completely, the new results for He and H_2 were remarkable, showing that Schrödinger's idea of postulating a wave equation in a two-electron configuration space was successful. From the good results one has then drawn the conclusion that the same type of wave equation should be valid also in a many-electron configuration space containing N particles. So far, no one has found any arguments against this rather drastic extrapolation from N = 2 to an arbitrary N, so we have all reasons to be optimistic and to believe that Dirac's famous 1929 prediction about quantum chemistry will become true.

2. Even the approximate solution of the Schrödinger equation is a rather difficult problem so, in addition to the simplest atoms and molecules, only a few other basic problems have been studied purely theoretically. Here I would like to mention the derivation of the van der Waals potential, the repulsive forces within ionic crystals, the Hume-Rothery rules for alloys, etc. - all based on various types of approximation.

3. A much more popular approach is represented by the semi-empirical theories obtained by unifying chemical and quantum-mechanical ideas. They are essentially devices for correlating one set of experimental data with another set, and many properties of e.g. the conjugated systems have been treated successfully in this way: the nature of the aromatic bond, the calculation of dipole moments, the directing power of substituents, the color and spectra, the addition problem, the stability of free radicals, and so on. Another successful attempt is provided by the crystal field theory. The importance of the contributions by Hückel, Pauling, Wheland, Mulliken, Sklar, Coulson, Longuet-Higgins, Platt and others in the former case and by Hartmann and the Copenhagen school in the latter case can hardly be overestimated.

4. One of the most fruitful methods for constructing the link between chemistry and electron physics was explored by the chemists themselves by starting from the ordinary chemical experiences and by translating and adapting the description to quantum-mechanical terminology. The best example is given by Pauling in his "Nature of the Chemical Bond", and we all know how successful this approach has been both qualitatively and quantitatively. Of course, it is very hard to decide whether the electronic interpretations gained in this way have a real background in nature or not, and the

discussion of this problem was actually the essential purpose of this meeting. I believe that, so far, we have not obtained a definite answer to this question and that the opinions are rather diversified among us.

Considering things quantitatively, one finds that the link between chemistry and the Schrödinger equation is still rather weak. It seems hence highly desirable to carry out much more pure calculations of atomic and molecular structure to get a more reliable basis both for the theory itself and for the other approaches built up analogously. The way from the Schrödinger equation to the "real truth", representing the behaviour of the system in nature, is definitely exceedingly long and, even in the treatment of the simple diatomic molecules, we are today in the stage of calculating molecular integrals for a very restricted basis, and it seems likely that the main difficulties still remain. The mathematical steps for solving the electronic Schrödinger equation have otherwise been rather standardized. The methods vary also with respect to the choice of basis. A more recent trend involves the transformation from an arbitrarily chosen basis to the natural spin-orbitals connected with the eigenvalue problem for the first-order density matrix and characterized by having maximum occupation numbers.

For a single molecule, it is now possible in principle to solve the Schrödinger equation to any degree of accuracy but, in practice, the problem is still very hard depending on the numerical difficulties involved. The development of the modern electronic computers has here been of almost revolutionary importance, since we may now study at least <u>small</u> molecules purely theoretically, but a great deal of additional efforts are definitely needed to speed up the progress within this important field. For large molecules, we are probably still confined to semi-empirical treatments or to methods breaking up the system into small "invariants", as has been stressed here previously by several speakers.

In conclusion we note that, in chemistry, one usually studies <u>series</u> of related compounds at the same time. Theoretically, it is hence important not only to solve the Schrödinger equation for a specific electronic system but also to study the change of the solution under varying conditions. The variation of the atomic Hartree-Fock functions with Z has, for instance, been studied in detail on intuitive grounds[x)], but the underlying mathematics treating the

---

x) See e.g. P.O. Löwdin, Phys. Rev. <u>94</u>, 1600 (1954) and references quoted there.

behaviour of the eigenfunction with varying parameters is perhaps not yet fully understood. Another important problem is the analytic character of the exact eigenfunctions and the nature of the energy spectrum and its discrete and continuous parts. In order to close the gap between chemistry and electron physics, studies in mathematics as well as in numerical technique may thus be of essential value.

Literature: P.O. Löwdin, "Present Situation of Quantum Chemistry", J. Phys. Chem. 61, 55 (1957); "Correlation Problem in Many-Electron Quantum Mechanics", Adv. Chem. Phys. (1958).

ON THE ONE-ELECTRON PICTURE AND THE QUESTION OF HYBRIDIZATION

Löwdin: .... It seems as if the theory of hybridization would follow a general pattern of behaviour characteristic for the entire quantum chemistry of today: an elementary theory may often be brought to give excellent agreement with the experimental results, whereas most refinements of the theory will then disturb the nice situation and cause disagreements:

Agreement
between theory
and experiment

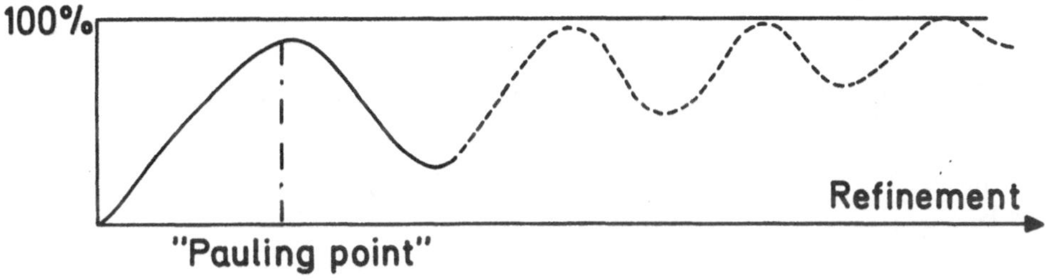

The peaks of the curve are usually obtained by using some extra empirical data, but the ultimate goal of the pure theory is, of course, to give the results desired by using only the values of e, m, h, and the atomic numbers of the nuclei involved. The minima of the curve reflect the fact that it is actually very difficult to "refine" a good semiempirical theory, and, in this connection, we all recall such remarkable things as the "nightmare of the inner shells", the "non-orthogonality catastrophe", and similar phenomena in the past. .....

ON NATURAL ORBITALS

......

Löwdin: Mulliken's remark "Give us a little bit more time and we will gain experience" applies also to natural spin orbitals for molecules.

Shull: What we want are the invariants in our calculations. For He the occupation numbers for natural orbitals are essentially invariant. For many-electron systems the corresponding invariants in terms of two-electron functions may represent the chemical bond.

Hjalmars: If this can be shown, I agree that the natural spin orbitals could be used. We want to build a large molecule from small pieces. Perhaps natural orbitals will be fairly constant and can be so used but it will take years to calculate them. In the meantime the analysis of s- and p-populations is still very useful, but should not be pushed too far.

Hall: The natural orbitals in general are not localized and therefore do not give a localized interpretation of part of a molecule. We want to find invariant localized interpretation of part of a molecule.

Löwdin: The natural spin orbitals can easily be obtained by diagonalizing the first order density matrix we derive from our wave function.

For example, Meckler's calculations on O_2 using Gaussian functions can be compared with other calculations by transforming to natural orbitals.

Hjalmars: There are so few calculations available exact enough to give the natural orbitals.

Shull: Even the crudest calculations have almost this invariance already.

McWeeny: Natural orbitals may be unique in principle but not in practice because of the truncated basis, which puts constraints on the form of the natural orbitals. I question if natural orbitals from Gaussian orbitals resemble those for Slater orbitals.

Shull: Yet the occupation numbers may be very close. Certainly for H_2 they are within 1 or 2%.

McWeeny: Coincidence! Meckler's calculation gives occupation numbers almost identical to those for Slater orbitals. Nevertheless

Meckler's total energy is much poorer and his functions are wrong at the nuclei. Occupation numbers merely determine general features of charge density: they may well be similar even for quite different functions.

Löwdin: I agree. Truncated sets are bad. We should all use infinite complete sets!

Pauling: It is just a hundred years since the carbon atom was first said by the chemists to be quadrivalent. Couper, in Paris, was the first to draw lines to represent bonds. The intensity of the wagging vibration of CH_4 gives a charge of 0.4 e on H^+. This implies 8% of ionic structure for the bond and 32% for the molecule. This may be a way of reinterpreting Voge's $(2s)^{1.38}$ $(2p)^{2.62}$ in terms of ionic character.

Mulliken: There is something in that - but I don't know how much.

....

Löwdin: Simple pictures are often more reasonable than more sophisticated ones.

Mulliken: Which ones!

Mulliken: But it is good to face the facts. For example, in Scherr's LCAO SCF calculations on N_2, the calculated dissociation energy fell from 8.0 to 1.2 eV when the inner shells were included properly. Scherr was disappointed, but I told him he should be glad that his calculation was now on a sound basis, even if still far from the accuracy needed before we are through.

Löwdin: Work in Uppsala on Al^{+3} (Fröman) shows that the difference between total experimental and HF-energies is about 22 eV. 11 eV of this is relativistic and even a 1% change should be chemically measureable.

ON CALCULATIONS

....

Pauling: Usually I am not very much interested in calculations because I know what the answers are going to be anyway. But I would like to ask if anyone can explain this? These are the bond energies in kcal/mole

C-C	85	C=C	125	C≡C 150
	45		45	
N-N	40	N=N	80	

and one predicts from this N≡N should be 150 - 45 = 105 whereas experimentally it is 225. I cannot see any reason for this to be so.

Mulliken: You may have a quadruple bond in N_2.

If you have a single bond which is twice as strong as normal, why not call it a double bond? In N_2 there is a very strong σ bond and this gives altogether in effect a quadruple bond. But one may reasonably ask, why don't we have quadruple bonds in a formal sense, for example in C_2?

ATOMS IN MOLECULES

Preuss (Summary of Talk): After a short discussion of the three different ways of quantum chemistry research: the half-empirical, theoretical- and half-theoretical methods, a method is suggested based on a half-theoretical treatment of a generalized form of the idea of "atoms in molecules". This formulation contains the methods of Moffitt and Bingel as special cases and give a description of the energies of molecules with many atoms. Applications of interactions between molecules are possible.

Pauling: Could this method be applied to H_3^+ where we do not know the answers?

There are two examples of a very similar method developed before the Heitler-London papers. These are the papers of Burrau on H_2^+ in 1926 and of Condon on H_2 in 1927. In this work the electronic repulsion was treated by an extrapolation from the experimental energy of He.

Löwdin: I like Heitler's idea that molecular properties should be expressed in terms of atomic wave functions, but Heitler's theory was built from exact atomic wave functions. It was not an orbital theory whereas Moffitt's is. Actually Moffitt seems to give up the proper definition of his basis and it is not clear what type of Hilbert space he is using. It would be very interesting to study the whole basis of Moffitt's method in greater detail. But is the method worthwhile? I do not object to semiempirical methods which give wonderful results with simple formulae, but I cannot see the meaning of semi-theoretical methods which give bad results with complicated formulae. Part of the same criticism applies also to Preuss' approach, even if it is described very elegantly here.

....

Hall: Would it not be better to think of ions in molecules instead of atoms? By using ions the electrons can be kept paired and difficulties in defining valence states will not arise. The greater part of the correlation energy will also be independent of nuclear distances and so can be cancelled, e.g. Löwdin's very good cohesive energies for ionic crystals. Similarly H_2 can be formed from H^- and H^+ with a continuous deformation of the orbitals and the correlation energy is independent of R. The rise in the correlation energy relative to two H-atoms is due to the discontinuity in the wave function.

Pauncz: The H^- wave function is very extended and surrounds both nuclei thus it includes much covalent character.

Mulliken: Covalent and ionic wave functions are not orthogonal and may even be almost identical. Thus for $(SO_4)^{--}$, wave functions built from those for S^{+6} and O^{--} may differ by only 10 to 20% (for a guess) from those based on S^{+4}, S^{+2}, S, etc. In a very stable molecule I believe the atoms are not still present to any clearly recognisable extent.

Pauling: The molecule HF can be treated using the structures H^+F^- and H - F but trouble starts if these are refined e.g. polarization of the ionic structure and resonance with the covalent structure are not independent refinements. This is why the empirical system of molecular structure has to remain simple. You can talk about partial ionic character of bonds if you do not talk about polarization at the same time.

LIGAND-FIELD THEORY DISCUSSION

Löwdin: This raises the general question of the relation between chemistry and quantum chemistry which perhaps may be taken up at this point. Everything is alright, if we stick to just one field. A characteristic thing about those who use successful models like the ligand-field theory is that, when they start refinements, they often argue about small details because they are not discussing what is going on in nature but perhaps more what is going on in their own minds. Such discussions can be very hot and controversal.

Kleiner (J. Chem. Phys. 20, 1784 (1952)) tried to investigate the Cr-ion by doing calculations on the perturbing effect of the neighbouring oxygens. The results were very bad and even gave the wrong sign for Δ_q.

This is the same as with other models: when we refine the theory, everything goes wrong. Why is it always so?

Mulliken: You are too pessimistic. Do you know that it always happens?

Löwdin: Actually we should remember the last two of Parr's "ten points",

9) Be optimistic!
10) Be patient!

Let me ask to what extent does an ion actually exist in one of these compounds? When Prof. Mulliken writes formulas with carbon atoms, does he mean normal C-atoms, deformed C-atoms (virial theorem) or C-atoms in valence states or what?

Mulliken: An atom means a nucleus and electrons.

Löwdin: How is the electronic cloud assigned to the nuclei?

Mulliken: This is arbitrary. There is a necessary uncertainty in very many of our chemical concepts. Is there any great use in trying to estimate this uncertainty?

Löwdin: If the theoretical people had been first in the field the question would not have arisen because they would probably never have invented bonds. But the chemists were there first and speak of specific atomic radii, electronegativities, covalent and ionic character, etc. They mean something by this, something concerning the electronic cloud, and it would be nice to know what they actually mean precisely.

Matsen: In his book, Pauling says: If the chemists had got there first there would not have been sp^3 but just tetrahedral orbitals.

Pauling: Yes, and s and p would be hybrids of the tetrahedral orbitals.

Löwdin: In the orbital picture of a molecule we can carry out linear transformations of the orbitals, and the wave function itself stands invariant. For ionization, one type of orbitals may be more convenient and, e.g. for internal rotation, another type, but the total wave functions are still identical. Conjugation involves a symmetry adaptation of the orbitals and can similarly be transformed away. Even in a more accurate treatment, if the occupation numbers are equal, the orbitals can be transformed, and there is no conjugation.

Mulliken: I don't agree with the last part of what you seem to be saying. If for example we use three tetrahedral hybrids (bent bonds) for a triple bond, the σ bonds in the corresponding $\sigma - \pi$ description will be quite different from our usually-assumed diagonal sp hybrids, and the total wave function is distinctly different, is it not? Likewise, conjugated and unconjugated wave functions differ appreciably, even though not radically, even if transformed to maximum similarity.

Löwdin: When we speak of bonds, we speak of assigning the electronic cloud to a pair of nuclei. Is this assignment not ambiguous?

Mulliken: In the MO-theory we may picture the bonds as σ- and π-bonds or we may use two or three equivalent bonds. In the VB-theory the bond can be described using paired electrons.

Pauling: In the MO-treatment a $\sigma + \pi$ description of a double bond is identical with the two bent bonds description, is this also true in the VB-theory?

I would like to propose the problem of finding the best set of AO:s to be used in a simple covalent VB wave function.

McWeeny: Are there orthogonality conditions imposed on the AO:s?

Löwdin: Slater has investigated this for H_2 and finds that with orthogonal AO:s there is no binding.

Mulliken: Pairs of electrons in localized MO:s really correspond better than VB pairs to Lewis' conception of electron-pair bonds.

Löwdin: The electronic cloud can be divided up into cells around each nucleus or into overlapping clouds associated with each nucleus. This ambiguity of description may physically not be so important except for the fact that the chemists have found properties of "atoms", such as atomic radii, electronegativities, etc. which are remarkably invariant from molecule to molecule. Actually we do not know what the chemists' atomic concept really means. I am coming back to this basic question again.

Mulliken: They can be <u>roughly</u> justified in terms of quantum theory, but only roughly (cf. Mulliken; J. chim. phys. <u>46</u>, 536 (1949)), and I do not have the impression that they are more than roughly invariant empirically either.

Löwdin: Ionic radii with Pauling's corrections for different surroundings are accurate to 0.001 Å. This is not just rough agreement.

Pauling: Even the exceptional structure of LiI can be explained by the mutual contact of the large I^- ions. This shows that there is something real about the assignment of a volume to each ion.

Löwdin: When Lundqvist calculated the Li^+H^- crystal (Arkiv Fysik 8, 177 (1954), he started out from the ions and afterwards looked at the electron density as an X-ray crystallographer would do and found it essentially covalent.

Lundqvist: It corresponded to about 25% ionic character.

Löwdin: When you start from an ionic structure and find it covalent there is something wrong with out concepts.

Mulliken: You can describe it with either concept.

Löwdin: They are not identical, but they overlap so much that our discussions are often meaningless.

Mulliken: H^- in the crystal is different from a free H^-.

Löwdin: How do we actually define H^- in a crystal?

Lindqvist: The experimental work by Brill-Grimm-Herman-Peters in 1939 and later by Witte-Wölfel (Summary in Rev. Mod. Phys. 30, 51 (1958)) indicates that the ionic radii obtained by considering the minimum points in the electron densities between the ions of some simple salts do not agree with the common scale of ionic radii used by crystallographers. There are great difficulties involved in the experiments and their evaluation, but the discrepancy must not be forgotten when we discuss the physical reality of ionic radii.

FORMAL CHARGES AND REACTIVITIES

Del Re: Reactivity may not wholly be a ground state phenomenon yet the formal charges usually give the situation for heterocyclics and substituted rings.

Pauling: It is an accident, but not a pure accident, that predictions are correct for some molecules.

McWeeny: The formal charges can only determine the earliest stage of the reaction. When there is no crossing of the energy curves the energy of the complex seems to be correlated with the formal charge but not otherwise. Both approaches involve a model of a reaction and neither is fully accurate (see for example R.D. Brown, Quart. Revs. 6, 63 (1952)).

Mulliken: I do not see any reason to question the reality of formal charges. It is only a matter of defining what we mean. There are no atoms there; all are fused together. Could the electron density distribution be used instead of atomic charges?

Pauling: I am not against the reality of atomic charges in molecules but just against the argument relating them to reactivi-ies. The atomic charges should be correlated with the real charges and related to dipole moments and quadrupole moments.

TOWARDS A STRATEGY FOR QUANTUM CHEMISTRY

by

George Hall

It is good to turn away, sometimes, from the technical details of a calculation on an atom, a molecule or a solid and to try to assess what quantum chemistry can hope to achieve in the future and how it should do so.

The basis of most of our calculations is the ordinary non-relativistic Schrödinger equation. The form of the full relativistic theory is not yet fully known but presumably in the future it, too will be used for molecular calculations. Because of the advances in electron computing, it is now possible to compute wavefunctions for molecules with sufficiently high accuracy to enable their properties to be predicted confidently.

It will never be possible, however, to calculate accurate wavefunctions for more than a small number of molecules. Even allowing for large improvements in computers and in mathematical methods, it seems reasonable to say that, if a molecule has three or more nuclei, other than H nuclei, more than 15 electrons and no high symmetry, wavefunctions of the James-Coolidge standard of accuracy will never be available. This standard, which allows for only 1/4 kc/m error in the total energy, is very high but it is necessary to ensure the accuracy of almost all the various properties calculated from the wavefunction. With a lower standard of accuracy, many more molecules could be calculated but the results would not be so reliable.

The conclusion is that, while the calculation of accurate

wavefunctions is of fundamental importance, it can never be the whole of the subject. Another method of procedure is to use the general theory to derive extrapolation formulae. These formulae apply to series of related molecules rather than to individual molecules. They can be used with parameters determined either by fitting experimental results or by extrapolation from accurate calculations on some of the simpler members of the series. An example of this method is the molecular orbital treatment of conjugated hydrocarbons. Such methods increase greatly the number of molecules which can be discussed and, by the use of perturbation methods, the number can be increased still further.

Most chemists, however, are interested not only in the numerical values of observables but in other less numerical concepts such as bonds, electronegativities and reactivities. None of these seems to have any relation to an accurate wave-function for a molecule. Nor, on the other hand, do they usually arise empirically by studying one molecule but only by comparing various related molecules. What is needed, then, is a method of finding the common properties of the wavefunctions of related molecules. This means that the extrapolation methods have a much closer relation to chemistry than the pure calculation method. To the accuracy of the molecular orbital theory, for example, a bond can be defined as an equivalent orbital and its properties computed. The theory can also be used to define and derive new concepts e.g. an alternant hydrocarbon. Another method of developing the subject is to use models. These are not concerned with the Schrödinger equation but with some simplification of it or abstraction from it. For this reason their assumptions and results cannot be checked by comparison with an accurate calculation, which is the vital safeguard of an extrapolation method.

Some models, such as ligand field theory, are suggested qualitatively by mathematical arguments. Others, such as resonance theory, have little mathematical justification but yet serve as very useful models of the chemist's concepts and experience. The most difficult to assess are the oversimplified models, such as free-electron theory, whose assumptions are much too simple but which can nevertheless provide good extrapolation formulae. Perhaps one of our most urgent tasks is to understand in terms of accurate wavefunctions why some of these models work so well.

Matsen: Man is strongly motivated to apply projection operators to his experience. These projection operators are correlations, ethical and metaphysical systems with which he hopes to project understanding out of his experience. Each man does not generally use the same operators but modifies them according to his own needs, past experience and his psychology. His strong motivation is based on his self-preserving instinct to predict the future and his impatience with boredom. Scientists differ from the rest only in the kind and intensity of their experiences. They have the same compulsions to project from their experience plus the additional one of professional self preservation.

Chemistry and part of physics differ from many fields of human activity in that there exists a commonly accepted axiomatic core from which it is believed that all experience in this area may be uniquely predicted. There are at present enormous technical difficulties in carrying out the program of prediction. The chemist is, however, compelled to continue to project his experiences whatever the status of the program. It is true that the existence of the axiomatic core and the hopes associated with it perturb the projections from experience as in fact the new experiences perturb the core. But the projection goes on. Conant relates how phlogiston theory was used for 130 years after all the necessary evidence was at hand to disapprove it. This was not blindness on the part of the scientists of that day but merely an expression of the great need for projection operators.

There is a possibility that a core of ideas in chemistry or in physics may develop independently of quantum mechanics. Since we believe that if a separate core should develop its connection with quantum mechanics could ultimately be established. The development of an alternate core could have two effects (1) it could enrich and enlarge our understanding,(2) it could impede the development of our understanding through the development of a language and attitude which must subsequently be modified. Pauling's remark that if chemists had got there first, they might have developed tetrahedral orbitals instead of s- and p-orbitals is perhaps not irrelevant here. Whichever the result the experimentalist must read his experiments and not be too perturbed by the quantum chemical prophets who are only too likely to say first, "that's impossible" and later, "we knew it all the time".

Griffel: Who do you think will develop another scheme?

Matsen: Before quantum mechanics, the nucleus of experience was surrounded by a sphere of ideas which spread out more or less symmetrically. This did serve the useful purpose of organizing the experimental material. Now, the sphere has tended to hybridize in the direction of quantum mechanics. Is there another core and a set of ideas around each core? At this point in the development of the subject we may hinder ourselves by thinking too much about Schrödinger's equation.

Our failure to develop a core of ideas would force the experimentalists themselves to develop such a core.

Sunner: Much of the best experimental work uses the method of comparative measurements. Is there any quantum mechanical way of treating molecules which are very similar so that the errors can be subtracted and only the differences calculated?

Mulliken: As we make more and more calculations on different simple molecules, I believe we will be able to make better and better extrapolations from these which will assist in understanding the experimental data on a wider variety of molecules. In experimental chemistry we consider a large number of cases and try to understand by comparing them. In quantum mechanics it is better to take a few cases and make accurate calculations.

Bastiansen: I divide science into real science, which deals with nature, and the study of models, not necessarily concerned with nature. Science is gathering information i.e. numbers. If our brains could memorize all the numbers we might not need theory and system.

Matsen: Even if we could, we, as human beings, would still demand a system.

Bastiansen: When you say you are interested in experimental results to high accuracy, do you mean it? Do you want them just to complete a table, or can you give a clear indication of the kind of problem they will solve?

Matsen: I think you will not be satisfied with raw data. You will want to project out some system of ideas e.g. by comparing related molecules and you will want to see these related back to a common axiomatic basis such as the Schrödinger equation.

Bastiansen: I think we experimentalists are doing this without knowing it.

Matsen: There are two separate schools of thought about this. You should not allow theorists to interfere too much.

Lindqvist: I agree with Matsen but I don't think that a new core will grow up.

It is an important safeguard to have the connection between the extrapolation methods and the pure theory. It might be worthwhile to look for a way to connect theory with some of the experimental relations which have been given for dipole moments and polarizabilities.

Pauling: It seems to me that what people are interested in is an understanding of nature. Nature includes the activities of an institute for quantum chemistry. The solution of the Schrödinger equation is a part of nature whether it is related to anything or not, in the same way as topology is a part of nature. Nevertheless, there is the possibility of expanding man's understanding by connecting these activities with what the chemist is doing. One of the things is to go to other moelcules either from the empirical or the theoretical side. Perhaps progress is more rapidly made from the theoretical side than the empirical by formulation of a generalization. A great many generalizations could be made by studying approximate solutions of the Schrödinger equation. There are so many chemical problems, that are not well understood, that I hope that the period of laying a sound framework of quantum chemistry will be approaching at least a temporary end. I hope that some of the activity in this field will be directed to the finding of new generalizations. What are the properties of the hydrogen-bond for example? I think we might well be able to predict some properties. This is so important in biology that I should like to see more theoretical work of high order done on the hydrogen-bond.

THE NON-ORTHOGONALITY PROBLEM AND ORTHOGONALIZATION PROCEDURES

Giuseppe Del Re

Cattedra di Chimica Teorica, Università di Napoli

Via Mezzocannone 4, 80134 Napoli, Italy

Historically, the non-orthogonality problem arose with the first attempts to pass from the qualitative quantum mechanical treatment of polyatomic molecules to more quantitative methods. Indeed, the very beginning of quantum chemistry was marked by the contradiction between considerations associating the strengths of the chemical bonds to the overlaps between atomic orbitals /1-4/ and treatments where the atomic orbitals were assumed to be orthogonal one to another /5/. The problem was at the same time a physical and a mathematical one, as will be illustrated below. Löwdin's work (/6/ and references given later) played a decisive role in formulating correctly and analyzing this problem.

1. PHYSICAL ORIGIN OF THE NON-ORTHOGONALITY PROBLEM

In connection with the physical aspects of the non-orthogonality problem, two questions may be asked. Has neglect of overlap a consequence on the general ability of a simple model to explain facts? Has overlap a physical meaning? The first question was the immediate stimulus for the non-orthogonality problem in its original form, the other question being more generally connected with the general topic "bases and models".

1a. Overlap and Simple Quantum Mechanical Models

The bond orbital model of a bond will be used both to introduce our notation and to illustrate a very important point. Given two atomic orbitals $|\mu\rangle$ and $|\nu\rangle$ associated to atoms M and N, and having an overlap integral $S = \langle\mu|\nu\rangle$ the single bond M-N is represented by a doubly occupied bond orbital $|b\rangle$

$$|b\rangle = c_{\mu b}|\mu\rangle + c_{\nu b}|\nu\rangle \qquad (1)$$

where the coefficients $c_{\mu b}$ and $c_{\nu b}$ form a column matrix $\mathbf{C}$ satisfying the equations

$$\mathbf{M}(\varepsilon_b)\ \mathbf{C}_b = 0 \qquad \mathbf{C}_b^{\dagger}\ \mathbf{S}\ \mathbf{C}_b = 1 \qquad (2)$$

Here the 'bond orbital' energy ε_b is the lower root of the secular equation associated with Eq. (2); the matrices $\mathbf{M}$ and $\mathbf{S}$, if $\mathbf{H}$ is the effective Hamiltonian matrix of the bond, are

$$\mathbf{M}(\varepsilon) = \begin{vmatrix} H_{\mu\mu} - \varepsilon & H_{\mu\nu} - \varepsilon S \\ H_{\mu\nu} - \varepsilon S & H_{\nu\nu} - \varepsilon \end{vmatrix} \qquad \mathbf{S} = \begin{vmatrix} 1 & S \\ S & 1 \end{vmatrix} \qquad (3)$$

respectively. In the same way one defines the root ε_a corresponding to the antibonding orbital, $|a\rangle$, which represents a repulsive one-electron state. The important point is that the dependence of the roots ε_a, ε_b on the value of overlap is such that the two roots appear to be symmetric with respect to the average $(\frac{H_{\mu\mu} + H_{\nu\nu}}{2})$ only when $\mathbf{S} = 0$. This is a general phenomenon for any number of basis orbitals. It becomes physically important mainly when the problem at hand involves occupied orbitals, as is the case with the radicals and ions of conjugated hydrocarbons in the π-electron Hückel method. In that method wide use was made of the symmetry of bonding and antibonding orbitals with respect to an appropriate 'non-bonding' energy /7/. Those results would lose much of their elegance and validity if overlap were not assumed to be zero. The most serious aspect of this can be illustrated in the case of the bond-orbital model by considering a negative ion - a three-electron system. The sum of the orbital energies of the three electrons as determined from Eq. (3) will never be positive if $\underline{S} = 0$, but can be as positive as desired if $\mathbf{S}$ is sufficiently close to unity - which can happen for certain δ bonds constructed with hybrid atomic orbitals. Therefore the overlap-neglecting one may lead to opposite conclusions regarding the very stability of the negative ion. This kind of difficulty is what is meant by the physical implications of the non-orthogonality problem. The latter, of course, would not subsist if the general theory did not admit in principle orthogonal bases. In fact, it does; even in a theory like that of vibronic transitions /8/ where the presence of the Franck-Condon factors seems to suggest that overlap is an essential ingredient of the theory, closer inspection shows that one could use orthogonal vibrational states in the same way as one uses orthogonal electronic states, basing the theory directly on the transition integrals, which the current theory treats as being proportional to the overlap integrals. The fact that in this problem it would be somewhat awkward to do without overlap is related to the significance of overlap itself in connection with the classical treatment of vibrations: it measures the average probability amplitude for the vibrating system to occupy the same positions in space in the states under consideration.

1b. Binding and Overlap

Intuitively, the interpretation just mentioned holds also for electron states like the atomic orbitals. This prompted Slater /1/ Mulliken /2/ Pauling /3/ to base the explanation of directed valency and other features of chemical binding on overlap. More generally, the whole MO-LCAO model of the electron distribution in a molecule rests on the atomic orbital concept at all levels of sophistication. And atomic orbitals form intrinsically non-orthogonal systems in the frame of the intuitive model which sees each atom entering the molecule grosso modo with its own orbitals. Even hybridization is seen at this stage as a zero-order mixing of (degenerate) hydrogenlike atomic orbitals, removing the symmetry but not changing the central nature of the force field to which those orbitals correspond /9/. On the other hand, the great successes of quantum chemistry, which led Coulson to say in 1959 that the best plums have already been plucked /10/, were largely based on neglect of overlap. Even later, neglect of overlap was a basic feature of models and methods; in particular the so-called PPP and the various CNDO methods /11,12/ are based on an assumption much stronger than orthogonality, - complete neglect of differential overlap. Even if this feature could be proven to be equivalent to physically more palatable approximations, like those introduced in the Hubbard Hamiltonian /13,14/, orthogonality of the basis would remain essential. Neglect of overlap also answers to criteria of simplicity and ease of interpretation in certain problems, like the interpretation of conjugation. There, the ansatz that two atomic orbitals are coupled just by the bond parameter (or 'hopping integral') $\langle\mu|\hat{H}|\nu\rangle$ ($\underline{\hat{H}}$ being the effective Hamiltonian) is sufficient to explain a number of facts, in spite of difficulties like those mentioned above in connection with level symmetry. In conclusion, a non-orthogonality problem exists in quantum chemistry because of the alternative between using overlap as an essential feature of a physical model and neglecting overlap entirely. The same situation is found in solid-state physics when the tight-binding model of solids is adopted. The connection between the two points of view and the mathematical implications of either assumption is a very important question. Löwdin's role in bringing those points to the light and in suggesting a very important solution can hardly be underestimated. It did not consist only in proposing the well known procedure which we shall widely refer to in the following /15/, but in encouraging discussions and studies on the problem, as the present author knows by personal experience. Moreover, Löwdin anticipated many conceptual and formal difficulties, as is proven by the very fact that most of the work on the question of overlap was developed after Löwdin's famous paper of 1950. That the non-orthogonality problem is still alive and should be further investigated is proven by very recent papers in the new fields of quantum chemistry and solid-state physics, e.g. some work by Grimley /16/ on the theory of chemisorption to which we shall refer in the following.

2. SYMMETRIC ORTHOGONALIZATION IN THE MO-LCAO MODEL

2a. Non-orthogonal Orbital Bases

The MO-LCAO model of molecules applies to the electrons of a molecule in the Born-Oppenheimer approximation. It studies at various levels of sophistication an independent particle model where the one particle states - the molecular orbitals $|k\rangle, |\ell\rangle, \ldots$ which form an orthonormal set $|k\rangle$ - are expressed as linear combinations of 'atomic orbitals' $|\mu\rangle, |\nu\rangle, \ldots$ - forming a system $|\chi\rangle$ of normalized but not necessarily orthogonal state vectors. In matrix notation, if $\underline{C}$ is a matrix having as many rows as the elements of $|\chi\rangle$ and as many columns as the elements of $|\mathbf{k}\rangle$, and the two sets are treated as row matrices, then

$$|\mathbf{k}\rangle = |\chi\rangle C. \tag{4}$$

The mathematical background of this simple relationship is rather formidable, as is well known from any textbook on quantum mechanics. Here it suffices to say that - regardless of whether the vectors in (4) form finite or infinite sets - the set $|\chi\rangle$ must be a basis of the vector space of the states for one particle, and thus it must be either complete in that space or be used just to represent a closed subspace of it. As is the case with any basis in a vector space, a matrix representation $\mathbf{S}$ of the metric tensor is associated with $|\chi\rangle$. The properties of $\mathbf{S}$ are that it should be symmetric and, by virtue of (4), positive definite. Following the standard procedure, we can define the conjugate space where $\langle\chi|$ is a basis, and define S as the matrix of the scalar products between pairs of elements of one of the elements being replaced by its conjugate

$$\mathbf{S} = \langle\chi|\chi\rangle \tag{5}$$

In the language of tensor algebra, $\mathbf{S}$ is the covariant form of the metric tensor on the basis $|\chi\rangle$. The mixed form, which must be the unit matrix $\mathbf{I}$, is obtained by introducing the basis

$$|\bar{\chi}\rangle = |\chi\rangle \mathbf{S}^{-1}:$$

$$\langle\bar{\chi}|\chi\rangle = \langle\chi|\bar{\chi}\rangle = \mathbf{I} \qquad (6)$$

while the contravariant form is

$$\langle\bar{\chi}|\bar{\chi}\rangle = \mathbf{S}^{-1}\langle\chi|\chi\rangle\mathbf{S}^{-1} = \mathbf{S}^{-1} \qquad (7)$$

The fact that we can associate with a given basis element two vectors, $|\mu\rangle$ and $|\mu\rangle$, is a very important point for many applications. It is sufficient here to write the time dependent Schrödinger equation in matrix form. Starting from

$$i\hbar \frac{\partial \psi\rangle}{\partial t} = \hat{H}|\psi\rangle \qquad (8)$$

and from the expansion of the state set $|\psi\rangle$ in the basis $|\chi\rangle$

$$|\psi\rangle = |\chi\rangle\mathbf{T} \qquad (9)$$

we find

$$i\hbar\mathbf{S}\frac{d\mathbf{T}}{dt} = \langle\chi|\hat{H}|\chi\rangle\mathbf{T} \qquad (10)$$

i.e.

$$i\hbar\frac{d\mathbf{T}}{dt} = \bar{\mathbf{H}}\mathbf{T}, \text{ with } \bar{\mathbf{H}} = \langle\bar{\chi}|\hat{H}|\chi\rangle \qquad (11)$$

($\bar{\mathbf{H}}$ corresponds to the mixed representation of the 'tensor' $\hat{H}$). Equation (11) is the form of the Schrödinger equation that corresponds to the eigenvalue equation

$$\mathbf{T}_{st}\mathbf{E} = \bar{\mathbf{H}}\mathbf{T}_{st}, \text{ i.e. } \mathbf{H}\mathbf{T}_{st} = \mathbf{S}\mathbf{T}_{st}\mathbf{E} \qquad (12)$$

where $\mathbf{T}_{st}$ now denotes the matrix T of equation (9) when $|\psi\rangle$ is the set of stationary states; $\mathbf{E}$ denotes the diagonal energy matrix; $\mathbf{H} = \langle\chi|\hat{H}|\chi\rangle$ is the covariant representation of $\hat{H}$. Thus, we must make use of two ket bases (and the two corresponding bra bases) in order to discuss consistently quantum mechanical problems on a non-orthogonal basis. A typical example of this situation is provided by the formulation of the Schrödinger equation in second quantization /16/. If $\mathbf{c}^+$ is the row of creation operators over the spinorbitals $|\underline{k}\sigma\rangle$ obtained from the basis $|\mathbf{k}\rangle$, $\hat{\mathbf{c}}$ the corresponding column of annihilation operators, the Hamiltonian operator can be written

$$\hat{H} = \hat{\mathbf{c}}^+\mathbf{H}_o\hat{\mathbf{c}} + \tfrac{1}{2}\hat{\mathbf{c}}^+\hat{\mathbf{W}}\hat{\mathbf{c}} \qquad (13)$$

where $\mathbf{H}_o$ is the (covariant) matrix representation of the core Hamiltonian and $\hat{\mathbf{W}}$ is a matrix whose elements are operators:

$$\hat{\mathbf{W}} = \sum_{n,p,\sigma} (\ell m|r_{12}^{-1}|np)\ \hat{c}^{+}_{n\sigma}\ \hat{c}_{p\sigma} \tag{14}$$

the numerical coefficients in the summation being the two-electron interaction integrals over the spinorbitals $|\ell\sigma\rangle$, $|m\sigma\rangle$ for one electron $|n\sigma'\rangle$, $|p\sigma'\rangle$ for the other. Equation (13) and (14) hold regardless of the choice of the one-particle spinorbital basis. However, there is an important difference in the meaning of the annihilation and creation operators depending on whether the basis is orthogonal or not. In the former case, one of such operators just creates or destroys an electron in the spinorbital to which it is associated. In the latter case the creation operator associated to the spinorbital $|m\sigma\rangle$ creates an electron in the spinorbital $|\bar{m}\sigma\rangle$ of the dual ket basis defined in Eq. (6). Another interesting consequence of non-orthogonality is that, if the Hamiltonian operator is used in its covariant matrix representation, the Greenian operator must appear in its contravariant representation to play the same role as with an orthogonal basis /16/.

2b. Symmetric Orthogonalization

- As is well known from tensor calculus, the choice of an orthogonal reference system suppresses all the complications just mentioned, since in that case the representation of the metric tensor is just the unit matrix, and the covariant, mixed and contravariant components coincide. The use of a non-orthogonal basis may be suggested by physical considerations, but certainly does not simplify the general theoretical scheme.

On the other hand, the physical significance of the basis, especially in view of approximations, is very important notwithstanding the fact that mathematically speaking complete bases are all equivalent.

A solution to this dilemma may come by referring to the most convenient and logically consistent orthogonal basis derived from the 'natural' non-orthogonal basis. The uniqueness of the choice is ensured by introducing an optimization requirement. The latter can be formulated in one of two ways. The more intuitive one is that no element of the new basis should be privileged with respect to the others ('symmetric' orthogonalization); the more formal one is that the new orthogonal state vectors should be as close as possible to the corresponding non-orthogonal ones. The latter requirement can be translated mathematically into the condition that the sum of the 'cosines' of the angles between corresponding old and new vectors be as high as possible subject to the condition that the state vectors forming the new basis be orthonormalized /17,18/.

Call $|\varphi>$ the new orthonormal basis. It will be related to $|\chi>$ through a non singular transformation Q

$$|\varphi> = |\chi> Q \quad (15)$$

We have to maximize the expression:

$$\mathrm{Tr}\ (<\varphi|\chi> + <\chi|\varphi>)\ \mathrm{Tr} \quad (16)$$

where M is a matrix of Lagrange multipliers. Substituting (15) into (16) we obtain

$$f(Q) = (Q^{+}S + SQ) - \mathrm{Tr}\ (MQ^{+}Q + Q^{+}QM^{+}) = \underline{\max} \quad (17)$$

Differentiating this expression with respect to Q^{+} and to Q we obtain - using the property of invariance of the trace with respect to cyclic permutations of the factors:

$$Q = SL \ , \quad L = + (M + M^{+})^{-1}. \quad (18)$$

L is thus a symmetric matrix to be determined by requiring that be an orthonormal basis:

$$<\varphi|\varphi> = Q^{+}S\ Q = LS^{3}L = I \therefore L = S^{-3/2} \quad (19)$$

Hence:

$$|\varphi> = |\chi> S^{-1/2} \quad (20)$$

This is the famous symmetric orthogonalization introduced by Löwdin in 1950 /15/.

Denoting the matrices in the new basis by a double bar, we have in particular:

$$\bar{\bar{H}} = S^{-\frac{1}{2}} H S^{-\frac{1}{2}}; \quad \bar{\bar{C}} = S^{\frac{1}{2}} C \quad (21)$$

where account has been taken of the fact that the set $|k>$ is given by

$$|\varphi> \bar{\bar{C}} = |\chi> C \quad (22)$$

The eigenvalue equation associated with $\bar{\bar{H}}$ is now, of course

$$\bar{\bar{H}}\,\bar{\bar{C}} = \bar{\bar{C}}\,E \quad (23)$$

and the Schrödinger equation is (cf. Eq. 11)!

$$i\hbar \frac{d\bar{\bar{T}}}{dt} = \bar{\bar{H}}\,\bar{\bar{T}}. \tag{24}$$

No problem arises in treating this kind of equation because the distinction between $\mathbf{H}$ and $\bar{\mathbf{H}}$, which was important with the non-orthogonal basis, is now removed.

2c. STO and LOAO's

We now compare the properties of the non-orthogonal orbitals in the case of STO's with Löwdin orbitals and with orthogonal orbitals obtained by the Schmidt procedure. The latter, as is well known, lies at the other extreme of Löwdin's procedure, in that it privileges one element of the original basis, which remains unchanged, and successively introduces more and more complicated corrections to the other elements of that basis. It depends on the order in which the elements are taken, and corresponds to building a triangular transformation matrix leading from the non-orthogonal to the orthogonal basis /19/.

The comparison in question can bear on two different aspects: computational and physical. The two are actually so strictly connected that we shall not attempt to keep a rigorous distinction between them in the following. The most important point concerning Löwdin's orbitals is a good example of this: the discussion of how localized they are. The linear transformation (20) leading from typical atomic orbitals to Löwdin ones adds to the atomic orbital associated with one center of a molecule contributions coming from the other centers - so that at first sight the orthogonalized orbitals are more diffuse than the original ones.

Actually Löwdin's orbitals are delocalized at least in so far as they have small cusps at the nuclei other than those to which they are associated. However, if standard measures of size are used, those orbitals appear to be more concentrated around the corresponding nuclei than the original STO's /20/. This result is intuitively acceptable if orthogonality is associated with separation in space; but this is not always the case, because orthogonality may be due to interference, as with molecular orbitals which extend over the same region of space. With σ systems this happens for LOAO's; the overlap between hybrids in a CH bond may reach values as high as 0.713 /21/. An important point is also that measures of delocalization depend on the particular operator used to measure the "size" of an orbital.

A particular point regarding the delocalized nature of the Löwdin orthogonalized orbitals is that cusps affect the kinetic energy more than other features of orbitals, and this might lead to difficulties when calculations not exactly invariant under a linear transformation are carried out. Doubts arose on the use of orthogonalized orbitals for such calculations /22/, but the conclusion was that the advantages of using orthogonalized orbitals and the optimal quality of Löwdin's ones justified use of the latter even with a 'small sacrifice in the energy' /23/.

The comparison of Löwdin's orbitals with orthogonalized bases was already discussed by Löwdin himself in his original papers, along with combined orthogonalization procedures /19/. There are cases when symmetric orthogonalization is not what one wants. The best example is orthogonalization of 2s orbitals to inner-shell 1s ones. It is clear that, if an STO 1s orbital is satisfactory for the K shell, it is the 2s STO of the same atom that must be made orthogonal to the 1s's by proper admixture of the latter.

2d. Orthogonalization of Linearly Dependent Bases

In his analyses of LCAO bases Löwdin has also treated the case of bases containing linear dependencies. In this connection the problem arises of orthogonalization when $\mathbf{S}$ is a singular matrix. The canonical orthogonalization proposed by Löwdin /19,24/ for this case can be derived from the optimization condition expressed by Eq. (17).

Let

$$\mathbf{N} = (\mathbf{M} + \mathbf{M}^+), \quad \mathbf{Q}\mathbf{N} = \mathbf{S} \tag{17a}$$

where (17) is in the form deriving immediately from (18) the matrix $\mathbf{N}$ must be a singular matrix because $\mathbf{S}$ is singular, and $\mathbf{L}$ cannot be defined. However, (17a) can be solved provided some condition on $\mathbf{Q}$ is imposed. In particular, require that $\mathbf{Q}$ should be a mxm matrix, m being the number of corresponding elements of the linearly dependent basis. If n is the number of the non-zero eigenvalues of $\mathbf{S}$, we write

$$\mathbf{Q}^+ \mathbf{S}\mathbf{Q} = \mathbf{I}^{(nn)} \tag{25}$$

where $\mathbf{I}^{(nn)}$ is an mxm matrix having n diagonal elements equal to one, zero elsewhere. A general solution of (25) is

$$\mathbf{Q} = \mathbf{U}\boldsymbol{\sigma}(-\tfrac{1}{2})\,\mathbf{T}, \quad \mathbf{U}^+\mathbf{U} = \mathbf{T}^+\mathbf{T} = \mathbf{I} \tag{26}$$

where $\mathbf{T}$ is arbitrary, $\mathbf{U}$ is the unitary mxm matrix which diagonalizes $\mathbf{S}$ to a matrix $\boldsymbol{\sigma}$. The matrix $\boldsymbol{\sigma}(-\frac{1}{2})$ is the matrix $\boldsymbol{\sigma}$ where only the non-zero elements are replaced by their inverse square roots.

Substituting (26) into (17a) and remembering that **N** is Hermitian we find

$$\mathbf{Q} = \mathbf{U}\,\boldsymbol{\sigma}\,(-\tfrac{1}{2})\,\mathbf{U}^{+} \tag{27}$$

This gives Löwdin's canonical orthogonalization if $\mathbf{U}^{+}$ is suppressed as well as the zero columns of $\boldsymbol{\sigma}(-\frac{1}{2})$, and **Q** becomes a rectangular matrix taking from $|\chi>$ to an orthonormalized basis with only n elements. Equation (27) sic et simpliciter is formally more satisfactory, but in fact defines a new linearly dependent set and demands an (arbitrary) choice of the n independent basis elements.

3. EXTENSIONS OF THE NON-ORTHOGONALITY PROBLEM

The impact of Löwdin's interest in the non-orthogonality problem was by no means limited to the introduction of symmetric orthogonalization. Actually he called attention more generally to the basis problem in the MO-LCAO method. Use of advanced techniques of matrix algebra was also one of his most important contributions to quantum chemistry and solid state physics. Thus it may correctly be said that many of the later developments of those studies are largely due to Löwdin's work, even when he did not contribute directly to the details. This is particularly true of much of the present author's work.

The developments of the non-orthogonality problem and basis studies are quite complicated and difficult to review, because they are intermingled with developments in computational chemistry. We shall mention here only three points which are particularly familiar to the present author and give some insight into the various facets of the question:

a) orthogonalization by contraction of the basis;
b) orthogonalization by non-linear transformation of the basis;
c) intermediate and partial Löwdin orthogonalization.

3a. Orthogonalization by Contraction of the Basis

Once the importance of the choice of the truncated basis in connection with orthogonality has been recognized, the search for extensions of the standard orthogonalization procedures becomes interesting. Normally, only linear transformations of a given truncated set are considered. However, one may wish to construct a truncated basis according to previously given specifications /25/. In that case one selects a convenient initial complete basis set $|\chi_{tot}>$ and extracts from it the required truncated basis. In particular, if $|\chi>$ must be an orthonormal basis where the individual elements have no cusps except on the atoms with which they are associated one may proceed as follows. The basis $|\chi_{tot}>$ may be one of STO's (Slater type orbitals) associated with the various atoms. This basis contains an infinite number of elements for each atom; yet in general, it is not complete, because the STO's do not form complete sets for the various atoms unless they are supplemented by the corresponding continua. If the continua are added, one may get overcompleteness for $|\chi_{tot}>$ /19,26/; if only one continuum is included,*e.g.* the one associated with the center of mass of the molecule taken as a van der Waals scattering center for free electrons, then the set is probably just complete, but it remains to be proved in general that the overlap matrix for the bound part of $|\chi_{tot}>$ is really non-singular, thus satisfying the required necessary and sufficient condition for linear independence /19/. With the above premises, let us assume that the starting basis has been chosen in the most convenient way for the present purpose (otherwise a linear transformation might be required), and in particular that a subset of STO's are associated with each atom. Calling X,Y,... the individual atoms, and denoting by Σ the spherical harmonic entering a given element, we can divide the basis into subsets denoted $|\chi_{tot}(X,\Sigma)>$. These subsets have zero overlap with other subsets associated with the same atom, but each contains in general several non-orthogonal elements associated with different principal quantum numbers. They can be truncated and contracted to a final orthonormal limited basis by a sequence of operations. First of all a truncation transforms $|\chi_{tot}>$ into a basis $|\chi_{ext}>$ which consists of finite subsets $|\chi_{ext}(X,\Sigma)>$, each associated with a single atom and a single spherical harmonic and containing $\underline{m}$ (X,Σ) elements. The final orthonormal atomic basis may then be constructed by contraction of the individual sets $|\chi_{ext}(X,\Sigma)>$ of $|\chi_{ext}>$ from $\underline{m}$ to $\underline{n}\underline{x}\underline{m}$ elements each so as to obtain final sets $|\chi(X,\Sigma)>$ consisting of $\underline{n}$ strictly atomic orbitals orthogonal to one another and to all the other atomic orbitals of the final basis $|\chi>$. Contraction is a combination of a linear transformation followed by it can be represented by a rectangular matrix $\mathbf{Q}$ which in our case consists of subblocks $\mathbf{Q}(X, \Sigma)$ having m rows and n columns. The equations defining these subblocks are

$$\mathbf{Q}^{+}(X', \Sigma')<\chi_{ext}(X', \Sigma')|\chi_{ext}(X, \Sigma)>\mathbf{Q}(X,\Sigma) = 0 \quad (25)$$

$$\text{for } X' \neq X;$$

$$\mathbf{Q}^{+}(X, \Sigma)<\chi_{ext}(X, \Sigma)|\chi_{ext}(X, \Sigma)>\mathbf{Q}(X, \Sigma) = \mathbf{I}^{(nn)} \quad (26)$$

where $\mathbf{I}^{(nn)}$ denotes the identity matrix of order $\underline{n}$ (X, Σ). If only one orbital of a given symmetry species per atom is desired, and only the nearest four neighbors are taken into account (25) and (26) correspond to 17 conditions per atom. Thus, the required basis

$|\chi_{ext}>$ is very large, and the choice of the initial atomic orbitals is a very delicate problem. Some aspects of the case when they consist of polynomials in r (distance of the electron from the nucleus) multiplying the standard exponentials with suitably chosen orbitals exponents and the appropriate spherical harmonics are treated in Ref. 27. The orbitals given by the above procedure, may be considered as 'Modified Atomic Orbitals' (MAO) in Mulliken's sense /28/. As has been mentioned, they have the atomic property of having cusps only on the corresponding atoms. This leaves the cusps on more than one center as a characteristic of molecular orbitals. On the other hand, among other things, the cusps appearing in the MAO's are far from corresponding to the contributions of individual atoms to the kinetic energy.

3b Orthogonalization by E-Dependent Transformation of the Basis

This particular procedure was introduced by the present author following discussions with prof. Löwdin, and was applied to the special case when the off-diagonal part of the Hamiltonian matrix was proportional to the overlap matrix /29/. It may be generalized to a form which provides better insight into the general non-orthogonality problem and into the very special qualities of Löwdin's orthogonalization.

Start as usual from the non-orthogonal basis $|\chi>$ and the Hamiltonian matrix $\mathbf{H}$. With $\mathbf{S}$ the overlap matrix, $\mathbf{C}$ the eigenvector matrix, and $\mathbf{E}$ the diagonal orbital-energy matrix, the secular equation is (cf. Eq. 12):

$$\mathbf{HC} = \mathbf{SCE} \tag{27}$$

Let a and b be two given scalars, and call $\mathbf{A}$ and $\mathbf{B}$ respectively the diagonal and the off-diagonal parts of $\mathbf{H}$. It is possible in general to define a third matrix $\mathbf{D}$, such that:

$$\mathbf{S} = \mathbf{D} + b\mathbf{H} - a\mathbf{A} \tag{28}$$

where, of course,

$$\text{diagonal part of } \mathbf{D} + (b-a)\mathbf{A} = \mathbf{I} \tag{29}$$

Substituting in (17) and shifting some terms to the left side with due caution with regard to commutativity gives

$$\mathbf{HC}(\mathbf{I} - b\mathbf{E}) = (\mathbf{D} - a\mathbf{A})\mathbf{CE}. \tag{30}$$

Applying Löwdin's procedure, we rewrite (30) in the form

$$(\mathbf{D} - a\mathbf{A})^{-\frac{1}{2}}\mathbf{H}(\mathbf{D} - a\mathbf{A})^{-\frac{1}{2}}(\mathbf{D} - a\mathbf{A})^{\frac{1}{2}}\mathbf{C}\mathbf{X} = (\mathbf{D} - a\mathbf{A})^{\frac{1}{2}}\mathbf{C}\mathbf{X}\mathbf{X}^{-1}\mathbf{E}(\mathbf{I} - b\mathbf{E})^{-1}\mathbf{X} \quad (31)$$

or

$$\bar{\bar{\mathbf{H}}}\bar{\bar{\mathbf{C}}} = \bar{\bar{\mathbf{C}}}\bar{\bar{\mathbf{E}}} \quad (32)$$

with

$$\bar{\bar{\mathbf{E}}} = \mathbf{X}^{-1}\mathbf{E}(\mathbf{I} - b\mathbf{E})^{-1}\mathbf{X}, \quad \bar{\bar{\mathbf{C}}} = (\mathbf{D} - a\mathbf{A})^{\frac{1}{2}}\mathbf{C}\mathbf{X} \quad (33a,b)$$

and the corresponding expression for $\bar{\bar{\mathbf{H}}}$.

The matrix $\mathbf{X}$, so far not specified, is required in order to satisfy the normalization condition. It must be diagonal (and thus commute with $\mathbf{E}$) in order that equation (32) be a regular diagonalization equation. It is immediate that the condition

$$\bar{\bar{\mathbf{C}}}^{+}\bar{\bar{\mathbf{C}}} = \mathbf{X}^{+}\mathbf{C}^{+}(\mathbf{D} - a\mathbf{A})\mathbf{C}\mathbf{X} = \mathbf{I} \quad \text{with } \mathbf{C}^{+}\mathbf{S}\mathbf{C} = \mathbf{I} \quad (34)$$

can be rewritten, remembering that $\mathbf{C}^{+}\mathbf{H}\mathbf{C} = \mathbf{E}$.

$$\mathbf{X}^{+}\mathbf{C}^{+}(\mathbf{S} - b\mathbf{H})\mathbf{C}\mathbf{X} = \mathbf{X}^{+}(\mathbf{I} - b\mathbf{C}^{+}\mathbf{H}\mathbf{C})\mathbf{X} = \mathbf{X}^{+}(\mathbf{I} - b\mathbf{E})\mathbf{X} = \mathbf{I}$$

and this equation is satisfied by

$$\mathbf{X} = (\mathbf{I} - b\mathbf{E})^{-\frac{1}{2}} \quad (35)$$

which is a diagonal matrix, as desired.

Two remarks are relevant in the present connection. First, equation (33b) is a generalization of Löwdin's procedure, because it becomes identical with it when $(\mathbf{D} - a\mathbf{A}) = \mathbf{S}$, which, by (28), means $b = 0$. Second, the fact that $\bar{\bar{\mathbf{C}}}$ is obtained by multiplying $\mathbf{C}$ by matrices on both sides implies that we cannot define a single basis to be associated with (32). In fact, for each eigenvalue $\underline{E}_j$ we have the basis

$$|\bar{\bar{\mathbf{X}}}^{(j)}\rangle = |\mathbf{X}\rangle(\mathbf{D} - a\mathbf{A})^{\frac{1}{2}}(1 - bE_j) \quad (36)$$

The various bases thus differ by scalar factors (the unit on a common reference axis). These scalar factors are functions of the original eigenvalues $\underline{E}_j$, which is why we speak of an $\underline{E}$-dependent basis, and they become identical only if $b = 0$, viz. for LOAO's. The other special case of (32) is when $\mathbf{D} = \mathbf{I}$ and $a = b$. Then, $\mathbf{D} - a\mathbf{A}$ is a diagonal matrix, and the transformation (36) does not delocalize the basis. This is the case treated in Ref. 29. Among other things, that work makes possible to reconcile the frontier-electron method of Fukui /30/ with the standard total charge treatment of reactivities. This lends some interpretational value to certain aspects of that orthogonalization procedure; more support is provided by the fact that it reconciles orthogonality of the eigenvector matrix **C** with the atomic character of the atomic orbitals.

On the other hand, the general case (32) does not fulfill another requirement implicit in Löwdin's orthogonalization procedure, viz. that the eigenvalue spectrum of the new representation of the Hamiltonian operator be the same as that of the old one. Equation (32) involves a slight generalization of that condition, because it admits the fuchsian parabolic transformation /31/ of the eigenvalues (33a). This may be taken as the transformation generally involved when overlap is neglected, the special values of a, **A** and b being determined by the invariants chosen in the specific cases. Note also that (34) removes in **E** the symmetry found in $\bar{\bar{\mathbf{E}}}$ (cf. the introduction).

3c. Intermediate and Partial Löwdin Orthogonalization

A third important application and extension of Löwdin's symmetric orthogonalization is found when one wants only certain blocks of the overlap matrix equal to zero. This may be the case, for instance, in a basis ensuring full mutual orthogonality of bond or group orbitals (vide infra). The problem has been treated by several authors /32,33/ and can be summarized as follows.

Let **S** be the overall overlap matrix and suppose that one wants to find a new basis $|\lambda\rangle$ whose overlap matrix is $\mathbf{S}_o$ (e.g., a block diagonal matrix extracted from **S** or even just a matrix which differs from **S** by some general correction matrix). We write:

$$|\lambda\rangle = |\chi\rangle \mathbf{Q}\ ,\ \mathbf{S} = \langle\chi|\chi\rangle = \mathbf{S}_o + \mathbf{S}_1, \tag{37}$$

and set

$$\mathbf{Q} = (\mathbf{I} + \mathbf{S}_o^{-1}\mathbf{S}_1)^{-\frac{1}{2}} = \mathbf{S}^{-\frac{1}{2}}\mathbf{S}_o^{\frac{1}{2}} \tag{38}$$

The basis $|\lambda\rangle$ then has an overlap matrix $\mathbf{S}_o$. This is proven by orthogonalizing $|\chi\rangle$ completely and multiplying by $\mathbf{S}_o^{\frac{1}{2}}$:

$$|\lambda\rangle = |\chi\rangle \mathbf{S}^{-\frac{1}{2}}\mathbf{S}_o^{\frac{1}{2}};\ \langle\lambda|\lambda\rangle = \mathbf{S}_o^{\frac{1}{2}}\mathbf{S}^{-\frac{1}{2}}\langle\chi|\chi\rangle\mathbf{S}^{-\frac{1}{2}}\mathbf{S}_o^{\frac{1}{2}} = \mathbf{S}_o.$$

The complete generality of this result, where the only special condition (non-singularity of inverted matrices) is automatically satisfied, opens the way to many applications, in particular to the problem of localization.

The matrix $\mathbf{S}_0^{-1}\mathbf{S}_1$ will normally be small, and consequently simple approximation formulas may be used with success.

A very useful and accurate procedure is Hartree's iteration formula /34/. If $\mathbf{D}$ is the given matrix, and $\mathbf{R}$ its inverse square root, the _n_-th approximation to $\mathbf{R}$ is

$$\mathbf{R}^{(n)} = \tfrac{1}{2}\mathbf{R}^{(n-1)}\left(3\cdot\mathbf{I} - \mathbf{D}\mathbf{R}^{(n-1)^2}\right) \qquad (39)$$

whose convergency is quite good /35/. Difficulties are anyway present every time $\mathbf{D}$ is a very large matrix and the computer is comparatively limited.

4. APPLICATIONS OF ORTHOGONALIZATION PROCEDURES

We shall consider here only three topics where orthogonalization procedures play an important role widely exceeding that of purely computational convenience. These are:

a) physical role of overlap;
b) localized description of molecules;
c) semi-empirical methods.

We shall refer explicitly to orbital treatments of molecules. The more general problem of overlap between many-electron states and between complete states of molecules and solids can be treated along the same lines; Section _4a_ applies without modification to overlapping quantum-mechanical states.

4a. Physical Role of Overlap

If the matrix representing the Hamiltonian operator of a physical system in a basis $|\psi\rangle$ is diagonal, the states of the basis are stationary states of the given system; but a necessary condition for this is that the basis be an orthonormal one. Thus, overlap plays the same role as the off-diagonal elements of the Hamiltonian matrix: it _couples_ the states one to another. The introduction of overlap into physical analyses of problems, however natural it may be, is physically redundant or amounts to a contribution to the forces between the systems described by the individual basis elements. Keeping in mind that physical equivalence between different mathematical descriptions is obtained only if the eigenvalue spectrum remains the same, and that a physically significant Hamiltonian matrix must be Hermitian, Löwdin's symmetric orthogonalization appears immediately as the most convenient technique for eliminating this overlap redundancy.

As an illustration take a two-state system like that already discussed in sec. 1. The physical problem one may be thinking of is two square wells located at a finite distance, each having one state in a given energy range and widely separated from the others. This is also a model of a bond.

Write the Hamiltonian matrix and the overlap matrix in the forms

$$\mathbf{H} = \begin{vmatrix} A & V \\ V & B \end{vmatrix} \qquad \mathbf{S} = \begin{vmatrix} 1 & S \\ S & 1 \end{vmatrix} \tag{40}$$

and assume B, A, V<0, B<A.

The standard interpretation of the above quantities is:

<u>A</u> is the energy of the given system in state $|1\rangle$;
<u>B</u> is the energy of the given system in state $|2\rangle$;
<u>V</u> is the coupling energy.

Introducing the further notation: $\Sigma = (A + B)/2$, $\Delta = (B-A)/2$, Σ may be taken as a zero point, and the two eigenvalues when $S = 0$ are $E_a, E_b = \pm\sqrt{\Delta^2 + V^2}$. Thus

$$\begin{aligned} D_a &= E_a - (-\Delta) = \sqrt{\Delta^2 + V^2} - |\Delta| \\ D_b &= E_b - (\Delta) = -\sqrt{\Delta^2 + V^2} + |\Delta| \end{aligned} \tag{41}$$

are the shifts of the energies of the two stationary states with respect to the diagonal elements of $\mathbf{H}$. In the case with overlap, the situation appears more complicated unless Löwdin's symmetric orthogonalization is applied. Then, with the original zero point, we have the orthogonalized Hamiltonian

$$\bar{\bar{\mathbf{H}}} = \mathbf{S}^{-\frac{1}{2}}\,\mathbf{H}\mathbf{S}^{-\frac{1}{2}} = \frac{1}{\cos^2\xi}\begin{vmatrix} \Sigma - \Delta\cos\xi - V\sin\xi & V - \Sigma\sin\xi \\ V - \Sigma\sin\xi & \Sigma + \Delta\cos\xi - V\sin\xi \end{vmatrix} \tag{42}$$

where $\xi = \arcsin S$.

This means that we have a shift in the arithmetic-mean zero-point which becomes

$$\Sigma' = \Sigma - V'\,S \tag{43}$$

with

$$V' = \frac{V - \Sigma\sin\xi}{\cos^2\xi} = \frac{V - \Sigma S}{1-S^2}; \quad (44)$$

and level separation with respect to A and B:

$$D'_a = \sqrt{\Delta^2(1-S^2) + V'^2} - |\Delta| + |V'|S \quad (45)$$

$$D'_b = -\sqrt{\Delta^2(1-S^2) + V'^2} + |\Delta| + |V'|S \quad (46)$$

Thus Löwdin's symmetric orthogonalization shows that the overlap-including problem is equivalent to a problem where the basis states are orthogonal but are shifted and coupled by $\underline{V}'$. Equation (46) might represent the distance ($\underline{R}$) dependent part of the orbital energy of H_2^+ and, in a semiclassical interpretation the corresponding force between the atoms would be (in this case $\Delta = 0$):

$$-\frac{1}{R^2} - \frac{\partial D_b}{\partial R} = -\frac{1}{R^2} + \frac{\partial|V'|}{\partial R}(1 + S) - |V'|\frac{\partial S}{\partial R} \quad (47)$$

Thus overlap plays a physical role by contributing to the forces of attraction between the two atoms.

4b. Localized Description of Molecules

The above considerations find an immediate application in connection with localization. Suppose one starts from the 'maximum localization criterion', which consists in constructing on the individual atoms sets of orthonormal hybrids each as close as possible to what would be obtained according to a local maximum overlap criterion. These hybrids give a new overlap matrix having blocks with large elements associated to bonds (or π systems)and many smaller elements associated with 'non-bonded interactions'. Physically speaking (See 4a) the latter elements are equivalent to couplings; it is not useful to keep them separated from other kinds of interactions because they complicate a localized-orbital model and formulas for many-electron energies on a bond orbital basis /36/. A partial symmetric orthogonalization removing the undesired small overlap elements is interesting; Eq. (37) provides the necessary mathematical formula.

The above subject can be attacked the other way round: instead of requiring orthogonality of localized orbitals as an aspect of localization, one may inquire whether current localization techniques lead to orthogonal orbitals. This problem has been briefly analyzed by Lévy /37/ who has remarked that nothing imposes that such orbitals should be orthogonal one to another; indeed, non-orthogonal orbitals may have certain advantages, like better convergency of configuration interaction /38/.

Lévy has found that a certain matrix **A** which depends on the localization criterion must satisfy Mulliken's relationship (off-diagonal elements = overlap x arithmetic mean of corresponding diagonal elements) for the localized orbitals to be orthogonal. The deviation from that relationship measures the degree of non-orthogonality. The orbitals localized according to Ruedenberg's criterion /39/ are found to be nearly orthogonal one to another at least in the case of benzene. This is in agreement with the remark that symmetrically orthogonalized atomic orbitals are more localized than regular atomic orbitals /20, 40/ (cf. sec. 2), and qualifies the localization criterion to which reference is made.

The case of Boys's criterion /41/ is different, because the resulting localized orbitals overlap much more. In conclusion, according to Lévy, there is no regular correlation between a posterior localization and orthogonality; all depends on the criterion adopted. Even contradictions may arise: if one tries to maximize the size of the orbitals, defined as the expectation value of the operator $\hat{r}^2 - \hat{r}\langle\hat{r}\rangle$ (with r distance of the electron from a suitable center) /42/, one ends up with using Boys's criterion. This is not compatible with Ruedenberg's criterion.

4c. Semi-Empirical Methods and the Non-Orthogonality Problem

We have already mentioned semi-empirical methods in connection with the non-orthogonality problem for the Hückel method and for a priori localized bond models. Here we consider only the intriguing zero-differential overlap (ZDO) approximation.

Three references /11, 40, 43/ give a general picture of the situation. Historically, it was probably Berthier /11/ who first elaborated the suggestion by Löwdin that the ZDO approximation could be justified as a linear approximation if orthonormalized atomic orbitals were used as a basis. A two-electron integral between atomic orbitals orthogonalized by the Löwdin procedure consists essentially of a combination of Coulomb integrals between atomic orbitals multiplied by overlap ($\underline{S}$) plus a correction containing the second order in $\underline{S}$, if one of the two electrons is associated in the given integrals with two different LOAO's.

The ZDO approximation is connected with that of the Mulliken approximation for the products of non-orthogonal atomic orbitals:

$$\mu\nu \simeq S_{\mu\nu} \frac{\mu\mu + \nu\nu}{2} \qquad (48)$$

which has been generalized by Cizek /44/, who replaced the arithmetic mean in (48) by a weighted average. The approximation becomes valid with much larger values of S, and the range of validity of the ZDO approximation is slightly increased. Another important topic where orthonormality plays a role is in connection with charges and electron populations. The basic question is whether populations defined according to Mulliken /45/ are invariant under Löwdin's transformation. It is easily proved that they are not. In fact, a Mulliken population is a diagonal element of the matrix

$$\mathbf{Q} = (\mathbf{C}\mathbf{n}\mathbf{C}^{+}\mathbf{S} + \mathbf{S}\mathbf{C}\mathbf{n}\mathbf{C}\quad)/2 \tag{51}$$

where $\mathbf{C}$ is the matrix of the coefficients of the molecular orbitals, $\mathbf{n}$ is the occupation number matrix, $\mathbf{S}$ as before the overlap matrix.

The transformation to Löwdin orbitals gives

$$\mathbf{Q} = (\mathbf{S}^{-\frac{1}{2}}\,\bar{\mathbf{C}}\mathbf{n}\bar{\mathbf{C}}^{+}\mathbf{S}^{\frac{1}{2}} + \text{transpose})/2 \tag{52}$$

In the 2x2 case, with the notation of sec. 4a, and

$$\mathbf{P} = \bar{\mathbf{C}}\mathbf{n}\bar{\mathbf{C}}^{+} \qquad \mathbf{n} = \begin{vmatrix} 1 & 0 \\ 0 & 0 \end{vmatrix} \tag{53}$$

we have (noting that $P_{11} + P_{22} = 1$)

$$\mathbf{Q} = \begin{vmatrix} \frac{1}{2} + \frac{1}{2\cos\xi}(P_{11}-P_{22}) & P_{12} \\ P_{21} & \frac{1}{2} + \frac{1}{2\cos\xi}(P_{22}-P_{11}) \end{vmatrix} \tag{54}$$

Thus, if $|P_{11} - P_{22}| > \cos\xi$, one of the two diagonal elements of $\mathbf{Q}$ will be negative (P_{11} and P_{22} are definite positive). The interpretation of those diagonal elements as electron populations becomes unsatisfactory /45/. Following Löwdin /19/, the Taylor expansion of $\mathbf{S}^{-\frac{1}{2}}$ and $\mathbf{S}^{\frac{1}{2}}$ indicates that in general the diagonal elements of $\mathbf{Q}$ coincide with those of $\mathbf{P}$ to the second order in $\mathbf{S}$ /25/.

The case of charges is an illustration of how the LOAO's can be used in the analysis of interpretational schemes of molecular wavefunctions.

5. CONCLUSION

The purpose of the preceding sections has been to give a brief summary of the non-orthogonality problem and its importance for quantum chemistry more than twenty-five years after Löwdin brought to light its importance and suggested his famous procedure.

There are many more points which could be discussed in this connection: for instance, the extension to the continuum, the application to configuration interaction, the question of bases for vibrational wave function calculations, etc. All such developments are of great importance in today's quantum chemistry, and they will certainly be the object of further research. We shall not discuss them here, but we wish to emphasize again that precisely the wealth of past and future studies connected with LOAO's is perhaps the major outcome of Löwdin's work on non-orthogonality; he did not confine himself to presenting a new result or a new application, but realized and communicated to the other scientists its importance for the whole field of quantum chemistry and solid-state theory. By doing so, he actually showed the way to many of us and promoted a long and fruitful line of progress. The best proof of the relevance of his contribution is that, after such a long time, new horizons are still opening up for the non-orthogonality problem and Löwdin's symmetric orthogonalization.

REFERENCES

1. J.C. Slater, Phys. Rev., 37, 481 (1931), 38, 325, 1109 (1932).

2. R.S. Mulliken, Phys. Rev., 41, 67 (1932).

3. L. Pauling, J. Amer. Chem. Soc., 53, 1367 (1931).

4. R.S. Mulliken, J. Amer. Chem. Soc., 72, 4493 (1950).

5. G.W. Wheland, J. Amer. Chem. Soc., 63, 2025 (1941); B.H. Chirgwin and C.A. Coulson, Proc. Roy. Soc. (London), A201, 196 (1950).

6. P.O. Löwdin, Adv. in Quant. Chem., 5, 185 (1970).

7. Cf. M.J.S. Dewar, J. Amer. Chem. Soc. 74, 3341-3357 (1952).

8. Cf.e.g. Th. Förster, in "Modern Quantum Cemistry", O. Sinanoglu ed. New York, Acad. Press, 1965, vol III, p. 93.

9. G. Del Re, Adv. Quant. Chem., 8, 95 (1974).

10. C.A. Coulson, Revs. Mod. Phys., 32, 170 (1960).

11. G. Berthier, Tetrahedron, 19, s.2 1 (1963).

12. G. Klopman, B.O. Leary, Fortschr. Chem. Forsch., 15, 445 (1970).

13. J. Koutecky, Chem. Phys. Lett., 1, 249 (1967).

14. J. Hubbard, Proc. Roy. Soc. (London), A276, 238; A277, 237 (1964).

15. P.O. Löwdin, Arkiv Mat. Astron. Fysik, 35A, n. 9 (1947); J. Chem. Phys. 18, 365 (1950).

16. T.B. Grimley, J. Phys. C: Solid St. Phys., 3, 1934 (1970).

17. B.C. Carlson and J.M. Keller, Phys. Rev., 105, 102 (1957).

18. G. Del Re, Theoret. Chim. Acta, 1, 188 (1963).

19. P.O. Löwdin, Adv. in Phys. 5, 1, (1956) See in particular pp. 40-56.

20. R. McWeeny, in Molecular Orbitals in Chemistry Physics and Biology (Eds. P.O. Löwdin, B. Pullman) New York Acad. Press (1969).

21. A. Veillard, G. Del Re, Theoret. Chim. Acta, 2, 55 (1964).

22. B.J. Duke, Theoret. Chim. Acta, 8, 87 (1967).

23. J.A. Chapman, D.P. Chong, Theoret. Chim. Acta, 10, 364 (1968).

24. P.O. Löwdin, J. Appl. Phys. (suppl.) 33, 251 (1962).

25. G. Del Re, Int. J. Quant. Chem., 1, 293 (1967).

26. H. Shull and P.O. Löwdin, J. Chem. Phys., 23, 1362, 1565 (1955).

27. A. Rastelli and G. Del Re, Int. J. Quant. Chem., 3, 543 (1969).

28. R.S. Mulliken, J. Amer. Chem. Soc., 88, 1849 (1966).

29. G. Del Re, Nuovo Cim. 17, 644 (1960).

30. K.Fukui, T. Youezawa, H. Shingu, J. Chem. Phys., 20, 722 (1952).

31. E. Goursat, Cours D'Analyses Mathématique Gauthiers-Villars, Paris 1949 (7ed.), col. 2 p. 65.

32. G. Del Re, Int. J. Quant. Chem., 7S, 193 (1973).

33. H. Kashivagi, F. Sasaki, Int. J. Quant. Chem., 7S, 515 (1973).

34. D.R. Hartree, Proc. Camb. Phys. Soc., 45, 230(1949).

35. G. Berthier, Tetrahedron, 19, s.2 1 (1963).

36. G. Del Re, in Localization and Delocalization in Quantum Chemistry (eds. O. Chalvet *et al*) Reidel, Dordrecht, 1976, Vol. II, p. 149-170.

37. B. Lévy, unpublished.

38. W. Mayer, J. Chem. Phys., 58, 1017 (1973).

39. C. Edmiston and K. Ruedenberg, Rev. Mod. Phys., 35, 457 (1963).

40. I. Fisher-Hjalmars, J. Chem. Phys. 42, 1962 (1965).

41. S.F. Boys and J. Foster, Rev. Mod. Phys. 32, 305 (1960).

42. L.M. Fal, S. Wolfe and I.C. Csizmadia, J. Chem. Phys., 59, 4047 (1973).

43. J.P. Dahl, Acta Chem. Scand., 21, 1244 (1967).

44. J. Cizek, J. Mol. Phys., 6, 19 (1963).

45. A. Julg, Fortsch. Chem. Forsch. (1975) in press and references therein.

BIORTHONORMAL BASES IN HILBERT SPACE

Luc Lathouwers

Dienst voor Teoretische en Wiskundige Natuurkunde

Rijksuniversitair Centrum Antwerpen, Antwerp, Belgium

INTRODUCTION

When Per-Olov Löwdin solved the non-othogonality problem, in his Ph.D. thesis[1] in 1947, he did this through an ingenious combination of mathematical tools and physical intuition. Like orthonormalisation, biorthonormalisation is a problem of both mathematical and physical importance. Covariant and contravariant representations, direct and reciprocal lattices, secular equations for non-orthogonal basis sets are just a few topics intimately related to the concept of biorthonormal bases. The full solution of the biorthonormality problem, i.e., an existence theorem for biorthonormalisation, has, to our knowledge, not been given. It is the intention of the present article to fill this gap. For this purpose we will use a theorem of R. Paley and N. Wiener[2] and the symmetric orthonormalisation [1], thus following Per-Olov Löwdin's example in combining mathematics and physics to solve a problem of interest to both fields.

1. MATHEMATICAL PRELIMINARIES

1.1. Hilbert Spaces

The definition of a Hilbert space H can be condensed into the following axioms:

I. H is a linear space.
II. H is a metric space in which the norm is defined by means of a scalar product.
III, H is complete, i.e., every Cauchy sequence in H converges to a limit in H.
IV. H is separable, i.e., the dimension of H is denumerably infinite.

We will consider the elements of H as abstract entities satisfying the above axioms. An important realisation of H is the sequential Hilbert space H_o consisting of those sequences of complex numbers $s = \{s_n\}$ such that

$$||\underline{s}||^2 = \underline{s}^+\underline{s} = \Sigma_{n=1}^{\infty} |s_n|^2 <+\infty \tag{1.1}$$

where $\underline{s}$ denotes an infinite column vector. The existence of H_o, for which one can show that I,....., IV are satisfied, guarantees that the axioms of abstract Hilbert space are consistent. Hilbert spaces are special cases of Banach spaces in which the norm is not necessarily defined by means of a scalar product. They are said to have "the best geometric properties" among all Banach spaces because one can introduce the concept of orthogonality. Furthermore, Hilbert spaces have special convergence features. One distinguishes between strong and weak convergence. A sequence of elements $\{f_n\}$ in H converges strongly to a limit f in H if $\lim_{n\to\infty} ||f_n - f|| = 0$. The sequence is said to be weakly convergent if $\lim_{n\to\infty} \langle g|f_n - f\rangle = 0$ for all g in H. It is easy to show that strong convergence implies weak convergence. The converse is false. A counterexample is given in ref. 16. In the following we will be interested in bases of H. A set of elements $\underline{\phi} = \{\phi_n\}$ in H is called a basis of H if every element f has a unique norm-convergent expansion

$$f = \Sigma_{n=1}^{\infty} \phi_n f_n \equiv \underline{\phi}\,\underline{f} \tag{1.2}$$

where $\underline{\phi}$ denotes an infinite row vector. The equality sign stands for norm-convergence which means

$$\begin{aligned} &\lim_{N\to\infty} ||f - \Sigma_{n=1}^{N} \phi_n f_n|| \\ &\equiv \lim_{N\to\infty} ||f - \underline{\phi}_N\, \underline{f}_N|| = 0 \end{aligned} \tag{1.3}$$

When a sequence bears a subscript N this will mean that all terms following the Nth one are replaced by zero.

1.2. Orthonormal Sets

A set of elements $\underline{\omega} = \{\omega_n\}$ is an orthonormal system (OS) in H if

$$\langle\underline{\omega}|\underline{\omega}\rangle = \underline{1} \tag{1.4}$$

where $\underline{1}$ denotes the infinite unit matrix, i.e., the unit operator in H_o. An OS for which $\underline{\omega}$ is also a basis is called an orthonormal basis (OB). In this case we can write

$$f = \underline{\omega}\, \underline{f}_\omega \qquad \underline{f}_\omega = \langle\underline{\omega}|f\rangle \qquad f = \underline{\omega} \langle \underline{\omega}|f\rangle \tag{1.5}$$

These equations can be summarised in the closure relation or resolution of the identity

$$I = |\underline{\omega}\rangle\langle\underline{\omega}| \tag{1.6}$$

I being the unit operator in H. Any OS satisfies Bessel's inequality

$$\underline{f}_\omega^+ \underline{f}_\omega = \leq ||f||^2 \tag{1.7.}$$

This inequality can be replaced by the equation, called Parseval's relation,

$$\underline{f}_\omega^+ \underline{f}_\omega = ||f||^2 \tag{1.8}$$

if and only if $\underline{\omega}$ is an OB.

1.3. Biorthonormal Sets

In order to obtain formulas for non-orthogonal bases, analogous to the onew we have just given for orthonormal sets, one can introduce the concept of biorthonormality. A pair of sequences $(\underline{\phi}\ \ \underline{\tilde{\phi}})$ is said to form a biorthonormal system (BS) if

$$\langle\underline{\tilde{\phi}}|\underline{\phi}\rangle = \langle\underline{\phi}|\underline{\tilde{\phi}}\rangle = \underline{1} \tag{1.9}$$

A biorthonormal basis (BB) is a pair $(\underline{\phi}, \underline{\tilde{\phi}})$, satisfying the biorthonormality relations (1.9), for which both $\underline{\phi}$ and $\underline{\tilde{\phi}}$ are bases of H. In analogy with (1.5) we have

$$f = \underline{\phi}\, \underline{f} \qquad \underline{f} = \langle\underline{\tilde{\phi}}|f\rangle \qquad f = \underline{\phi}\, \langle\underline{\tilde{\phi}}|f\rangle \tag{1.10a}$$

$$f = \underline{\tilde{\phi}}\, \underline{\tilde{f}} \qquad \underline{\tilde{f}} = \langle\underline{\phi}|f\rangle \qquad f = \underline{\tilde{\phi}}\, \langle\underline{\phi}|f\rangle \tag{1.10b}$$

which can be condensed in the closure relation

$$I = |\underline{\phi}\rangle\langle\tilde{\underline{\phi}}| = |\tilde{\underline{\phi}}\rangle\langle\underline{\phi}| \qquad (1.11)$$

The question now arises whether a BS satisfies an inequality of Bessel's type and whether BB's can be characterized uniquely among BS's by a generalised Parseval relation. The following derivation is similar to the classical proof of Bessel's inequality.

We try to approximate an element $\underline{f}$ as closely as possible in terms of the finite sets $\underline{\phi}_N$ or $\tilde{\underline{\phi}}_N$, i.e., we want to find the best coefficients $\underline{f}_N$ and $\tilde{\underline{f}}_N$ such that the distances $||f - \underline{\phi}_N\, \underline{f}_N||$ and $||f - \tilde{\underline{\phi}}_N\, \tilde{\underline{f}}||$ are minimal. It can be proven that the optimal coefficients are given by

$$\underline{f}_N = \langle\tilde{\underline{\phi}}_N|f\rangle \qquad \tilde{\underline{f}}_N = \langle\underline{\phi}_N|f\rangle \qquad (1.12)$$

Substitution in the above norms yields the generalised Bessel inequality

$$\underline{f}_N^+\, \tilde{\underline{f}}_N = \tilde{\underline{f}}_N^+\, \underline{f}_N \leq ||f||^2 \qquad (1.13)$$

Furthermore it has been shown elsewhere[18] that a BS is a BB if and only if the generalised Parseval relation

$$\underline{f}^+\, \tilde{\underline{f}} = \tilde{\underline{f}}^+\, \underline{f} = ||f||^2 \qquad (1.14)$$

holds. We refer to the excellent book of F. Riesz and B.Sz. Nagy[17] for further details about functional analysis in Hilbert spaces

---

Footnote:
Notice that $(\phi_N, \tilde{\phi}_N)$ is an N dimensional biorthonormal set and not the truncation of $(\phi, \tilde{\phi})$ after N terms.

---

2. ORTHONORMALISATION AND BIORTHONORMALISATION

In order to obtain bound state solutions to the Schrödinger equation

$$H\, \Psi = E\, \Psi \qquad (2.1)$$

one often expands Ψ in a basis ϕ

$$\Psi = \underline{\phi}\ \ \underline{C} \qquad (2.2)$$

In practice this expansion will be truncated after a finite number of terms

$$\Psi \tilde{=} \underline{\phi}_N \, \underline{C}_N \tag{2.3}$$

thus leading to an N dimensional secular equation

$$(\underline{H}_N - E\,\underline{\Delta}_N)\,\underline{C}_N = 0 \tag{2.4}$$

which is an approximation to the full secular equation

$$(\underline{H} - E\,\underline{\Delta})\,\underline{C} = 0 \tag{2.5}$$

Computationally it is convenient to reduce equations (2.4) and (2.5) to simpler ones having a diagonal metric. This can be accomplished by orthonormalisation of the set $\underline{\phi}_N$ and the basis $\underline{\phi}$. We refer to ref. 3 for the N dimensional case and concentrate ourselves to the orthonormalisation of bases of H. According to the variational principle the Schrödinger equation is equivalent to the secular equation (2.5). Hence one should be able to derive Ψ from the eigenvectors $\underline{C}$ and vice versa. Equation (2.2) solves half this problem. The counterpart necessitates the existence of a biorthonormal basis $(\underline{\phi}, \tilde{\underline{\phi}})$.

Indeed, applying the closure relation (1.11) to the Schrödinger equation one obtains

$$(\langle\underline{\phi}|H|\phi\rangle - E\,\langle\underline{\phi}|\phi\rangle)\,\langle\tilde{\underline{\phi}}|\Psi\rangle = 0 \tag{2.6}$$

which reduces to the secular equation provided one makes the identification

$$\underline{C} = \langle\tilde{\underline{\phi}}|\Psi\rangle \tag{2.7}$$

which completes the connection between Ψ and $\underline{C}$. Thus for (2.1) and (2.5) to be equivalent there should exist a basis $\tilde{\underline{\phi}}$ which, together with the original basis $\underline{\phi}$ forms a BB. The procedure to obtain $\tilde{\underline{\phi}}$ from $\underline{\phi}$ will be called biorthonormalisation. The primary goal of this article is to prove an existence theorem for the biorthonormalisation procedure.

2.1. Orthonormalisation

For a linear transformation $\underline{A}$ to transform a non-orthogonal basis $\underline{\phi}$ into an orthonormal set $\underline{\omega}$

$$\underline{\omega} = \underline{\phi}\,\underline{A} \tag{2.8}$$

it should satisfy the equation

$$\langle \underline{\omega} | \underline{\omega} \rangle = \underline{1} = \underline{A}^{+} \, \underline{\Delta} \, \underline{A} \tag{2.9}$$

If the OS $\underline{\omega}$ is to be a basis there should exist a matrix $\underline{B}$ such that

$$\underline{\phi} = \underline{\omega} \, \underline{B} \tag{2.10}$$

Combination of (2.8) and (2.10) gives

$$\underline{A} \, \underline{B} = \underline{B} \, \underline{A} = \underline{1} \tag{2.11}$$

i.e., a matrix $\underline{A}$ which orthonormalises a basis $\underline{\phi}$ to an OB $\underline{\omega}$ must have a unique left and right inverse. It then follows from (2.10) that

$$\langle \underline{\phi} | \underline{\phi} \rangle = \underline{\Delta} = \underline{B}^{+} \, \underline{B} \tag{2.12}$$

Conversely if $\underline{A}$ obeys (2.9) and has a unique left and right inverse $\underline{\omega}$ is an OB since

$$f = \underline{\phi} \, \underline{f} = \underline{\omega} \, \underline{B} \, \underline{f} = \underline{\omega} \, \underline{f}_{\omega} \tag{2.13a}$$

and

$$\underline{f}_{\omega}^{+} \, \underline{f}_{\omega} = \underline{f}^{+} \, \underline{B}^{+} \, \underline{B} \, \underline{f} = \underline{f}^{+} \, \underline{\Delta} \, \underline{f} = ||f||^{2} \tag{2.13b}$$

such that Parseval's relation holds.

2.2. Biorthonormalisation

Given a non-orthogonal basis $\underline{\phi}$ the following questions arise

I. Does there exist a set $\underline{\tilde{\phi}}$ such that $(\underline{\phi}, \underline{\tilde{\phi}})$ is a BS?

II. Does there exist a basis $\underline{\tilde{\phi}}$ such that $(\underline{\phi}, \underline{\tilde{\phi}})$ is a BB?

The first question is easily answered with yes by combining two famous theorems. It follows from a corollary to the Hahn-Banach theorem[4] that for a basis $\underline{\phi}$ there exists a set of functionals $\underline{F} = \{F_n\}$ for which

$$\underline{F} \, (\underline{\phi}) = \underline{1} \tag{2.14}$$

On the other hand, according to Riesz' theorem[15], one can find for every set of functionals $\underline{F}$ a unique set $\underline{\tilde{\phi}}$ in H such that

$$\underline{F} \, (\underline{\phi}) = \langle \underline{\tilde{\phi}} | \underline{\phi} \rangle \tag{2.15}$$

Clearly $(\underline{\phi}, \underline{\tilde{\phi}})$ is a BS. The second question is a much more difficult one. So far only sufficient conditions on the original basis $\underline{\phi}$ have been given in order that a conjugated basis $\underline{\tilde{\phi}}$ exists. We will, in the following, answer question II with yes without imposing any restrictions on the initial basis $\underline{\phi}$.

The biorthogonalisation process is unique. Indeed, suppose that both $(\underline{\phi}, \underline{\tilde{\phi}}_1)$ and $(\underline{\phi}, \underline{\tilde{\phi}}_2)$ are BB's. Subtracting the biorthonormality relations

$$\begin{aligned} \langle\underline{\phi}|\underline{\tilde{\phi}}_1\rangle &= \underline{1} \\ \langle\underline{\phi}|\underline{\tilde{\phi}}_2\rangle &= \underline{1} \end{aligned} \tag{2.16a}$$

we obtain

$$\langle\underline{\phi}|\underline{\tilde{\phi}}_1 - \underline{\tilde{\phi}}_2\rangle = \underline{0} \tag{2.16b}$$

and since $\underline{\phi}$ is a basis it follows that $\underline{\tilde{\phi}}_1 = \underline{\tilde{\phi}}_2$. We will consider two methods for biorthonormalisation.

<u>Method I</u>

If $(\underline{\phi}, \underline{\tilde{\phi}})$ is a BB we have

$$\underline{\tilde{\phi}} = \underline{\phi}\ \langle\underline{\tilde{\phi}}|\underline{\tilde{\phi}}\rangle \tag{2.17a}$$

$$\underline{\phi} = \underline{\tilde{\phi}}\ \langle\underline{\phi}|\underline{\phi}\rangle \tag{2.17b}$$

Taking the scalar product to the left of (2.17a) and (2.17b) with $\underline{\phi}$ and $\underline{\tilde{\phi}}$ respectively, it follows that

$$\underline{\Delta}\ \langle\underline{\tilde{\phi}}|\underline{\tilde{\phi}}\rangle = \langle\underline{\tilde{\phi}}|\underline{\tilde{\phi}}\rangle\ \underline{\Delta} = \underline{1} \tag{2.18}$$

Thus the metric matrix of $\underline{\tilde{\phi}}$ is nothing but the inverse of $\underline{\Delta}$

$$\underline{\Delta}^{-1} = \langle\underline{\tilde{\phi}}|\underline{\tilde{\phi}}\rangle \tag{2.19}$$

Substitution of (2.19) in (2.17a) yields

$$\underline{\tilde{\phi}} = \underline{\phi}\ \underline{\Delta}^{-1} \tag{2.20}$$

We can therefore state: The biorthonormalisation of a basis is equivalent to the inversion of its metric matrix.

The straightforward inversion of $\underline{\Delta}$ is a non-trivial problem since, contrary to the finite dimensional case, there exists no concept of determinants for infinite matrices. A possiblity to solve the problem is to define $\underline{\Delta}$ as an operator on a suitable sequence space such that the inverse operator exists. If the metric matrix would be a bounded operator on H_o one could use Toeplitz' theorem[5]. This theorem says that, if $\underline{\Delta}$ is bounded, $\underline{\Delta}^{-1}$ exists and is bounded if and only if $\underline{\Delta}$ is bounded away from zero, i.e.,

$$\underline{\Delta} \geq \varepsilon > o \tag{2.21}$$

A procedure to define $\underline{\Delta}$ as a bounded operator on H_o will be given in section 3.

Formally one can immediately write down the solution to the biorthonormalisation problem since if we set

$$\underline{\tilde{\phi}} = \underline{\omega}\ \underline{A}^{+} \tag{2.22}$$

we have

$$<\underline{\tilde{\phi}}|\underline{\phi}> = <\underline{\omega}\ \underline{A}^{+}|\underline{\omega}\ \underline{B}> = \underline{A}\ <\underline{\omega}|\underline{\omega}>\ \underline{B} = \underline{1} \tag{2.23}$$

However, there is no guarantee that the elements of $\underline{\tilde{\phi}}$, as defined through (2.22), are situated in H. One has to impose the additional condition that the metric matrix $<\underline{\tilde{\phi}}|\underline{\tilde{\phi}}>$, i.e. $\underline{\Delta}^{-1}$, exists

$$<\underline{\tilde{\phi}}|\underline{\tilde{\phi}}> = \underline{\Delta}^{-1} = \underline{A}\ \underline{A}^{+} \tag{2.24}$$

Therefore the validity of the formal solution (2.22) amounts to the existence of Δ^{-1}. Since $\underline{\Delta}^{-1}$ is a metric matrix it is sufficient that its diagonal elements exist

$$\Delta^{-1}_{nn} = ||\tilde{\phi}_n||^2 = (\underline{A}\ \underline{A}^{+})_{nn} = \Sigma^{\infty}_{m=1} |A_{nm}|^2 < +\infty \tag{2.25}$$

We conclude that the biorthonormalisation problem admits the solution (2.22) provided the rows of the orthonormalisation matrix $\underline{A}$ converge modulus squared.

Method II

Equations (2.10) and (2.22) suggest the following theorem. If $\underline{\omega}$ is an OB and T a linear, bounded and invertible operator (isomorphism) on H a BB can be generated according to the following mapping scheme

$$\underline{\phi} \ \underset{T}{\overset{T^{-1}}{\rightleftarrows}}\ \underline{\omega}\ \underset{(T^{-1})^{+}}{\overset{T^{+}}{\leftrightarrows}}\ \underline{\tilde{\phi}} \tag{2.26}$$

The resulting pair ($\underline{\phi}$, $\underline{\tilde{\phi}}$) is a BS

$$<\underline{\tilde{\phi}}|\underline{\phi}> = <(T^{-1})^{+}\ \underline{\omega}|T\omega> = <\underline{\omega}|T^{-1}T\ \underline{\omega}> = \underline{1} \tag{2.27}$$

and it is also a BB since for all f in H we have

$$\begin{aligned} f &= T(T^{-1}f) = T\ \underline{\omega}\ <\underline{\omega}|T^{-1}\ f> \\ &= T\ \underline{\omega}\ <(T^{-1})^{+}\ \underline{\omega}|f> = \underline{\phi}\ <\underline{\tilde{\phi}}|f> \end{aligned} \tag{2.28}$$

The expansion of f in $\underline{\tilde{\phi}}$ can be obtained in a similar way. According to this theorem a basis can be biorthonormalised by orthonormalising it to an OB $\underline{\omega}$ related to $\underline{\phi}$ by an isomorphism. The drawback of this method, closely related to the possible non-convergence of (2.25), is that not every orthonormalisation procedure will give rise to an OB $\underline{\omega}$ related to the initial basis by an isomorphism. In their joint publication[2] Paley and Wiener have given a sufficient condition for the intermediate OB $\underline{\omega}$ to lead to a linear, bounded and invertible operator on H. In a form convenient for our purposes the Paley-Wiener theorem (PW th.) reads:

If, for a given basis $\underline{\phi}$, there exists an orthonormal basis $\underline{\omega}$ such that

$$||(\underline{\omega} - \underline{\phi})\ \underline{a}||^{2} < \underline{a}^{+}\ \underline{a} \tag{2.29}$$

for all finite sequences of complex numbers $\underline{a}$, there exists a basis $\underline{\tilde{\phi}}$ such that $(\underline{\phi}, \underline{\tilde{\phi}})$ is a BB.

A proof of the PW th., given by Nagy[6], can be found in ref. 16. It illustrates nicely how method II works.

Roughly the PW th. says that biorthogonalisation is possible if there exists an OB "sufficiently near" to $\underline{\phi}$. There is an extensive mathematical literature about theorems of the PW type and many equivalent formulations have been given[7]. In order to obtain a more tractable form of the PW condition (2.29) we can combine the theorem with an orthonormalisation procedure. Substitution of (2.10) in (2.29) yields

$$\begin{aligned} ||(\underline{\omega} - \underline{\phi})\ \underline{a}||^{2} &= ||\underline{\omega}\ (\underline{1} - \underline{B})\ \underline{a}||^{2} = \\ &= \underline{a}^{+}\ (\underline{1} - \underline{B}^{+})\ <\underline{\omega}|\underline{\omega}>\ (\underline{1} - \underline{B})\ \underline{a} = \\ &= \underline{a}^{+}\ \underline{a} + \underline{a}^{+}\ \underline{\Delta}\ \underline{a} - 2\ \mathrm{Re}\ (\underline{a}^{+}\ \underline{B}\ \underline{a}) \end{aligned} \tag{2.30}$$

The PW condition is thus equivalent to

$$\underline{a}^{+}\ \underline{\Delta}\ \underline{a} < 2\ \mathrm{Re}\ (\underline{a}^{+}\ \underline{B}\ \underline{a}) \tag{2.31}$$

which is an inequality depending solely on the metric of the initial basis $\underline{\Delta}$ and the orthonormalisation procedure used to obtain $\underline{\omega}$, i.e., the matrix $\underline{B}$.

3. THE METRIC MATRIX AS AN OPERATOR ON THE SEQUENTIAL HILBERT SPACE

Let $\underline{\bar{\phi}}$ be a basis of H such that $||\bar{\phi}_n|| = 1$. Its metric matrix $\underline{\bar{\Delta}} = \langle\underline{\bar{\phi}}|\underline{\bar{\phi}}\rangle$, considered as an operator on H_o, has, in general, a restricted domain consisting of those sequences $\underline{x} = \{x_n\}$ for which $||\underline{\bar{\Delta}}\,\underline{x}||$ is finite. On this domain $\underline{\Delta}$ is Hermitian and positive definite. As for the existence of $\underline{\bar{\Delta}}^{-1}$, there is no guarantee that $\underline{\bar{\Delta}}$ is invertible. In order to systematise the problem we propose the following renormalisation procedure. We multiply each element ϕ_n with a non-zero complex number s_n such that the sequence $\underline{s} = \{s_n\}$ is in H_o. The resulting set $\underline{\phi}$, of elements $\phi_n = \bar{\phi}_n\, s_n$, is a basis if and only if $\underline{\bar{\phi}}$ is a basis[7]. The metric matrix of the renormalised basis $\underline{\Delta} = \langle\underline{\phi}|\underline{\phi}\rangle$ now has some interesting properties.

Using Schwartz' inequality one easily verifies that

$$\Sigma^{\infty}_{m,\, n = 1}|\Delta_{mn}|^2 \leq (\underline{s}^+\underline{s})^2 < +\infty \tag{3.1}$$

which, according to von Neumann[8], proves that $\underline{\Delta}$ is a completely continuous operator on H_o, i.e., an operator which transforms a weakly convergent sequence into a strongly convergent sequence. There is even more, since

$$\mathrm{Tr}\,(\underline{\Delta}) = \Sigma^{\infty}_{n = 1}\,\Delta_{nn} = \underline{s}^+\,\underline{s} < +\infty \tag{3.2}$$

$\underline{\Delta}$ is also a finite trace operator. In view of these properties the renormalised metric matrix has the following spectral characteristics. $\underline{\Delta}$ has a set of strictly positive eigenvalues which we order in a decreasing sequence

$$d_1 \geq d_2 \geq \cdots \qquad \cdots > o \tag{3.3}$$

The corresponding normalised eigenvectors form an OB in H_o. They can be arranged columnwise in a unitary matrix $\underline{U}$ such that the eigenvalue problem for $\underline{\Delta}$ can be written as

$$\underline{\Delta}\,\underline{U} = \underline{U}\,\underline{d} \tag{3.4}$$

where $\underline{d}$ is the diagonal matrix of the eigenvalues (3.3). Multiplying (3.4) to the right with $\underline{U}^+$ one obtains the spectral resolution of $\underline{\Delta}$

$$\underline{\Delta} = \underline{U}\ \underline{d}\ \underline{U}^{+} \tag{3.5}$$

Taking the trace of this expression yields

$$\mathrm{Tr}\ (\underline{\Delta}) = \Sigma_{n=1}^{\infty} d_n < + \infty \tag{3.6}$$

Consequently the eigenvalues cluster to zero

$$\lim_{n \to \infty} d_n = o \tag{3.7}$$

This means that Toeplitz' theorem is not valid for a renormalised metric matrix since $\underline{\Delta}$ cannot be bounded away from zero. Therefore the possible inverse Δ^{-1} will certainly be an unbounded operator on H_o. In general a power of $\underline{\Delta}$ can be defined[3] through a function of $\underline{\Delta}$.

$$f\ (\underline{\Delta}) = \underline{U}\ f\ (\underline{d})\ \underline{U}^{+} \tag{3.8}$$

Being bounded, Hermitian and positive definite $\underline{\Delta}$ has a unique positive square root $\underline{\Delta}^{+\frac{1}{2}}$ given by

$$\underline{\Delta}^{+\frac{1}{2}} = \underline{U}\ \underline{d}^{+\frac{1}{2}}\ \underline{U}^{+} \tag{3.9}$$

The negative powers, $\underline{\Delta}^{-\frac{1}{2}}$ and $\underline{\Delta}^{-1}$, which are of crucial importance in orthonormalisation and biorthonormalisation, will, if they exist, be given by

$$\underline{\Delta}^{-\frac{1}{2}} = \underline{U}\ \underline{d}^{-\frac{1}{2}}\ \underline{U}^{+} \tag{3.10a}$$

$$\underline{\Delta}^{-1} = \underline{U}\ \underline{d}^{-1}\ \underline{U}^{+} \tag{3.10b}$$

As demonstrated above, and as can also be verified directly, their domains will be subspaces of H_o.

4. BIORTHONORMALISATION VIA SCHMIDT ORTHONORMALISATION

The most straightforward way to orthonormalise a basis $\underline{\phi}$ is the Schmidt procedure or successive orthonormalisation[9]. It is based on the recursion formula

$$\omega_{n+1} = \frac{\phi_{n+1} - \Sigma_{m=1}^{n} \omega_m \langle \omega_m | \phi_{n+1} \rangle}{|| \phi_{n+1} - \Sigma_{m=1}^{n} \omega_m \langle \omega_m | \phi_{n+1} \rangle ||} \tag{4.1}$$

since $\omega_1, \omega_2, \ldots \quad \ldots, \omega_n, \phi_{n+1}$ are linearly independent, the denominator in (4.1) is never zero and the process can be continued at infinitum. Hence Schmidt's procedure allows us to orthonormalise ϕ by a matrix transformation

$$\underline{\omega} = \underline{\phi}\ \underline{T} \tag{4.2}$$

where $\underline{T}$ is an upper triangular matrix. Mathematically the successive orthonormalisation is quite satisfactory since all other orthonormal bases are related to $\underline{\omega}$ by unitary transformations. Practically and physically, however, the method suffers from serious drawbacks[3].

For the formal solution (2.22) to be valid the rows of $\underline{T}$ should converge modulus squared. This condition can usually not be checked since the matrix $\underline{T}$ is generated column by column through (4.1) such that no algoritm for the rows of $\underline{T}$ is known. It is possible to carry out the Schmidt procedure by successive inversion[3] of the finite order metric matrices $\underline{\Delta}_N$. However, for the infinite case this reduces once more to the existence of $\underline{\Delta}^{-1}$.

Alternatively one can try to apply method II, i.e., substitute the left and right inverse $\underline{T}^{-1}$ of T in the PW condition (2.31) which gives

$$\underline{a}^{+}\ \underline{\Delta}\ \underline{a} < 2\ \mathrm{Re}\ (\underline{a}^{+}\ \underline{T}^{-1}\ \underline{a}) \tag{4.3}$$

It is highly unlikely that this inequality will be satisfied for all finite $\underline{a}$. Indeed the left hand side of (4.3) is positive definite while there is no reason that T^{-1} should have a positive real part. Moreover both conditions, the convergence criterion on the rows of $\underline{T}$ and the PW condition, will depend upon the order in which the elements of $\underline{\phi}$ are orthonormalised.

We can conclude that the Schmidt procedure, for physical as well as for computational applications, is not suited for biorthonormalisation. As an example one can take the Hilbert space L^2 (-1, +1) and the basis $\underline{x} = \{x^n\}$.

5. BIORTHONORMALISATION VIA LÖWDIN ORTHONORMALISATION

The Löwdin orthonormalisation[1] or symmetric orthonormalisation is defined for the finite dimensional case by the formula

$$\underline{\omega}_N = \underline{\phi}_N\ \underline{\Delta}_N^{-\frac{1}{2}} \tag{5.1}$$

Contrary to the Schmidt procedure, the elements of $\underline{\phi}_N$ are treated simultaneously and on an equivalent basis. The method has been very successful in molecular and solid state applications since it preserves symmetry[10]. The extension of (5.1) to the infinite dimensional case requires some care since we do not know whether $\underline{\Delta}^{-\frac{1}{2}}$ exists. However, by virtue of the renormalisation procedure, we can perform the following three-step orthonormalisation of $\underline{\phi}$

$$\underline{\phi} \;\underset{\underline{U}^{+}}{\overset{\underline{U}}{\rightleftarrows}}\; \underline{\Xi} \;\underset{\underline{d}^{+\frac{1}{2}}}{\overset{\underline{d}^{-\frac{1}{2}}}{\rightleftarrows}}\; \underline{\chi} \;\underset{\underline{U}}{\overset{\underline{U}^{+}}{\rightleftarrows}}\; \underline{\omega} \qquad (5.2)$$

The first step $\underline{\Xi} = \underline{\phi}\,\underline{U}$ is an orthogonalisation since $\langle\Xi|\Xi\rangle = \underline{d}$. The second step $\underline{\chi} = \underline{\Xi}\,\underline{d}^{-\frac{1}{2}} = \underline{\phi}\,\underline{U}\,\underline{d}^{-\frac{1}{2}}$ normalises the set $\underline{\Xi}$. The basis $\underline{\chi}$ is known as the canonically orthonormalised set and has proved to be very useful in the case of approximate linear dependence[11-13]. The last step $\underline{\omega} = \underline{\chi}\,\underline{U}^{+}$ rotates the OB $\underline{\chi}$ to a new OB $\underline{\omega}$ which we define as the symmetric orthonormal basis in the infinite dimensional case. It is clear that $\underline{\omega}$ is related to the initial basis $\underline{\phi}$ by the positive square root $\underline{\Delta}^{+\frac{1}{2}}$ since

$$\underline{\phi} = \underline{\omega}\,\underline{U}\,\underline{d}^{+\frac{1}{2}}\,U^{+} = \underline{\omega}\,\underline{\Delta}^{+\frac{1}{2}} \qquad (5.3)$$

We have thus succeeded in defining the symmetric OB without explicitly using the matrix $\Delta^{-\frac{1}{2}}$ whose existence is not yet proven. Substitution of (5.3) in the PW condition gives

$$\underline{a}^{+}\,\underline{\Delta}\,\underline{a} < 2\,\underline{a}^{+}\,\underline{\Delta}^{+\frac{1}{2}}\,\underline{a} \qquad (5.4)$$

Observe that now both sides are positive definite. It is easy to show that (5.4) is satisfied if

$$d_n < 4 \qquad (5.5)$$

This rather strange condition that the eigenvalues of the metric matrix should be strict smaller than 4 results from a series of sufficient conditions. Probably most of the inequalities we have written can be weakened. We can now perform the final step to the existence theorem for biorthonormalisation since in order to satisfy (5.5) one merely has to choose the renormalisation sequence such that

$$\mathrm{Tr}\,(\underline{\Delta}) = \Sigma_{n=1}^{\infty}\, d_n = \underline{s}^{+}\,\underline{s} < 4 \qquad (5.6)$$

which implies that all eigenvalues are smaller than 4. One could simply set $\phi_n = \Phi_n/n$ for which $\mathrm{Tr}\,(\underline{\Delta}) = \pi^2/6 < 4$. The PW condition can thus always be satisfied through a suitable renormalisation procedure. The existence of the BB $(\underline{\phi}, \underline{\tilde{\phi}})$ is now established and one can change the normalisation of both $\underline{\phi}$ and $\underline{\tilde{\phi}}$ such that the biorthonormality relation $\langle\underline{\phi}|\underline{\tilde{\phi}}\rangle = \underline{1}$ remains invariant. Consequently we can formulate the following general existence theorem:

Conjugated to every basis $\underline{\phi}$ of a Hilbert space there exists a basis $\underline{\tilde{\phi}}$ which, together with the original set, forms a biorthonormal basis.

From this theorem and the previously obtained result that biorthonormalisation is equivalent to inversion of the metric matrix we obtain the corollary:

Every metric matrix $\underline{\Delta} = \langle\underline{\phi}|\underline{\phi}\rangle$ has a unique inverse given by the metric matrix of the biorthonormal basis.

The fact that these theorems originated through a combination of the PW th. and the symmetric orthonormalisation is not so surprising. Indeed the sets $\underline{\omega}_N$ have the property of minimising the Tr $(\langle\underline{\omega}_N - \underline{\phi}_N|\underline{\omega}_N - \underline{\phi}_N\rangle)$ i.e., they are the N dimensional OB's which "resemble" the sets $\underline{\phi}_N$ "as closely as possible"[14].

6. EPILOGUE

It seems highly unlikely that, in the course of more than 60 years of functional analysis in Hilbert spaces, nobody arrived at the above results. However a reasonably thorough search of the literature did not bring these theorems to light. The reason is probably that mathematicians do not go beyond Schmidt's procedure whereas it has now become clear that the symmetric orthonormalisation has the desired features for biorthonormalisation.

This work was performed during the author's stay at the Uppsala Quantum Chemistry Group (October 1974 - May 1975) and supported by the Svenska Institutet (Sweden) and the Interuniversitair Instituut voor Kernwetenschappen (Belgium).

REFERENCES

1. P.O. Löwdin, "A Theoretical Investigation into some Properties of Ionic Crystals", Thesis, Uppsala 1948, Almqvist and Wiksells.

2. R. Paley and N. Wiener, "Fourier Transforms in the Complex Domain", Am. Math. Soc. (1934) 100.

3. P.O. Löwdin, "Advances in Quantum Chemistry V", Academic Press (1970).

4. S. Banach, "Theorie des opérations linéaires", Monografje Matematyczne (1932) 80 th. 5.

5. O. Toeplitz, Prace Math.-Fiz. 22 (1911) 113.
E. Hellinger and O. Toeplitz, Math. Ann. 69 (1910) 289.

6. B. Sz.-Nagy, Duke Math. J. 14 (1947) 975.

7. I. Singer, "Bases in Banach Spaces I", Springer (1970), this book contains an extensive bibliography on basis problems upto 1970.

8. J. von Neumann, "Mathematical Foundations of Quantum Mechanics", Princeton U.P. (1955) 184.

9. E. Schmidt, Math. Ann. 63 (1907) 433.

10. J.C. Slater and G.F. Koster, Phys. Rev. 94 (1954) 1498.

11. P.O. Löwdin, Adv. Phys. 5 (1956) 1.

12. J. Nordling, J. Mol. Spectr. 13 (1964) 57.

13. L. Lathouwers, accepted for publ. in Intern. J. Quant. Chem.

14. B.C. Carlson and J.M. Keller, Phys. Rev. 105 (1957) 102.

15. F. Riesz, Comptes Rendus 144 (1934) 34.

16. F. Riesz and B. Sz.-Nagy, "Functional Analysis" F. Unger Publ. Co., New York.

17. P.O. Löwdin, Phys. Rev. A139 (1965) 357.

18. L. Lathouwers, submitted for publ. in Intern. J. Quant. Chem.

ENERGY WEIGHTED MAXIMUM OVERLAP (EWMO)

Jan Linderberg[†], Yngve Öhrn[††], and Poul W. Thulstrup[†]

[†]Department of Chemistry, Aarhus University, DK-8000 Aarhus C, Denmark

[††]Department of Chemistry, University of Florida Gainesville, Florida 32611, U.S.A.

INTRODUCTION

Löwdin emphasized the importance of the correct treatment of overlap between atomic functions in molecular and solid-state theory[1]. He pioneered the rigorous handling of overlap in the famous calculations on ionic crystals[2] and in molecular theory[3]. His name is today attached to a symmetrical orthogonalization procedure[1], and he has formulated and discussed the general problem of orthogonalization of a basis[4,5]. We find it appropriate to dedicate to him the following account of a very simple calculational method for molecular electronic structure determinations, particularly one electron spectra and properties, based on interatomic overlap.

The arguments leading to the Energy Weighted Maximum Overlap (EWMO) model are outlined in the next section. Remaining sections are devoted to reviews of applications and some previously unpublished results.

ATOMIC AND MOLECULAR ORBITALS

The formalism, which was first published in "Propagators in Quantum Chemistry"[6], is reconsidered here in the same notation. The electron propagator in the energy representation, $\langle\langle\psi(\xi);\ \psi^{\dagger}(\xi')\rangle\rangle_E$, is explored through its matrix elements in a basis of atomic spin orbitals, $\{u_s(\xi)\}$:

$$G_{sr}(E) = \langle\langle a_s;\, a_r^\dagger \rangle\rangle_E. \tag{1}$$

The set of annihilation operators $\{a_s\}$ is defined in terms of the field operator $\psi(\xi)$ as

$$a_s = \int d\xi u_s^*(\xi)\psi(\xi), \tag{2}$$

and the set of creation operators $\{a_s^\dagger\}$ is defined by the adjoint equation. They satisfy the anticommutation relations

$$\{a_s, a_r\}_+ = 0,\ \{a_s, a_r^\dagger\}_+ = \delta_{sr} + S_{sr},\ \{a_s^\dagger, a_r^\dagger\}_+ = 0, \tag{3}$$

where Löwdin's choice of definition for the overlap integral is used:

$$S_{sr} = \int d\xi u_s^*(\xi) u_r(\xi) - \delta_{sr} \tag{4}$$

We will use normalized orbitals such that $S_{rr} = 0$. The eigenstates $|n\rangle$ and the eigenvalues E_n of the many-electron Hamiltonian over the Fock space are used in writing the spectral density function of the propagator matrix elements as[6)]

$$\begin{aligned} A_{sr}(\varepsilon) &= \lim_{\delta\to 0} \{G_{sr}(\varepsilon+i\delta) - G_{sr}(\varepsilon-i\delta)\}(i/2\pi) \\ &= \Sigma_n \{\langle 0|a_s|n\rangle\langle n|a_r^\dagger|0\rangle\delta(\varepsilon-E_n+E_0) + \\ &\quad \langle 0|a_r^\dagger|n\rangle\langle n|a_s|0\rangle\delta(\varepsilon-E_0+E_n)\}. \end{aligned} \tag{5}$$

The reference state $|0\rangle$ is here taken to be the N electron ground state and from Eq. (5) it follows that the spectral density functions have contributions from states where an electron or hole is added to the reference state. The propagator matrix is singular at those energies where the spectral density function is non-vanishing and we conclude that a singularity at $E = \varepsilon_k$ corresponds to an energy difference such that $E_n(N\pm 1) = E_0(N)\pm\varepsilon_k$. The upper (lower) sign is used if ε_k is larger (smaller) than a separation parameter μ. This parameter equals the chemical potential in the case that the system is infinite and we consider the zero temperature limit. The spectral density function (5) contains the necessary information for the calculation of the photoelectron spectrum from the system[7)]. Hartree-Fock theory gives singularities at the orbital energies ε_k and the resultant photoelectron spectrum is the one found by Koopmans' theorem[8)].

Approximate forms of the propagator matrix can be studied with advantage in terms of moment expansions. These are written compactly as

$$G_{sr}(E) = \Sigma_j\ E^{-j-1}\int d\varepsilon\varepsilon^j A_{sr}(\varepsilon) = \Sigma_j E^{-j-1}\langle(a_s^{(j)},a_r^\dagger)_+\rangle \tag{6}$$

where $a_s^{(j)} = (a_s^{(j-1)}, H)$, $a_s^{(0)} = a_s$, and H is the appropriate many-electron Hamiltonian. The average values are evaluated for the reference state $|0\rangle$.

The first term in the expansion (6) is

$$E^{-1}\langle(a_s,a_r^\dagger)_+\rangle = (\delta_{sr} + S_{sr})/E, \tag{7}$$

while the second term can be expressed as

$$E^{-2}\langle((a_s,H),\ a_r^\dagger)_+\rangle = E^{-2}\langle(a_s,\ (H,a_r^\dagger))_+\rangle = f_{sr}/E^2, \tag{8}$$

where f_{sr} is a matrix element of the Fock operator[6] in the basis $(\bar{u}_s(\xi))$. It is seen to be hermitian and might be brought to a diagonal form together with the metric matrix of Eq. (7). The transformation involves the formation of canonical molecular orbitals. Similarly we have a diagonal form in the limit of separated atoms when canonical atomic self-consistent orbitals are used as a basis. The corresponding limiting form of the matrix elements, $f_{sr}\to\delta_{sr}\ W_s$, offers the interpretation of the atomic parameters (W_s) as orbital energies.

The knowledge of the first moments of the spectral density functions allows us to put forth an effective Hamiltonian in the limit of separated atoms as

$$H = \Sigma_A H_A,\ \text{with}\ H_A = \Sigma_s W_s(A)a_s^\dagger a_s, \tag{9}$$

that is, a sum of atomic Hamiltonians H_A given in diagonal form as a sum of one-electron terms. The <u>assumption</u> of our model is that the effective Hamiltonian form (9) applies also at interatomic distances appropriate for molecular systems. The effect of overlap is included in the relations (3). It may be argued that the effective Hamiltonian is correct to "first order in overlap" by comparison to the energy formulas for the cohesive energy in the Hartree-Fock approximation as given by Löwdin[2,4]. The form (9) admits the calculation of all higher moments of the spectral density function and gives the explicit form for the propagator matrix:[6]

$$\underline{G}\ (E) = (E(\underline{1}+\underline{S})^{-1}-\underline{W})^{-1}. \tag{10}$$

The diagonal matrix $\underline{W}$ has elements, W_s, which will all be negative when the spin orbital basis is limited to occupied atomic self-consistent orbitals. Then we can write $W_s = -\chi_s^2$ with χ_s as a positive parameter. The poles of the matrix $\underline{G}(E)$ in Eq. (10) will occur at the characteristic values of the secular determinant

$$D(\varepsilon) = ||(\varepsilon + \chi_s^2)\delta_{sr} + \chi_s S_{sr} \chi_r|| = 0. \tag{11}$$

This can be considered as a diagonalization of the overlap matrix, where each basis spin orbital is weighted by its parameter χ_s, and this is the reason for the term Energy Weighted Maximum Overlap (EWMO). A straight diagonalization of the overlap matrix leads to the so-called Maximum Overlap molecular orbitals[9]. The term Modified Extended Hückel Model (MEHM) was originally used at Aarhus[10] but is abandoned as being less descriptive.

The matrix with elements $f'_{sr} = -\chi_s (\delta_{sr} + S_{sr})\chi_r = W_s\delta_{sr} - S_{sr}(W_sW_r)^{\frac{1}{2}}$ is a representative of the Fock matrix in a basis which can be characterized as being energy weighted Löwdin orthogonalized[11]. The relation to the original basis is brought out in the expression for the molecular orbitals, $\omega_k(\xi)$, in terms of the unitary matrix (x_{rk}), which diagonalizes the matrix $\underline{f}'$, and the new basis of orthonormal spin orbitals $\{\hat{u}_s(\xi)\}$:

$$\omega_k(\xi) = \Sigma_s \hat{u}_s(\xi) x_{sk} = \Sigma_s u_s(\xi)(W_s/\varepsilon_k)^{\frac{1}{2}} x_{sk}. \tag{12}$$

The role of energy weighting is seen in the last expression in Eq. (12) where the positive square root is used of the ratio between the atomic and molecular orbital energies. We notice that only one diagonalization is required to determine the molecular orbitals and yet no neglect of overlap is necessary.

The charge and bond order matrix is defined in the EWMO model from the orthonormal set and can conveniently be calculated[6,12] from the Coulson integral formula[13]

$$(2\pi i)^{-1} \int_C dE \left(E\underline{1} + \underline{\chi}(\underline{1} + \underline{S})\underline{\chi}\right)^{-1} \tag{13}$$

where C encloses those molecular orbital energies that correspond to occupied orbitals. Thus one finds that the formal charge in spin orbital s, which is equal to the orbital population according to Mulliken[14], is

$$q_s = \Sigma_k n_k |x_{sk}|^2, \tag{14}$$

where n_k are molecular spin orbital occupation numbers.

The resulting formal charges in the various orbitals is generally different in the molecule and the separated atoms and it was judged to be of interest to attempt a self-consistent iteration determination of the form of the orbitals. This kind of thinking led Slater to suggest a Hyper-Hartree-Fock method where fractional occupation numbers are used[15]. We find the ensemble average procedure more satisfactory[16] and define a density operator for each atom as

$$\rho = \Pi_s^A \left(1-q_s+(2q_s-1)a_s^\dagger a_s\right). \tag{15}$$

The average value of the complete atomic Hamiltonian with respect to this density operator is then made stationary to variations in the radial factor of the spin orbitals and these are then solved from the Euler equations. The ensemble (15) does not correspond to a fixed value of the number of electrons in the system and may be called a "grand canonical" ensemble although no thermodynamic quantities are implied. Other forms of ensemble averages were introduced by Löwdin[17] and has been revived by Goscinski[18]. It is now possible to find basis orbitals that are consistent with the resultant molecular charge distribution through an iterative procedure, and some of the results to be presented have been obtained that way. A slight, but computationally decisive, modification of the general density operator (15) has been used; the formal charges have been averaged over spin and magnetic quantum numbers in order to ensure that the density operator commutes with spin and angular momentum. This simplification could possibly cause some significant changes when the atoms have a definitely nonspherical environment as in linear or planar molecules. Work is being pursued[19] with the general density operator (15).

APPLICATIONS

We show in Table 1 electron binding energies (the negative of molecular orbital energies) as calculated with the EWMO method for the SF_6 molecule both with and without the iterative procedure. Comparisons are made with the Multiple Scattering Xα method, various calculations based on Gaussian orbitals, and experimental results. The self-consistent formal charges arrived at from the EWMO calculation are +1.50 on the sulfur and -.25 on each of the flourines in atomic charge units. The EWMO calculations reported here were performed by numerical integration of the atomic central field equations for the orbitals[20] and the overlap integrals were calculated by a method of smooth interpolation and direct integration of the numerically given atomic orbitals according to a procedure introduced by P.W. Thulstrup[21].

A range of tetrahedral molecules have also been considered with the nonself-consistent procedure[22] and as an example of these results we show in Fig. 1 the orbital energies for the

Table 1

Electron binding energies for SF_6 (R_{S-F} = 2.948 au, octahedral) in eV.

	ESCA[a]	HF[b]	HF[c]	MSXα[d]	MSXα[e]	EWMO[f]	EWMO[g]
$3e_g$		13.3	19.4	17.5	13.8	12.5	18.0
	16						
$1t_{1g}$		14.0	18.2	15.9	12.1	12.2	17.8
$1t_{2u}$	17.3	15.3	19.4	16.8	13.1	13.3	19.1
$5t_{1u}$	18.7	15.9	19.0	16.8	13.0	13.1	18.4
$1t_{2g}$	19.9	17.6	22.2	18.7	15.0	15.7	21.8
$5a_{1g}$	22.9	19.9	29.6	26.7	23.0	25.9	27.5
$4t_{1u}$	27.0	23.0	24.7	21.8	18.1	20.7	23.5
$2e_g$	39.3	41.6	45.4	35.6	31.8	38.1	41.3
$3t_{1u}$	41.2	45.5	46.9	36.5	32.8	48.5	47.5
$4a_{1g}$	44.2	49.7	50.4	39.3	35.5	62.6	58.3

[a]From reference 35.

[b]Hartree-Fock results from reference 36, without d-functions on sulfur.

[c]Hartree-Fock results quoted in reference 36, with d-functions on sulfur and a very extensive basis.

[d]Transition state from reference 37.

[e]"Muffin-tin" eigenvalues from reference 37.

[f]Iterative self-consistent results.

[g]Non-iterative calculation.

highest occupied levels of the tetrabromides of carbon, silicon, germanium, and tin in comparison to values deduced from photoelectron experiments[23]. Spanget-Larsen has considered planar molecules and used analytical orbitals in studies of azines[24] and azanaphthalenes[25]. He found it possible to correlate his results convincingly with photoelectron spectra.

Several applications of the EWMO model have added to the understanding of problems in electron spin resonance experiments. Quadrupole coupling constants were obtained by Byberg and Spanget-Larsen[26] from the asymmetry of the charge distribution around in the valence shell according to the assumptions of Townes and Dailey[27]. The method was then used[28] to determine the bond distance in the radical $C\ell O_3$. The result was later confirmed by neutron diffraction measurements[29]. A simple solvation model has been used by Pedersen and Spanget-Larsen[30] to study solvent sensitivities of the hyperfine interaction constants in substituted benzosemiquinones. These satisfactory results spurred an attempt to calculate the g-tensor in the spin Hamiltonian for radicals[11]. It was then assumed that the atomic Hamiltonians were given in the diagonal form (9) for a basis of atomic spin orbitals in the strong coupling form characterized by the total angular momentum j and its projection on the external magnetic field $\underline{B}$. The spin orbitals were also functions of the field through the London phase integral recipe[31]. This causes a fine structure of the molecular spin orbital levels, which varies with the field direction since the overlap integrals do. Thus it is possible to derive an effective g-tensor from calculations with different strengths and directions of $\underline{B}$. The calculations were performed as finite perturbations but derivative forms can also be developed. The results were quite satisfactory. Some recent results are presented in Table 2. A combination of the thinking behind the quadrupole coupling constants and the magnetic field effects has prompted an attempt to calculate chemical shifts in nuclear magnetic resonance spectra that presently is being pursued[32].

CONCLUSION

The EWMO model offers a simple description of molecular electronic structure derived from an atomic spin orbital picture without the need for parametrization or assumptions about negligible differential overlap. The calculational effort is less than that required in Hoffman's scheme[33] which otherwise should be judged to be at the same level of sophistication, and one matrix diagonalization provides molecular orbitals, their energies, and an atomic basis of orthogonal orbitals, which serves for the definition of formal charges and bond orders. The self-consistent version, which is based on ensemble average Hartree-Fock atomic orbitals, seems to be rapidly convergent.

Table 2

Comparison of some calculated and observed g-values.

Species		EWMO	Obs.	
HCO	g_{xx}	1.9961	1.9960	Geometry
	g_{yy}	2.0023	2.0027	relatively
	g_{zz}	2.0042	2.0041	certain
	g_{xx}	2.0033	2.004	Geometry
SO_3^-	g_{yy}	2.0033	2.004	chosen as
	g_{zz}	2.0024	2.004	in $C\ell O_3$
	g_{xx}	2.0023	2.0019	Geometry
SO_2^-	g_{yy}	2.0129	2.012	from SO_2.
	g_{zz}	2.0033	2.0057	
	g_{xx}	2.0119	2.0079	Geometry
CS_2^-	g_{yy}	1.9702	1.9661	semiemperical
	g_{zz}	2.0023	1.9993	1.55Å, 141°.
	g_{xx}	2.0404	2.0299	Geometry from
NO_3	g_{yy}	2.0404	2.029	NO_3^- (D_{3h}).
	g_{zz}	2.0061	1.998	

Overlap integrals have been used extensively to estimate matrix elements, notably by Mulliken[34], but such attempts have generally started from explicit representations in configuration space and proceeded through integral approximations. The arguments put forth in order to establish the EWMO model are no more rigorous, but they rest on propagator concepts and are complementary to the ones based on Hamiltonian operators. Only the electron propagator is considered, and it is thus not feasible to discuss total energy expressions and so-called particle-hole excitations within the present EWMO model. It might be feasible to estimate the Hartree-Fock energy of a molecule from the knowledge of the EWMO density matrix which is easily calculated and where no approximation has been introduced in the treatment of overlap. This would be an immediate generalization of Löwdin's cohesive energy method for closed shell systems in the atomic orbital representation[4], and his integral evaluation techniques could be readily implemented. Such efforts have not been undertaken since while we have learnt from Löwdin that a correct treatment of overlap is essential, we have also been taught by him that electronic correlation is significant. It seems to us to be possible to develop the propagator approach further in order to include the dominant correlation effects associated with chemical bonding.

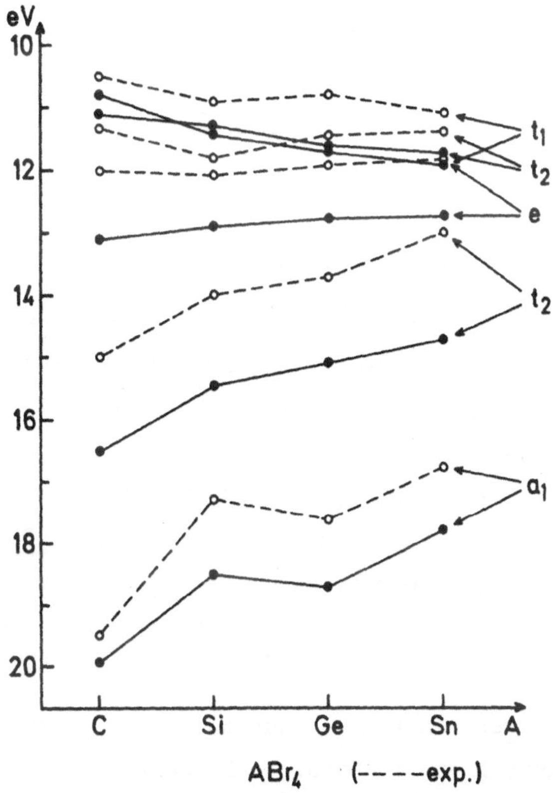

Figure 1. Valence shell photoionization energies from experiments (---------------) and EWMO calculations (————————) for some tetrabromides.

REFERENCES

1. P.O. Löwdin, J. Chem. Phys., 18 365 (1950).
2. P.O. Löwdin, "A Theoretical Investigation into some Properties of Ionic Crystals". Almquist and Wiksell, Uppsala, Sweden (1948).
3. P.O. Löwdin, J. Chem. Phys., 19 1570 (1951).
4. P.O. Löwdin, Adv. Phys., 5 40 ff (1956).
5. P.O. Löwdin, Adv. Quantum Chem., 5 185 (1970).
6. J. Linderberg and Y. Öhrn, "Propagators in Quantum Chemistry", Academic Press, London (1973).
7. G. Purvis and Y. Öhrn, J. Chem. Phys., 60 4063 (1974).
8. T. Koopmans, Physica, 1 104 (1933).
9. R.J. Bartlett and Y. Öhrn, Theoret. Chim. Acta. 21 215 (1971).
10. J. Spanget-Larsen, Computer Program MIEHM, QCPE 246, Quantum Chemistry Program Exchange, Indiana University, Bloomington, Indiana (1974).
11. E. Dalgaard and J. Linderberg, Intern. J. Quantum Chem., S9 (1975).
12. J. Linderberg, Chem. Phys. Letters, 5 134 (1970).
13. C.A. Coulson, Proc. Camb. Phil. Soc., 40 201 (1940).
14. R.S. Mulliken, J. Chem. Phys., 23 1833 (1955).
15. J.C. Slater et al, Phys. Rev., 184 672 (1969).
16. S.F. Abdulnur et al, Phys.Rev. A, 6 889 (1972).
17. P.O. Löwdin, Phys. Rev., 97 1490 (1955).
18. O. Goscinski et al, J. Phys. B, 8 11 (1975).
19. Y. Öhrn et al,to be published.
20. The program was put at our disposal by C.F. Fisher and slight changes were introduced to accommodate fractional occupation numbers.
21. P.W. Thulstrup, Intern. J. Quantum Chem., 9 789 (1975).
22. P.W. Thulstrup and Y. Öhrn, to be published.
23. J.C. Green et al, Phil. Trans. Roy. Soc. Lond., A268 111 (1970).
W.E. Bull et al, Inorg. Chem., 9 2474 (1970).
P.J. Basset and D.R. Lloyd, J. Chem. Soc. (A), 641 (1971).
24. J. Spanget-Larsen, J. Electr. Spectry., 2 33 (1973).
25. J. Spanget-Larsen, J. Electr. Spectry., 3 369 (1974).

26. J. Byberg and J. Spanget-Larsen, Chem. Phys. Letters, 23 247 (1973).

27. C.H. Townes and B.P. Dailey, J. Chem. Phys., 17 782 (1949).

28. J. Byberg, Chem. Phys. Letters, 23 414 (1973).

29. F.K. Larsen, Private Communication.

30. J.A. Pedersen and J. Spanget-Larsen, Chem. Phys. Letters, 35 41 (1975).

J. Spanget-Larsen and J.A. Pedersen, J. Mag. Res., 18 383 (1975).

31. F. London, J. Phys. Radium, 8 397 (1937).

32. E. Dalgaard and J. Linderberg, to be published.

33. R. Hoffmann, J. Chem. Phys., 39 1397 (1963).

34. R.S. Mulliken, J. Chim. Phys., 46 497, 675 (1949).

35. K. Siegbahn et al, "ESCA Applied to Free Molecules", North Holland Publishing Co., Amsterdam (1969).

36. F.A. Gianturco et al, Chem. Phys. Letters, 10 269 (1971).

37. J.W.D. Connolly and K.H. Johnson, Chem. Phys. Letters, 10 616 (1971).

THE CALCULATION OF THE EXCHANGE PARAMETER $J = \frac{1}{2}(E_{singlet} - E_{triplet})$ FOR TWO EQUIVALENT ELECTRONS USING CANONICAL MOLECULAR ORBITALS

R.D. Harcourt

Department of Chemistry, University of Melbourne

Parkville, Victoria 3052, Australia

INTRODUCTION

In atomic and molecular quantum mechanics, an important realm of investigation is concerned with the calculation of the energy separation between orbital configurations that represent singlet (S = 0) and triplet (S = 1) spin states. The simplest cases to consider involve two singly occupied orbitals for the triplet state. In theories of ferro- and antiferromagnetism, these orbitals are sometimes referred to as the magnetic orbitals (1). The spin Hamiltionian for the two magnetic electrons may be written as $\hat{H}_{spin} = -2J\hat{s}_1 \cdot \hat{s}_2$, in which $\hat{s}_1 \cdot \hat{s}_2 = -3/4$ for S = 0 and +1/4 for S = 1. The singlet-triplet energy separation is then given by $^1E - ^3E = 2J$, in which J is called the "exchange integral" or the exchange parameter (2, 3). Ferromagnetic or antiferromagnetic ground-states (with parallel or antiparallel spins for the two magnetic electrons) pertain when $J > 0$ and $J < 0$ respectively. For H_2, the derivation and analysis of the expression

$$J = (\langle\Phi|\hat{H}\hat{P}|\Phi\rangle - \langle\Phi|\hat{P}|\Phi\rangle\langle\Phi|\hat{H}|\Phi\rangle)/(1 - \langle\Phi|\hat{P}|\Phi\rangle)^2$$

with $\Phi = s_a(1)s_b(2)$ has been described by Löwdin (4).

In this paper, we shall summarize some elementary procedures that may be used to calculate J for the following pairs of <u>equivalent</u> magnetic orbitals:

a) Two overlapping atomic orbitals (χ_a and χ_b) centred on two atomic nuclei. Example: The 1s atomic orbitals of two hydrogen atoms.

b) Two orthogonal 2-centre molecular orbitals located around the same pair of atomic nuclei. Examples: The π^* orbitals of O_2, S_2 and SO; the bonding π-orbitals of B_2.

c) Two overlapping 2-centre molecular orbitals (ψ_L^* and ψ_R^*) involving

i) three atomic nuclei, one of which is common to both molecular orbitals. Examples: The angular MXM linkages of binuclear Cu(II) and other transition-metal ion complexes. (The X ligands are anions such as O^{2-}, OH^- and the halides).

ii) four atomic nuclei (two for each molecular orbital). Examples' $(NO)_2$ and the $M\overline{X \qquad X}M$ linkages of binuclear Cu(II) complexes. The $\overline{X \qquad X}$ ligands are usually anions such as carboxylates (as in the copper acetate dimer), in which the X atoms are oxygen atoms.

In Figure 1, we display sets of atomic orbitals that are used to construct the equivalent orbitals ψ_L^* and ψ_R^* for each of the examples for c). Later, we shall show that for these systems, the ψ_L^* and ψ_R^* are MX antibonding orbitals.

In various guises, the theory of J for the examples of a) and b) is well-known, but it is helpful to re-describe it here for the purpose of showing its relationship to the theory for c). The latter provides a molecular orbital interpretation of aspects of superexchange, i.e. magnetic effects which arise from the delocalization of ligand lone-pair electrons into singly-occupied transition-metal ion orbitals (1,5-9).

For each of a) - c), we shall initially construct two canonical molecular orbitals (ψ_k^* and ψ_m^*) from the relevant atomic orbitals. One One consequence of doing this is that the same type of expression for J then pertains for each of a) - c). Where appropriate, we shall express the molecular orbital integrals that occur in J in terms of integrals over atomic orbitals. We shall then examine the origin of the sign of J when INDO (intermediate neglect of differential overlap) assumptions are made. These assumptions invoke zero differential overlap except with respect to 1-centre exchange and ionic integrals ($(\mu\mu'|\mu\mu')$ and $(\mu\mu|\mu\mu')$) (10,11).

In this work, we shall assume that the same set of atomic orbital exponents pertains for each pair of singlet and triplet states. Therefore, we shall not take account of recent considerations concerning the origin of the singlet-triplet energy separation (12). The geometries of both spin states are also assumed to be the same, and any overlap that may occur between the magnetic atomic orbitals (e.g. those of Figure 1) and other atomic orbitals is neglected.

When orthogonal magnetic orbitals (atomic or molecular) are used, the importance of configuration interaction (or "covalent-ionic resonance") for the calculation of J has been stressed by Löwdin (4), Slater (13), McWeeny (14) and others. The theory presented here takes account of this requirement.

EXPRESSION FOR J IN TERMS OF CANONICAL MOLECULAR ORBITALS

Here, we shall designate the two singly-occupied molecular orbitals of the S = 1 configuration as ψ_k and ψ_m. We shall assume that these canonical orbitals are orthonormal eigenfunctions of the Hartree-Fock operator $\hat{F}$ for a closed-shell S = 0 configuration, i.e.

$$\langle\psi_i|\psi_j\rangle = \delta_{ij}, \text{ and } \hat{F}\psi_i = \varepsilon_i\psi_i \tag{1}$$

for i = k and m. The ε_i are the molecular orbital energies.

Using ψ_k and ψ_m, and two electrons, we may construct the four molecular orbital configurations* of Eq. (2) in which R represents

* The spin quantum numbers for $^3\Phi_3$ are $S = S_z = 1$. We have not written down the two other degenerate configurations with $S_z = 0$ and -1. Salem and Rowland (16) have given some consideration to the energies for the configurations of Eq. (2).

all of the doubly occupied canonical molecular orbitals that are common to each configuration.

$$^1\Phi_1 = |R\psi_k^\alpha\psi_k^\beta|; \quad ^1\Phi_2 = |R\psi_m^\alpha\psi_m^\beta|;$$
$$^3\Phi_3 = |R\psi_k^\alpha\psi_m^\alpha|; \quad ^1\Phi_4 = (|R\psi_k^\alpha\psi_m^\beta| + |R\psi_m^\alpha\psi_k^\beta|)/\sqrt{2} \tag{2}$$

If we express all energies realtive to that of $^1\Phi_1$, and assume that ψ_k and ψ_m are eigenfunctions of the Hartree-Fock operator for this configuration, then the energies of $^1\Phi_2$ and $^3\Phi_3$ are given by Eqs. (3) and (4) (15). We shall assume that A > 0, i.e. when A > 0, $^1\Phi_1$ has a lower energy than $^1\Phi_2$. For the molecular systems that we shall examine, the symmetry of $^1\Phi_4$ prevents it from interacting with the S = 0 configurations $^1\Phi_1$ and $^1\Phi_2$.

$$^{1}E_2 \equiv A = 2\varepsilon_m - 2\varepsilon_k - 4J_{km} + J_{kk} + J_{mm} + 2K_{km} \qquad (3)$$

$$^{3}E_3 = \varepsilon_m - \varepsilon_k - J_{km} \equiv \tfrac{1}{2}A + J_{km} - \tfrac{1}{2}(J_{kk} + J_{mm}) - K_{km}$$

$$\equiv {}^{1}E_4 - 2K_{km} \qquad (4)$$

In Eqs. (3) and (4), the molecular orbitals coulomb and exchange integrals are defined by Eqs. (5) and (6). The molecular orbital energies are defined by Eq. (7), in which $\hat{H}^{o}$ is the core Hamiltonian, and $\hat{R}$ ($\equiv \Sigma_i(2\hat{J}_i - \hat{K}_i)$) is constructed from all of the doubly occupied molecular orbitals for $^{1}\Phi_1$ except for ψ_k.

$$J_{ij} = \langle\psi_i(1)\psi_j(2)|e^2/r_{12}|\psi_i(1)\psi_j(2)\rangle \equiv (ii|jj) \qquad (5)$$

$$K_{ij} = \langle\psi_i(1)\psi_j(2)|e^2/r_{12}|\psi_j(1)\psi_i(2)\rangle \equiv (ij|ji) \qquad (6)$$

$$\varepsilon_i \equiv \langle\psi_i|\hat{F}|\psi_i\rangle = \langle\psi_i|\hat{H}^{o} + \hat{R} + 2\hat{J}_3 - \hat{K}_3|\psi_i\rangle \qquad (7)$$

The wave-function for the lowest-energy S = 0 state involves one of the linear combinations* of Eq. (8). Relative to that of $^{1}\Phi_1$, its energy ($^{1}E_G$) is given by Eq. (9), in which A is calculated

---

*An alternative formulation for $^{1}\psi$ of Eq. (8) is $^{1}\psi = (|R\psi_+^{\alpha}\psi_-^{\beta}| + |R\psi_-^{\alpha}\psi_+^{\beta}|)/\sqrt{2}$, with $\psi_+ = (\psi_k + \ell\psi_m)/(1 + \ell^2)^{\frac{1}{2}}$, $\psi_- = (\psi_k - \ell\psi_m)/(1 + \ell^2)^{\frac{1}{2}}$, and $\ell = (-C_m/C_k)^{\frac{1}{2}}$. For the MXM and MX XM linkages of this paper, we shall assume that ψ_3 and ψ_4 of Eq. (29) are the canonical molecular orbitals ψ_k and ψ_m. If we set λ = μ in these orbitals, then this formulation for $^{1}\psi$ is equivalent to the Löwdin alternant molecular orbital wave-function (18) for the two magnetic electrons, which is a generalization of the Coulson-Fischer wave-function (19).

---

from Eq. (3), and $\langle^{1}\Phi_1|\hat{H}|^{1}\Phi_2\rangle = K_{km}$

$$^{1}\psi = C_1\,{}^{1}\Phi_1 + C_2\,{}^{1}\Phi_2 \qquad (8)$$

$$^{1}E_G = \tfrac{1}{2}\{A - (A^2 + 4K_{km}^2)^{\frac{1}{2}}\} \approx -A^2/8K_{km} \text{ for small A} \qquad (9)$$

By subtracting Eq. (4) from Eq. (9), we obtain Eq. (10) for the exchange parameter J (17). This expression is quite general, and independent of the number and type of non-magnetic electrons that pertain to each of ${}^1\Phi_1$, ${}^1\Phi_2$ and ${}^3\Phi_3$, provided we assume that the S = 1 state is to be represented by only a single configuration, namely ${}^3\Phi_3$. This then represents the simplest approach to the calculation of J. More elaborate treatments will involve extensive configuration interaction studies for both the singlet and the triplet states.

$$J = \tfrac{1}{2}\{-J_{km} + \tfrac{1}{2}(J_{kk} + J_{mm}) + K_{km} - \tfrac{1}{2}(A^2 + 4K_{km}^2)^{\frac{1}{2}}\} \qquad (10)$$

It may be proved that $-J_{km} + \frac{1}{2}(J_{kk} + J_{mm}) \geq 0$ (20). The inequality therefore generates a ferromagnetic contribution to J. On the other hand, $K_{km} - \frac{1}{2}(A^2 + 4K_{km}^2)^{\frac{1}{2}} \leq 0$, and therefore when $A > 0$, this expression generates antiferromagnetic coupling of the electron spins in the ground-state. The sign of J is therefore determined by whichever of these two terms has the larger magnitude in Eq. (10).

If we assume that Eq. (10) gives a reasonable approximation to an exact expression for J (based for example on separate multi-configuration interaction treatments for both the singlet and triplet states), it is of interest to analyse Eq. (10) for the systems referred to in the Introduction*. We shall do this in the following Sections. But it should be noted that when $J > 0$ (i.e.,

*After we had completed this work, Hay, Thibeault and Hoffmann (31) reported a derivation and examination of the J of Eq. (10) which is valid when the doubly-occupied molecular orbitals of R in Eq. (2) are not considered. Their expression for A in Eq. (10) is $2H^o_{mm} + J_{mm} - 2H^o_{kk} - J_{kk}$, with $H^o_{ii} = \langle\psi_i|\hat{H}^o|\psi_i\rangle$. Other workers (32, 33) have also provided a similar type of derivation of Eq. (10). When the orbitals of R are considered, substitution of Eq. (7) into Eq. (3) shows that the term $2(R_{mm} - R_{kk})$ with $R_{ii} = \langle\psi_i|\hat{R}|\psi_i\rangle$ must be added to the above expression for A. Our derivation of Eq. (10) (an outline of which we have reported previously in (17), where $J \equiv E_{triplet} - E_{singlet}$) takes account of this term, whose contribution has significance for the INDO analysis of J for angular MXM linkages (cf Eqs. (32), (38) and (41)). Hoffmann et al have approximated their expression for A to $2(\varepsilon_k - \varepsilon_m)$, in which the ε_i are the energies of ψ_k and ψ_m based on their S = 1 configuration. Their subsequent studies of binuclear complexes are based on their extended Hückel calculations of $\varepsilon_k - \varepsilon_m$.

the ground-state is the S = 1 configuration ${}^3\Phi_3$), the relative energies of ${}^1\Phi_4$ of Eq. (2) and the lower-energy ${}^1\psi$ of Eq. (8) may need to be examined. However, when $J < 0$, the ground-state S = 0 wave-function must be obtained from Eq. (8). This is because $^1\Phi_4$ lies above $^3\Phi_3$ by an amount equal to $2K_{km}$.

J FOR TWO OVERLAPPING ATOMIC ORBITALS

From the equivalent atomic orbitals χ_a and χ_b centered on two atomic nuclei, we may form the orthogonal canonical orbitals of Eq. (11), in which $S_{ab} = \langle\chi_a|\chi_b\rangle > 0$. The $^1\psi_G$ of Eq. (8) now corresponds to the Weinbaum wave-function (19) for H_2, for which $J < 0$ for $0 < R_{ab} < \infty$. In terms of integrals over atomic orbitals ($a = \chi_a$, $b = \chi_b$) the J_{ij} and K_{ij} of Eq. (10) may be expressed as Eqs. (12) - (15).

$$\psi_k = (\chi_a + \chi_b)/(2 + 2S_{ab})^{\frac{1}{2}}, \quad \psi_m = (\chi_a - \chi_b)/(2 - 2S_{ab})^{\frac{1}{2}} \tag{11}$$

$$J_{kk} = \tfrac{1}{2}\left((aa|aa) + (aa|bb) + 2(ab|ab) + 4(aa|ab)\right)/(1 + S_{ab})^2 \tag{12}$$

$$J_{mm} = \tfrac{1}{2}\left((aa|aa) + (aa|bb) + 2(ab|ab) - 4(aa|ab)\right)/(1 - S_{ab})^2 \tag{13}$$

$$J_{km} = \tfrac{1}{2}\left((aa|aa) + (aa|bb) - 2(ab|ab)\right)/(1 - S_{ab}^2) \tag{14}$$

$$K_{km} = \tfrac{1}{2}\left((aa|aa) - (aa|bb)\right)/(1 - S_{ab}^2) \tag{15}$$

To compare the J of Eq. (10) with the Heitler-London expression (1,4) of Eq. (16), we shall express the $^1\Phi_1$ and $^1\Phi_2$ configurations of Eq. (2) as Eqs. (17) and (18), for which $^1\psi_{cov}$ (≡ Heitler-London S = 0 wave-function) and $^1\Phi_{ion}$ are defined by Eqs. (19) and (20). By expanding the Slater determinant for $^3\Phi_3$ we obtain the $^3\psi_{cov}$ of Eq. (21), which is the Heitler-London wave-function for the S = 1 state.

$$J = (\langle ab|\hat{H}|ba\rangle - S_{ab}^2 \langle ab|\hat{H}|ab\rangle)/(1 - S_{ab}^4) \tag{16}$$

$$^1\Phi_1 = \{(1 + S_{ab}^2)^{\frac{1}{2}}/(1 + S_{ab})\}(^1\psi_{ion} + {}^1\psi_{cov})/\sqrt{2} \tag{17}$$

$$^1\Phi_2 = \{(1 + S_{ab}^2)^{\frac{1}{2}}/(1 - S_{ab})\}(^1\psi_{ion} - {}^1\psi_{cov})/\sqrt{2} \tag{18}$$

$$^1\psi_{cov} = (|\chi_a^\alpha \chi_b^\beta| + |\chi_b^\alpha \chi_a^\beta|)/(2 + 2S_{ab}^2)^{\frac{1}{2}} \tag{19}$$

$$^1\psi_{ion} = (|\chi_a^\alpha \chi_a^\beta| + |\chi_b^\alpha \chi_b^\beta|)/(2 + 2S_{ab}^2)^{\frac{1}{2}} \quad (20)$$

$$^3\Phi_3 \equiv {}^3\psi_{cov} = |\chi_b^\alpha \chi_a^\alpha|/(1 - S_{ab}^2)^{\frac{1}{2}} \quad (21)$$

By expressing $^1\psi_{cov}$ as a linear combination of $^1\Phi_1$ and $^1\Phi_2$, we obtain Eq. (22) as the energy of $^1\psi_{cov}$ relative to $^1\Phi_1$. The energy of $^3\psi_{cov}$ is given by Eq. (4) and on subtracting Eq. (4) from Eq. (22), we obtain Eq. (23) as the Heitler-London Valence-bond (HLVB) expression for J expressed in terms of integrals over the canonical molecular orbitals ψ_k and ψ_m. On substituting the LCAO forms of these orbitals into Eq. (23), this expression for H_2 reduces to that of Eq. (16).

$$^1E_{cov} = \{\tfrac{1}{2}(1 - S_{ab})^2 A - (1 - S_{ab}^2)K_{km}\}/(1 + S_{ab}^2) \quad (22)$$

$$J_{HLVB} = \{-J_{km} + \tfrac{1}{2}(J_{kk} + J_{mm}) + (2S_{ab}^2 K_{km} - S_{ab}A)/(1 + S_{ab}^2)\} \quad (23)$$

The J of Eq. (10) may now be expressed as Eq. (24).

$$J = J_{HLVB} + \tfrac{1}{2}\{S_{ab}A + (1 - S_{ab}^2)K_{km}\}(1+S_{ab}^2)-1/4(A^2+4K_{km}^2)^{\frac{1}{2}} \quad (24)$$

If the overlap integral S_{ab} is neglected, then Eqs. (12) - (15) give $-J_{km} + \frac{1}{2}(J_{kk} + J_{mm}) \overset{ab}{=} 2(ab|ab)$, and therefore J_{HLVB} $J_{HLVB} = (ab|ab) > 0$. If zero differential overlap is assumed, then $J_{HLVB} = 0$. On the basis of these considerations, Löwdin (4), Slater (13), McWeeny (14) and others have discussed how it is not possible to obtain antiferromagnetic coupling and bonding for H_2 by using orthogonal orbitals and the Heitler-London model. It is then necessary to include the "polar" or ionic wave-function ($^1\psi_{ion}$ of Eq. (20)) which influences the energy of $^1\psi_{cov}$ but not that of $^3\psi_{cov}$ (4,13,14). When this is done, and zero differential overlap is assumed, we obtain $J = \frac{1}{2}\{K_{km} - \frac{1}{2}(A^2 + 4K_{km}^2)^{\frac{1}{2}}\} < 0$, thereby accounting for bonding and the correct sign for J. (When zero differential overlap is invoked for H_2, $A = -4\langle\chi_a|\hat{H}^0|\chi_b\rangle \equiv -4\beta_{ab}$).

J FOR TWO ORTHOGONAL 2-CENTRE MOLECULAR ORBITALS

Here, ψ_k and ψ_m are a pair of orthogonal equivalent molecular orbitals, the LCAO forms of which are given by Eq. (25). The χ_a and

χ_b, and χ_c and χ_d, are pairs of equivalent atomic orbitals centred on two atomic nuclei, and $S_{ab} = \langle\chi_a|\chi_b\rangle \equiv \langle\chi_c|\chi_d\rangle = 0$.

$$\psi_k = (\chi_a + \kappa\chi_d)/(1 + 2\kappa S_{ad} + \kappa^2)^{\frac{1}{2}},$$
$$\psi_m = (\chi_b + \kappa\chi_c)/(1 + 2\kappa S_{ad} + \kappa^2)^{\frac{1}{2}} \quad (25)$$

In Eq. (25), the parameter κ is either > 0 or < 0 according to whether ψ_k and ψ_m are bonding or antibonding, and $S_{ad} \equiv S_{bc}$.

To simplify the J of Eq. (10), we proceed as follows. Because ψ_k and ψ_m are also equivalent orbitals, $J_{kk} = J_{mm}$, and A = 0. (The latter result is obtained by noting that, due to the equivalence of ψ_k and ψ_m, $\langle\psi_k|\hat{R}|\psi_k\rangle = \langle\psi_m|\hat{R}|\psi_m\rangle$ and $\langle\psi_k|\hat{H}^o|\psi_k\rangle = \langle\psi_m|\hat{H}^o|\psi_m\rangle$ in the ε_i of Eq. (7). Therefore, for the A of Eq. (3), we obtain $\varepsilon_m - \varepsilon_k = 2J_{km} - J_{kk} - K_{km}$, and hence A = 0). By substituting these results into Eq. (10), we obtain Eq. (26).

$$J = \tfrac{1}{2}(J_{kk} - J_{km}) \quad (26)$$

Because $J_{12} < \frac{1}{2}(J_{11} + J_{22})$, and $J_{11} = J_{22}$, it follows that J > 0, i.e. the ground-state is a triplet spin-state, as is the case for O_2, S_2, SO and B_2.

On substituting the L.C.A.O. molecular orbitals of Eq. (25) into Eq. (26), and assuming zero differential overlap, we obtain Eq. (27) for J, which is > 0.

$$J = \tfrac{1}{2}\{((aa|aa) - (aa|bb) + 2\kappa^2((aa|dd) - (aa|cc)) + \kappa^4((dd|dd) - (cc|dd))\}/(1 + \kappa^2)^2 \quad (27)$$

When A = 0, ${}^1\psi_G$ of Eq. (8) is given by ${}^1\psi_G = (|R\psi_k^{\alpha}\psi_k^{\beta}| - |R\psi_m^{\alpha}\psi_m^{\beta}|/\sqrt{2}$

For the linear molecules of this Section, ψ_k and ψ_m are equivalent π-type molecular orbitals. When ψ_k and ψ_m are real orbitals (i.e. either π_x and π_y, or π_x^* and π_y^*), ${}^1\psi_G$ and ${}^1\Phi_4$ of Eq. (2) are degenerate components of a ${}^1\Delta$ state. Relative to the S = 1 ground-state wave-function (${}^3\Phi_3$ of Eq. (2)), the energy of ${}^1\Phi_4$ is $2K_{km}$. Therefore the identity of Eq. (28) pertains. (A demonstration of this equality may be provided by expressing ψ_k and ψ_m in terms of the complex orbitals π_+ and π_- (or π_+^* and π_-^*) and noting that due to symmetry, integrals of the type $(\pi_+\pi_-|\pi_+\pi_-)$ are zero (22).)

$$J = \tfrac{1}{2}(J_{kk} - J_{km}) \equiv K_{km} \quad (28)$$

J FOR MX—XM AND ANGULAR MXM-LINKAGES

For the MXM and MX—XM linkages of Figure 1, the canonical molecular orbitals are those of Eq. (29), in which λ and μ are parameters ≥ 0. Because we shall be invoking INDO assumptions, overlap integrals have been omitted from the normalizing constants.

$$\begin{aligned}
\psi_1 &= \{\chi_c + \chi_d + \lambda(\chi_a + \chi_b)\}/\{2(1 + \lambda^2)\}^{\frac{1}{2}} \\
\psi_2 &= \{\chi_d - \chi_c + \mu(\chi_a - \chi_b)\}/\{2(1 + \mu^2)\}^{\frac{1}{2}} \\
\psi_3 &= \{\lambda(\chi_c + \chi_d) - (\chi_a + \chi_b)\}/\{2(1 + \lambda^2)\}^{\frac{1}{2}} \\
\psi_4 &= \{\mu(\chi_d - \chi_c) - (\chi_a - \chi_b)\}/\{2(1 + \mu^2)\}^{\frac{1}{2}}
\end{aligned} \qquad (29)$$

Both ψ_3 and ψ_4 are antibonding with respect to pairs of adjacent M and χ atomic orbitals (i.e. χ_a and χ_d, and χ_b and χ_c), and therefore these are the magnetic molecular orbitals ψ_k and ψ_m for our discussion.

The canonical molecular orbitals for configuration $^1\Phi_1$ of Eq. (2) involve $\lambda \neq \mu$. Because of this, an analysis of Eq. (10) for J is algebraically complicated (23). We shall outline one procedure that may be used to do this in a later section of this paper. Here, we shall make the simplying approximation that $\lambda = \mu = \kappa$. For carboxylate X—X-type linkages, the results of calculations suggest that this may be a good approximation (17,24a); this is not necessarily the case for MXM-type linkages (24a). However, the assumption that $\lambda = \mu = \kappa$ is satisfactory to take account of magnetic effects arising from one important difference between the two types of linkages, namely that the ligand orbitals χ_c and χ_d are respectively orthogonal and overlapping in MXM and MX—XM.

When $\lambda = \mu = \kappa$, the molecular orbitals of Eq. (29) may be expressed as in Eq. (30), in which ψ_L, ψ_R, ψ_L^* and ψ_R^* are pairs of equivalent 2-centre molecular orbitals that are respectively MX bonding and MX antibonding. These 2-centre orbitals are defined in Eq. (31).

$$\begin{aligned}
\psi_1 &= (\psi_L + \psi_R)/\sqrt{2};\ \psi_2 = (\psi_L - \psi_R)/\sqrt{2}; \\
\psi_3 &= (\psi_L^* + \psi_R^*)/\sqrt{2};\ \psi_4 = (\psi_L^* - \psi_R^*)/\sqrt{2};
\end{aligned} \qquad (30)$$

$$\psi_L = (\chi_d + \kappa\chi_a)/(1+\kappa^2)^{\frac{1}{2}};\ \psi_R = (\chi_c + \kappa\chi_b)/(1+\kappa^2)^{\frac{1}{2}};$$
$$\psi_L^* = (\kappa\chi_d - \chi_a)/(1+\kappa^2)^{\frac{1}{2}};\ \psi_R^* = (\kappa\chi_c - \chi_b)/(1+\kappa^2)^{\frac{1}{2}} \quad (31)$$

Using the ψ_3 and ψ_4 of Eq. (30), the J_{km} and K_{km} of Eqs. (3) and (10) may be evaluated from Eqs. (12) - (15) by replacing χ_a and χ_b with ψ_L^* and ψ_R^*, and setting $<\psi_L^*|\psi_R^*> = 0$ in the denominators of these integrals. The $\varepsilon_m - \varepsilon_k$ of Eq. (10) may be expressed in terms of ψ_L^* and ψ_R^* in the following manner:

$$\varepsilon_m - \varepsilon_k = -2<\psi_L^*|\hat{F}|\psi_R^*> \equiv <\psi_L^*|\hat{H}^o + \hat{R} + 2\hat{J}_3 - \hat{K}_3|\psi_R^*>$$
$$= -2(<\psi_L^*|\hat{H}^o + \hat{R}|\psi_R^*> + 2(\psi_3\psi_3|\psi_L^*\psi_R^*) - (\psi_3\psi_L^*|\psi_3\psi_R^*))$$
$$= -2(<\psi_L^*|\hat{H}^o + \hat{R}|\psi_R^*> + (\psi_L^*\psi_L^*|\psi_L^*\psi_R^*)$$
$$+ 3/2\ (\psi_L^*\psi_R^*|\psi_L^*\psi_R^*) - \tfrac{1}{2}(\psi_L^*\psi_L^*|\psi_R^*\psi_R^*)) \quad (32)$$

The resulting expressions for A and J become those of Eqs. (33) and (34), in which for convenience only, we have approximated $K_{34} - \frac{1}{2}(A^2 + 4K_{34}^2)^{\frac{1}{2}}$ to $-A^2/8K_{34}$.

$$A = -4(<\psi_L^*|\hat{H}^o|\psi_R^*> + <\psi_L^*|\hat{R}|\psi_R^*> + (\psi_L^*\psi_L^*|\psi_L^*\psi_R^*)) \quad (33)$$

$$J = (\psi_L^*\psi_R^*|\psi_L^*\psi_R^*) - A^2/16K_{34} \quad (34)$$

The $\hat{R}$ operator of Eq. (33) is constructed from the ψ_1 and ψ_2 molecular orbitals of Eq. (30), and therefore $<\psi_L^*|\hat{R}|\psi_R^*>$ of Eq. (33) may be expressed firstly as Eq. (35) and then as (36).

$$<\psi_L^*|\hat{R}|\psi_R^*> = \sum_{i=1,2}(2(\psi_L^*\psi_R^*|\psi_i\psi_i) - (\psi_L^*\psi_i|\psi_R^*\psi_i)) \quad (35)$$

$$= 4(\psi_L^*\psi_R^*|\psi_L\psi_L) - 2(\psi_L^*\psi_L|\psi_R^*\psi_L) \quad (36)$$

On substituting the 2-centre molecular orbitals of Eq. (31) into the integrals of Eqs. (33) - (36), and invoking INDO-type assumptions, we obtain Eqs. (37) - (39).

$$<\psi_L^*|\hat{H}^o|\psi_R^*> = (\beta_{ab} - 2\kappa\beta_{ac} + \kappa^2\beta_{cd})/(1+\kappa^2) \quad (37)$$

$$\langle\psi_L^*|\hat{R}|\psi_R^*\rangle = 2\kappa^2(cc|cd)/(1+\kappa^2)^2 \quad (38)$$

$$(\psi_L^*\psi_R^*|\psi_L^*\psi_R^*) = \kappa^4(cd|cd)/(1+\kappa^2)^2 \quad (39)$$

For the MXM linkages, the atomic orbitals χ_c and χ_d are orthogonal, and therefore $\beta_{cd} = 0$ in Eq. (37).

The resulting expressions for J and A are those of Eq. (40) and (41)

$$J = \kappa^4(cd|cd)/(1+\kappa^2)^2 - A^2/16K_{34} \quad (40)$$

$$A = -4\{\beta_{ab} - 2\kappa\beta_{ac} + \kappa^2(\kappa^2+2)\ (cc|cd)/(1+\kappa^2)\}/(1+\kappa^2) \quad (41)$$

For the ligand orbitals χ_c and χ_d of the MXM linkage of Figure 1, the ionic integral $(cc|cd)$ is non-zero when these orbitals are hybrid atomic orbitals with some s character. If each orbital is written as $\chi_\mu = (p_\mu + \alpha s)/(1+\alpha^2)^{\frac{1}{2}}$ in which α is a hybridization parameter, then for $\alpha < \infty$, it may be proved that $(cc|cd) < (cd|cd)$. If χ_c and χ_d are orthogonal p orbitals when $\sphericalangle$ MXM $= 90^\circ$, then for this angle, $\beta_{ac} = 0$ in Eq. (41). For the $\overline{\text{MX}\quad\text{XM}}$ linkage of Figure 1, the exchange and ionic integrals $(cd|cd)$ and $(cc|cd)$ are 2-centre integrals, and therefore they are set equal to zero when INDO assumptions are made. The resulting expression for J becomes that of Eq. (42).

$$J = -(\beta_{ab} - 2\kappa\beta_{ac} + \kappa^2\beta_{cd}^2)/\{(1+\kappa^2)^2K_{34}\} \quad (42)$$

a). Angular MXM Linkages

Examination of Eqs. (40) and (41) shows that the orthogonality of the ligand orbitals (through $(cd|cd)$) generates ferromagnetism, whereas both 2-centre overlap (through β_{ab} and β_{ac}) and the s-character of the ligand orbitals (through β_{ac} and $(cc|cd)$) generate antiferromagnetism. For the molecular orbitals of Eq. (30), the parameter κ gives a measure of the extent to which the ligand lone-pair electrons delocalise into the singly-occupied metal-ion orbitals of Figure 1. Therefore, κ may be referred to as a "superexchange" parameter,* with values ranging from 0 to ∞. With these values

*When $\kappa = 0$ and $\kappa = \infty$, the molecular orbitals of Eq. (30) generate the following INDO atomic orbitals charges $(P_{\mu\mu})$ for each of the molecular orbital configurations of Eq. (2). $\kappa = 0$: $P_{aa} = P_{bb} = 1$, $P_{cc} = P_{dd} = 2$; $\kappa = \infty$: $P_{aa} = P_{bb} = 2$, $P_{cc} = P_{dd} = 1$. For the canonical

Eqs. (40) and (41) give $J = -\beta_{ab}^{2}/K_{34} < 0$, and $J = (cd|cd) - (cc|cd)^{2}/K_{34} > 0$, thereby generating antiferromagnetism and ferromagnetism respectively. (When $\kappa = \infty$, $K_{34} = (cd|cd)$. These expressions for J are the same as those obtained previously for pairs of equivalent overlapping atomic orbitals and orthogonal π-orbitals. For the latter, $\kappa = \infty$ in Eq. (27), together with the identity of Eq. (28), gives $J = \frac{1}{2}((dd|dd) - (cc|dd)) \equiv (cd|cd)$; $(cc|cd) = 0$ for p orbitals.)

For $\kappa = 0$ and $\kappa = \infty$, the atomic orbital populations and spins for the ground-state valence-bond structures are displayed in Table 1. They are (I) with antiparallel spins, (II) and (III) for $\kappa = 0$, and (IV) with parallel spins for $\kappa = \infty$. When $0 < \kappa < \infty$, these structures participate in resonance with either seven other S = 0 structures or five other S = 1 structures. For the S = 0 state, valence bond considerations similar to these have been described elsewhere for N_2O_4 and $Cu_2(RCO_2)_4$ with two RCO_2^- ligands (17,25).

Table 1

Electron distributions for valence-bond structures

	χ_a	χ_b	χ_c	χ_d		χ_a	χ_b	χ_c	χ_d
(I)	↑	↓	↑↓	↑↓					
(II)	↑↓		↑↓	↑↓	(V)	↑↓	↑↓	↑	↓
(III)		↑↓	↑↓	↑↓	(VI)	↑↓	↑↓	↑↓	
(IV)	↑↓	↑↓	↑	↑	(VII)	↑↓	↑↓		↑↓

Both ferro- and antiferromagnetic Cu(II) hydroxy- and alkoxo-bridge dimers have been reported (26-28), and the simplest form of the INDO theory which we have described shows that both types of magnetic behaviour are possible.

b). MX——XM Linkages

For the MX——XM linkages, Eq. (42) gives $J < 0$ for all values of κ. Consequently the INDO theory predicts antiferromagnetic coupling of the two magnetic electrons, in agreement with what has been observed for a large number of Cu(II) carboxylates (29). For $\kappa = 0$

---

molecular orbitals of Eq. (29), the λ and μ are respectively super-exchange parameters for $^{1}\Phi_1$ and $^{1}\Phi_2$ of Eq. (2).

and $\kappa = 0$ and $\kappa = \infty$, Eq. (42) gives $J = -\beta_{ab}^2/K_{34}$ and $J = -\beta_{cd}^2/K_{34}$. The latter value reflects the difference between $\overline{MX \quad XM}$ and MXM linkages, namely that arising from the overlap of χ_c and χ_d for $\overline{MX \quad XM}$ and the orthogonality of χ_c and χ_d for MXM. For $\kappa = \infty$, the ground-state $S = 0$ valence-bond structures have the electron distributions of (V), (VI) and (VII) for $\overline{MX \quad XM}$ linkages.

Examination of Eq. (42) shows that the relative importance of metal-metal, metal-ligand and ligand-ligand bonding for antiferromagnetic coupling is determined by both the magnitudes of the core resonance integrals (and hence the overlap integrals S_{ab}, S_{ac} and S_{cd}) and the superexchange parameter κ. For copper acetate, with $d_{x^2-y^2}$ metal orbitals forming a Cu-Cu δ-bond, and sp^2 oxygen orbitals, the results of some INDO molecular orbital calculations (24a) suggest that OO bonding is more important than CuCu bonding for generating antiferromagnetism.

In Eqs. (41) and (42), the sign of the coefficient of the core resonance integral β_{ac} is opposite to that of β_{ab} and (in Eq. (42)) β_{cd}. Therefore, if the overlap integrals S_{ab}, S_{cd} and S_{ac} have the same signs, the antiferromagnetic contribution to J arising from the M-M and (in Eq. (42)) the X-X overlap will be reduced by the (non-neighbour) M-X overlap. These overlaps will only reinforce each other when the sign of S_{ac} is opposite to those of S_{ab} and S_{cd}.

J FOR BINUCLEAR SYSTEMS WITH TWO X OR $\overline{X \quad X}$ LIGANDS

In the previous Section, we used the simplifying assumption that $\lambda = \mu = \kappa$ in the molecular orbitals of Eq. (29). Here, we shall provide a brief outline of one analysis of J that takes account of the essential non-equality of λ and μ in these canonical molecular orbitals. In doing so, we shall simultaneously give consideration to the presence of two X or $\overline{X \quad X}$ type ligands, (as in $Cu_2(OH)_2^{2+}$ and $Cu_2(RCO_2)_4$). For these D_{2h} symmetry M_2X_2 or M_2X linkages, there are six canonical molecular orbitals. If, as was done in the previous Section, we assume that the antibonding MX molecular orbitals are singly occupied in the $S = 1$ configuration, then ψ_k and ψ_m are now given by Eq. (43), in which χ_a, χ_d and χ_f are "L" moiety orbitals and χ_b, χ_c and χ_e are "R" moiety orbitals.

$$\psi_k = \{\tfrac{1}{2}\lambda(\chi_c + \chi_d + \chi_e + \chi_f) - (\chi_a + \chi_b)/\sqrt{2}\}/(1 + \lambda^2)^{\frac{1}{2}}$$
$$\psi_m = \{\tfrac{1}{2}\mu(\chi_d - \chi_c - \chi_e + \chi_f) - (\chi_a - \chi_b)/\sqrt{2}\}/(1 + \mu^2)^{\frac{1}{2}} \qquad (43)$$

The superexchange parameters λ and μ may be expressed , as in Eq. (44), in terms of a Hückel electronegativity parameter (h) for

---

In Hückel molecular orbital theory, the energies of the D_{2h} molecular orbitals of Eq. (43) may be expressed in terms of the Hückel coulomb and resonance integrals (α_X and β_M) for the X ligand orbitals and the atomic orbitals for the Mχ bonds of Figure 1. We write $\alpha_m = \alpha_X + h\beta_M$, $\beta_{MM} = k\beta_{MX}$, and solve the secular equations $(\alpha_\mu - \varepsilon)c_\mu + \Sigma\, \beta_{\mu\nu} c_\nu = 0$. The electronegativity and resonance integral parameters h and k are chosen so that the Hückel and Hartree-Fock matrix elements commute i.e. so that they generate the same sets of eigenvectors (30).

---

a metal-ion orbital, and a Hückel resonance integral parameter (k) for the MM bond (25a). To make any convenient simplifications when INDO-type assumptions are introduced, it is necessary to assume that k is small, and disregard k^2 and higher powers of k. When this is done, λ and μ may be expressed as in Eq. (45), with κ and ρ defined in Eq. (46). The resulting INDO-type expressions for J and A are then given by Eqs. (47) and (48) (23).

$$\lambda = (\{(h + k)^2 + 8\}^{\frac{1}{2}} + h + k)/2\sqrt{2},$$

$$\mu = (\{(h - k)^2 + 8\}^{\frac{1}{2}} + h - k)/2\sqrt{2} \qquad (44)$$

$$\lambda = \kappa + \rho k,\ \mu = \kappa - \rho k \qquad (45)$$

$$\kappa = \{(h^2 + 8)^{\frac{1}{2}} + h\}/2\sqrt{2};\ \rho = \kappa^2/\{\sqrt{2}(1 + \kappa^2)\}. \qquad (46)$$

$$J = \tfrac{1}{2}\kappa^3(\kappa(cd|cd) + 4\rho k(cc|cd))/(1 + \kappa^2)^2 - A^2/16K_{km} \qquad (47)$$

$$\begin{aligned} A = &-4\{\beta_{ab} - 2\sqrt{2}\kappa\beta_{ac} + \kappa^2(\beta_{cd} + \beta_{de})\}/(1 + \kappa^2) \\ &-8\kappa^2(1 + 3\kappa^2/4)(cc|cd)/(1 + \kappa^2) \\ &+2\rho\kappa\{2\gamma_{ab} + 2(\kappa^2 - 1)\gamma_{ac} - \kappa^2(\gamma_{cd} + \gamma_{de})\}/(1 + \kappa^2)^3 \\ &+\rho\kappa^3 k(3(cd|cd) + 4(cc|cd))/(1 + \kappa^2)^3 \end{aligned} \qquad (48)$$

In Eq. (48), the $\gamma_{\mu\nu}$ ($\equiv(\mu\mu|\nu\nu)$) are the coulombic repulsion integrals for the atomic orbitals.

For complexes with $\overline{X \qquad X}$-type ligands, the (cd|cd) and (cc|cd) integrals are set equal to zero when zero differential overlap is invoked, and therefore antiferromagnetism is predicted from Eq. (47). As was remarked above, Eqs. (47) and (48) are only

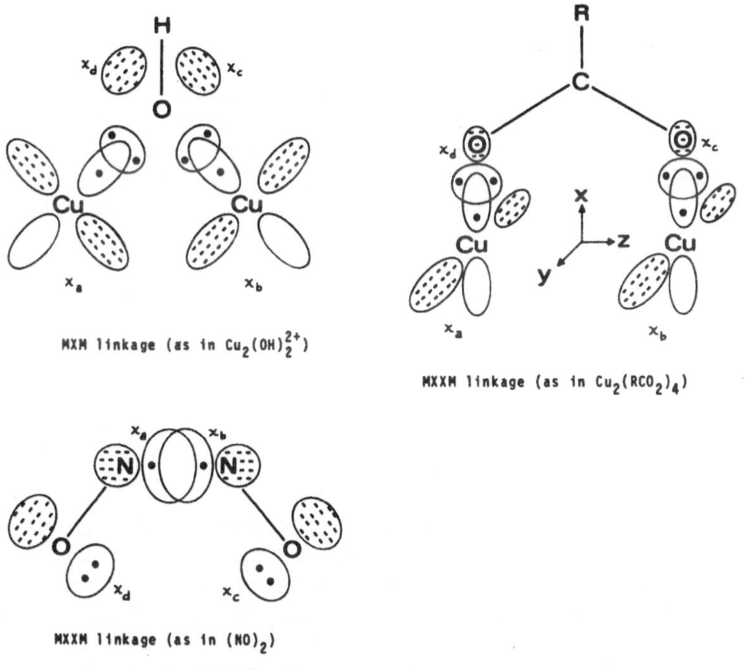

Figure 1. Atomic orbitals for 6-electron 3-centre and 6-electron 4-centre bonding (-ve lobes are shaded).

appropriate when k is small. This occurs for RCO_2^- ligands, with the ligand atomic orbitals (χ_c and χ_d, or χ_e and χ_f) located on non-adjacent atoms (17,24a). Indeed, for the copper carboxylates, it is a good approximation to set k = 0. However, for copper pyrazolate dimers $Cu_2(pz)_2^{2+}$, the two nitrogen atomic orbitals of each pyrazolate ligand are located on adjacent atoms, and their overlap can be appreciable (24b). The results of some preliminary INDO calculations suggest that k may not necessarily be small (24b), and therefore deductions made from Eq. (47) may not be reliable.

For complexes with two X-type ligands, β_{cd} = 0 in Eq. (48). For MXM bondangles rather greater than 90°, the overlap integral S_{de}, and hence the resonance integral β_{de} of Eq. (48), may have appreciable magnitude (24a). When this is the case, som significant responsibility for antiferromagnetic interactions may be ascribed to the χ_d - χ_e-type overlap. As is the case for $Cu_2(pz)_2^{2+}$, k for these types of complexes may not be sufficiently small for the approximations of Eq. (45) to be reliable, and higher powers of k would then be required for the analysis of J. It may also be noted that ψ_k in particular is not necessarily given by Eq. (43); the results of INDO calculations (24a) indicate that the MX non-bonding molecular orbital $\frac{1}{2}(\chi_c + \chi_d - \chi_e - \chi_f)$ may have a higher energy than has ψ_k of Eq. (43). However, all treatments indicate the relevance of the orthogonality of each pair of ligand orbitals for ferromagnetism.

CONCLUSION

By invoking configuration interaction for the S = 0 state, we have provided an analysis of the origin of the sign of the exchange parameter J for several types of molecular systems. The basic formula for J (namely Eq. (10)) has both a ferromagnetic and an antiferromagnetic component, at least one of which does not reduce to zero when INDO-type assumptions are made; 2-centre overlap and 1-centre exchange are still able to generate either antiferro- or ferromagnetic coupling of the electron spins.

ACKNOWLEDGEMENT

It is with great pleasure that I present this paper to honour Professor P.O. Löwdin on the occasion of his 60th birthday. I wish to thank Dr. A.C. Hurley, C.S.I.R.O. Division of Chemical Physics, and Mr. Peter Taylor, Department of Theoretical Chemistry, University of Sydney, for reading the manuscript and making very helpful comments.

BIBLIOGRAPHY

1. A.P. Ginsberg, Inorg. Chim. Acta Revs., 5, 45 (1971).
2. W. Heisenberg, Z. Physik, 38, 411 (1926).
3. P.A.M. Dirac, Proc. Roy. Soc. London, A123, 714 (1929).
4. P.O. Löwdin, (a) Revs. Modern Phys. 34, 80 (1962); (b) J. Applied Phys. Suppl., 33, 251 (1962); (c) International Summer School Lectures, Uppsala.
5. H.A. Kramers, Physica, 1, 182 (1934).
6. P.W. Anderson, in "Magnetism", (Ed. G.T. Rado and H. Suhl Academic Press, New York, 1963) Vol. 1. p.99.
7. J.B. Goodenough, Phys. Rev., 100, 564 (1955).
8. J. Kanamori, J. Phys. Chem. Solids, 10, 87 (1959).
9. R.L. Martin, in "New Pathways in Inorganic Chemistry", (Eds. E.A.V. Ebsworth, A.G. Maddock and A.G. Sharpe, C.U.P., 1968) p. 175.
10. J.A. Pople, D.L. Beveridge and P.A. Dobosh, J. Chem. Phys., 47, 2026 (1967).
11. R.N. Dixon, Mol. Phys., 12, 83 (1967).
12. J.P. Colpa, Mol. Phys., 28, 581 (1974) and references therein.
13. J.C. Slater, (a) J. Chem. Phys., 19 220 (1951); (b) Revs.Modern Phys., 25, 199 (1953).
14. R. McWeeny, Proc. Roy. Soc. London, A233, 63, 306 (1954).
15. J.N. Murrell and K.L. McEwen, J. Chem. Phys., 25, 1143 (1956).
16. L. Salem and C. Rowland, Angewandte Chem., Int. Ed. 11, 92 (1972).
17. R.D. Harcourt, Australian J. Chem., 27, 2065 (1974).
18. P. O. Löwdin, Phys. Rev., 97, 1509 (1955).
19. C.A. Coulson and I. Fischer, Phil. Mag., 40, 386 (1949).
20. C.C.J. Roothaan, Revs. Modern Phys. 23, 161 (1951).
21. S. Weinbaum, J. Chem. Phys., 1, 593 (1933).
22. A.C. Hurley and P. Taylor, private communication.
23. R.D. Harcourt, unpublished results.
24. (a) R.D. Harcourt and G. Martin, submitted for publication; (b) R.D. Harcourt and M. Leigh-Brown, unpublished results.
25. R.D. Harcourt, (a) Theor. Chim. Acta, 2, 437 (1964); errata, 4, 202 (1966); (b) ibid, 6, 131 (1966); (c) Int. J. Quantum

Chem., 4, 173 (1970); J. Mol. Structure, 9, 221 (1971).

26. A.T. Casey, Australian J. Chem. 25, 2311 (1972).

27. K.T. McGregor, N.T. Watkins, D.L. Lewis, R.F. Drake, D.J. Hodgson, and W.E. Hatfield, Inorg. Nucl. Chem. Letters, 9, 423 (1973).

28. C.G. Barraclough, R.W. Brookes and R.L. Martin, Australian J. Chem., 27, 1843 (1974).

29. R.W. Jotham, S.F.A. Kettle and J.A. Marks, J. Chem. Soc. (Dalton) 428 (1972).

30. R.D. Brown and R.D. Harcourt, Australian J. Chem. 16, 737 (1963).

31. P.J. Hay, J.C. Thibeault and R. Hoffmann, J. American Chem. Soc., 97, 4884 (1975).

32. I.G. Dance, Inorg. Chim. Acta, 9, 77 (1974).

33. C.K. Jørgensen, in "Modern Aspects of Ligand Field Theory" (North Holland, Amsterdam, 1971)p. 317.

IMPORTANCE OF OVERLAP IN THE ANALYSIS OF WEAK EXCHANGE INTERACTIONS BY PERTURBATION METHODS

R. Block and L. Jansen

Institute of Theoretical Chemistry, University of Amsterdam, Amsterdam, The Netherlands

INTRODUCTION

The accurate determination of exchange interactions, i.e. those which arise as a consequence of permutation symmetry of the Hamiltonian for a system considered, is one of the most difficult tasks of the quantum theory of molecules and solids. On the other hand, the general importance of exchange interactions can hardly be overestimated: starting with Heitler and London's (1927) pioneering work on the hydrogen molecule, it has become clear that they often play an essential role in problems of molecular stability and conformation, of the cohesive energy of insulators and metals, magnetic ordering in solids with paramagnetic cations, etc. Their accurate evaluation is, consequently, a conditio sine qua non for any satisfactory theoretical analysis of properties of condensed matter.

Let us first mention a few characteristic properties of exchange interactions: a) they decrease in strength very rapidly (say, exponentially,) with increasing distance between two interacting atoms or molecules, invariably becoming repulsive at the smallest separations; b) exchange interactions are typically non-additive. For example, the interaction energy between three hydrogen atoms is not the same as that for three isolated pairs of hydrogen atoms. This was demonstrated already in 1929 by London (1) for the system H_3 and pointed out later by Margenau (2) in connection with short-range repulsions between closed-shell atoms.

An interesting model system on which to study exchange interactions in more detail is that composed of two hydrogen atoms A and

B and one helium atom C (the so-called "three-center, four-electron sstem"). If we denote the total (exchange) energy of interaction for the system by E(ABC), the corresponding quantities for the isolated pairs (AB), (AC) and (BC) by E(AB), E(AC) and E(BC), respectively, then

$$\Delta E(ABC) \equiv E(ABC) - \{E(AB) + E(AC) + E(BC)\}$$

is the non-additive component of the exchange energy. In view of the short-range character of exchange, ΔE(ABC) depends sensitively on the geometry of the triangle formed by the triplet (ABC). Since the two hydrogen atoms A and B are the "interesting" particles of the system in view of their reactivity, ΔE(ABC) is usually taken to define the "indirect" exchange between A and B. Likewise, we might define the quantity ΔE(ABC) + E(AB) to denote the "effective" exchange between the two hydrogen atoms; it still contains the direct exchange contribution E(AB) for the isolated pair of hydrogen atoms.

The a priori potential significance of a non-additivity quantity such as ΔE(ABC) is twofold. First, it is structure-sensitive with respect to molecular conformation or local symmetry in solids; second, indirect exchange may have a considerably longer range than direct exchange interactions. In 1934, Kramers (3) was the first to develop a theory of indirect exchange interactions between the cations of a paramagnetic solid via the diamagnetic anions (e.g. between M^{2+}-cations via the O^{2-}-anions in solid MnO). The first experimental indications that such effects might be appreciable, just became available at that time (4). In his analysis, Kramers employed a perturbation approach, starting from free-ion wave-functions, "adapted" to the crystalline field, in zeroth order. He illustrated the indirect exchange schematically by considering a system of two cations and one anion as the fundamental unit of superexchange. In the simplest possible representation of this system, Kramers used the three-center, four-electron model, i.e. the unpaired electrons on either cation are replaced by one "effective" electron, those on the anion by two spin-paired effective electrons. However, in Kramers' scheme the first non-zero indirect exchange resulted from third order in the perturbation. The conclusion was that either indirect exchange should be negligibly small, or that perturbation theory was not an appropriate tool for its evaluation.

The first possibility was soon proven to be invalid. In 1946, de Klerk (5) obtained further evidence for indirect exchange between paramagnetic cations (the name "superexchange" is now commonly used for this effect (6) from measurement of the specific heat of copper potassium sulfate at temperatures below 1 K. The specific heat was much larger than that due to isolated unpaired spins on the Cu^{2+}-ions, pointing to some kind of indirect coupling mechanism between them (direct exchange is very small, even between nearest cations). As magnetostatic coupling is much too weak at the distance between neighboring Cu^{2+}-ions, the indirect coupling must be of exchange type. In 1949, Shull and collaborators (7) published the first results of a series of neutron-diffraction analyses of non-conducting 3d-solids at low temperatures. The evidence of magnetic ordering of spins on different paramagnetic cations was final and conclusive proof for existence of the super-exchange phenomenon. Since then, magnetic order and superexchange interactions have formed a major domain of experimental and theoretical solid-state research and have found many technological applications (ferrites).

In 3d-solids, indirect exchange is of the order of 10^{-2} cV in magnitude, i.e. several orders weaker than conventional chemical bonding. It is thus to be expected that a perturbation analysis should be appropriate in evaluating indirect exchange, and the question arises as to the cause of the failure in Kramers' approach. As mentioned before, Kramers starts from free-ion wavefunctions in zeroth order of perturbation, "adapted" in some way to the crystalline field. Then, he neglects overlap between the zeroth-order functions, not only that between cations, but also between cations and anions. With this approximation, it is immediately obvious that indirect exchange can only occur as a higher-order effect, i.e. arising from matrix elements containing the perturbation operator more than once, since this operator consists of terms not involving the position vectors of more than two electrons. On the other hand, if overlap is not neglected, then indirect exchange in principle already occurs in the lowest perturbation order. It is, thus, not a priori plausible to expect that Kramers' third-order result is even qualitatively correct.

Anderson (8) gave an explicit formulation of Kramers' analysis (1950). Aware of convergence and orthogonality difficulties, he later developed (1959) (9) a quite different, very ingenious, approach to the problem of indirect exchange in solids with paramagnetic cations. Since we here meet with another approximation which may in principle seriously affect the reliability of numerical results, a brief outline of Anderson's method is given next. Anderson's approach is very similar to the one he used in the problem of localized magnetic moments in 3d-metals, often called the "s-d mixing model" (10,11).

In the case of a non-conducting solid with paramagnetic cations, the evaluation of exchange interactions involving unpaired electrons on different cations proceeds in two steps. First, Anderson considers one magnetic electron in the field of all the anions (ligands) and of all the other cations in the solid, while leaving out exchange between the unpaired electron considered and those of the other cations, in a Hartree-Fock scheme, i.e. in a one-electron approximation. If we start from a state in which the spins of all unpaired electrons are parallel (ferromagnetic case), then the Hartree-Fock operator for the electron considered has the periodicity of the (cation-) lattice, so that the solutions are running Bloch waves. These can then be transformed to Wannier functions, localized around the cation positions; by their definition, Wannier functions centered at different lattice points are orthogonal to one another, which eliminates one of the difficulties of Kramers' VB-approach. In the second step, Anderson takes into account exchange interactions for two electrons on different Wannier orbitals, taking those orbitals as zeroth-order wavefunctions. A complication arises here, because of the fact that the Wannier functions are not eigenfunctions of the complete Hartree-Fock operator for each electron. Instead, the operator must be split into a part diagonal in the basis of Wannier functions, plus a "delocalization operator" (it mixes into a given Wannier function those located at all other lattice points). This latter (one-electron) part of the Hartree-Fock operator for each electron is then combined with the electron-electron interaction to form the perturbation. In first order, the delocalization part of the perturbation gives zero contribution to the energy, whereas electron-electron exchange favors parallel alignment of their spins ("potential exchange"). Both results are a consequence of orthogonality of Wannier functions at different lattice points. In second order, only the delocalization part of the perturbation is taken into account. A delocalization (mixing) of the two orbitals can, however, take place only if the spins of the two electrons are antiparallel, in view of the Pauli principle. Thus, delocalization, accompanied by a lowering in one-electron energy ("kinetic exchange") requires antiparallel spins of the two electrons concerned.

Attractively enough, the Anderson analysis leaves the net result of the two competing effects ("potential exchange" favoring parallel spins, "kinetic exchange" only in the case of antiparallel spins) open. It is, however, clear that an actual calculation on this basis is hardly feasible. Furthermore, whereas in Kramers' VB-model the orbitals of magnetic and ligand electrons occur equally explicitly in the formalism, in Anderson's method the effect of the ligands is replaced by a smeared-out exchange potential. This tends to over-emphasize delocalization of the magnetic orbitals.

The important problem of explaining, or predicting, most stable magnetic patterns for solids with different paramagnetic cations at low temperatures, can, unfortunately, not be solved along these lines. Semi-empirical rules governing superexchange phenomena must then fill the gap between theory and experiment; they were developed by Goodenough (12) and Kanamori (13)(so-called "Goodenough-Kanamori rules"; for a clear review, and a concise formulation of these rules, see Anderson (14)). Still, it must be said that, with the large supply of accurate experimental data now available, even semi-empirical theory lags far behind the experimental facts.

The above considerations were primarily given with the aim of illustrating some of the complexities inherent in an accurate theoretical analysis of "indirect" exchange. From Kramers' analysis it can be concluded that neglect of non-orthogonality of basis functions in a perturbation approach to superexchange has a pronounced effect on the order of perturbation in which non-zero indirect exchange first appears. Anderson's Hartree-Fock method for constructing zeroth-order one-electron wavefunctions, on the other hand, clearly overemphasizes the role of the magnetic electrons in indirect exchange. In addition, such a formalism is far too complicated to be applied in actual calculations.

In the next Sections we will discuss the results of an evaluation of indirect exchange based on a Kramers-type model of effective electrons, in which non-orthogonality of the basis wave-functions is rigorously taken into account. Such a model was first applied to the problem of crystal stability of rare-gas and simple ionic solids (15), following Löwdin's pioneering research on the crystal stability of alkali halides and the deviations of their elastic constants from the Cauchy relations (16). In addition, the model has been used to evaluate superexchange interactions in 3d- or 4f-solids at low temperatures (17,18), to analyze stability of rare-gas halides (19) and to give an interpretation of rotational barriers in simple organic and inorganic molecules (20). A comprehensive review of such applications has appeared recently (21).

For the sake of completeness, it should be remarked here that indirect exchange is a very common phenomenon, not limited to non-conducting solids and to molecules, as we have thus far considered. Indirect exchange interactions between localized spins of paramagnetic atoms (cations) in a metal via conduction electrons are known as the RKKY (Ruderman-Kittel-Kasuya-Yosida)-effect; they form the basis of much of our present-day interpretation of magnetism in metals and alloys (22). Conversely, indirect exchange between conduction electrons via a localized electron spin leads, in a Heisenberg effective spin-Hamiltonian formalism, to so-called "spin-flip" or "dynamical" scattering of conduction electrons, accounting for the observed sudden rise in resistivity of many metals and alloys upon decreasing the temperature; this phenomenon is called the Kondo effect (23). Recently, Witkowski (24)has suggested that also hydrogen bonding may be interpreted in terms of indirect exchange, namely between non-bonded electrons via the chemical bond. Witkowski's formal analysis closely resembles that of the BCS-theory of superconductivity, although the BCS indirect electron-electron coupling is not of exchange type. The fact that two electrons forming a Cooper pair have their spins opposite results from the fact that the wavefunctions of these electrons can overlap in orbital space, increasing their indirect interaction via lattice vibrations (phonons).

EFFECTIVE-ELECTRON MODEL WITH OVERLAP

As in Kramers' analysis, we start the perturbation analysis of indirect exchange interactions for the system considered from the free-ion wavefunctions; however, their overlap is now explicitly taken into account. Also, we assume that the exchange part of the interactions may reliably be calculated without considering spatial symmetries of the Hamilton operator for the ions in the crystal. In other words, we will not consider any crystalfield effects. As a further approximation, we adopt an effective-electron model: electrons with unpaired spins on a cation are replaced by one effective electron, those on a diamagnetic anion are replaced by two, spin-paired, effective electrons. A justification for this assumption may be given by noting that the magnetic patterns observed with (non-conducting) 3d-solids neither depend on the number of unpaired electrons per cation, nor on the type of anion. In addition, the measured temperatures (Néel tempertures) of the transition from the magnetically ordered to the disordered (paramagnetic) state of the different solids are all of the same order of magnitude (a few hundred K). For the same reason, we neglect the possible effect of the net ionic charges on indirect exchange. On the basis of the above approximations, we start from electrically neutral "ions" with spherically symmetric zeroth-order wavefunctions for the effective electrons. The perturbation method is of the Rayleigh-Schrödinger type, adapted to exchange (25).

Consider now a system of two paramagnetic cations A and B, separated by a diamagnetic anion C. According to the assumptions mentioned above for the three-center, four-effective electron model, the ions A and B are each described by a half-filled s-shell (one unpaired electron), whereas ion C is described only by one closed shell of s-type (two spin-paired electrons). The corresponding functions are denoted by ϕ_A, ϕ_B and ϕ_C, respectively. We do not assume orthogonality between the spherically symmetric shell orbitals. In accordance with the model, the three centers will further be called "atoms". In terms of the simple atomic wavefunctions, the correct zeroth-order wavefunction describing the singlet and triplet groundstates of the coupled system can then be written as

$$\psi_\sigma^{ABC} = A\phi_A(1)\phi_B(2)\phi_C(3)\phi_C(4)\sigma \equiv A\phi\sigma.$$

A is the antisymmetrizer for the total system and σ denotes the triplet or singlet spin-eigenfunction of the squared totalspin operator S^2, i.e.

$$\sigma(\text{singlet}) = (\tfrac{1}{2})\{\alpha(1)\beta(2) - \alpha(2)\beta(1)\}\{\alpha(3)\beta(4) - \alpha(4)\beta(3)\}$$

and

$$\sigma(\text{triplet}) = (\tfrac{1}{2})\{\alpha(1)\beta(2) + \alpha(2)\beta(1)\}\{\alpha(3)\beta(4) - \alpha(4)\beta(3)\}.$$

It can be shown (25) that, in good approximation, the associated first-order interaction energy $E_\sigma^{(1)}$ (only terms linear in the perturbation Hamiltonian) is given by

$$E_\sigma^{(1)} = \frac{(\phi\sigma, VA\phi\sigma)}{(\phi\sigma, A\phi\sigma)} . \qquad (1)$$

Here, V is the perturbation Hamiltonian for the system ABC in which the labels of the electrons on the individual atoms are fixed.

In order to obtain a convenient expression for $E_\sigma^{(1)}$, a symmetrical double coset (SDC) decomposition (26) of the projector A can be carried out with respect to the invariance group K of the function ϕ, i.e. the group formed by the unit operator and the permutation P_{34} interchanging the electron labels 3 and 4. The transformation properties of the spinfunctions follow at once from their definitions. We generate all permutations P in the antisymmetrizer A,

$$A = \sum_P (-1)^P P \qquad (P \in S_4)$$

as a linear combination of symmetric double cosets of the form

$$\sum_i \frac{1}{\delta_i} K \lambda_i K,$$

where δ_i is the repetition number of the elements in the SDC with generator λ_i. By substitution of this expansion for the antisymmetrizer into (1), and making use of the transformation properties of the spinfunctions, the expression for $E_\sigma^{(1)}$ is reduced to

$$E_\sigma^{(1)} = \frac{\sum_i (-1)^{\lambda_i} (\phi, V\lambda_i\phi)(\sigma, \lambda_i\sigma) \cdot \frac{1}{\delta_i}}{\sum_i (-1)^{\lambda_i} (\phi, \lambda_i\phi)(\sigma, \lambda_i\sigma) \cdot \frac{1}{\delta_i}} . \qquad (2)$$

It can be shown (26) that the SDC generators can always be chosen as pure interchanges between the different closed- and open-shell electron indices; the cycles in an SDC generator based on this choice do not contain more than one electron label of each shell. For further details of the theory of the SDC decomposition, the properties of the SDC generators and the calculation of the repetition numbers, we refer to the literature (26). In order to obtain an expression for the first-order interaction energy in terms of an effective spin-Hamiltonian, we use the Dirac identity in spin space,

$$P^{\sigma}_{ij} = \tfrac{1}{2} + 2\, s_i \cdot s_j ,$$

where P^{σ}_{ij} denotes the permutation operator acting on the indices of the two electrons $\underline{i}$ and $\underline{J}$; s_i and s_j are the operators of their spin angular momenta, measured in units of $\hbar$. By applying this operator identity in spin space to the generators λ_i, it is easily found that each matrix element $(\sigma,\lambda_i\sigma)$ can be expressed as the expectation value of an operator

$$a + b\langle s_1 \cdot s_2\rangle ,$$

where a and b denote constants (27). The results are listed in Table I.

Table I

λ_i	δ_i	$(-1)^{\lambda_i} \cdot \frac{1}{\delta_i} \quad (\sigma,\lambda_i\sigma)$
e	2	1/2
13	1	- 1/2
23	1	- 1/2
12	2	$-1/4 - \langle s_1\cdot s_2\rangle_\sigma$
123	1	$1/4 + \langle s_1\cdot s_2\rangle_\sigma$
132	1	$1/4 + \langle s_1\cdot s_2\rangle_\sigma$
13,24	1	$1/4 - \langle s_1\cdot s_2\rangle_\sigma$

As a consequence, expression (2) now assumes the form

$$E^{(1)}_\sigma = \frac{d_1 + d_2\langle s_1\cdot s_2\rangle_\sigma}{d_3 + d_4\langle s_1\cdot s_2\rangle_\sigma} \equiv \frac{c_1 + c_2\langle s_1\cdot s_2\rangle_\sigma}{1 + c_4\langle s_1\cdot s_2\rangle_\sigma} , \qquad (3)$$

where d_1 and d_2 are linear combinations of orbital matrix elements of the operator V; d_3 and d_4 are linear combinations of overlap integrals. Expression (3) can be expanded as the series

$$E_\sigma^{(1)} = c_1 + (c_2 - c_4 c_1) \langle s_1 \cdot s_2 \rangle_\sigma \cdot \{1 - c_4 \langle s_1 \cdot s_2 \rangle_\sigma + c_4^2 \langle s_1 \cdot s_2 \rangle_\sigma^2 \ldots\}. \quad (4)$$

It should be noted that, due to the fact that there are only two unpaired electrons, the first-order interaction energy can always be expressed as the expectation value of a purely bilinear spin-Hamiltonian of the form (28)

$$E_\sigma^{(1)} = M + N \langle s_1 \cdot s_2 \rangle_\sigma,$$

where

$$M = (1/4)\,(3\,E^{(1)}\,(\text{triplet}) + E^{(1)}\,(\text{singlet}))$$

and

$$N = E^{(1)}(\text{triplet}) - E^{(1)}\,(\text{singlet}).$$

However, in an analysis of the results from model calculations, it is useful to derive an expression for the interaction energy which can be interpreted as a sum of contributions of the various SDC generators (the pure **inter-atomic permutations**). Since, in these calculations, it turns out that the quadratic and higher-order terms in the scalar product $s_1 \cdot s_2$ of expression (4) contribute only 1-2% to the total value of $E_\sigma^{(1)}$, the interaction energy can be written, in good approximation, as

$$E_\sigma^{(1)} = c_1 + (c_2 - c_4 c_1) \langle s_1 \cdot s_2 \rangle_\sigma \quad . \quad (5)$$

With respect to the average energy c_1 of the triplet and singlet level, defined through the constants d_1 and d_3, it is possible to split the coefficient of the expectation value $\langle s_1 \cdot s_2 \rangle_\sigma$ as

$$\sum_i (c_2^i - c_4^i c_1),$$

with c_2^i and c_4^i defined as d_2^i/d_3 and d_4^i/d_3, respectively; d_m^i is the contribution of the generator λ_i to d_m. Note that there is no splitting of the constants d_1 and d_3 in terms of SDC-generator contributions, i.e. the term $c_2^i - c_4^i c_1$ is interpreted as the contribution of the generator λ_i to the coefficient $(c_2 - c_4 c_1)$, with respect to the spin-independent quantities d_1 and d_3 in expression (3).

In our model calculations each shell is described by a Slater function $\phi(r) = r^{n-1} \exp(-pr)$, with r the distance to the nucleus and p a variable parameter. Computations of the interaction energy splittings with n = 1, 2, 3 and 4 yield very similar results (29). Therefore, we limit ourselves to 1s-type Slater functions (n = 1) for all orbitals. Furthermore, we only consider those geometric configurations in which the two cation-anion distances are equal; the opening angle θ, at the site of the anion C, varies from 40° to 180°. The cations A and B, with fixed distance R_{AB} = 7 au, are assumed to be of the same kind. The choice of the orbital parameters p (i.e. $p_{cation} = 1\ au^{-1}$; $p_{anion} = 0.75\ au^{-1}$) determines the orbital extensions. The p-values roughly correspond to those for the unit Ni-Cℓ-Ni in an application of the effective-electron model to the analysis of 180°-superexchange (18).

RESULTS AND DISCUSSION

In Figure 1 we present the contributions to the coefficient of the scalar product $s_1 \cdot s_2$ in the first-order energy expression (5) from the SDC-generators λ_4, λ_5, λ_6 and λ_7 (see Table I), as a function of the opening angle θ at the anion. The values of the parameters p and of the cation-anion distance $R_{AC} = R_{BC}$ are those given at the end of the previous Section. The orbital matrix elements involving the permutations P_{123} and P_{132} are identical for the isosceles configurations considered; these contributions are added in the figure. A positive coefficient implies that the singlet state of the coupled system is favored, whereas a negative coefficient implies triplet coupling. The unit of energy is 10^{-6}au.

From the figure we conclude that first-order indirect exchange between the cations favors antiparallel coupling of their spins, except for a region of weak ferromagnetic alignment around $\theta = 80^{\circ}$. The permutations $P_{13,24}$ and P_{12} always lead to antiparallel coupling, whereas P_{123} and P_{132} always favor a parallel alignment. It is further obvious from the figure that direct cation-cation exchange is negligibly small; for large values of the opening angle θ the coupling is strongly antiferromagnetic, in agreement with experiments (18).

Of particular interest for comparison with experiment is the narrow θ-range where, according to the model, weak ferromagnetic coupling occurs. In the following Figure 2 the results are replotted on a larger scale, between $\theta = 60^{\circ}$ and 95°.

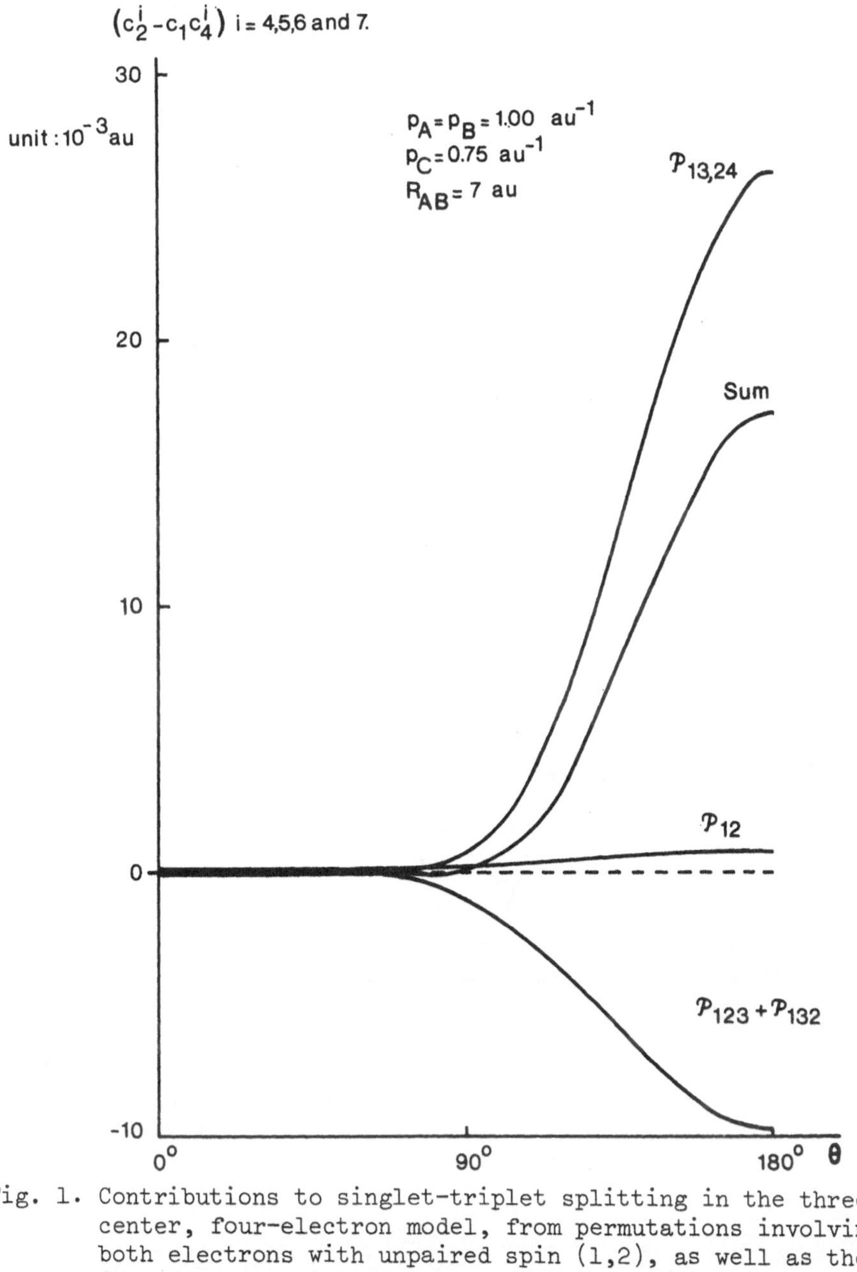

Fig. 1. Contributions to singlet-triplet splitting in the three-center, four-electron model, from permutations involving both electrons with unpaired spin (1,2), as well as their sum, as a function of the opening angle θ at the anion. Parameters P_A, P_B (cations) and P_C (anion) of the 1s-Slater-orbitals are given. The unit of energy is 10^{-3} au. Positive values imply singlet, negative values triplet stability.

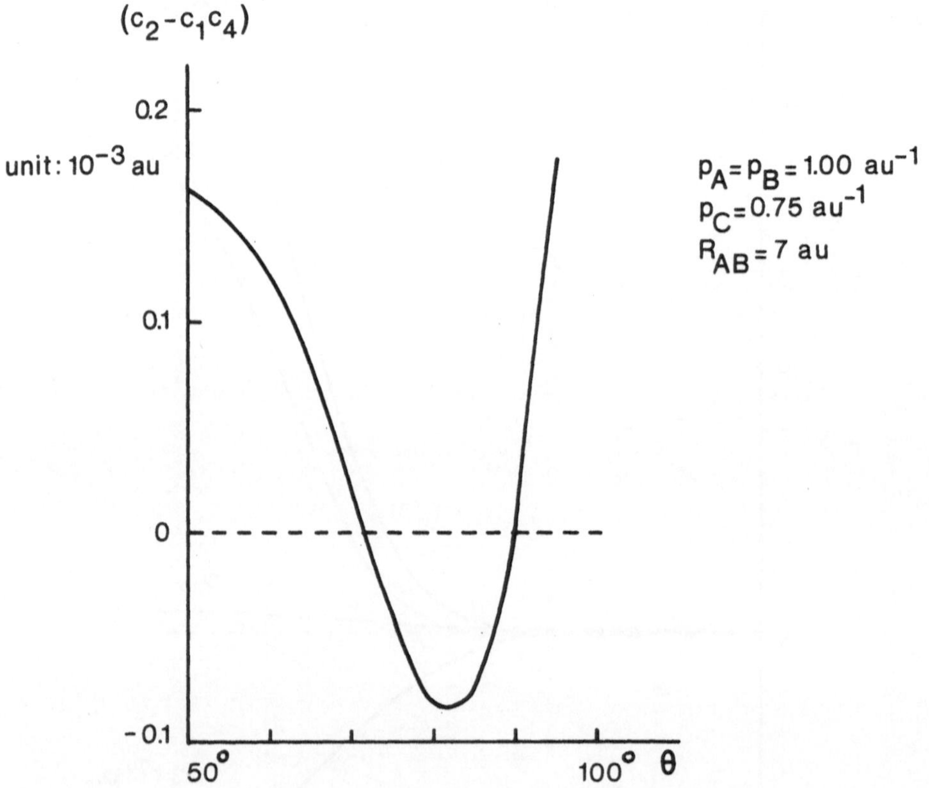

Fig. 2. Larger-scale drawing of the sum-curve of Fig. 1, in the range $\theta = 50^\circ$ to 100°. Unit of energy is 10^{-3}au.

The second crossing from parallel to antiparallel coupling occurs at approximately $\theta = 90^o$; in that region, the change in the coupling constant with θ is very pronounced. Considering, in particular, the results for $\theta = 85^o$ ($-78 \cdot 10^{-6}$ au) and for $\theta = 95^o$ ($+178 \cdot 10^{-6}$ au) and translating the energy of interaction in terms of a Heisenberg Hamiltonian $C - 2\, J_{AB}\, S_A \cdot S_B$, where S_A and S_B are the total-spin operators for the cations, we find for a unit Ni-Cℓ-Ni (2 unpaired 3d-electrons per cation), corresponding approximately to the values used for the parameters, a variation in J_{AB}/k between +3 and -7 K. Experimentally, it is known that the sign of J_{AB} does indeed sensitively depend on the value of θ just in this region (18,30). The accuracy with which a simple model of three centers and four electrons, in an exchange-perturbation approach, reproduces the experimental findings, is surprising indeed. It is further to be noted that the angle-dependence of the indirect exchange is, in the present model, due solely to permutation symmetry of the zeroth-order wavefunction in the non-orthogonal basis of atomic wavefunctions. If non-orthogonality had been neglected then, of course, indirect exchange would have appeared only in higher order, as with Kramers' model.

It is easy to show why, in a Heitler-London formalism, the non-orthogonal basis gives rise to a θ-range of triplet stability between $\theta = 180^o$ (singlet) and $\theta = 0^o$ (singlet). Obviously, we may always (Schmidt)-orthogonalize the cation wavefunctions to that of the anion, since this leaves the total wavefunction unchanged. If we now decrease the angle θ, starting from $\theta = 180^o$ (singlet stability), there will be a value of the opening angle θ at which the Schmidt-orbitals for the cations are orthogonal (at this value, the overlap between the atomic wavefunctions for the cations is equal to the square of the cation-anion overlap). We then have three orthogonal orbitals for the system; the only first-order exchange splitting is due to direct exchange between electrons on these Schmidt-orbitals, which favors parallel alignment of their spins. Upon decreasing θ further, there finally remains only the direct exchange between electrons on the non-orthogonal atomic wavefunctions ϕ_A and ϕ_B, again favoring the singlet state.

As is well known, the stability of the hydrogen molecule, in a Heitler-London approach, is due principally to an exchange contribution involving the electron-proton interaction, for non-orthogonal wavefunctions. Without this contribution, the energy of interaction for the system is repulsive and favors the triplet state. The indirect exchange via a closed-shell anion, however, is not based on the presence of the cation nuclear charges, but is mediated through non-zero overlap with the anion wavefunction. To show this, we plot in the following Figure 3 again the singlet-triplet splitting, this time omitting the nuclear charges. For $\theta = 0^o$, the interaction energy amounts to $+194 \cdot 10^{-6}$ au.

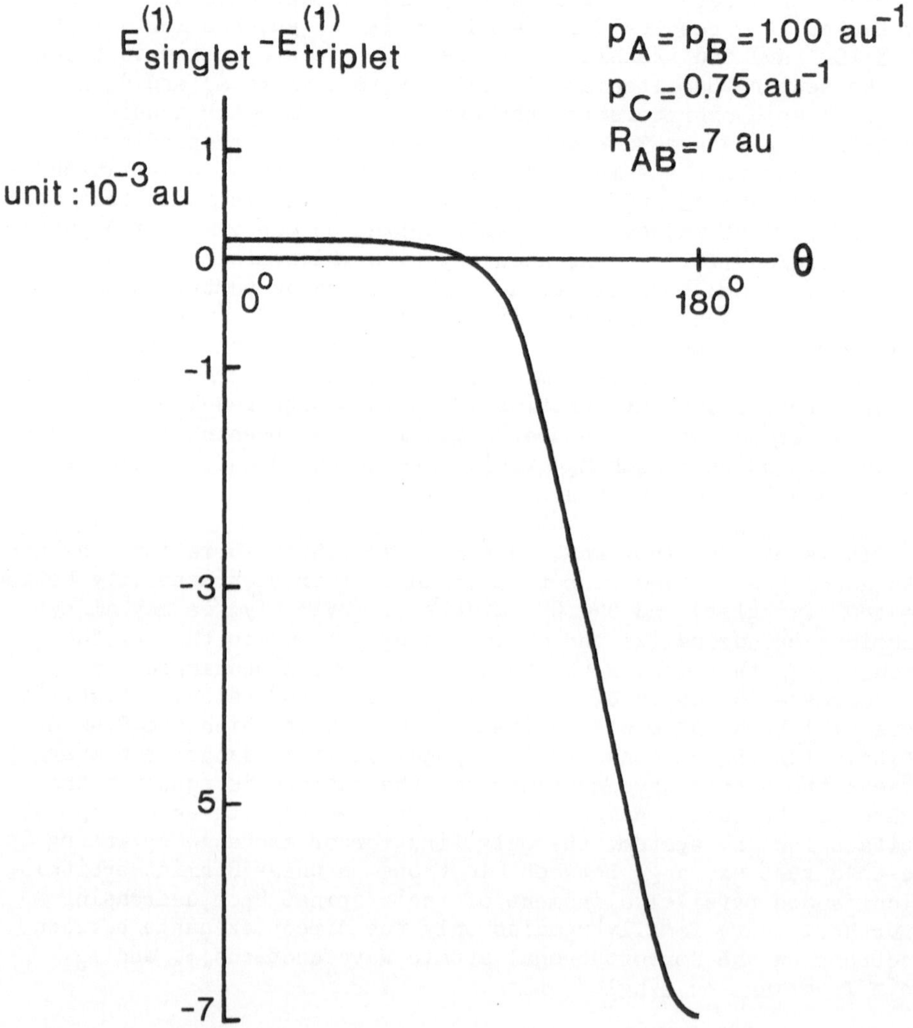

Fig. 3. First-order singlet-triplet splitting, $E^{(1)}_{singlet} - E^{(1)}_{triplet}$, in the three-center, four-electron model, omitting the nuclear charges of the two cations, as a function of the opening angle θ at the anion. Limiting value of splitting at $\theta = 0^o$ is $+194\cdot10^{-6}$ au. Positive values imply triplet, negative values singlet stability.

As is seen from the figure, indirect exchange persists in the absence of the cation nuclei, although there is now only one crossing point with the axis separating triplet from singlet stability. The θ-range for which the singlet configuration is the more stable one, remains very nearly the same when the nuclear charges have been omitted. The possibility that such exchange correlation effects may lead to singlet pairing of electrons also in metals, is being investigated.

A qualitative difference between the present analysis of superexchange and the one based on conventional models (e.g. Kramers' model) is that we ascribe the origin of the indirect exchange to singlet-triplet splitting in first order of the perturbation series, using a non-orthogonal basis, whereas the Kramers'-, Anderson- and other models are based on mixed free-cation-ligand wavefunctions in an assumed, or enforced, orthogonal basis. Contributions to the singlet-triplet splitting arising from excited free-ion states can actually be implicitly taken into account in the present model by including the second-order interaction energy. To illustrate this, we consider for simplicity the expressions for first- and second-order energies, $E^{(1)}$ and $E^{(2)}$, in Rayleigh-Schrödinger perturbation theory without exchange. Analogous results hold when exchange is taken into account, with the zeroth- and first-order wavefunctions, $\psi^{(o)}$ and $\psi^{(o)}$, adapted to the permutation symmetry of the system and with the perturbation V redefined such that it is permutation-invariant [25].

In RS-perturbation theory, the following equality holds:

$$E^{(1)} + E^{(2)} = (\psi^{(o)} + \psi^{(1)}/2, V(\psi^{(o)} + \psi^{(1)}/2))$$
$$-(1/4)(\psi^{(1)}, V\psi^{(1)}), \qquad (6)$$

which shows that $E^{(1)} + E^{(2)}$ can be written as the expectation value of V with the wavefunction $\psi^{(o)} + \psi^{(1)}/2$ if the second term in (6) (containing V three times if $\psi^{(1)}$ is re-expressed in terms of unperturbed functions) may be neglected relative to the first. We are in that case left with a "first-order" expression for an "adapted" groundstate wavefunction which includes contributions from excited states. In the three-center, four-electron model, a slight change in the parameters of the effective-electron orbitals may be taken to simulate this modification, again assuming that permutation symmetry is the principal cause of indirect interactions. Consistency of this approach requires that the parameters for the same cation (anion) and different anions (cations) must be approximately the same; this is indeed found in the applications of the model to superexchange in paramagnetic solids [17,18]. Similar considerations hold with respect to rotational barriers in molecules (19).

To summarize, we stress again the critical importance of taking into account non-orthogonality of basis wavefunctions in a perturbation approach to weak indirect exchange interactions, occurring in a large number of molecular and solid-state phenomena. If non-orthogonality is ignored, then indirect exchange becomes, incorrectly, only a higher-order effect in the series. If orthogonality is enforced, as in Anderson's analysis, then the formalism, correct in a Hartree-Fock approximation, becomes practically unmanageable. A three-center, four-electron model with overlapping zeroth-order wavefunctions was found to yield, in exchange perturbation theory, results in surprisingly good agreement with experiment. Finally, it was established that indirect exchange persists in the absence of nuclear charges of the cations in superexchange.

The above general results apply as well to indirect exchange other than superexchange effects. For example, the weak indirect exchange between conduction electrons in a metal via a localized unpaired electron, of great interest in magnetic metals, cannot be evaluated in a standard second-quantization scheme (which requires orthogonal basis functions) if the wavefunctions of the conduction electrons are assumed to be plane waves in zeroth order. Moshinsky and Seligman (31) have recently shown how a second-quantization formalism for non-orthogonal bases can be developed. Also, the "non-orthogonality problem" in valence bond theory, of long a serious stumbling block in applications to systems containing more than a few electrons, may now be approaching a practical solution (32).

REFERENCES

1. F. London, Z. Elektrochemie 35 (1929) 552; see also D.C. Mattis, The Theory of Magnetism, Harper and Row (New York, 1965) p. 40 ff.

2. H. Margenau, Rev. Modern Phys. 11 (1939) 1.

3. H.A. Kramers, Physica 1 (1934) 182.

4. J. Becquerel, W.J. de Haas and J. van den Handel, Physica 1 (1934) 383.

5. D. de Klerk, Physica XII (1946) 513.

6. The origin of the name "superexchange" is somewhat obscure. Kramers, in his original article (3), calls the effect "indirect exchange", but in a 1939-publication (Réunion d'Etudes sur le Magnétisme (Strasbourg, 1939) 3, 45) speaks of (in French) "superéxchange". Apparently, then, the name was invented during the period 1934-39.

7. C.G. Shull and J.S. Smart, Phys. Rev. 76 (1949) 1256.

8. P.W. Anderson, Phys. Rev. 79 (1950) 350.

9. P.W. Anderson, Phys. Rev. 115 (1959) 2.

10. P.W. Anderson, Phys. Rev. 124 (1961) 41.

11. P.W. Anderson and A.M. Clogston, Bull.Am.Phys.Soc. 6 (1961) 124.

12. J.B. Goodenough, Phys. Rev. 100 (1955) 564; J. Phys. Chem. Solids 6 (1958) 287.

13. J. Kanamori, J. Phys. Chem. Solids 10 (1959) 87.

14. P.W. Anderson, in Solid State Physics, edited by F. Seitz and D. Turnbull, Academic Press (New York, 1963) Vol. 14, p. 99.

15. For a review, see e.g. L. Jansen and E. Lombardi, Disc. Faraday Soc. 40 (1965) 78.

16. P.-O. Löwdin, A Theoretical Investigation into Some Properties of Ionic Crystals, Almqvist and Wiksell (Uppsala, 1948); Repts. Progr. Phys. 5 (1956) 1.

17. R. Ritter, L. Jansen and E. Lombardi, Phys. Rev. B8 (1973) 2139; L. Jansen, R. Ritter and E. Lombardi, Physica 71 (1974) 425.

18. L.J. de Jongh and R. Block, Physica 79B (1975) 568.

19. E. Lombardi, R. Ritter and L. Jansen, Int. J. Quantum Chem. 7 (1973) 155;
E. Lombardi, L. Pirola, G. Tarantini, L. Jansen and R. Ritter, ibid 8 (1974) 335.

20. E. Lombardi, G. Tarantini, L. Pirola, L. Jansen and R. Ritter,

J.Chem.Phys. 61 (1974) 894.

21. L. Jansen, in Crystal Structure and Chemical Bonding in Inorganic Chemistry, edited by C.J.M. Rooymans and A. Rabenau, North-Holland Publ. Co (Amsterdam, 1975) p. 205.

22. K. Yosida, Phys. Rev. 106 (1957) 893; for a critical evaluation of applications see, e.g. H.J. Zeiger and G.W. Pratt, Magnetic Interactions in Solids, Clarendon Press (Oxford, 1973) Chapter 8; F. van der Woude and G.A. Sawatzky, Phys. Reports 12C (1974) 336, or K.H.J. Buschow, J. Chem. Phys. 61 (1974) 4666.

23. For a review, see e.g. A.J. Heeger, in Solid State Physics, edited by F. Seitz and D. Turnbull, Academic Press (New York, 1969) Vol. 23, p. 283.

24. A. Witkowski, Mol. Phys. 29 (1975) 1441.

25. L. Jansen, Phys. Rev. 162 (1967) 63; W. Byers Brown, Chem. Phys. Letters 2 (1968) 105.

26. See e.g. P. Kramer and T.H. Seligman, Nuclear Phys. A136 (1969) 545; A186 (1972) 49.

27. R. Block, Physica 73 (1974) 312.

28. P.-O. Löwdin, Revs. Mod. Phys. 34 (1962) 80.

29. E. Lombardi, G. Tarantini, R. Block, R. Roël, G. ter Maten, L. Jansen and R. Ritter, Chem. Phys. Letters 12 (1972) 534.

30. J. Als-Nielsen, R.J. Birgenau and H.J. Guggenheim, Phys. Rev. B6 (1972) 2030.

31. M. Moshinsky and T.H. Seligman, Ann. Phys. 66 (1971) 311.

32. J.M. Norbeck and R. McWeeny, Chem. Phys. Letters 34 (1975) 206.

INELASTIC SCATTERING OF PHOTONS FROM IONIC CRYSTALS AND EFFECTS OF OVERLAP

K.-F. Berggren

Linköping University, Dept of Physics and Measurement Technology
S-581 83 Linköping, Sweden

Abstract: The Compton scattering experiment has recently been developed as a sensitive probe for electron wave functions in atoms, molecules and solids. The experimental technique in general and the basic theoretical assumptions are discussed. The Compton profile of ionic crystals is evaluated on the basis of a tight-binding (Heitler - London) model and Löwdin's expansion in terms of overlap integrals. Comparison between experimental and theoretical results for LiF and LiH is made.

I. INTRODUCTION

Due to improved instrumentation there has been a revival of interest in recent years in the study of electron momentum distributions by inelastic scattering of photons - Compton scattering - which was originally pioneered in the 1920's and 1930's /1,2/. A typical experiment consists of measuring the spectrum of radiation arising when an essentially monochromatic beam is scattered by an atom, molecule or solid into a fixed angle of scattering. With the condition that the photon energy loss is much greater than the electron binding energy, the scattering electron's initial momentum along the scattering vector uniquely determines the wavelength of the scattered photon. Hence, measurement of the spectrum of Compton scattered radiation is equivalent to determination of a particular projection of the scatterer's electronic momentum distribution. More specifically, within the impulse approximation (scattering by single, independent electrons, neglect of electron binding energy, plane wave final electron states) the differential cross section is simply related to the Compton profile, defined as an integral over the initial momentum distribution with the integration extending over a plane perpendicular to the scattering vector. Compton scattering is thus closely related to other experimental techniques used to obtain information about momentum distributions. When thermal positrons annihilate in matter, the angular correlation between the decay photons is determined by the initial momentum distribution of electrons /3/. Unfortunately, the need to have accurate positron wave functions makes it difficult to obtain reliable momentum distributions by this method. Another technique involves (e, 2e)-reactions /4-6/. In this case the momentum distribution can be determined directly for electrons in individual orbitals, but so far the method is limited to atoms and molecules only.

In the field of Compton scattering experiment is ahead of theory and a wealth of experimental data about various atoms, molecules and solids are accumulating fast. This provides an interesting challenge to solid state theorists and quantum chemists to test their wave functions. An ideal calculation would go as follows. Assuming a certain electronic configuration, or in general a certain wave function $\psi(\bar{r}_1, \bar{r}_2, \ldots\ldots \bar{r}_n)$, the momentum transform $\psi(\bar{p}_1, \bar{p}_2, \ldots\ldots \bar{p}_n)$ is taken and the corresponding momentum density $n(\bar{p})$ and Compton profile are evaluated. Comparison with experimental results then indicates whether one must revise the assumptions made about the electronic state and repeat the procedure until one, in principle at least, obtains agreement with experiment. Ideally, deciding which of several wavefunctions gives the "best" Compton profile should be a simple matter of comparing theory with experiment, either by eye or with a more quantitative statistical test. However, such comparisons have so far required extreme care, since experimental data may be reported with a 1 % statistical uncertainty at the peak of the profile, but may be subject to systematical errors of several times that magnitude because of multiple scattering, neglect of residual instrument functions, relativistic corrections and other effects. In this connection ionic crystals are of particular interest, as we indicate below.

In his thesis "A Theoretical Investigation into Some Properties of Ionic Crystals" Löwdin demonstrated that the ground state of many of the ionic crystals may be viewed as a simple superposition of closed shell ions which overlap each other only very weakly /7/. By expanding physical quantities in terms of small overlap integrals Löwdin was able to explain on the basis of quantum mechanics the cohesive and elastic properties of a series of ionic crystals. In this respect Löwdin's thesis represents the first large scale calculation for such systems and is a milestone on the way to our present understanding of the solid state. In the present article we intend to show how recent Compton scattering experiments may be tested in the light of Löwdin's Ansatz. The virtue of such a test lies in the fact that we know, because of Löwdin's work, that the initial wave function is of high accuracy; hence the differential cross section and the Compton profile can be evaluated with confidence and by comparison with experiments a good test of the various assumptions made in the Compton scattering technique is obtained.

In Section II we describe in general terms the derivation of the differential cross section and the Compton profile by means of time-dependent perturbation theory. The presentation is rather detailed in order to bring out clearly all the approximations made to obtain the Compton profile. In Section III we briefly describe the experimental technique. Section IV deals with the evaluation of Compton profiles of ionic crystals. In particular we will show how Löwdin's expansion in terms of small overlap integrals gives rise to a directional dependence of the Compton profile. The close agreement between theory and recent measurements on the angular dependence of LiF indicates that the Compton scattering technique has now developed as a sensitive probe of electron wave functions. Thus it is remarkable that overlap effects as originally elaborated by Löwdin in his thesis are nowadays amenable to direct sensitive measurements. Section V, finally, contains a short summary.

II. INELASTIC SCATTERING OF LIGHT AGAINST BOUND STATE SYSTEMS

a) Nonrelativistic Theory

Consider a nonrelativistic electron system and a free radiation field specified by the vector potential $A(\bar{x},t)$ /8/. The interaction Hamiltonian between the bound electrons and the radiation field is then obtained by the standard prescription $\bar{p} \rightarrow \bar{p} - e\bar{A}/c$, i.e.

$$H_{int} = \sum_i \{ \frac{e^2}{2mc^2} \bar{A}(\bar{r}_i,t) \cdot \bar{A}(\bar{r}_i,t) - \frac{e}{2mc} (\bar{p}_i \cdot \bar{A}(\bar{r}_i,t)$$

$$+ \bar{A}(\bar{r}_i,t) \cdot \bar{p}_i)\} = H_1 + H_2 \qquad (1)$$

The summation is over the various electrons that participate in the interaction. The probability that the electron system makes a transition from the unperturbed initial state A to a final state B under the action of H_1 is then determined by the matrix element $< B, \bar{k}', \alpha' | H_1 | A, \bar{k}, \alpha >$ in which the incoming photon is defined by the wave vector $\bar{k}$ and polarization α, and the scattered photon by $(\bar{k}', \alpha')$. Inserting the quantized radiation field

$$\bar{A}(\bar{r},t) = \sum_{\bar{k},\alpha} c\sqrt{\hbar/2\omega V}\{ a(\bar{k},\alpha)\cdot \bar{e}(\alpha) \exp(i\bar{k}\cdot\bar{r} - i\omega t)$$

$$+ a^{+}(\bar{k},\alpha)\cdot \bar{e}(\alpha)\exp(-i\bar{k}\cdot\bar{r} + i\omega t)\} \quad (2)$$

where V is the volume, ω the frequency, $\bar{e}(\alpha)$ transverse polarization vectors, and $a^{(+)}(\bar{k},\alpha)$ an annihilation (creation) operator, the matrix element becomes

$$< B,\bar{k}'\alpha'|H_1|A, \bar{k}\alpha >$$

$$= \sum_i (e^2\hbar/2mV\sqrt{\omega\omega'})\, \bar{e}(\alpha)\cdot\bar{e}(\alpha')\exp(-i(\omega-\omega')t)<B|\exp(i(\bar{k}-\bar{k}')\cdot\bar{r}_i|A>. \quad (3)$$

Here primed quantities refer to the scattered photon. The corresponding transition amplitude is according to ordinary time dependent perturbation theory

$$c_1(t) = \frac{1}{i\hbar}\int_0^t dt'< B, \bar{k}'\alpha'|H_1|A, \bar{k}\alpha > \exp(i(E_B-E_A)\, t'/\hbar) \quad (4)$$

The interaction term H_2 in Eq. (1) contributes to the inelastic scattering only in second perturbation theory. The transition amplitude is therefore

$$c_2(t) = \frac{1}{(i\hbar)^2}\sum_I \int_0^t dt'' \int_0^{t'} dt'< B|H_2(t'')|I>\exp(i(E_A-E_I)t''/\hbar)$$

$$\times < I\,|H_2(t')|\, A>\exp(i(E_I-E_A)t'/\hbar) \quad (5)$$

where the summation is over all intermediate states $|I>$. If the incoming photon is monochromatic and the outgoing, scattered photon is characterized by $(\bar{k}',\omega',\alpha')$ we then have

$$c_2(t) = -\,(e^2/2m^2\hbar\sqrt{\omega\omega'})$$

$$\times \sum_I\sum_i\{<B|\exp(-i\bar{k}'\cdot\bar{r})\bar{p}_i\cdot\bar{e}(\alpha')|I><I|\exp(i\bar{k}\bar{r}_i)\bar{p}_i\cdot\bar{e}(\alpha)|A>/$$

$$(E_I-E_A-\hbar\omega) + <B|\exp(i\bar{k}\bar{r}_i)\bar{p}_i\cdot\bar{e}(\alpha)|I><I|\exp(-i\bar{k}'\bar{r}_i)\bar{p}_i\cdot\bar{e}(\alpha')|A>/$$

$$(E_I-E_A + \hbar\omega')\}$$

$$x(-i\hbar)\int_0^t dt''\exp(i(E_B-E_A+\hbar\omega'-\hbar\omega)t''/\hbar). \qquad (6)$$

The scattering processes associated with H_1 and H_2 are illustrated in Fig. 1.

The differential cross section of a photon into the solid angle segment (Ω, $\Omega + d\Omega$) and frequency intervall (ω', $\omega' + d\omega'$) is

$$\frac{d^2\sigma}{d\omega' d\Omega} = \frac{(\omega' V)^2}{(2\pi)^3\hbar c^4}\sum_B |c_1(t) + c_2(t)|^2 /t \qquad (7)$$

With the expression for $c_1(t)$ and $c_2(t)$ inserted and letting $t \to \infty$ we obtain

$$\frac{d^2\sigma}{d\omega' d\Omega} = r_0^2\frac{\omega'}{\omega}\sum_B\Big|\sum_i\{\bar{e}(\alpha)\,\bar{e}(\alpha') < B|\exp(i(\bar{k}-\bar{k}')\bar{r}_i)|A>$$
$$-\sum_I\Big[<B|\exp(-i\bar{k}'\bar{r}_i)\bar{p}_i\bar{e}(\alpha')|I><I|\exp(i\bar{k}\bar{r}_i)\bar{p}_i\bar{e}(\alpha)|A>/m(E_I-E_A$$
$$-\hbar\omega) + <B|\exp(i\bar{k}\bar{r}_i)\bar{p}_i\bar{e}(\alpha)|I><I|\exp(-i\bar{k}'\bar{r}_i)\bar{p}_i\bar{e}(\alpha')|A>/m(E_I$$
$$-E_A+\hbar\omega')\Big]\Big\}\Big|^2 \times \delta(E_B - E_A + \hbar\omega' - \hbar\omega) \qquad (8)$$

where r_0 stands for the classical radius of the electron,

$$r_0 = \frac{e^2}{4\pi mc^2} \simeq \frac{1}{137}\,\frac{h}{mc} \simeq 2{,}82 \times 10^{-13}\ \text{cm} \qquad (9)$$

This expression for the differential cross section is general, but as such rather intractable. We shall therefore make the usual simplification corresponding to the experimental situation /9/. If the energy of the incident and scattered photon is much higher than the ionization energy (but smaller than $2mc^2$) the process becomes similar to the scattering of free electrons. The momentum of the bound electrons is small compared to $\hbar\bar{k}$ and $\hbar\bar{k}'$ and consequently the momentum associated with the intermediate states in the second term is $\sim\hbar\bar{k}$ and in the third term $\sim\hbar\bar{k}'$. Then

$$E_I - E_A - \hbar\omega \simeq \hbar\omega\ (\hbar\omega/2mc^2 - 1) \simeq -\hbar\omega \qquad (10)$$

and similarly for the third term. In practice the photon energy is not greater than $\sim$ 160 keV. We may therefore neglect the two

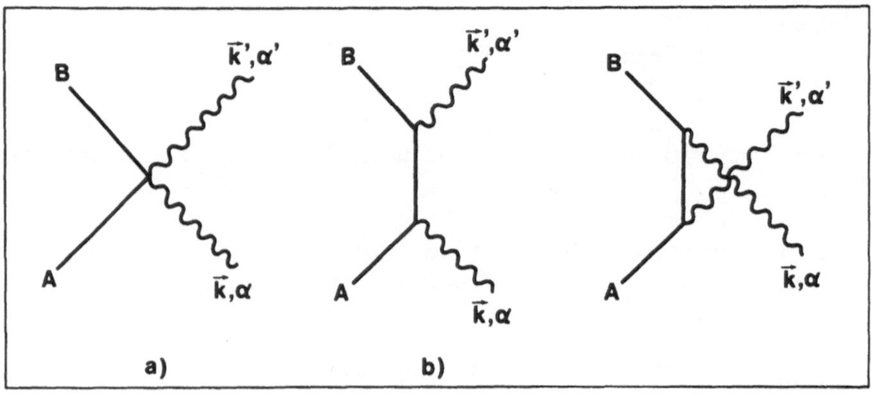

Fig. 1. Space-time diagrams for scattering of photons. Diagram a) corresponds to the A^2-term in the Hamiltonian and b) and c) to the second order processes associated with the $(\bar{p}\cdot\bar{A} + \bar{A}\cdot\bar{p})$ term.

last terms in Eq. (8). Furthermore, if we assume that the excited electron goes into the plane wave state $\exp(i\bar{p}'\bar{r}/\hbar)$ and that the initial state is a Slater determinant the simplified cross section reads

$$\frac{d^2\sigma}{d\omega' d\Omega} \simeq r_0^2 \frac{\omega'}{\omega} \frac{1 + \cos^2(\theta)}{2(2\pi\hbar)^3} \int d\bar{p}' \sum_i \left| \int d\bar{r} \exp(i(-\bar{p}'/\hbar + \bar{k} - \bar{k}')\bar{r}) \chi_i(\bar{r}) \right|^2 \times \delta(p'^2/2m - \varepsilon_i + \hbar\omega' - \hbar\omega). \tag{11}$$

Here ε_i is the orbital energy of the electron before scattering. The scattering angle θ comes from averaging over the polarizations. In Eq. (11) the scattering refers to individual electrons. A more complete expression relates the differential cross section to the dynamic structure factor. Eq. (11) can be further simplified by assuming that the energy transfer $\Delta\hbar\omega = \hbar\omega - \hbar\omega'$ is so large that we may neglect the potential energy in E_A and simply write $\varepsilon_i \simeq p^2/2m$. From the conservation of momentum then follows

$$\delta(p'^2/2m - \varepsilon_i - \Delta\hbar\omega) \simeq \delta(\bar{p}\cdot\bar{q}\hbar/m + q^2\hbar^2/2m - \Delta\hbar\omega) \tag{12}$$

where $\bar{q} = \bar{k} - \bar{k}'$ is the scattering vector. Introducing the momentum transform of $\chi_i(\bar{r})$,

$$\chi_i(\bar{p}) = (2\pi\hbar)^{-3/2} \int d\bar{r}\, \chi_i(\bar{r}) \exp(-i\bar{p}\bar{r}/\hbar) \tag{13}$$

we obtain the differential cross section

$$\frac{d^2\sigma}{d\omega' d\Omega} = r_0^2 \frac{1 + \cos^2\theta}{2} \frac{\omega'}{\omega} \int d\bar{p}' \sum_i | \chi_i(\bar{p})|^2 \delta(\bar{p}\bar{q}\hbar/m + q^2\hbar^2/2m - \Delta\hbar\omega). \tag{14}$$

If the z-axis is assumed to be along the direction of $\bar{q}$ we finally obtain

$$\frac{d^2\sigma}{d\omega' d\Omega} = r_0^2 \frac{1 + \cos^2\theta}{2} \frac{\omega'}{\omega} \frac{m}{\hbar q} \iint dp_x \, dp_y \sum_i | \chi_i(\bar{p}) |^2_{p_z}. \tag{15}$$

The integral in Eq. (15) is to be evaluated over a plane perpendicular to the scattering vector $\bar{q}$ and at a constant value of p_z given by

$$p_z \simeq -\hbar q/2 + m\Delta\omega/q. \tag{16}$$

The integral in Eq. (15) is usually referred to as the Compton profile,

$$J(p_z) = \iint dp_x \, dp_y \, n(\bar{p}) \tag{17}$$

where we have introduced the momentum density

$$n(\bar{p}) = \sum_i | \chi_i(\bar{p})|^2 \tag{18}$$

Consequently

$$\frac{d^2\sigma}{d\omega' d\Omega} = r_0^2 \frac{1 + \cos^2\theta}{2} \frac{\omega'}{\omega} \frac{m}{\hbar q} J(p_z). \tag{19}$$

For isotropic systems or for polycrystalline solids and systems such as liquids or gases in which rotational averaging is essentially complete, the experimentally accessible quantity is not the directional Compton profile $J(p_z)$ but its spherical average $J(q)$, where (with $q \equiv |p_z|$)

$$J(q) = 2\pi \int_q^\infty dp \, p < n(\bar{p}) > \tag{20}$$

and $< n(\bar{p}) >$ is the spherical average of $n(\bar{p})$.

Consequently a measurement of the differential cross section gives direct information about the Compton profile and the momentum distribution of the unperturbed electron system. $J(p_z)$ is of central importance and represents the form in which the experimentalists usually present their results.

For later reference we return for a moment to the complete expression for the differential cross section in Eq. (8). With the assumption that the energy transfer is so large that both final and intermediate states are plane waves the cross section becomes

$$\frac{d^2\sigma}{d\omega' d\omega} = r_0^2 \frac{\omega'}{\omega} \int d\bar{p}' n(\bar{p})\ X\ \delta\ (p'^2/2m - \varepsilon_i - \Delta\hbar\omega) \qquad (21)$$

where the X-factor is given by

$$X = \{\bar{e}(\alpha)\cdot\bar{e}(\alpha')+(\bar{e}(\alpha)\cdot\bar{p})(\bar{e}(\alpha')\cdot\bar{p}')/\kappa-(\bar{e}(\alpha')\cdot\bar{p})(\bar{e}(\alpha)\cdot\bar{p}')/\kappa'\ \}^2 \qquad (22)$$

The nominators κ and κ' are simple algebraic expressions involving momenta and frequencies. Also in a more complete theory the differential cross section is thus related to the momentum distribution of the target, although the relationship is more complicated. As mentioned above we have good reasons to drop the two last terms in Eq. (22).

b) Relativistic Corrections

The nonrelativistic theory may be sufficient in x-ray work. To ensure, however, that the impulse approximation is valid the trend has been to use γ-rays in Compton scattering experiments. Typical cources in use are ^{241}Am(59.54 keV) and ^{123m}Te(159.0 keV), but sources with higher energies are currently under consideration. It has therefore been found necessary to consider the relativistic differential cross section for inelastic photon scattering from bound systems. Although the general principles of such a calculation are well known an actual evaluation of the cross section is quite difficult due to complicated matrix elements and the large set of intermediate states to be summed over /10/. The reason for this complexity is that the Dirac equation is a first order differential equation whereas the Schrödinger equation is of second order. Consequently one has to resort to second order perturbation theory in order to describe the relativistic Compton process. The three different graphs in Fig. 1 are then replaced by the two graphs in Fig. 2. Clearly there is no longer a natural distinction between the $\bar{A}^2$- and $\bar{p}\cdot\bar{A}$-contributions, as in the nonrelativistic theory. This is unfortunate since it was demonstrated above that the $\bar{A}^2$- term is the dominant one and that the $\bar{p}\cdot\bar{A}$ -terms may be dropped. Obviously a relativistic expression for the cross section

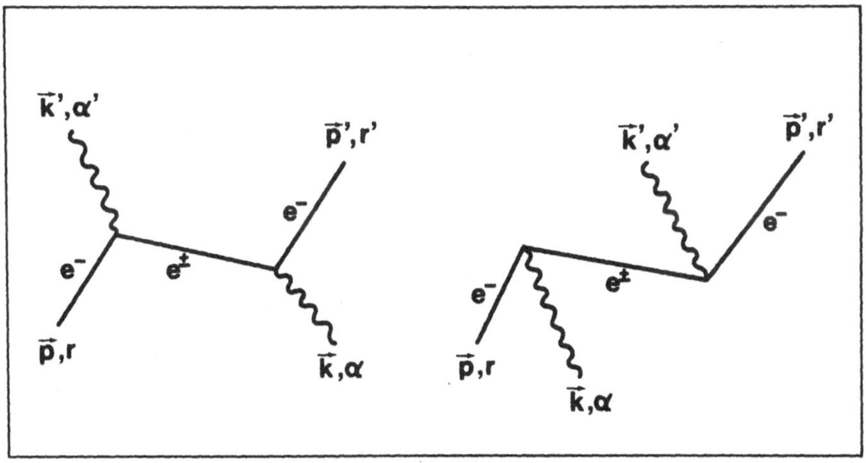

Fig. 2. Feynman diagrams for photon-electron scattering. The symbols $\bar{p}^{(\prime)}$, $r^{(\prime)}$ denote the momentum and spin of the electron and $\bar{k}^{(\prime)}$, $\alpha^{(\prime)}$ the wave vector and polarization of the photon before (after) scattering. The different particle states can be either electron (e^-) or positron (e^+)-states.

has to approach the general expression in Eq. (8) in the nonrelativistic limit. Hence we conclude that the relativistic theory is "overdoing things". Without going into details we will describe how to circumvent this difficulty.

Eisenberger and Reed /11/ and Manninen et al. /12/ have considered an intuitive approach which recently has been elaborated by Ribberfors /13,14/. In brief it goes as follows. The inelastic cross section for colliding beams of electrons and photons is available in a closed analytical form. If the energy transfer is so large that binding effects can be neglected we may view the scattering of photons from a bound electron state as the scattering against a stationary wave packet composed of plane wave states (or a collection of electron beams), characterized by the probability function $n(\bar{p})$. With a properly revised flux factor the desired relativistic cross section may then be written as an integral over each free particle scattering event times the weight factor n(p). In this way one obtains (with $\hbar = c = 1$)/15/

$$\frac{d^2\sigma}{d\omega' d\Omega} = \frac{m^2 r_o^2 \omega'}{2\omega} \int d\bar{p}\; n(\bar{p})\, \frac{X(K,K')}{EE'}\, \delta(E + \omega - E' - \omega') \qquad (23)$$

where the scattering kernel X is defined as /14/

$$X = \frac{1}{2}\left(\frac{K}{K'}+\frac{K'}{K}\right)-1+2\{\bar{e}(\alpha)\cdot\bar{e}(\alpha')+(\bar{e}(\alpha)\cdot\bar{p})(\bar{e}(\alpha')\cdot\bar{p}')/K-(\bar{e}(\alpha')\cdot\bar{p})$$

$$(\bar{e}(\alpha)\cdot\bar{p}')/K'\}^2 \tag{24}$$

Here E and E′ are relativistic free particle energies and

$$K = E\omega - \bar{p}\cdot\bar{k}$$

$$K' = E\omega' - \bar{p}\cdot\bar{k}' \tag{25}$$

Except for the first term, which appears because of relativistic kinematics, Eq. (24) is similar to Eq. (22). In a loose way we may say that the terms containing the $e \cdot p$ - products correspond to the second order contributions in the nonrelativistic theory. For the same reason as above we may therefore neglect them and write

$$X \simeq \frac{1}{2}\left(\frac{K}{K'}+\frac{K'}{K}\right) - 1 + 2\{\bar{e}(\alpha)\cdot\bar{e}(\alpha')\}^2 \tag{26}$$

This step simplifies the algebra considerably, and it turns out the differential cross section is to a good approximation given by the expression

$$\frac{d^2\sigma}{d\omega' d\Omega} \simeq \frac{m\, r_o^2\, \omega'}{2\omega q} J(p_z)\, \bar{X} \tag{27}$$

where

$$\bar{X} = \frac{1}{2}\left(\frac{R}{R'}+\frac{R'}{R}\right) - 1 + 2\{\bar{e}(\alpha)\cdot\bar{e}(\alpha')\}^2 \tag{28}$$

and

$$R = \omega(m - (\omega-\omega' \cos\theta)\, p_z/q)$$

$$R' = R - \omega\omega'(1-\cos\theta) \tag{29}$$

Instead of the approximate expression in Eq. (16) for p_z the relativistic relation

$$p_z = (m(\omega-\omega') - \omega\omega' (1-\cos\theta))/q \tag{30}$$

is to be used. Eq. (27) is important since it allows a definition of the Compton profile also in a relativistic context /14/.

We emphasize that the relativistic expression in Eqs. (23) and (28) are based on intuitive arguments. A strict derivation of the differential cross section for scattering of photons from bound state systems is yet to be found. In the case of large atomic numbers and high incident radiation energies there are indications that such a cross section will not be related to the Compton profile in a simply way /16/.

III. EXPERIMENTAL TECHNIQUE

As mentioned in Section IIb the trend in recent years has been to use γ-rays rather than x-rays. In the case of γ-rays the experimental arrangement is in principle quite simple, as illustrated in Fig. 3. The radiation from an radioactive source (normally ^{241}Am or ^{123m}Te) is scattered from the sample in a vacuum chamber through an angle θ (usually $\gtrsim 150^\circ$). The spectrum of the scattered radiation is recorded by a solid state detector. A typical example of raw experimental data is shown in Fig. 4.

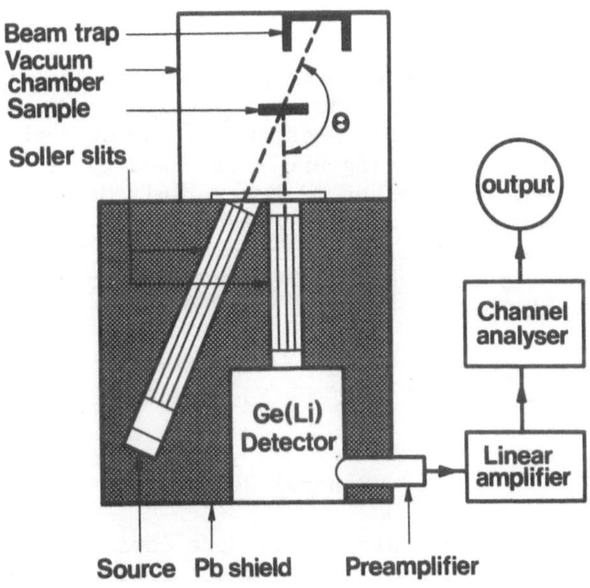

Fig. 3. Schematic drawing of the experimental arrangement.

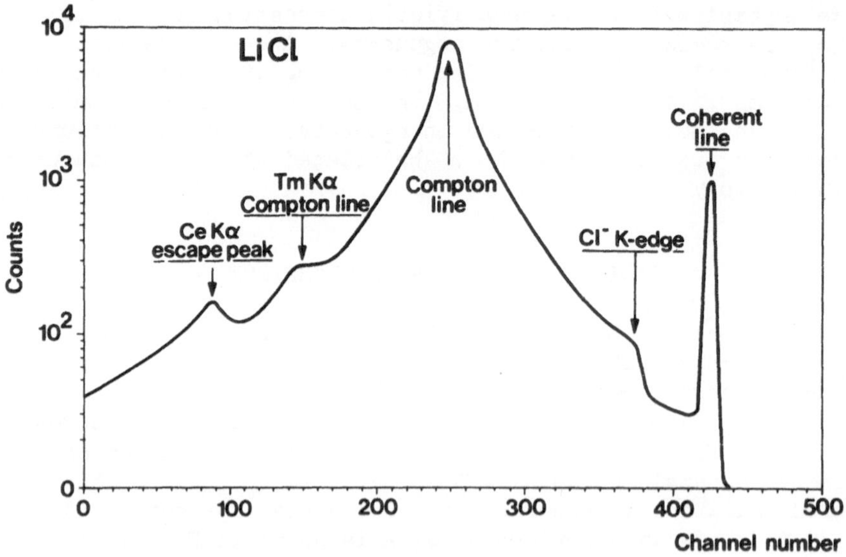

Fig. 4. Raw experimental data for a LiCl sample (from ref. 17). The channel width corresponds to an energy of 60.7 eV.

In order to extract a Compton profile from the experimental data corrections have to be made for the background, the energy dependence of the Compton cross section and the absorption of γ-rays. Effects of multiple scattering are eliminated by either measuring foils of different thicknesses or by explicit Monte-Carlo calculations of the scattering process. Sometimes the data are also corrected for the instrumental broadering by means of a deconvolution procedure. Finally the results are converted to an electron momentum scale through Eq. (30).

IV. COMPTON SCATTERING FROM IONIC CRYSTALS

a) General Description

As mensioned in the introduction the ionic crystals can be viewed as a superposition of closed shell ions which in most cases overlap each other only very weakly. For some time now a series of calculations of total energies, lattice constants /7,18/, form factors /19/, etc. have indeed achieved substantial agreement with experiments with such a description. This indicates, furthermore, that electronic correlation may be less important in describing ground state properties. We are therefore dealing with a rather ideal case; the total wave function is simple and accurate, and hence the Compton profile can be computed with high precision and with relative ease. The simple ionic crystals should thus offer a beautiful opportunity for studying the various assumptions which are incorporated in the Compton scattering technique. In the remaining part of this section we shall describe in some detail recent calcularions for LiF. This system has the advantage that the number of electrons per ion pair is small and that the overlap is indeed small. Finally we shall deal with LiH which has a somewhat unique position among ionic crystals. In this case the overlap is not small, a fact which gives rise to considerable solid state effects in the Compton cross section. Throughout the section we shall use atomic units $\hbar = m = e^2 = 1$.

Let

$$\chi_\alpha^g(\bar{r}) = \chi_\alpha(\bar{r} - \bar{R}_g) \tag{31}$$

be undistorted free ion solutions centered at the different lattice sites $\bar{R}_g$. Index α denotes the quantum numbers (nlm). The total crystal wave function is hence the single determinant

$$\Phi(\bar{r}_1, \bar{r}_2, \ldots \bar{r}_N) = \frac{1}{\sqrt{N!}} \det \{\chi_\alpha^g (\bar{r})\} \tag{32}$$

and the corresponding one-particle density matrix /7,18/

$$\hat{\rho} = 2 \sum_{\alpha\beta}^{gh} |\chi_\alpha^g\rangle (\underset{\sim}{\Delta}^{-1})_{\alpha\beta}^{gh} \langle \chi_\beta^h| . \tag{33}$$

Here the g- and h- summations refer to all occupied states α and β at sites $\bar{R}_g$ and $\bar{R}_g$, respectively. The factor of two comes from spin and $\underset{\sim}{\Delta}^{-1}$ is the inverse of the overlap matrix. Except for LiH the off-diagonal elements in $\underset{\sim}{\Delta}$ are usually small and therefore Löwdin's /7,18/ expansion

$$\underset{\sim}{\Delta}^{-1} = (\underset{\sim}{1} + \underset{\sim}{S})^{-1} = \underset{\sim}{1} - \underset{\sim}{S} + \underset{\sim}{S}^2 + \ldots \tag{34}$$

can be used. To second order in $\underset{\sim}{S}$ we then have the properly normalized density matrix

$$\hat{\rho} = 2 \sum_{\alpha}^{g} |g,\alpha\rangle\langle g,\alpha| - 2 \sum_{\alpha\beta}\sum^{gh} |g,\alpha\rangle S^{g,h}_{\alpha,\beta} \langle h,\beta| + $$

$$+ 2 \sum_{\alpha,\beta}^{g} |g,\alpha\rangle (\underset{\sim}{S}^2)^{g,g}_{\alpha,\beta} \langle g,\beta| \tag{35}$$

where

$$S^{g,h}_{\alpha,\beta} = \langle g,\alpha|h,\beta\rangle - \delta_{g,h}\,\delta_{\alpha,\beta} \tag{36}$$

and $|g,\alpha\rangle$ corresponds to the orbital $\chi_\alpha(\bar{r} - \bar{R}_g)$.

b) The Momentum Density and the Compton Profile

The momentum distribution of the crystal is in general defined as

$$n(\bar{p}) = \langle\bar{p}|\hat{\rho}|\bar{p}\rangle \tag{37}$$

where $|\bar{p}\rangle$ stands for a plane wave state normalized as $\langle\bar{p}|\bar{p}'\rangle = \delta(\bar{p}-\bar{p}')$. The momentum density of an ion, for example a F^--ion in LiF, is then from Eqs. (35) and (37)

$$n_{F^-}(\bar{p}) = 2\sum_{\alpha\beta}\sum^{00} [\delta_{\alpha\beta} + (\underset{\sim}{S}^2)^{0,0}_{\alpha,\beta}]\langle\bar{p}|\alpha\rangle\langle\beta|\bar{p}\rangle$$

$$-2\sum_{\alpha\beta}\sum^{0h}\langle\bar{p}|\alpha\rangle\, S^{0,h}_{\alpha,\beta}\,\langle\beta|\bar{p}\rangle \exp(-i\bar{p}\cdot\bar{R}_h) \tag{38}$$

where

$$\langle\bar{p}|\alpha\rangle = (2\pi)^{-3/2}\int d\bar{r}\, \exp(i\bar{p}\cdot\bar{r})\, \chi^{o}_{\alpha}(\bar{r}) \tag{39}$$

defines the momentum transform of an orbital at the central site <u>0</u>. There is no restriction in chosing the ion as the central one as in Eq. (38).

Using Eq. (38), it is then convenient to write the Compton profile as

$$J_{F^-}(p_z) = J^0_{F^-}(p_z) + J^1_{F^-}(p_z) + J^2_{F^-}(p_z) \tag{40}$$

where

$$J^{0}_{F^-}(p_z) = 2\iint dp_x dp_y \sum_{\alpha}^{0} | \langle \bar{p}|\alpha\rangle |^2 \tag{41}$$

$$J^{1}_{F^-}(p_z) = 2\iint dp_x dp_y \sum_{\alpha\beta}^{0h}\sum \langle \bar{p}|\alpha\rangle S^{0,h}_{\alpha,\beta} \langle\beta|\bar{p}\rangle \exp(-i\bar{p}\cdot\bar{R}_h) \tag{42}$$

$$J^{2}_{F^-}(p_z) = 2\iint dp_x dp_y \sum_{\alpha\beta}^{0} (\underset{\approx}{S}^2)^{0,0}_{\alpha,\beta} \langle\bar{p}|\alpha\rangle\langle\beta|\bar{p}\rangle \tag{43}$$

Similar expressions hold for the Li^+-ion, so that the total profile is

$$J(p_z) = J_{Li^+}(p_z) + J_{F^-}(p_z) \tag{44}$$

In the present elementary version of the tight binding model J^0 clearly represents a superposition of free ions whereas J^1 and J^2 represent the solid state effects. Both J^0 and J^2 are independent of the direction of scattering vector. The directional dependence of $J(p_z)$ is thus linear in $\underset{\approx}{S}$.

c) Isotropic Profiles of Simple Ionic Crystals

In the case of polycrystalline systems the calculation of the Compton profile is much faciliated because the angular integrations can easily be carried out. We refer to Aikala *et al.*/17/ for mathematical details.

The first calculation of the Compton profile of an ionic crystal by means of the expansion in Eq. (34) was performed by Berggren /21/. The profile of LiF was computed using only the two first terms in the expansion. Qualitatively solid state effects can be understood in this way but a proper normalization requires that also $\underset{\approx}{S}^2$-terms are included. Such calculations have been performed for LiF /17,20/, LiCl /17,20/, LiBr /17/, NaCl /17,22/, NaF /17,23/, NaBr /17/, KF /17/, KCl /17/, KBr /17/, and MgO /24/. Theoretical and experimental results for LiF are displayed in Fig. 5.

For the series of ionic crystals mentioned above one notes that already a plain superposition of free ion solutions (neglect of overlap) gives a much better fit to experimental data than a superposition of free atom solution. Thus the Compton scattering

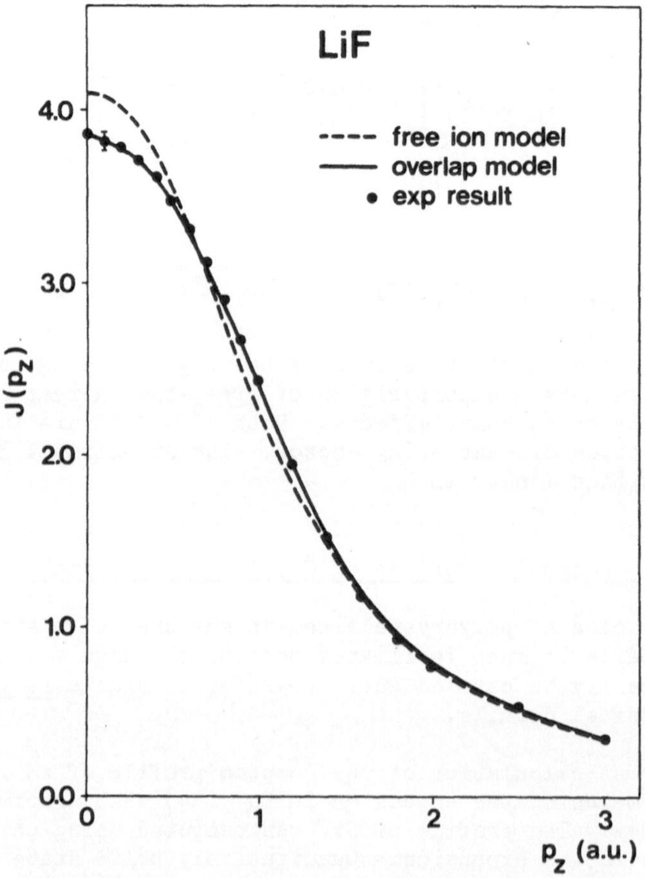

Fig. 5. Experimental (circles) and theoretical Compton profiles for polycrystalline LiF. The broken curve refers to a superposition of free ion solutions and the solid curve to the overlap model. (From ref. 17).

experiment demonstrates in a quick and easy way the ionic character of these crystals. Since the electrons are tightly bound to the ions the inclusion of overlap corrections have only a small effect but is important to bring measured and computed values into agreement with each other. As demonstrated by Fig. 5 this agreement is perfect in the case of LiF. Also for the other ionic crystals theory and experiments are remarkably close, although it seems that some problems about multiple scattering are still to be solved /17/.

d) Anisotropic Compton profiles of LiF

Although the remarkable agreement in the case of LiF (Fig. 5) indicates the correctness of both the experimental and theoretical machinary we will now proceed to a much severer test. We will thus deal with the delicate features of the directional dependence of the Compton profile.

For some time now Compton profiles and anisotropic momentum distributions in covalent bonded solids have been available /25, 25/. In these crystals relatively large anisotropies exist. This naturally raises the question as to the degree of anisotropy which exists in highly ionic crystals. Although Compton profiles have been reported on a number of ionic crystals only Weiss /27/ has previously looked for anisotropy. His conclusion for LiF was that to within 2 % there was no significant anisotropy. Due to the greater precision attainable using γ-rays LiF has recently been remeasured in order to place smaller error limits on the measurements /28,29/. Still another reason for the remeasurement has been that a Hartree-Fock calculation of the directional Compton profiles by Euwema et al. /30/ appeared in the literature. In the experiments /28,29/ the measurements were made with the scattering vector parallel to the <100>, <110>, and <111> axes and with the use of 159 keV γ-rays. After data processing the profiles were subtracted and smoothed with a digital filter. In order to give an impression of the statistical fluctuations the difference between J_{100} and J_{111} before smoothing is shown in Fig. 6. The anisotropy is readily apparent and is symmetric for positive and negative values of p_z. Furthermore, Fig. 6 indicates that no spurious effects are present.

Calculations of the directional Compton profiles have been performed as outlined in Section IVb /28,29/. Three different sets of orbitals for the Li^+ and F^--ions were used, namely Clementi's free ion orbitals /31/, Kunz /32/ free ion orbitals and Kunz' orbitally relaxed crystal orbitals. The calculations are straightforward but somewhat laborious. Therefore we shall not give the details here, as they may be found in ref. 29. We remark,

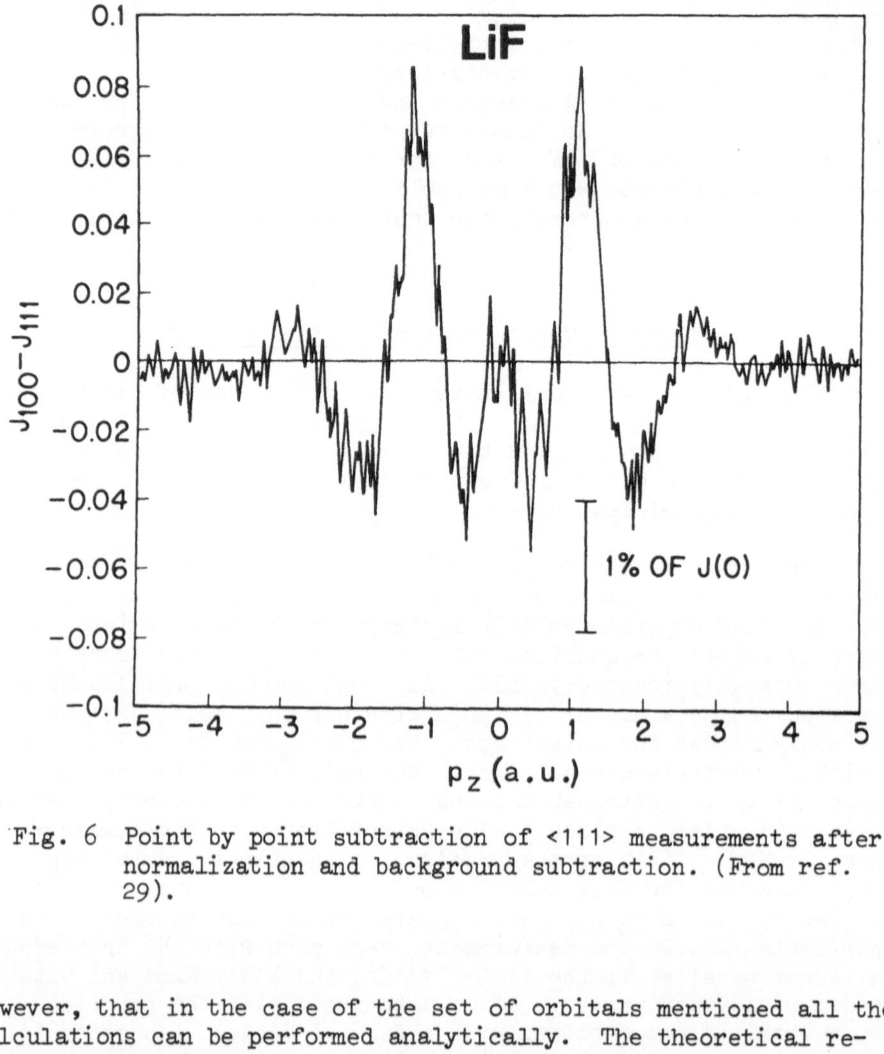

Fig. 6 Point by point subtraction of <111> measurements after normalization and background subtraction. (From ref. 29).

however, that in the case of the set of orbitals mentioned all the calculations can be performed analytically. The theoretical results for the difference $J_{100} - J_{111}$ are shown in Fig. 7. There is obviously an overall qualitative agreement between the anisotropies as calculated from the three different choices of ionic orbitals. The numerical differences that result from Clementi's and Kunz' free ion solutions merely reflect the use of slightly different basis functions. Ideally the curves corresponding to these two wave functions should of course coincide. The differences resulting from Kunz' two sets of orbitals are, on the other hand, real. Their smallness indicates, however, that relaxation due to the crystal environment is not an important effect in LiF. The same features were also found for the remaining anisotropies. Before the theoretical results are compared with experiments they

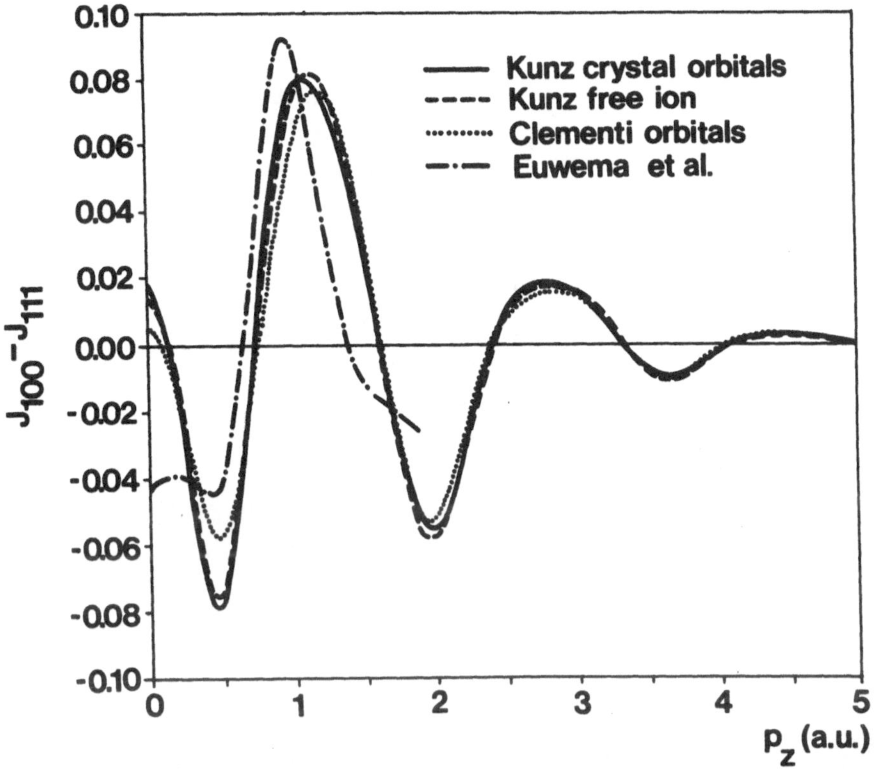

Fig. 7. Unsmeared difference between <100> and <111> Compton profiles as obtained from different wave functions. The dotted curve refers to Clementi's free ion solutions, the dashed curve to Kunz' free ion solutions, and the full drawn curve to Kunz' orbitally relaxed crystal solutions. The results of Euwema et al. are denoted by the dash-dot curve (From ref. 29).

should, however, be convoluted with the experimental resolution function. (We feel that it is better to convolue the theory with the resolution function than try to remove the effect of finite resolution from the experimental data.) After convolution the differences in Fig. 7 become even less significant. For this reason we have only considered Kunz' crystal orbitals in the final comparison with experiments in Fig. 8. It is clear, however, that the results of Euwema et al. /30/ differ distinctly from ours before as well as after convolution.

Considering the smallness of the anisotropies in LiF the

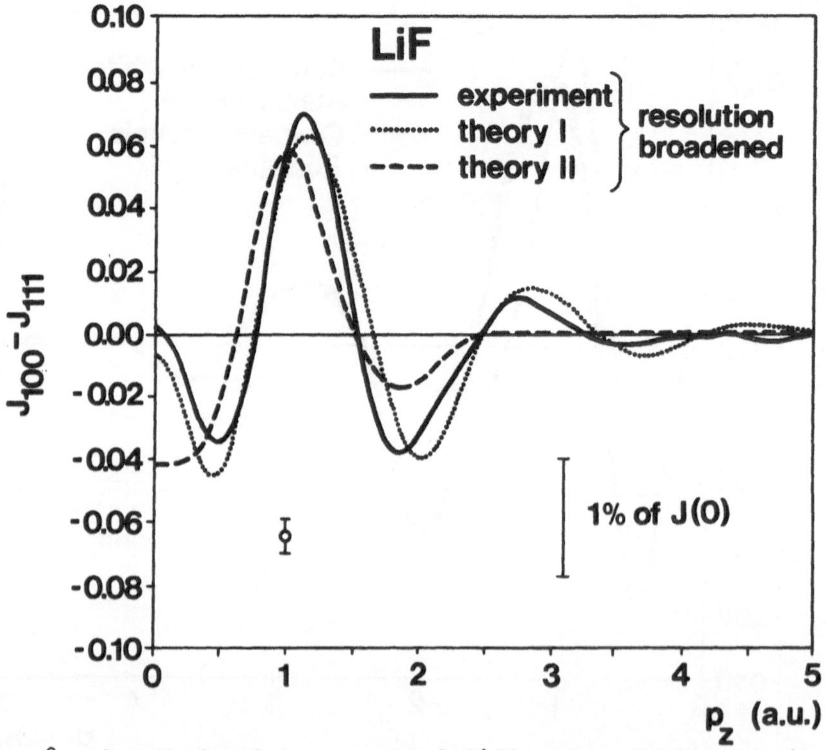

Fig. 8. Smoothed and $\pm p_z$ averaged difference of the experimental data for <100> and <111> directions shown in Fig. 6 (solid line). The theory of Euwema et al. smeared with the resolution function is shown as a dashed line (II). The present theory using Kunz' crystal orbitals and smeared in the same way is shown as a dotted line (I). (From ref. 29.)

agreement in Fig. 8 between theory and experiments is quite satisfactory. It is remarkable that this agreement is brought about by a mere orthogonalization of the ionic orbitals as prescribed by Löwdin long ago /7,18/. One may now ask why the theory of Euwema et al. /30/, which is quite successful in many other respects, fails to predict the correct anisotropy in Compton scattering. A possible reason may be their choice of a Gaussians basis. The difficulties of properly describing the tails of wave functions with such a basis are well known. A further difficulty is their use of Gaussian lobes rather than proper p-functions. Besides that Gaussian contractions appropriate to free atoms were used, whereas contractions appropriate to Li^+ and F^- should have been a better choice as remarked by Euwema et al. This fact resulted in certain difficulties in predicting the correct binding energy. Only when a separate Hartree-Fock calculation, using the same contracted basis together with some additional longer-range Gaussians, was performed for the free Li^+ and F^- ions did a correct binding energy obtain. The lack of these longer-range Gaussians in the original basis, and perhaps the use of Gaussians in general should be reflected in an inadequate description of the low momentum region.

Another interesting result of the work described above is that one can now understand the apparent paradox between the charge density anisotropy, as measured by elastic x-ray scattering, and the momentum density anisotropy, as measured by Compton scattering /28/. The most recent conclusion drawn from an analysis of elastic x-ray scattering measurements /33-36/ is that the charge is spherically symmetrical about each ion /36/. It is clear from the Compton measurements that the momentum density is not isotropic. Since Compton scattering, unlike elastic x-ray scattering, is related to both the off-diagonal and diagonal elements of the single particle density matrix, one would expect it to yield more complete information about these anisotropies. Both the experimental and theoretical results indicate that this is indeed the case. The explanation lies in the dependency of the elastic scattering factors on only the diagonal elements of the single particle density matrix. This results in a "smearing out" of anisotropies, as compared with Compton scattering. For any form of anisotropic charge density the anisotropy in the Compton profile is orders of magnitude larger than that for the form factor for smaller values of q, and the smaller the magnitude of the anisotropy, the larger the q values at which the most significant part of the form factor anisotropy occurs. In the case where the anisotropy results from overlapping charges, the anisotropy of the Compton profile goes linearly with the overlap, as mentioned in Section IVb, whereas the form factor anisotropy goes effectively worse than the square of the overlap /28,29/. These circumstances suggest the use of the Compton scattering technique to measure anisotropies in electron distributions.

e) Compton profiles of LiH

Because of its unique position as the simples heteronuclear crystal LiH has been an attractive subject for theoretical investigations /37/. On the basis of its general physics-chemical properties LiH is usually referred to as a simple salt made up to $Li^+(1s)^2$ and $H^-(1s)^2$ ions. When for example molten LiH is electrolyzed, hydrogen is liberated at the positive electrode. Consequently it has been natural to apply the Born-Mayer theory of ionic solids /38/ or the quantum mechanical Heitler-London method with free-ion wave functions /39,40/. By treating the overlap integrals between the free-ion 1s functions as small quantities and including them to first order only Hylleraas /39/ obtained theoretical values of the cohesive energy and the equilibrium lattice constant which were in good agreement with experiments. Hylleraas' calculations were later re-examined by Lundqvist /40/ who made the following observation. Expansions based on the overlap integrals as the parameters of smallness are not allowed in the case of LiH, and in fact means that Hylleraas' results correspond to the first term in a divergent series. The same type of difficulties affects the calculations by Morita and Takahasi /41/ and Kunz /42/. The problem associated with expansions based on overlap integrals is easily understood if we consider the dimensions of the H^--ion. The classical ionic radius is ~ 2Å which is of the same order of magnitude as the nearest H^- - H^- separation. In the crystal the H^--ions therefore penetrate appreciably into regions beyond the nearest neighbours. In fact this diffuseness of the anion wave functions have made various itinerant electron models quite attractive /18,43,44/. Consequently the electronic structure of the LiH crystal may be a bit more interesting and difficult to describe theoretically than what the simple chemical notation Li^+H^- implies.

Obviously the Compton scattering technique offers an interesting opportunity to investigate momentum distributions and to test the quality of different models for the electronic structure of LiH. The first measurements on LiH were performed by Philips and Weiss /45/, who recorded $J(p_z)$ with the x-ray scattering vector along the <100>, <110>, and <210> directions. The anisotropy was found small but the Compton profile was not measured with sufficient statistical accuracy or for enough crystal orientations to allow the determination of a unique model of the momentum-density asymmetry. The measurements were soon followed by two theoretical calculations of the isotropic profile, one based on an itinerant Wigner-Seitz approach /46/ and another /47/ based on the tight-binding model discussed above but with a proper inversion of the overlap matrix in Eq. (33). Of these two models the tight-binding model was found to give the best agreement with experiments. Although there was an overall qualitative agreement between experiment and theory, the experimental profile was somewhat broader than what theory predicted. For example, at $p_z = 0$ theory was off by ~ 5 %, an error which is unacceptable by present standards. For this reason Felsteiner et al, /48/ remeasured polycrystalline LiH using 60 keV γ-rays instead of x-rays. They also performed a theoretical calculation of the isotropic profile by means of Kunz' crystal orbitals /42/. Their results did not, however, remove the discrepancy between theory and experiment; rather it was accentuated. Recent calculations by Grosso et al, /49/ also reconfirm this.

The dilemma with the Compton profile of LiH has remained until today. An experimental difficulty is that LiH is highly hygroscopic. Therefore Paakkari /50/ has recently remeasured polycrystalline LiH with 60 keV γ-rays and made precautions that the water content of the sample was as low as possible. Indeed he found a sharper profile which agrees well with the earlier theoretical predictions /47/. The agreement is even perfect if theory is convoluted with the experimental resolution function as shown in Fig. 9. We have therefore found it appropriate to extend the previous Heitler-London calculations to include also directional effects with the hope that the results would inspire to new measurements /51/.

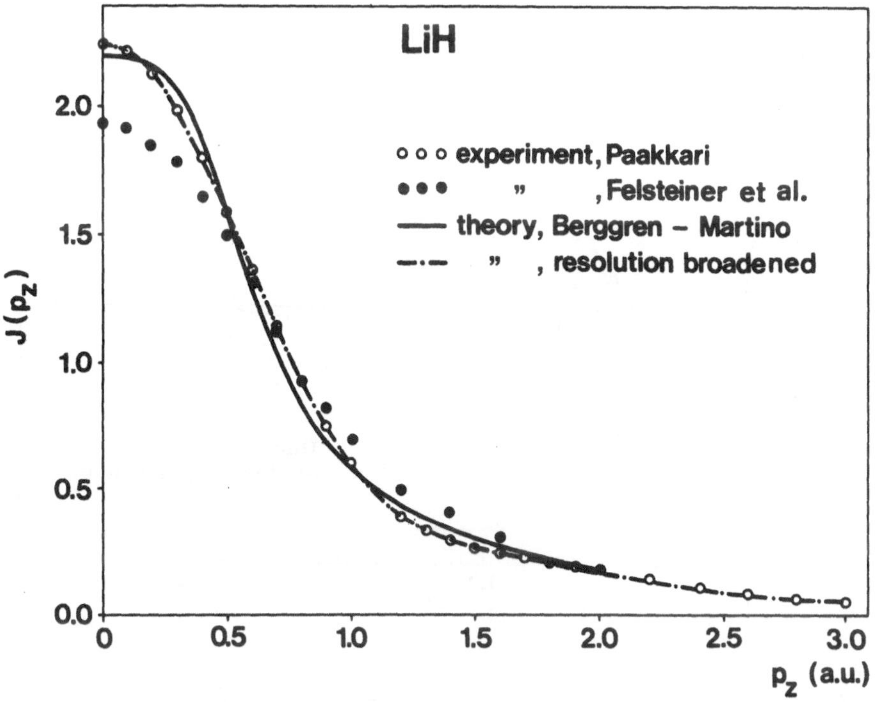

Fig. 9. Isotropic Compton profile of LiH. Experimental results are taken from refs. 48 (Felsteiner *et al.*) and 50 (Paakkari). Theoretical values are from ref. 47, (Berggren and Martino). The experimental data of Phillips and Weiss (ref. 45) are somewhat closer to Paakkari's.

If we let the ionic orbitals simply be 1s-type orbitals with orbitals exponents α_{Li^+} and α_{H^-} it is an easy matter work out the directional dependence of $J(p^z)$ /51/. For numerical work the orbital exponents $\alpha_{Li^+} = 3 - 5/16 = 2.6875$ and $\alpha_{H^-} = 0.7208$ were chosen. The value for α_{Li^+} follows from minimization of the total energy of a free Li^+-ion. The value for α_{H^-} was determined by Lundqvist /40/ from a minimization of the crystal energy keeping α_{Li^+} constant. The summations over the lattice in Eq. (33) were carried out to the fourth shell. By then the contributions seemed small enough to allow for a termination of the series. In order to be consistent with the restricted summations the inverse of the

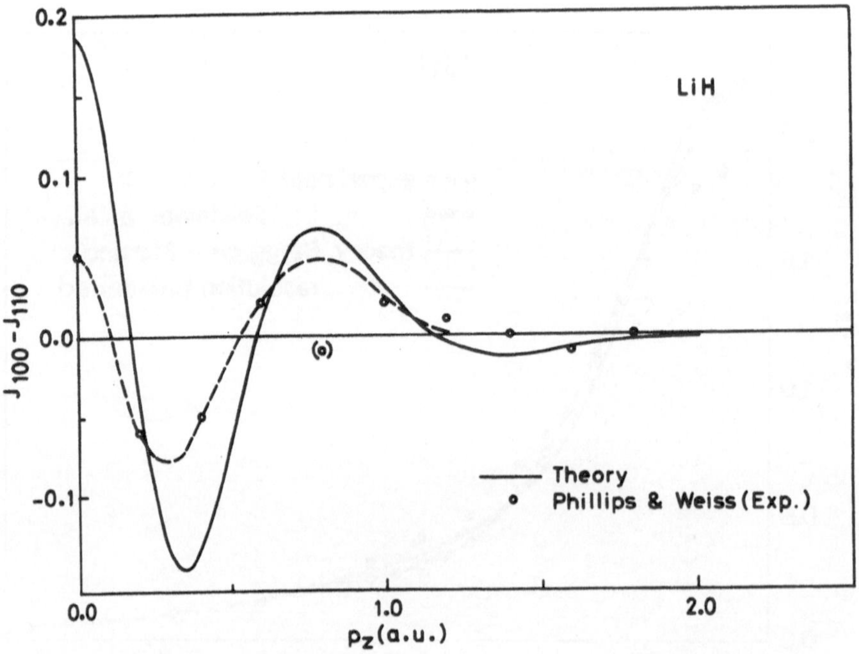

Fig. 10. Difference between <100> and <110> Compton profiles of LiH. The full drawn curve refers to theory (Berggren, ref. 51) and the circles to experiments (Phillips and Weiss, ref. 45). The measured value within parenthesis is believed to be inaccurate due to statistical fluctuations. The dashed curve is an interpolation by free hand.

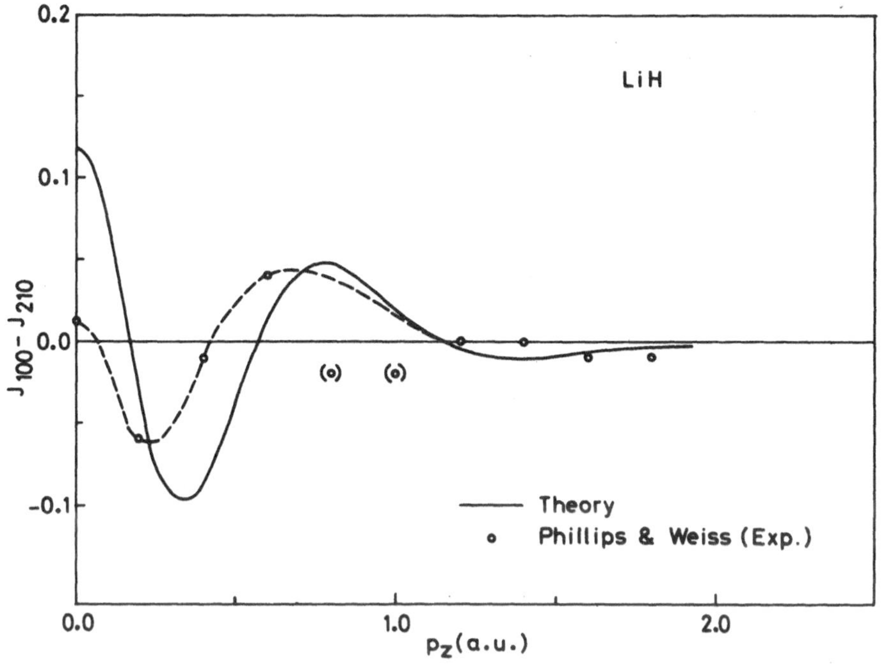

Fig. 11. Difference between <100> and <210> Compton profiles of LiH. The notations are the same as in Fig. 10.

overlap matrix was determined for a cluster containing the central ion, either Li^+ or H^-, and four shells. In the expression for the Compton profile the diagonal terms, nearest neighbour interactions and $H^- - H^-$ interactions were found to dominate by far. For small values of p_z the largest contribution to the anisotropy derives from the $H^- - H^-$ interactions. For larger values of p_z first and second shells contribute about equally. The computed anisotropies are shown in Figs. 10 - 12.

The calculated anisotropies are remarkably large, about twice the size found for LiF /28,29/. At $p_z = 0$ $J_{100} - J_{110}$ amounts to ~ 10 % of the total profile which is as large as the anisotropy in diamond /52/. As mentioned above Phillips and Weiss /45/ have measured the <100>, <110> and <210> profiles. In spite of the limited statistical accuracy there seems to be at least a qualitative agreement between theory and experiment for $J_{100} - J_{110}$ and $J_{100} - J_{210}$ for $p_z \lesssim 0.8$ a.u. (Figs. 10 and 11) and $J_{110} - J_{210}$ for $p_z \gtrsim 1.0$ a.u. (Fig. 12). An inspection of Figs. 10 - 12 makes us

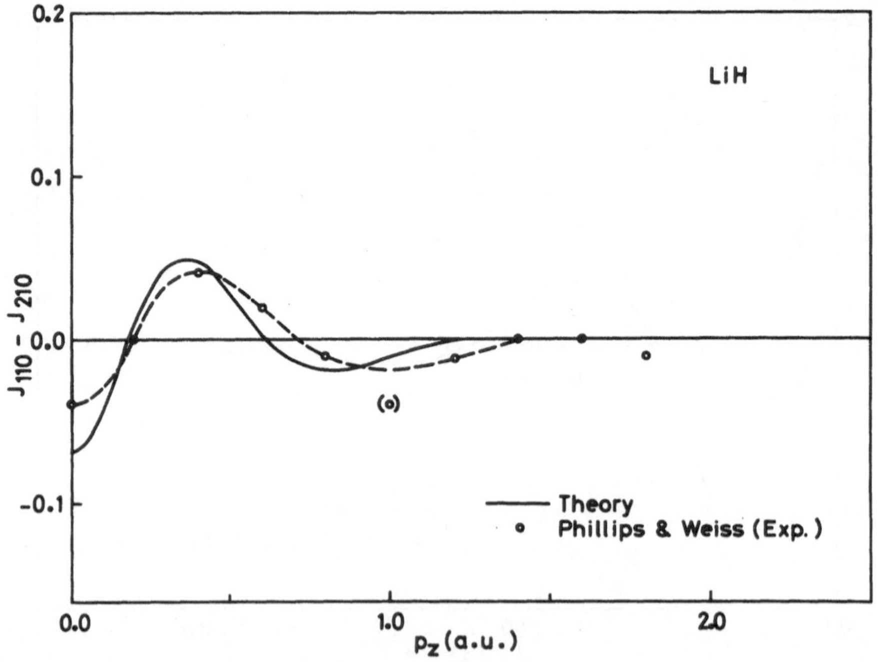

Fig. 12. Difference between <110> and <210> Compton profiles of LiH. The notations are the same as in Fig. 10.

believe that the measured values for J_{100} (0.8) and J_{210} (1.0) are inaccurate. We therefore suggest that the directional Compton profile of LiH should be remeasured in order to improve the statistical accuracy.

V. SUMMARY

With some ionic crystals as examples we have illustrated the present theoretical capabilities of predicting Compton profiles. We have seen that a detailed comparison between theory and experiments is nowadays possible, which thus sheds light on the quality of theoretical wave functions as well as on experimental techniques. It is particularly satisfying that recent refinements in these techniques are beginning to offer independent experimental evidence that available wave functions in ionic crystals are indeed of high accuracy. Thus Löwdin, as far back as 1948 in his thesis, did work which was instrumental in the coming of age of a new experimental technique in the last decade.

REFERENCES

1. For a review, see M. Cooper, Advan. Phys. 20, 453 (1971).

2. B. Williams (Editor), "The Compton Effect", (McGraw-Hill, book to be published).

3. A.T. Stewart and L.O. Roelling (Editors), "Positron Annihilation", (Academic Press, N.Y., 1967).

4. E. Weigold, S.T. Hood, and P.J.O. Teubner, Phys. Rev. Lett. 30, 475 (1973).

5. I.E. McCarthy, A. Ugbabe, E. Weigold and P.J.O. Teubner, Phys. Rev. Lett. 33, 459 (1974).

6. S. Dey, I.E. McCarthy, P.J.O. Teubner, and E. Weigold, Phys. Rev. Lett. 34, 782 (1975).

7. P.O. Löwdin, "A Theoretical Investigation into Some Properties of Ionic Crystals" (Almqvist & Wiksells, Uppsala, 1948).

8. The first part of Section III a is rather close to the presentation by J.J. Sakurai, "Advanced Quantum Mechanics", (Addison-Wesley, 1967), Sec. 2-5.

9. W. Heitler, "The Quantum Theory of Radiation", (Oxford University Press, 1954), 3rd edition, p. 192.

10. A.I. Achieser and W.B. Berestezki, "Quantenelektrodynamik", (B.G. Teubner Verlagsgesellschaft, Leipzig 1962).

11. P. Eisenberger and W.A. Reed, Phys. Rev. B 9, 3237 (1974).

12. S. Manninen, T. Paakkari, and K. Kajantie, Philos. Mag. 29, 167 (1974).

13. R. Ribberfors, Phys. Rev. B, 12, 2067 (1975).

14. R. Ribberfors, Phys. Rev. B, 12, 3136 (1975)

15. An approach similar to the one in refs. 11 - 14 has been considered in connection with radiation transport theory (B.R. Wienke, Nuclear Science and Engineering, 52, 247 (1973); J. Quant. Spectrosc. Radiat. Transfer, 15, 151 (1975)).

16. R. Ribberfors, "Relativistic differential cross sections for Compton scattering of photons from bound electron systems", Report LiH-IFM-R-40 (Linköping University, Sept. 1975).

17. T. Paakkari, E.-L. Kohonen, O. Aikala, K. Mansikka, and S. Mikkola, Phys. Fenn. 9, 207 (1974).

18. P.O. Löwdin, Philos. Mag. Suppl. 5, 1 (1956).

19. O. Aikala and K. Mansikka, Phys. Kondens. Materie, 11, 243 (1970); ibid. 14, 105 (1972).

20. O. Aikala, V. Jokela, and K. Mansikka, J. Phys. C: Solid State Phys., 6, 1116 (1973).

21. K.-F. Berggren, Solid State Commun., 9, 861 (1971).

22. S. Manninen, T. Paakkari, O. Aikala, and K. Mansikka, J. Phys. C: Solid State Phys., 6, L 410 (1973).

23. O. Aikala, K. Mansikka, L. Ekström, and K.-F. Berggren, Philos. Mag., 28, 997 (1973).

24. K. Mansikka and O. Aikala, Ann. Univ. Turkuensis AI, 162, 43 (1973).

25. W.A. Reed and P. Eisenberger, Phys. Rev. B, 6, 4596 (1972).

26. W.A. Reed, P. Eisenberger, K.C. Pandey, and L.C. Snyder, Phys. Rev. B, 10, 1507 (1974).

27. R.J. Weiss, Philos. Mag., 21, 1169 (1970).

28. W.A. Reed, P. Eisenberger, F. Martino, and K.-F. Berggren, Phys. Rev. Lett., 35, 114 (1975).

29. K.-F. Berggren, F. Martino, P. Eisenberger, and W.A. Reed, Phys. Rev. B, Feb. (1976) (in press).

30. R.N. Euwema, G.G. Wepfer, G.T. Surratt, and D.L. Wilhite, Phys. Rev. B, 9, 5249 (1974).

31. E. Clementi, IBM J. Res. Dev., 9, 2 (1965).

32. A.B. Kunz, Phys. Rev. B, 2, 2224 (1970).

33. W.H. Zachariasen, Acta. Cryst. A24, 324 (1968).

34. R.C.G. Killean, J.L. Lawrence and V.C. Sharma, Acta. Cryst. A28, 405 (1972).

35. C.J. Howard and R.G. Khadake, Acta. Cryst. A30, 296 (1974).

36. C.J. Howard and B. Dawson, International Crystallography Conf., Melbourne, Australia, 1974.

37. W.H. Mueller, J.P. Blackledge, and G.G. Libowitz (Eds.), "Metal Hydrides" (Academic Press, New York, 1968).

38. See e.g., T.R.P. Gibb in "Progress in Inorganic Chemistry", Ed. F.A. Cotton (Interscience Publishers, Inc., New York, 1962), Vol. III, pp. 315-522.

39. E.A. Hylleraas, Z. Phys., 63, 771 (1930).

40. S.O. Lundqvist, Arkiv for Fysik, 8, 177 (1954).

41. A. Morita and K. Takahashi, Prog. Theor. Phys., 19, 257 (1958).

42. A.B. Kunz, phys. status solidi, 36, 301 (1969); see also A.B. Kunz and D.J. Mickish, Phys. Rev. B, 11, 1700 (1975).

43. W. Brandt, L. Eder, and S.O. Lundqvist, Phys. Rev., 142, 165 (1966).

44. K.-F. Berggren, J. Phys. C: Solid St. Phys., 2, 802 (1969).

45. W.C. Phillips and R.J. Weiss, Phys. Rev., 182, 923 (1969).

46. W. Brandt, Phys. Rev. B, 2 561 (1970).

47. K.-F. Berggren and F. Martino, Phys. Rev. B, 3, 1509 (1971).

48. J. Felsteiner, R. Fox, and S. Kahane, Phys. Rev. B, 6, 4689 (1972).

49. G. Grosso, G. Pastori Parravicini, and R. Resta, phys. stat. sol. (to be published).

50. T. Paakkari (private communication).

51. K.-F. Berggren, "Anisotropic Compton scattering in crystalline LiH", Report LiH-IFM-IS-40 (Linköping University, March 1975).

52. W.A. Reed and P. Eisenberger, Phys. Rev. B, 6, 4596 (1972).

Note added in proof: Recently O. Aikala (J. Phys. C: Solid St. Phys., to be published) has extended the calculations of the directional Compton profiles of LiH using the same method as in ref. 51. The overlap matrix has been inverted for a larger number of neighbours, which results in smaller anisotropies for $p_z \lesssim 0.8$. The qualitative features are in agreement with the results in ref. 51.

SOME COMMENTS ON THE QUANTUM-MECHANICAL TREATMENT OF DEFECTS IN IONIC CRYSTALS

Jean-Louis Calais

Quantum Chemistry Group
University of Uppsala
Box 518, S-751 20 UPPSALA, Sweden

I. INTRODUCTION

Per-Olov Löwdin's thesis[1] "A theoretical investigation into some properties of ionic crystals", constitutes a landmark in the theory of the electronic structure of ionic crystals. By combining mathematical ingenuity and an enormous amount of numerical work without the help of electronic computers Löwdin was able to get not only very good agreement between experimental and theoretical values for a number of physical parameters for nearly all the alkali halides, but also and of greater importance to introduce a new physical concept. His calculations of cohesive energies and equilibrium lattice constants gave a quantum mechanical background and "explanation" of the Born-Mayer theory. But his quantum mechanical treatment went far beyond the Born-Mayer model, since it contained ingredients, which could explain the deviations from the Cauchy relations for the elastic constants. It is particularly interesting to notice that the effective many-ion forces, which are responsible for these deviations, actually came out of the calculations as a result of the careful treatment of the overlap between the ions. Löwdin's alkali halide work shows what in a favourable case can be achieved by an "ab initio" calculation.

For many years Löwdin's work seemed to be the definitive word in the quantum mechanical treatment of ionic crystals, and it is only rather recently that some papers have appeared, in which one has tried to go beyond some of the approximations used by Löwdin. The present situation has recently been summed up by the author[2]. In this connection we notice in particular Ra's[3] procedure for obtaining a more rapidly convergent power series expansion of the inverse of the overlap matrix and the method proposed by Kohn[4,5] for the direct calculation of Wannier functions.

Löwdin's investigations as well as the more recent calculations based on his method dealt exclusively with perfect ionic crystals. A corresponding quantum mechanical treatment of "static" properties of defects of various kinds in ionic crystals is certainly much more difficult and nothing of that kind seems to have been attempted, as can be seen for example by the following quotation from a recent survey article by Lidiard[6]: "Although there exist orbital theories of the cohesion of ionic crystals (Löwdin 1956; Lundqvist 1957) which give important insight and guidance into the nature of ionic bonding in solids, in calculations of lattice relaxation we use semi-empirical models." It is the purpose of this paper to try to tie up some loose ends which should be of importance for a detailed quantum mechanical analysis of the defect problem. Per-Olov Löwdin has always stressed the importance of considering quantum chemistry and solid state physics as two aspects of the same field, and since the problem of defects in solids forms an excellent bridge between these two areas, a discussion of this problem seems appropriate in a volume dedicated to him.

II. SEMI-EMPIRICAL PROCEDURES

Lidiard[6] has recently given a survey of the present state of the theories of defects in crystalline solids. Semi-empirical and empirical methods dominate the field of ionic crystals for which we are here primarily interested in the distortion and polarization of the crystal in the neighbourhood of the defect. The polarization around a defect in an ionic crystal is traditionally treated by means of the "semi-discrete" method of Mott and Littleton[7]. Only the nearest neighbours of the defect are treated as discrete and the remainder of the crystal is considered as a dielectric continuum. Kanzaki[8] has introduced a procedure for calculating the distortion around a defect, which is based entirely on the discrete nature of the lattice. This method - sometimes called lattice statics - is quite general, provided an expression (semi-empirical or ab initio) for the potential energy of the perfect lattice is known. The defect is replaced by fictitious external forces, which create the same distortions as the real defect. For further aspects on these and other methods used we refer to the reviews by Lidiard[6] and Hardy-Flocken[9].

The intensive semi-empirical work on defects, which is summed up in these reviews, is of course necessary and extremely valuable. But the question is also whether time is not now ripe for a quantum mechanical attack on these problems. Apart from the general interest of basing semi-empirical procedures on a quantum mechanical ground, there also seems to be a definite need of methods that can do better than those utilized so far[9]. Dick and Das[10,11,12,13] have made en extensive set of calculations of lattice distortions, field gradients and other quantities in alkali halide solid solutions. The results were found to be very sensitive to the various semi-empirical parameters used. The field gradients in the latest of these papers were good only to within a factor of two, despite the fact that the shell model was used and a very large number of neighbours were taken into account. Quigley and Das[14] stress the desirability of a quantum mechanical treatment and in particular point out the need for an improved procedure for describing the interaction between an impurity ion and its neighbours.

III. HEITLER-LONDON TREATMENT OF A SUBSTITUTIONAL IMPURITY IN AN IONIC CRYSTAL. EXTENSIONS OF THIS METHOD

Previous quantum mechanical calculations of cohesive properties of perfect ionic crystals can be characterized as Heitler--London approximations or slight improvements thereof[2]. The Wannier functions of the crystals were approximated by orthogonalized atomic orbitals of the free ions, but in some of the more recent investigations radial deformations of the outermost orbitals of the negative ions were also introduced.

As pointed out by Lundqvist[15], the Heitler-London approximation does not account for any polarization of the ions. It might therefore seem vain to attempt any kind of defect calculation within that scheme. It would certainly be highly unrealistic to do so for a vacancy or an interstitial. For a substitutional impurity with the same charge as the host ion it could be regarded as a possible first step, which should be extended to include polarization later on. There would be essentially two advantages with a Heitler-London calculation for such a defect. First of all it would allow us to study in a relatively easy way the dependence of the geometry on interactions. Secondly we could use the experience we have from similar calculations for perfect crystals.

Following Löwdin[16] we write the total energy for an ionic crystal with a substitutional impurity as (atomic units for all quantities with the Hartree as the unit for energy are used)

$$E_d = \frac{1}{2}\sum_{g,h}{}' \frac{Z_g Z_h}{R_{gh}} + \sum_{g\neq o}\frac{nZ_g}{R_{go}} - \frac{1}{2}\int \Delta\gamma_d(x_1|x_1')dx_1 - \sum_{g\neq o} Z_g \int \frac{\gamma_d(x_1|x_1)}{r_{1g}}dx_1$$

$$- Z_o\int\frac{\gamma_d(x_1|x_1)}{r_{10}}\,dx_1 + \int\frac{\Gamma_d(x_1x_2|x_1x_2)}{r_{12}}\,dx_1dx_2. \qquad (1)$$

Here γ_d and Γ_d are the (exact or approximate) first and second order density matrices for the crystal, which satisfy

$$\int\gamma_d(x|x)dx = N_d; \quad \int\Gamma_d(x_1x_2|x_1x_2)dx_1dx_2 = \binom{N_d}{2}, \tag{2}$$

where N_d is the total number of electrons in the system. The impurity ion is placed at the origin and its charge is assumed to satisfy

$$Z_o - N_o = \pm(Z_g - N_g), \tag{3}$$

where Z_g is the nuclear charge and N_g the number of electrons of ion g. The number $n = N_o - N_+ = Z_o - Z_+$ (assuming that the ion at the origin is a cation) is the difference in the nuclear charges of the impurity and the host ion.

For a closed shell system like an ionic crystal with or without substitutional impurities, the Hartree-Fock level is a natural approximation. This means

$$\gamma_d(x_1|x_1') = \bar{\rho}_d(x_1,x_1'); \tag{4a}$$

$$\Gamma_d(x_1x_2|x_1'x_2') = \frac{1}{2}\Big[\bar{\rho}_d(x_1,x_1')\bar{\rho}_d(x_2,x_2') -$$

$$- \bar{\rho}_d(x_1,x_2')\bar{\rho}_d(x_2,x_1')\Big] \tag{4b}$$

In analogy with the perfect crystal, we approximate the Fock-Dirac matrix by the expression

$$\bar{\rho}_d = |\Phi\rangle\Delta^{-1}\langle\Phi|, \tag{5}$$

where $|\Phi\rangle$ is the set of atomic orbitals used for the free ions and Δ their overlap matrix. This overlap matrix is no longer cyclic, but we can still calculate its inverse as a power series expansion in $\mathbf{S} = \Delta - \mathbf{1}$; there the procedure proposed by Ra[3] should be preferable.

Then we subtract from expression (1) the total energy of those free ions which make up the crystal with the impurity. The result can be written - in analogy with the perfect crystal case - as

$$E_{d,coh} = E_{elstat} + E_{exch} + E_S \tag{6}$$

E_{elstat} is the Madelung energy with corrections for the fact that the ions are not point charges, E_{exch} is the exchange energy for all the ion pairs, and in E_S are collected all those terms which depend on the overlap integrals.

Now $E_{d,coh}$ should be minimized with respect to the "geometrical" parameters, so as to yield the optimum nuclear configuration around the impurity. It is valuable to study how the three terms in Eq. (6) depend separately on the nuclear configuration, and in particular a study of E_S, which contains the effective many-ion forces, should be of great interest.

Such calculations are quite feasible, but since the form (6) has to be calculated for a large number of nuclear configurations it will be necessary to have an efficient integral program available.

An extension of the Heitler-London treatment, which should give better approximations of the Wannier functions than the orthogonalized atomic orbitals (OAO) has been proposed by Bauer[17,18]. He pointed out that in that approximation of the Hartree-Fock equations for the crystal which leads to the OAO's, the first term neglected is the cubic part (for a perfect crystal) of the lattice potential. Inclusion of this potential will mix the ordinarily used ground states orbitals of the free ions with excited orbitals. Bauer has shown that this cubic potential is essential to explain the density distribution in the silver halides which is considerably different from that of the alkali halides. - As extensions of the OAO-approximation we can also regard the work by Kunz[19] and his group (for further references see Calais[2]), based on the Adams[20] - Gilbert[21] localized orbitals.

Such extensions of the original Heitler-London approximation will hopefully more and more approach the Hartree-Fock solutions for the crystal with a defect. The localized version of these Hartree-Fock orbitals would then be what is now beginning to be called the generalized Wannier functions. This concept and its possible importance for defects in ionic crystals will be treated in the next section.

IV. GENERALIZED WANNIER FUNCTIONS

For perfect crystals Kohn[4,5] has developed a variational procedure to calculate Wannier functions (WF) directly. The total energy of a filled band is written as a functional of the Wannier functions and minimized with respect to them. - Kohn and Onffroy[22] have introduced - in a one-dimensional case - the concept of generalized

Wannier functions (GWF), which play the same role in a non-periodic system as the Wannier functions do for a periodic one.

Gay and Smith[23)] have analyzed the generalized Wannier functions (GWF) in the three-dimensional case. These functions can be thought of as unitary transformations of the eigenfunctions of the effective one-electron Hamiltonian for the non-periodic problem (e.g. a crystal with a defect), which are of a localized nature. They can also be expanded in terms of the Wannier functions of the corresponding periodic system. Gay and Smith show explicitly the structure of the different transformation matrices.

A very interesting aspect of the GWF's is seen in connection with the Koster-Slater method[24)] for treating defects in solids (for a simple derivation see Löwdin[25)]). In that method the eigenfunctions of the perturbed Hamiltonian are expanded in terms of the WF's of the perfect crystal. The equation is transformed to an integral equation by means of the Green's function of the perfect crystal and finally to a difference equation for the coefficients of the WF's in the eigenfunctions. - This expansion in terms of the WF's runs over many sites and many bands. If one expands instead in terms of GWF's, which are adapted to the perturbed crystal, the one-band expansion becomes exact.

A variational principle similar to the one for the WF's exists also for the GWF's. In the neighbourhood of the impurity the GWF's contain contributions from WF's of many bands, but further away from the defect they become identical to one of the WF's.

The GWF's would seem to provide an ideal tool both for formulating defect problems and possibly also for numerical work. They have been used successfully for surface problems[26,27,28)], but as far as we know not for any problem concerned with defects in ionic crystals.

It seems probable that in the future both WF's and GWF's will play an important role not only as a formal tool, as they have been essentially confined to so far, but also for explicit numerical work. In this connection it is worth recalling the important results on localization and asymptotic behaviour of WF's obtained by Kohn[29)], Bulyanitsa and Svetlov[30)], des Cloizeaux[31,32,33)], the Stuttgart group[34-36)] and Kertész and Biczó[37)].

V. CONCLUSION

The purpose of this paper was to indicate that with the experience we have of calculations of cohesive properties of perfect ionic crystals and with the new concepts and procedures which are becoming available, the prospects for meaningful quantum mechanical calculations of the electronic structure of ionic crystals with defects look reasonably bright. Such calculations can be regarded as meaningful if they lead to new concepts which help us to interpret and understand exerimental results, so as to give us more insight into the nature of the chemical bonds in these system. Very much work remains to be done indeed, in this nearly virgin territory: actual calculations for many different cases, analysis of such calculations and comparison with previous semi-empirical work and with experimental data.

ACKNOWLEDGEMENTS

The author would like to use this opportunity to express his great indebtedness and warm gratitude to Per-Olov Löwdin for many years of close collaboration. He is also grateful to Dr. Osvaldo Goscinski for a critical reading of the manuscript.

REFERENCES

1. P-O Löwdin, Thesis, Almqvist & Wiksell, Uppsala 1948.
2. J-L Calais, Int. J. Quantum Chem. 9S, 497 (1975).
3. Ø. Ra, Int. J. Quantum Chem. 10, 5, 57 (1976).
4. W. Kohn, Phys. Rev. B7, 4388 (1973).
5. W. Kohn, Phys. Rev. B10, 382 (1974).
6. A.B. Lidiard in "Orbital Theories of Molecules and Solids", ed. by N. March, Clarendon Press, Oxford 1974.
7. N.F. Mott, M.J. Littleton, Trans. Faraday Soc. 34, 485 (1938).
8. H. Kanzaki, J. Phys. Chem. Solids 2, 24 (1957).
9. J.P. Hardy, J.W. Flocken, Critical Reviews in Solid State Sciences 1, 605 (1970).
10. B.G. Dick, T.P. Das, Phys. Rev. 127, 1053 (1962).
11. T.P. Das, B.G. Dick, Phys. Rev. 127, 1063 (1962).
12. T.P. Das, Phys. Rev. 140, A1957 (1965).
13. B.G. Dick, Phys. Rev. 145, 609 (1966).
14. R.J. Quigley, T.P. Das, Phys. Rev. 164, 1185 (1967).
15. S.O. Lundqvist, Arkiv Fysik 12, 263 (1957).
16. P-O Löwdin, Adv. Phys. 5, 1 (1956).
17. R. Bauer, phys. stat. sol. (b) 50, 225 (1972).

18. R. Bauer, J. Nonmetals 1, 257 (1973).

19. A.B. Kunz, phys. stat. sol. (b) 36, 301 (1969).

20. W.H. Adams, J. Chem. Phys. 34, 89 (1961); 37, 2009 (1962).

21. T.L. Gilbert in "Molecular Orbitals in Chemistry, Physics, and Biology", ed. P-O Löwdin, Acad. Press 1964.

22. W. Kohn, J. Onffroy, Phys. Rev. B8, 2485 (1973).

23. J.G. Gay, J.R. Smith, Phys. Rev. B9, 4151 (1974).

24. G.F. Koster, J.C. Slater, Phys. Rev. 95, 1167 (1954).

25. P-O Löwdin, J. Mol. Spectroscopy 14, 119 (1964).

26. J.R. Smith, J.G. Gay, Phys. Rev. Lett. 32, 774 (1974).

27. S.C. Ying, J.R. Smith, W. Kohn, Phys. Rev. B11, 1483 (1975).

28. J.G. Gay, J.R. Smith, Phys. Rev. B11, 4906 (1975).

29. W. Kohn, Phys. Rev. 115, 809 (1959).

30. D.S. Bulyanitsa, Y.E. Svetlov, Soviet Phys.-Solid State 4, 981 (1962).

31. J. des Cloizeaux, Phys. Rev. 129, 554 (1963).

32. J. des Cloizeaux, Phys. Rev. 135, A685 (1964).

33. J. des Cloizeaux, Phys. Rev. 135, A698 (1964).

34. B. Lix, phys. stat. sol. (b) 44, 411 (1971).

35. G. Kögel, phys. stat. sol. (b) 44, 577 (1971).

36. E. Krüger, phys. stat. sol. (b) 52, 215 (1972); 52, 519 (1972).

37. M. Kertész, G. Biczó, phys. stat. sol. (b) 60, 249 (1973).

PROPERTIES OF COMPRESSED ATOMS FROM A SPHERICAL CELLULAR MODEL

Anders Fröman

National Defence Research Institute

104 50 STOCKHOLM, Sweden

INTRODUCTION

When pressure is applied to matter the volume per atom will decrease and, as conjectured by P.W. Bridgman long ago, changes will occur in the electron structure and hence in several solid-state properties. Static and in particular dynamic high pressure experiments on the pure elements /1-6/ have revealed that solid-solid transitions are not exceptional and that several properties e.g. compressibilities, electrical conductivity and melting temperatures do not change with pressure in a simple and regular fashion. The prime examples of this are perhaps the transitions in Ce /7/ and Cs /8,9/ at 7 kbar and 42 kbar respectively where a large discontinuous decrease of the volume occurs without changes in the lattice structure. These transitions have been denoted electronic, a concept which Drickamer and Frank /2/ interpret as a pressure-induced shift to a new or greatly modified electronic ground state of a system.

When the volume available to an atom is changed it is quite obvious that the electron structure will also change. The extreme situations are the infinitely compressed state where the free electron gas is the appropriate description and the free atom with well-separated, discrete one-electron energies. At normal, zero-pressure conditions it is wellknown that the solid-state properties of the elements show a characteristic variation with the atomic number reflecting the importance of the electron structure which forms the basis for the periodic system /10/. If one wants to get an over all picture of the possible pressure-induced changes in the electron structures one thus has to employ a model which is capable of describing the characteristic atomic properties. It is the purpose of this note to present such a model and to give some results of calculations carried out 5 to 10 years ago.

THE MODEL

The Thomas-Fermi model and extensions of it are adequate at very high compressions but they lead to properties which are smooth functions of the atomic number. Unless one makes additional assumptions the statistical models are therefore not suitable for the present problem. Similarly, a model based directly on the free atom such as e.g. the tight-binding picture will freeze the electron structure which is anticipated to change. The model chosen is the spherical cellular model (SCM) which may be looked upon as an atomic model with spherical boundary conditions representing a finite atom in a condensed phase. Alternatively, it may be considered to be the sphericalized version of the Wigner-Seitz cellular method. There are many possible choices of boundary conditions. The one used in this investigation is the following.

Within the spherical volume with radius R the crystal orbital, ψ, is expanded in spherical harmonics with

$$\psi(\vec{r}) = \sum_{l,m} S_{lm}(\theta\phi)\phi_l(\varepsilon,r)B_{lm} \tag{1}$$

where $\phi_l(\varepsilon,r)$ is the radial wave-function at the energy ε. If for a given reduced wave-vector $\bar{k}$ one inserts the expansion (1) in the Wigner-Seitz boundary conditions applied to the endpoints of a diameter, one obtains the following set of equations after multiplying with S^*_{LM} and integrating over the angles: ($L = 0,1\ldots L_{MAX}$)

$$\sum_{l=M}^{L_{max}} B_{lM}(k)\phi_l(\varepsilon,R)\beta(l,L,M;kR)\{1 - (-1)^L\} = 0$$

$$\sum_{l=m}^{L_{max}} B_{lM}(k)\phi_l'(\varepsilon,r)\beta(l,L,M;kR)\{1 + (-1)^L\} = 0 \tag{2}$$

where $\beta(l,L,M;kR) = C_{norm}\int_{-1}^{+1} dx\; P_l^{|M|}(x)P_L^{|M|}(x)\exp(-ikRx)$.

The system (2) determines the bandstructure $\varepsilon_i(k)$ where $i = (nLM)$

is the band index. At k = 0 one obtains the common approximate conditions for the bottom and top of the band, $\phi_l'(R) = 0$ and $\phi_l(R) = 0$, respectively. The method /11/ is very similar to the one used by Gandelman /12/ but two differences are worth noting. Firstly we have in general used the full Slater $\rho^{1/3}$ exchange in the radial Schrödinger equation and secondly we have made the calculation self-consistent in the restricted sense described below.

With a given charge density and potential, one can compute the band-structure and crystal orbitals fulfilling the boundary condition. The occupied states are determined and from the corresponding crystal orbitals one may then define l-occupation numbers, N_l, by the following relations where Z is the atomic number and i is the band index.

$$N_{li} = \int^{\text{occupied}} dk\, k^2\, |B_{li}(k)|^2 \qquad l \lesssim L_{max} \tag{3a}$$

$$Z = \sum_{l,i} N_{li} = \sum_{l} \{N_l + N_{l\,core}\} \tag{3b}$$

In the core states the occupation numbers, $N_{l\,core}$, are integers but for the conduction bands and the closed shells, which have been influenced by the boundary, the N_l's are in general non-integers. A new charge density and potential can then be constructed using these l-occupation numbers and, for simplicity, radial solutions which satisfy $\phi_l' = 0$ for the corresponding l-value and with the proper number of nodes. One may then iterate until the l-occupation numbers used in the charge density are equal to the ones obtained from the resulting, occupied crystal orbitals.

This method of obtaining approximate self-consistent solutions is straightforward for simple metals in which case the choice of starting potential is not crucial and the calculated properties are in general not sensitive to vatiations in the N_l's. For transition and, in particular, rare-earth elements the situation is different. This is illustrated in figure 1 where for Ce at R = 4.2 a.u. the resulting 4f-occupation number is given as function of the number of 4f-electrons assumed in the potential. The reason for the extreme sensitivity appears to be that the effective 4f-potential has a well which is strong enough to bind the electrons in a 4f-core state if N_{4f} is less than 2 and which is too weak to bind the 4f-electrons if N_{4f} is larger than 3.5. This is indicated in figure 1 showing that q, the atomic volume times the electron density at the boundary, is small for localized 4f-states. In figure 2 the self-consistent l-occupation numbers for Ce at R = 3.8 are given as functions of a

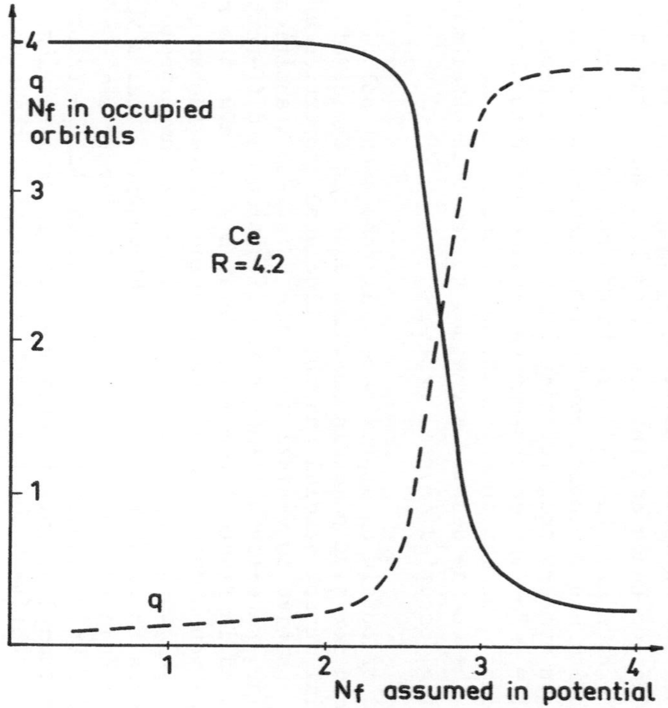

Figure 1. The f-occupation number and q (atomic volume times the electron density at the boundary) versus the number of 4f-electrons assumed when constructing the potential of Ce at R = 4.2 a.u.

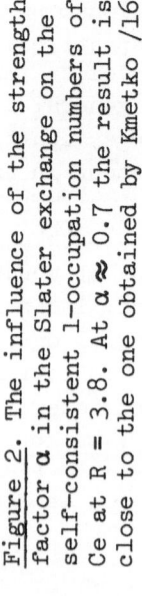

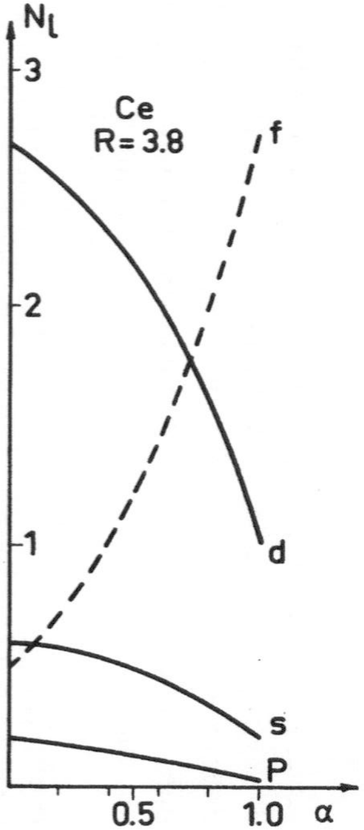

Figure 2. The influence of the strength factor α in the Slater exchange on the self-consistent l-occupation numbers of Ce at R = 3.8. At $\alpha \approx 0.7$ the result is close to the one obtained by Kmetko /16/.

strength parameter, α, multiplying the Slater exchange. Again one can note that the N_l and in fact also other properties are sensitive to the potential in the case of Ce. The use of more accurate exchange correlation potential may thus be needed.

The extensive use of l-occupation numbers in this work has two reasons. From a computational point of view it has turned out to be a convenient way of achieving self-consistency and thereby circumventing the ambiguity in the choice of potential. Initially the potential was obtained from the Thomas-Fermi solution for the finite atom, but particularly for non-simple metals this gives quite different results from the ones obtained with the self-consistent potential. The other reason is of a more physical nature. The transitions in Ce and Cs mentioned above have been interpreted to indicate a $4f \rightarrow 5d$ and $6s \rightarrow 5d$ transfer, respectively. By focusing the attention to the l-occupation numbers one may hopefully obtain an insight into the pressure-induced rearrangements of the electrons.

RESULTS AND DISCUSSION

The Equation-of-State

Initially this investigation was undertaken in an attempt to calculate the equation-of-state of solids at zero temperature. A sensitive test of any model developed for this purpose is to calculate the zero-pressure density of the elements in their condensed phases. In this work the pressure has not been obtained from a numerical differentiation of the volume-dependent total energy or from the virial theorem but has been calculated with the following formula /11,12/:

$$P = P_{cell} + P_{cell,cell} \tag{4a}$$

where
$$P_{cell} = \frac{1}{S} \int_S dS \left\{ \frac{\partial^2}{\partial n \partial n'} - \frac{\partial^2}{\partial n^2} \right\} \gamma\,(1'1) \tag{4b}$$

$$P_{cell,cell} = \frac{1}{3V} \sum_{\vec{R}_g \neq 0} \left\{ \frac{Z^2}{2|\vec{R}_g|} - 2Z \int_V d\tau_1 \frac{\gamma(11)}{|\vec{r}_1 - \vec{R}_g|} + \int_V d\tau_1 \int_V d\tau_2 \frac{2\gamma(11)\gamma(22) - \gamma(12)\gamma(21)}{|\vec{r}_1 - \vec{r}_2 - \vec{R}_g|} \right. \tag{4c}$$

$$\approx \frac{1}{3V^{4/3}} \{\lambda q^2 - \mu(q)\} = P_c + P_x \qquad (4d)$$

Here V and S are the volume and surface, respectively, of the atomic cell, $\vec{R}_g$ a lattice vector, n the outward normal unit-vector and γ the first-order spin-free density matrix. If one assumes free electrons with a density equal to the density at the atomic boundary, one obtains the approximate expression in eq. (4d). Here q is the atomic volume times the electron density at the boundary , λ is a constant depending on the lattice structure, and the function $\mu(q)$ is derived from the exchange part of $P_{cell,cell}$ using plane waves.

The zero-pressure atomic volumes obtained from the condition $P_{cell} + P_{cell,cell} = 0$ are given in figure 3, where we also present the experimental values /10/ and the results from two statistical models, the Thomas-Fermi-Dirac (TFD) and the Thomas-Fermi with quantum corrections (TFC) /13/. The simple Thomas-Fermi method does not give finite volumes. For simple metals - alkalis, alkaline earths, and noble metals - the spherical cellular model gives reasonable results provided one includes the $P_{cell,cell}$ term. This term is negative and corresponds to the exchange pressure in an electron gas. Incidentally, this contribution to the pressure is not taken into account in the results presented for the statistical methods. The present model also gives a qualitatively correct description of the small volumes associated with the transition elements (d-shells) but fails for the rare-earth metals and more reasonably so with this metallic appraoch for the inert gases and the elements in the right part of the periodic table. Considering α, the strength factor in the Slater exchange, as a parameter, it is possible to improve the situation in particular for the d- and f-elements. For instance, the correct zero-pressure volume of γ-Ce is obtained with α approximately equal to 0.7 and 0.2 using the conditions $P_{cell} = 0$ and $P_{cell} + P_{cell,cell} = 0$, respectively. If one considers compressed states of the rare-earths and several other elements (e.g. K, Ti, Cs, Ba) where the d-occupation number increases upon compression, the contribution from P_{cell} can be negative in a volume region and in addition $P_{cell,cell}$ gives a large negative contribution. Consequently, one obtains in these cases unstable states which are completely unphysical. This situation can only partly be remedied by lowering α. A possible cause for these deficiencies in the model is that the boundary conditions are not appropriate for the d-electrons and that the approximate treatment of $P_{cell,cell}$ is not accurate enough when localized d- and f-electrons play an important role.

The different contributions to the total pressure are given in

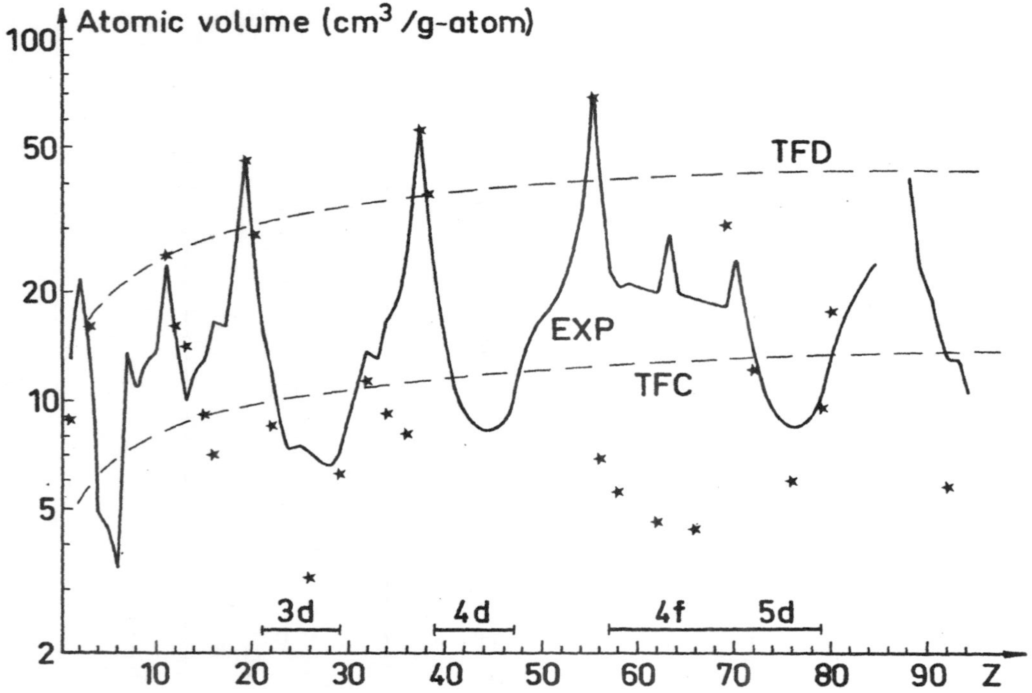

Figure 3. The zero-pressure, low temperature atomic volume of the elements. The statistical models give the curves denoted TFD and TFC. The SCM results (✱) are determined from $P_{cell} + P_{cell,cell} = 0$ using $\alpha = 1$. If P in the SCM calculations is zero at more than one finite volume the largest one is given (for K, Ca, Ti, Sr, Cs and Tm).

Table 1. Contributions (in kbar) to the total pressure of Na, Ce, and Pb at different compressions

Element	Vo/V	Pcell,cond	Pcell,core	Pcell	Px	Pc	P
Na	0.80	13.0	-0.2	12.8	-19.1	0.3	-6.0
(R_o=3.99)	0.99	31.7	-0.7	31.0	-27.1	0.4	4.3
(α=1)	1.48	101	-4	97	-50	1	48
	2.35	354	-28	326	-102	2	226
	4.60	1880	-240	1640	-300	10	1350
	7.94	6750	-780	5970	-750	20	5240
Ce	0.86	235	-57	178	-117	6	67
(R_o=3.81)	1.01	281	-86	195	-149	8	54
(α=0)	1.29	318	-137	181	-215	12	-22
	1.69	371	-159	212	-362	24	-126
	2.05	500	-106	394	-533	39	-100
Pb	0.89	26	-4	22	-60	3	-35
(R_o=3.66)	1.49	745	-45	700	-249	12	463
(α=1)	2.78	5700	-430	5270	-870	60	4460
	3.54	11070	-570	10500	-1410	110	9200
	4.59	23200	+800	24000	-2400	200	21800
	6.11	45400	5500	50900	-4200	500	47200

Table 1 for the elements Na, Ce, and Pb which are typical examples. As already emphasized the exchange part of $P_{cell,cell}$ is important at normal densities and small compressions but is only a correction factor at large compressions. The P_{cell} term has two parts, one associated with the conduction electrons and one with the core. Upon compression the latter part first gives a negative contribution and then a rapidly increasing positive one representing the strong repulsion between closed shells. The negative part is somewhat surprising and may be due to the way the different contributions are grouped in the pressure formula (4). However, from a physical point of view it is not unreasonable that, with the boundary condition on a finite atom, the core-state energies will initially decrease due to a mechanism similar to the one causing binding in the alkali metals. When the core is not treated with frozen orbitals but adjusted automatically to the potential, changes will certainly occur upon compression, possibly giving a different balance between the bottom and top of the band.

In figure 4 the calculated pressure-density curves for Pb, Al, and Li are compared with experiments and the TFC model /13/. The results are in fair agreement with experiments in these somewhat favourable cases and represent an improvement on the results from the statistical models. In order to have reliable results, however, one needs more elaborate and careful band calculations which are feasible to carry out /14/.

Electron Structure at Normal Densities

If one applies the spherical cellular method to the free-electron gas /15/, i.e. a constant potential, the l-occupation numbers will coincide with the exact ones obtained from a decomposition in spherical harmonics of plane waves provided the number of electrons per atom, n, is less than two. At larger n, deviations will occur, in particular N_d will be considerably smaller in the model when $2 < n < 10$, N_p considerably larger when $4 < n < 11.5$ and N_s larger ($2 < n < 6.5$) and smaller ($6.5 < n$). This can be traced back to the bandstructure at finite volumes which is not the free electron parabola, but exhibits bandgaps and at the reduced wave vector equal to zero gives the states in the order s, d, s and p (degenerate), g with increasing energy. Thus for the alkali and alkaline-earth metals the deficiences of the present model are expected to be less important, but for elements with more than two electrons outside the ion core the model is likely to introduce some unphysical features.

In Table 2 the l-occupation numbers are presented for the normal density of the elements. In this table there are also included

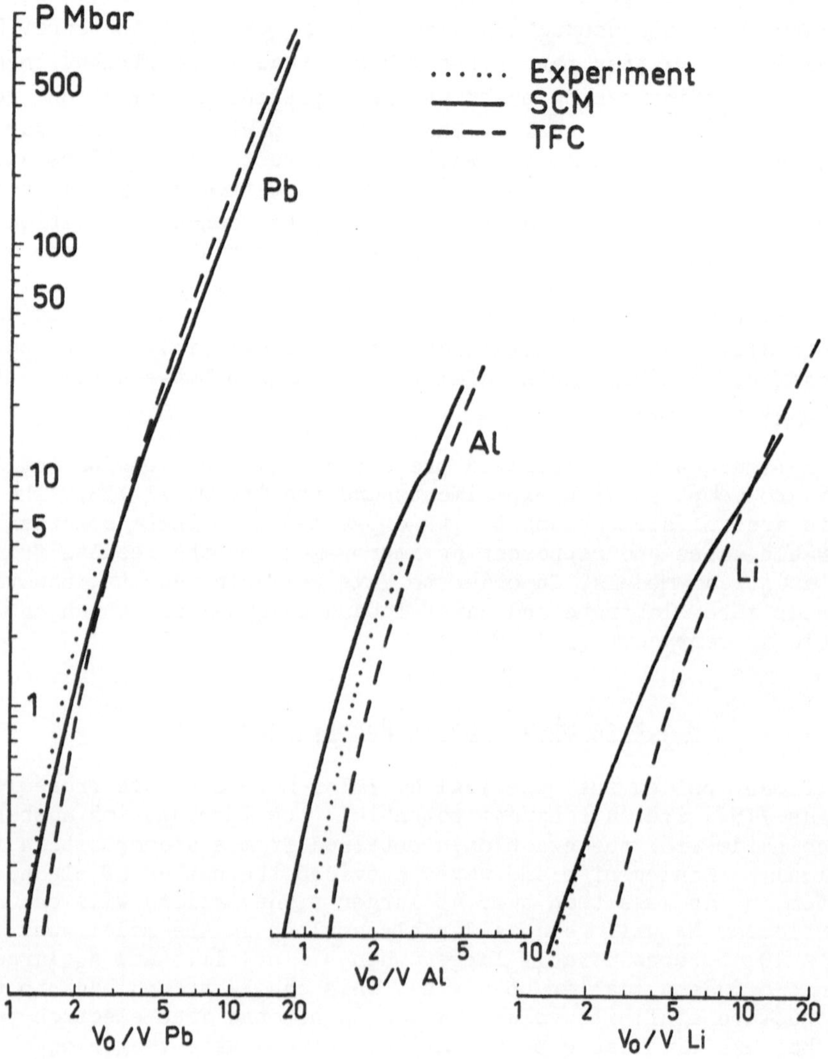

Figure 4. The equation-of-state at zero temperature for Pb, Al and Li. The statistical model results (TFC) are from reference 13 and the experimental ones from ref. 3 (Al and Pb) and 21 (Li).

Table 2. l-occupation numbers and some other electronic properties of the leements at the experiment normal density as calculated with the spherical cellular model.

Element	N_s	N_p	N_d	N_f	q	$\frac{g(\epsilon_F)}{g(\epsilon_F)_{exp}}$
El.gas n=1	0.65	0.30	0.04	0.00		
3Li	0.50	0.47	0.03	0.00	1.07	0.31
3Li*	0.51	0.46	0.03	0.00		
11Na	0.61	0.34	0.04	0.00	1.04	0.85
19K	0.62	0.31	0.07	0.00	1.08	0.91
37Rb	0.63	0.28	0.08	0.01	1.10	0.43
55Cs	0.62	0.25	0.12	0.01	1.14	0.85
55Cs*	0.62	0.21	0.16	0.00		
29Cu	0.60	0.45	9.91	0.03	1.65	0.69
79Au	0.58	0.59	9.68	0.09	2.31	0.33
El.gas n=2	1.04	0.78	0.17	0.02		
4Be	0.57	1.30	0.13	0.00	2.00	
12Mg	0.83	0.97	0.18	0.01	2.04	0.99
20Ca	0.75	0.65	0.55	0.01	1.99	2.03
20Ca*	0.85	0.63	0.51	0.01		
38Sr	0.81	0.66	0.51	0.02	2.12	~0.9
38Sr*	0.88	0.50	0.60	0.02		
56Ba	0.65	0.27	0.98	0.10	2.00	1.81
56Ba*	0.67	0.33	0.98	0.03		
El.gas n=3						
" (exact)	1.30	1.28	0.37	0.05		
" (SCM)	1.53	1.19	0.17	0.07		
13Al	1.50	1.23	0.17	0.07	2.50	0.84
22Ti	0.60	0.47	2.88	0.04	2.26	1.88
72Hf	0.66	0.46	2.77	0.09	3.18	0.76
26Fe	0.41	0.25	7.26	0.06	1.76	1.11
26Fe*	0.60	0.62	6.72	0.06		
58Ce α=1	0.23	0.02	1.02	2.73	1.91	5
58Ce α=0	0.59	0.20	2.68	0.50	4.17	0.66
62Sm α=1	0.26	0.05	0.37	7.32	1.35	8
62Sm α=0	0.66	0.30	2.15	4.82	3.95	2.4
66Dy	0.34	0.08	0.22	11.35	1.26	40
69Tm	0.56	0.21	0.17	13.98	1.56	0.06
76Os	0.44	0.31	7.00	0.15	3.32	0.47
82Pb	1.86	1.84	0.15	0.10	2.64	5
82Pb*	1.59	1.95	0.35	0.11		

(See notes on p. 190).

Notes to Table 2: A* identifies the results of Kmetko /16/. There seems to be a misprint for N_s of Ba. The value 0.67 is given instead of 0.81 in /16/.

For the electron gas with the density, n, equal to 1 and 2 the exact and SCM results are identical.

Unless specified $\alpha = 1$ in the SCM calculations.

q is defined as the atomic volume times the electron density at the boundary.

The experimental density-of-states at the Fermi energy, $g(\varepsilon_F)$, are deduced from the electronic specific heat γ /10/.

results for the electron gas and from the APW calculations of Kmetko /16/, who used optimized α's. One can first note that the alkali metals are indeed free electron like with Li as a notable exception. In the noble metals (Cu and Au) the p-component is relatively larger and q is quite different from 1 indicating a contribution from the filled d-shell, which is in fact influenced by the boundary. None of the alkaline earth elements are from this point of view free electron like though q is close to 2. With increasing atomic number the l-occupation numbers in the alkali series exhibit a reasonably smooth behaviour which is not present in the alkaline-earth series. In fact some physical properties such as cohesive energy and electronic specific heat γ appear to be more irregular in the alkaline-earth series than in the alkali series. In table 2 the electron density-of-states at the Fermi energy, $g(\varepsilon_F)$, are compared with the experimental ones /10/ obtained from γ without introducing any corrections for e.g. enhancement. $g(\varepsilon_F)$ varies considerably among the elements and it is gratifying that the present model with its crude bandstructure does not give larger deviations.

For the other elements included in table 2 the available data do not permit any general conclusions on systematic trends. It is, however, worth noting that Ti and Hf are quite similar and that, as expected, Al is free electron like though the exact and the SCM results differ in the case of 3 free electrons.

If one takes into account that Kmetko /16/ has used different α-values and that the choice of α in many cases influences the occupation numbers as illustrated in table 2 and figure 2, there is a fair agreement between the APW and SCM results. This suggests that some relevance may be attached to the changes in the N_l's occurring upon compression and that the results, e.g. potential and electron configuration, from a spherical cellular calculation may be used as a start for a proper bandstructure calculation.

Electron Structure of Compressed States

In figure 5 the volume dependence of the d-occupation numbers of the alkali elements are displayed. It is quite obvious that the free-electron character present at normal densities is lost for the heavier elements. In the case of Cs the d-component increases rapidly at $V/V_o < 0.5$ and simultaneously the s- and p-components decrease, N_s and N_d being equal at $V/V_o = 0.5$. This behaviour is also found in Yamashita's KKR calculation /17/ which, however, gives a slightly less pronounced effect. The Cs transition at $V/V_o \approx 0.45$ was characterized in Sternheimer's cellular calculation /8/ as a sudden 6s-5d electron rearrangement. The present and Yamashita's work indicate a continuous change in the fractional occupation numbers and a crossing of the bottoms of the 6s- and 5d-band at $V/V_o = 0.45$. In analogy with Cs one would expect similar transitions in Rb and K at roughly a 4-fold compression. For Na and Li N_d is small even at a ten-fold compression. One is then approaching the state when the L- and K-shell, respectively, are not core-states but bands which eventually will mix with the conduction band to give a free electron behaviour and as consequence an increased d-character.

It was noted above that the alkaline earth metals at normal densities are not free-electron like and behave in an irregular way. The volume dependence of the d-occupation numbers presented in figure 6 exhibits that this situation prevails. Ca and Sr are quite similar and different from Ba. In the case of Be and Mg the d-component is fairly small as in Li and Na. This is due to the fact that at low atomic numbers the d- and f-states are far from being bound ones. If one also takes into account the s-, p- and f-component (not shown in figure) the picture is not changed - Ca and Sr are similar and Ba exceptional. No transitions of the Cs-type are known in the alkaline earth metals but shock-wave experiments /3/ have shown that a kink exists in the Hugoniot curve of Ca and Sr at $V/V_o \approx 0.45$ indicating a rather abrupt decrease in the compressibility. This change appears to be associated with the large d-component which is present in Ba already at its normal density. In the case of compressed Ca a rapid increase in N_d is also found in the APW calculation by McCaffrey et al. /18/ but again the present model seems to give too large N_d's. A similar redistribution of the l-occupation numbers also occurs in Ti and Hf as illustrated in figu-

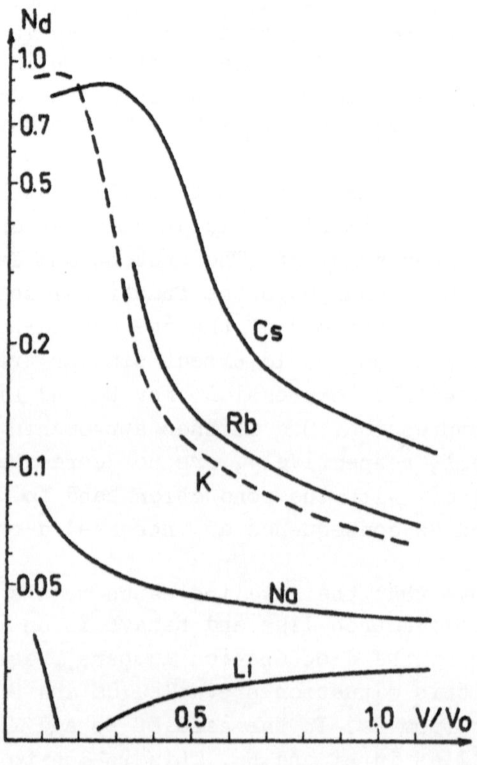

Figure 5. The volume dependence of the d-occupation numbers of the alkali metals. The experimental V_o is used.

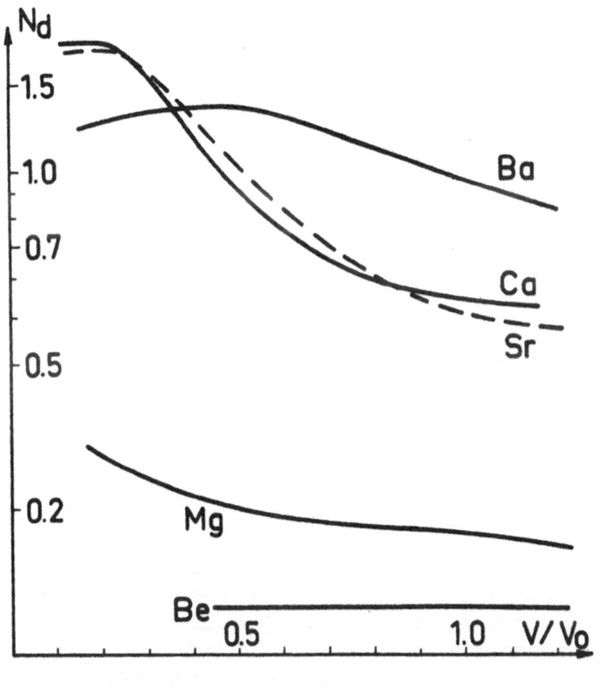

Figure 6. The volume dependence of the d-occupation numbers of the alkaline earth metals. The experimental V_o is used.

re 7. A kink in the Hf Hugoniot is in fact observed at $V/V_o = 0.75$ and by a similarity argument the SCM results suggest that it would occur at $V/V_o = 0.5$ for Ti. This change in the conduction band appears to be a mechanism for the Hugoniot kinks different from the onset of closed-shell repulsions suggested by Gust and Royce /19/.

A particularly interesting result is obtained for Al. At $V/V_o \approx 0.3$ there is a rather rapid change from a free electron distribution to a distribution with markedly increased d-component and decreased s-component (See figure 8). The mechanism is not a crossing of electron states as in Cs but a flattening of $\varepsilon_{3d}(k)$ and eventually a shift of the bottom of the 3d-band from the boundary of the spherical Brillouin zone to the center ($k = 0$) of the zone /11/. The calculated equation-of-state shown in figure 4 has an irregularity at the corresponding compression. The similar behaviour found in the experimental Hugoniot /3/ occurs, however, at a lower compression. At the transition the electronic Grüneisen γ_G, defined as $\gamma_G = \frac{\partial \log g(\varepsilon_F)}{\partial \log V}$, fluctuates, suggesting that the assumption of a constant or slowly varying γ_G may not be valid when reducing, at high compressions, the Hugoniot to isothermals. In fact the SCM results show that for many elements, where the model is reasonably reliable, the Grüneisen γ_G has a rather complicated dependence on the compression. This may have a second-order influence on the statistical equation-of-state in addition to the primary shell-structure effect discussed by Zink /20/.

The l-analysis as applied to Cs indicates that SCM mechanism at the 42-kbar transition supports the conventional 6s → 5d explanation, which of course cannot be taken literally to mean a transfer of a whole electron. On the other hand the explanation of the Ce-transition is not at all supported by the SCM results. In the compression region of interest and using $\alpha = 1$ the N_f and N_d are approximately 2.7 and 1.0 respectively and both increase slightly upon compression. With $\alpha = 0$ N_d is 2.7 and approximately constant whereas N_f is 0.5 at the normal density and increases upon compression. Even if the notion of a 4f → 5d transition is interpreted to mean a fractional change as in Cs, the SCM results give the opposite picture suggesting that mechanisms not describable with a one-electron band model may be involved in the Ce transition /22/.

CONCLUSIONS

The spherical cellular model has been used in an investigation of some 30 elements covering for each element a considerable volume range. The results presented in this note are a selection showing

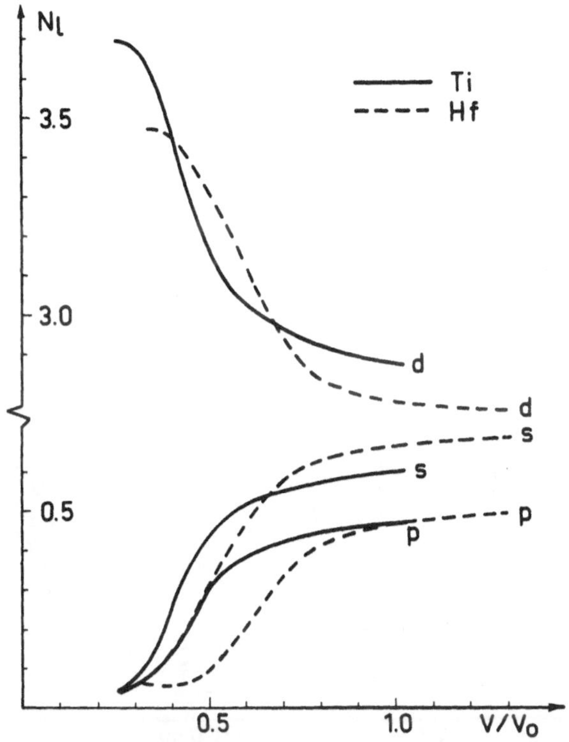

Figure 7. The volume dependence of the l-occupation numbers of Ti and Hf. The experimental V_o is used.

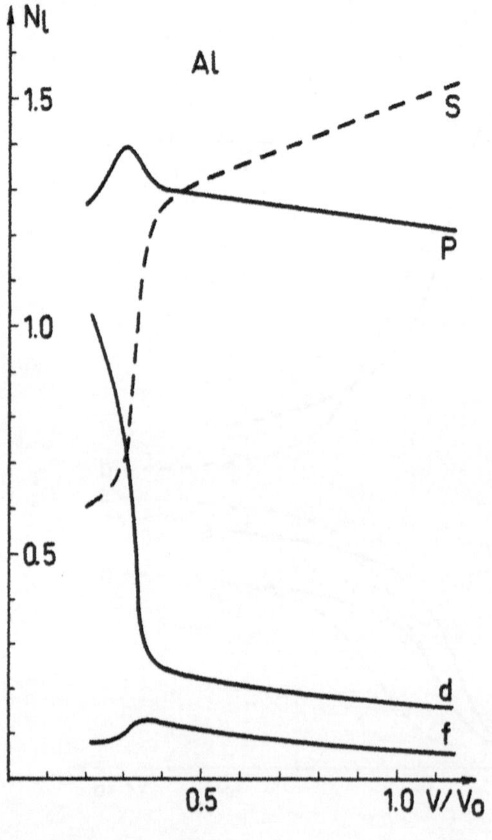

Figure 8. The volume dependence of the l-occupation numbers of Al. The experimental V_o is used.

cases where the model works well and where it fails. As expected the model accounts in a qualitative way for the characteristic variations with the atomic number not reproducible with statistical models. The spherical cellular model gives, however, some unphysical results for compressed states of elements with more than two electrons outside the core. This is mainly due to the chosen boundary conditions which certainly can be altered so that more realistic bandstructures are obtained. No conclusive evidence is found indicating that the assumption of a spherical atomic volume is basically wrong but probably one needs for some elements a more accurate potential and pressure formula and for heavy elements a relativistic treatment.

Some attention has been paid to the volume dependence of the l-occupation numbers which seems to agree reasonably well with the ones obtained from proper crystal-calculations. Characteristic variations in these fractional occupation numbers occur and there appears to be a definite possibility to correlate these variations with some physical and chemical properties of the solids. This is and extension of the concept of integral electron transfer.

One main advantage of the model is the large computational simplifications due to the spherical symmetry. In the past this has been of great importance but with the increasing capability of computers and the development of efficient methods for accurate band-calculations this will be less important in the future. For exploratory studies, however, the spherical cellular model such as the one used in this work or modified versions of it may still be useful in investigations of the high pressure properties of matter, a field which probably contains phenomena not yet discovered.

POSTSCRIPT

The work presented in this paper was initiated in 1961 while I was associated with the Uppsala Quantum Chemistry Group. It is a great pleasure to dedicate this brief summary of unpublished results to Pelle Löwdin, my teacher at Uppsala. Though the primitive methods used may not be characteristic of a Löwdin student, in retrospect it is quite clear to me that his attitude to and knowledge of numerical techniques and basic concepts in the quantum theory of matter have greatly influenced the general course of this investigation.

REFERENCES

1. P.W. Bridgman, "Physics of High Pressure", G. Bell and Sons, Ltd., London (1949).
2. H.G. Drickamer and C.W. Frank, "Electronic Transitions and the High Pressure Chemistry and Physics of Solids", Chapman and Hall, London (1973).
3. L.V. Al'tshuler and A.A. Bakanova, Soviet Physics Uspekhi 11, 678 (1969).
4. W.J. Carter, J.N. Fritz, S.P. Marsh and R.G. McQueen, J. Phys. Chem. Solids 36, 741 (1975).
5. J.F. Cannon, J. Phys. Chem. Ref. Data 3, 781 (1974).
6. A. Jayaraman, Phys. Rev. 139, A690 (1965).
7. A.W. Lawson and T.-Y. Tang, Phys. Rev. 76, 301 (1949).
8. R. Sternheimer, Phys. Rev. 78, 235 (1950). See also ref. 17 and F.W. Averill, Phys. Rev. B4, 3315 (1971), B6, 3637 (1972).
9. H.T. Hall, J.D. Barnett and L. Merrill, Science 146, 1297 (1964).
10. K.A. Gschneider, Jr., Solid State Physics (ed. F. Seitz and D. Turnbull), 16, 275 (Academic Press, N.Y., 1964).
11. K.F. Berggren and A. Fröman, Arkiv för Fysik 39, 355 (1969).
12. G.M. Gandel'man, Soviet Physics JETP 16, 94 (1963); 24, 99 (1967).
    A.I. Voropinov, G.M. Gandel'man and V.G. Podval'nyi, Soviet Physics Uspekhi 13, 56 (1970).
13. N.N. Kalitkin, Soviet Physics JETP 11, 1106 (1960).
14. See e.g. F.W. Averill (loc cit);
    M. Ross and K.W. Johnson, Phys. Rev. 2, 4709 (1970)
    J.F. Janak, V.L. Moruzzi and A.R. Williams, Phys. Rev. B12, 1257 (1975)
    W.E. Rudge, Phys. Rev. 181, 1033 (1969).
15. K.F. Berggren and A. Fröman, Report 174 (1966) from the Quantum Chemistry Group, Uppsala University, and other unpublished material.
16. E.A Kmetko, U.S. National Bureau of Standards Spec. Publ. 323 p. 67 (1970).
17. J. Yamashita and S. Asano, J. Phys. Soc. Japan 29, 264 (1970).
18. J.W. McCaffrey, D.A. Papaconstatopoulos and J.R. Anderson, Solid State Communications 8, 2109 (1970).

19. W.H. Gust and E.B. Royce, Phys. Rev. B8, 3595 (1973).
See also G. Leman, talk presented at the 5th International Conference on High Pressure Physics and Technology, Moscow, May 26-31, 1975.
R. Grover and B.J. Alder, J. Phys. Chem. Solids 35, 753 (1974).

20. J.W. Zink, Phys. Rev. 176, 279 (1968).

21. M.H. Rice, J. Phys. Chem. Solids 26, 483 (1965).

22. See e.g. B. Johansson, Phil. Mag. 30, 469 (1974).

STATIC AND DYNAMIC CORRELATIONS IN MANY-ELECTRON SYSTEMS

In honour of Professor P.O. Löwdin's 60th birthday

Stig Lundqvist

Inst. för teoretisk fysik och mekanik, Chalmers Tekniska

Högskola, Fack, S-402 20 Göteborg 5, Sweden

I. INTRODUCTION

My first contact with correlation problems dates back to my first year as a graduate student in theoretical physics at Uppsala. My interest in theoretical physics was very much stimulated by Per-Olov Löwdin's inspiring lectures and I started as a graduate student in 1949 just one year after he had published his remarkable thesis on the properties of ionic crystals. This breakthrough in the large-scale application of quantum mechanical methods to treat complicated solid state problems created for natural reasons a great excitement and had a strong influence on the research activities at the department over a long period. Professor Ivar Waller had already in the early forties been doing some pioneering work in the lattice dynamics of KCl together with M. Iona and took a major interest in looking into the possibilities to study the lattice dynamics of ionic crystals starting from the quantum mechanical theory of the cohesive properties. Related to this idea was some very early work that he and Per-Olof Fröman, now professor in Uppsala, did to develop the theory of neutron diffraction as a method to studying phonons in solids. Alf Sjölander continued this line of research and later made fundamental contributions to the field. In this connection, however, I would like to recall some very early actual calculations he did on the scattering by an ionic solid, which I believe was done quite some time before any experimental data existed.

There were many intense discussions about various extensions and new applications of the new theoretical ideas and Per-Olov Löwdin himself was of course the leader in developing new techniques and ideas to treat other problems. One of the problems which was always coming up in the discussions was the problem about electron correlations. One of the puzzling features was why the results for ionic solids agreed so well with experiments even though all effects of correlations were neglected, whereas the correlation effects are of major importance for the cohesion of metals.

I was involved during my first year in a project together with Per-Olof Fröman to introduce correlation effects in a calculation of the cohesive properties of crystalline LiH. The idea was to extend the one-electron scheme by introducing two-electron wave functions for the Li^+ and H^- ions respectively, in which correlations were explicitly introduced. Although the project seemed reasonable in the start we soon ran into so many problems that we had to give it up and I had to start with the considerably less glamorous task to make an improved calculation for LiH using the one-electron scheme developed by Löwdin.

The discussions about correlation effects in ionic crystals led to the reasonable assumption that correlation was essentially a localized atomic phenomenon which was not changed in any significant way when forming a solid. The rather large error in the total energy introduced by using one-electron theory would then essentially cancel out when one subtracts the terms in the energy formula which give the energy of the free ions or atoms. Although certainly not of universal validity the assumption seemed to work rather well for ionic solids.

Per-Olov Löwdin's own work on various aspects of the correlation problem is well known to all workers in the field of electron structure. The rigorous formulation of the theory in terms of density matrices has laid the foundation for further progress. My own lasting interest in some aspects of the correlation problem goes back to the many stimulating discussions I had with Per-Olov Löwdin and members of his group during my years in Uppsala. As a small contribution to this volume in honour of his sixtieth birthday I would like to present some remarks on correlation problems in connection with recent work going on in our group.

II. REMARKS ABOUT GROUND STATE CORRELATIONS

The concept of an exchange-correlation hole was introduced already in the pioneering work by Wigner and Seitz /1/ about cohesion in alkali metals. Using the method of density matrices, Löwdin /2/ gave a definition for arbitrary systems of the exchange-correlation hole and the corresponding contribution to the energy and this work has been of great importance for the later developments. For this discussion we need only the diagonal parts of the one- and two-particle density matrices, i.e. the density

$$\rho(\underline{x}_1) = N\int|\Psi|^2 d\underline{x}_2 \ldots\ldots\ldots d\underline{x}_N \tag{1}$$

and

$$\Gamma(\underline{x}_1,\underline{x}_2) = \frac{N(N-1)}{2} \int|\Psi|^2 d\underline{x}_3 \ldots\ldots\ldots d\underline{x}_N \tag{2}$$

where $\underline{x}$ stands for the space and spin coordinates of an electron. Subtracting out the classical part of the Coulomb interaction energy one obtains for the exchange and correlation contribution the formula

$$\begin{aligned} E_{xc}^{Coul} &= \int d\underline{x}_1 d\underline{x}_2 \frac{e^2}{R_{12}} \left(\Gamma(\underline{x}_1,\underline{x}_2) - \tfrac{1}{2}\rho(\underline{x}_1)\,\rho(\underline{x}_2)\right) \\ &= \tfrac{1}{2}\int d\underline{x}_1 d\underline{x}_2 \frac{e^2}{R_{12}} \rho(\underline{x}_1) \left(\frac{2\Gamma(\underline{x}_1,\underline{x}_2)}{\rho(\underline{x}_1)} - \rho(\underline{x}_2)\right) \\ &= \tfrac{1}{2}\int d\underline{x}_1 d\underline{x}_2 \frac{e^2}{R_{12}} \rho(\underline{x}_1)\, \rho_{xc}(\underline{x}_1,\underline{x}_2) \end{aligned} \tag{3}$$

where

$$\rho_{xc}(\underline{x}_1,\underline{x}_2) = \frac{2\Gamma(\underline{x}_1 x_2)}{\rho(\underline{x}_1)} - \rho(\underline{x}_2) \tag{4}$$

The interpretation of the last form of E_{xc}^{coul} is that the integrand represents the interaction between an electron at $\underline{x}_1$ and charge distribution $\rho(\underline{x}_1)$ with a hole charge distribution around $\underline{x}_1$ which has the value $\rho_{xc}(\underline{x}_1\underline{x}_2)$ at the point $\underline{x}_2$. This function is related to the pair distribution function in the ground state through the formula

$$\rho_{xc}(\underline{x}_1,\underline{x}_2) = \rho(\underline{x}_2)\left(g(\underline{x}_1,\underline{x}_2) - 1\right) \tag{5}$$

From the definition it follows that

$$\int\rho_{xc}(\underline{x}_1,\underline{x}_2)d\underline{x}_2 = -1 \tag{6}$$

i.e. that the hole charge displaced around the electron always corresponds to exactly one electron. For a system of non-interacting fermions ρ_{xc} describes the Fermi hole and only electrons with parallell spins repel each other. When one considers the effects of the Coulomb repulsion between the electrons, one will have a hole in the distribution of electrons having spins opposite to the spin at $\underline{x}_1$. Because of the important condition in Eq. (6) restricting the total displaced charge to be exactly one unit, the contributions to the hole from electrons with parallell spins must be reduced in the corresponding amount relative to the pure exchange hole for independent fermions.

The brief resumé given here follows essentially the treatment given by Löwdin /2/. There is a minor technical difference in that he separated the hole into a pure exchange hole (as in Hartree-Fock theory) and a correlation hole. Since the exchange hole exhausts the "sum rule" (6) the correlation hole must always have zero charge.

The formula (3) gives only the Coulomb interaction part of the total exchange and correlation energy and one will have to add the contribution from the kinetic energy and from one-body potentials. This was done by Löwdin by identifying the correlation contribution to the first order density matrix and calculating the corresponding contributions to the energy. An alternative procedure is to calculate the total exchange and correlation energy by using the method, first introduced by Pauli, to integrate over the coupling constant to obtain the total interaction energy. This gives the formula

$$E_{xc} = \tfrac{1}{2}\int d\underline{x}_1 \rho(\underline{x}_1) \int_0^{e^2} d\lambda \int d\underline{x}_2 \frac{1}{R_{12}} \rho(\underline{x}_2) \{g(\underline{x}_1\underline{x}_2,\lambda) - 1\} \qquad (7)$$

where $g\ (\underline{x}_1\underline{x}_2\lambda)$ describes the exchange-correlation hole in a system having the density $\rho(\underline{x})$ and coupling constant λ (the physical value being e^2). Thus we are back at the same form as (3) except for the additional integration over the coupling constant.

The exchange and correlation hole is obviously the key quantity to understand ground state correlations. For an electron liquid or a simple metal the hole is spherical or nearly so and describes the screening cloud around the electron. At a metal surface the hole will be highly asymmetrical and tends to be located in the surface plane when $\underline{x}_1$ recedes from the surface. The asymmetric polarization cloud at the surface will in the classical limit correspond to the image force. A highly asymmetric exchange-correlation hole will in general be characteristic of all systems with strong density gradients such as transition metals, semiconductors, solid surfaces and atoms. Unfortunately very few actual mappings of the hole have been carried out and therefore the detailed structure in various types of systems is not very well known.

In all equations of motion determining one-electron properties in a many-electron system there appears a generalized potential describing the interaction with the other electrons. This potential is at least qualitatively related to the exchange-correlation hole. The best known approximate theory is the Hartree-Fock theory where the non-local exchange potential directly derives from the exchange hole. The natural spin orbitals introduced by Löwdin satisfy a

Schrödinger equation with a non-local potential related to the exchange-correlation hole in the ground state. The density functional scheme developed by Hohenberg, Kohn and Sham /3/ uses a local potential for exchange and correlation which is defined as the functional derivative of E_{xc} with regard to the local density. The one-electron equations describing physical states where one has added or subtracted one electron describe the interaction with other electrons through the non-local and energy-dependent self-energy of an electron.

These schemes are, however, seldom used in actual applications. One usually simplifies the problem by using local potentials of a simple form and the most popular procedure is to introduce the local density approximation. The best known approximations are the following:

(a) The Dirac-Kohn-Sham exchange potential /3/ which is equal to the exchange contribution μ_x to the chemical potential in a uniform electron gas

$$v_x = \mu_x = -\frac{e^2}{\pi}(3\pi^2\rho(\underline{x}))^{1/3} \tag{8}$$

(b) The Slater exchange potential /4/ where one takes an average over the Fermi sea which results in the formula

$$v_x^S = \frac{3}{2}\mu_x \tag{9}$$

(c) The so-called Xα-method introduced by Slater, in which one introduces a parameter α and writes

$$v_{x\alpha} = \frac{3}{2}\alpha\mu_x \tag{10}$$

It should be remarked that the Xα-method is not connected with the density functional approach. There are no free parameters in the density functional approach.

The local density result for exchange and correlation can be found from the calculations and interpolation formulas for an electron liquid, described e.g. in ref. /5/. The result can be written in the form

$$v_{xc} = \beta(\rho)v_x \tag{11}$$

where $\beta(\rho)$ is a slowly varying function of the density and v_x is the exchange potential given by Eq. (8).

In the high-density limit there is only exchange so that $\beta(\rho)$ starts from the value 1 and increases monotonically towards values in the range 1.2 - 1.3 for metallic densities.

The density functional approach has been extended to spin-polarized systems and the local density approach has been studied by von Barth and Hedin /6/ and by Gunnarsson and B.I. Lundqvist /7/. They derived a local potential depending on the local spin polarization from the properties of a spin-polarized interacting electron liquid. Applications of the method to chemical binding in small molecules /8/, cohesion in metals /9/, band theory for magnetism in transition metals /10/ etc., have been very successful.

The applications of the local density approach have on the whole been extremely successful even though the conditions for the use of the approximation, e.g. slowly varying density, have not been fulfilled. In some applications one notices variations by a factor of 10 or more over the range of the exchange-correlation hole. It is indeed hard to see why the method works at all and further theoretical work to develop the theory of the non-uniform electron gas is badly needed. Expansions in terms of the gradient of the density would probably not be the solution, since the condition on the total displaced charge in the exchange-correlation hole would be violated. The density functional approach seems to be the simplest theory to consider for further theoretical developments, since the exchange-correlation potential is a local potential, whereas other exact formulations require a non-local potential. One should therefore aim at constructing a more general local potential which depends also on the density distribution over a region around the point. Such potentials have been mentioned over and over again in the literature but very little of actual work has been done. The asymmetry of the exchange-potential hole in cases with strong density gradients would seem to complicate the treatment. An important observation was recently made by Gunnarsson and B.I. Lundqvist /7/, who noted that only the spherical average over the exchange-correlation hole $\rho_{xc}(\underline{x}_1,\underline{x}_2,\lambda)$ will contribute to the exchange-correlation energy E_{xc}. One might therefore obtain a good description of many properties by considering only the spherical average of the hole when constructing a more general local potential, although the non-spherical parts no doubt will be important in some problems. In such a spherical approximation there are various possibilities to construct a more general local potential. One would choose a spherical hole normalized to the correct displaced charge and may impose any further condition which may seem reasonable. One such possible condition is that the result should be exact to leading order in the limit of small density variations around a uniform density. Work along such lines is in progress and will soon be published /11/.

III. DYNAMIC CORRELATION EFFECTS

The effects of correlation in the ground state of atomic systems are important for an accurate quantitative theory but, particularly for closed shell systems, correlations mainly give fairly modest correction terms to the results of the one-electron approximation. Correlation energies are generally rather small and the changes in quantities like diamagnetic susceptibilities, static polarizabilities, average excitation potentials etc., when correlation effects are taken into account, often show some interesting trends but are numerically fairly small. However, when one considers phenomena at high frequencies, the situation is dramatically changed. The effects of dynamical correlations are often large and may even give rise to new qualitative features and new physical phenomena which cannot be explained at all in the one-electron picture. An example of such effects is the giant dipolar resonances which seem to occur in many heavy elements. Theoretical work by Wendin /12/ and by Amusia and coworkers /13/, who have applied many-body techniques to calculate the photoabsorption cross section in the soft X-ray region, shows excellent agreement with experiments.

The conceptually simplest case of dynamical correlations occurs in the photoemission of electrons at high energies studied with the ESCA method. In the limit of high velocity of the ejected photoelectron one can consider the process as an instantaneous creation of a hole in the case of a particular atom in the system. The energy spectrum seen is that of the N-1 electron system, which we simply refer to as the "hole spectrum". This is indeed only an approximation which assumes that the electron suffers no energy losses in its path from the atom to the detector. This is a good approximation for atoms in gases but in solids the effects of ineleastic scattering of the outgoing electron are often important.

The sudden creation of a localized charged hole in the system acts as a probe to study the dynamics of the electrons. The simplest case is that of a strongly localized hole which is well isolated from neighbouring shells both with regard to energy separation and the spatial extension of the orbitals. In this case the hole acts essentially as an external Potential to which the other electrons respond.

In the basic description of the photoelectron spectrum there are the following two main features:

(a) There is a relaxation shift of the energy level of the core electron which is due to the adjustment of the outer electrons to the additional attractive potential due to the hole. This corresponds in the normal situations to the lowest state of the remaining N-1 electrons in the presence of the hole.

(b) The photoelectron may leave the atom in an excited state and such states appear in the photoelectron spectrum as shake-up or shake-off lines or continua. In all simple applications where the hole simply acts as an external potential, the conservation of angular momentum requires that the excited states will correspond to monopole transitions.

For the theoretical representation of the spectrum it is convenient to introduce the spectral weight function $A(\underline{x},\underline{x}',\varepsilon)$, ε being the energy, which is a powerful generalization of the density matrix to problems concerning one-electron excitations of the system. In the case of localized holes created by a high energy excitation, it is practical to introduce a set of suitable one-electron orbitals $u_n(\underline{x})$ and define the corresponding matrix $A_{nn'}(\varepsilon)$. In the simplest case, where we have a well separated core state we can with good approximation neglect non-diagonal terms and obtain the explicit formula

$$A_{nn}(\varepsilon) = \frac{|\mathrm{Im}\Sigma_{nn}(\varepsilon)|}{\left(\varepsilon - \varepsilon_n - \mathrm{Re}\Sigma_{nn}(\varepsilon)\right)^2 + \left(\mathrm{Im}\Sigma_{nn}(\varepsilon)\right)^2} \tag{12}$$

Here ε_n represents the unperturbed hole energy i.e. the orbital energy in the ground state and $\Sigma_{nn}(\varepsilon)$ is the self-energy which corresponds to all the interactions between the core hole and the other electrons. The resonances in the spectrum are found by solving the equation

$$\varepsilon = \varepsilon_n + \mathrm{Re}\Sigma_{nn}(\varepsilon) \tag{13}$$

and the width of the resonance is determined by the imaginary part of the self-energy.

There are some important properties of the spectral function, that have the form of sum rules and are of key importance in applications of the theory. For simplicity we supress the space and quantum labels in the next couple of pages.

(a) The spectrum is normalized to unity

$$\int\frac{d\omega}{2\pi}\, A(\omega) = 1 \tag{14}$$

(b) The energy averaged over the spectrum is equal to the energy of the system, when we suddenly remove the core electron without allowing the other electrons to relax. This corresponds to the sudden approximation in perturbation theory. The other electrons are left in what we call the corresponding energy by ε_{frozen} and the second sum rule states that

$$\int \frac{d\omega}{2\pi} \omega A(\omega) = \varepsilon_{frozen} \tag{15}$$

This rule says that the average energy over the spectrum is equal to the energy of the system in a frozen configuration after the hole is created. This sum rule was first found by B.I. Lundqvist /14/ and was later verified on general grounds by Langreth /15/. A different formulation was later given by Doniach /16/. The often quoted paper by Manne and Åberg is an approximate and specialized version of the results given in ref. /14/.

In the normal cases the lowest state of the system with a core hole is the state corresponding to complete relaxation around the hole which has the energy ε_r. The difference $\varepsilon_{frozen} - \varepsilon_r$ is the relaxation shift. The second sum rule tells us that the excited states of the N-1 electron system must exactly balance the relaxation shift. Doniach /16/ has stated the sum rule in a different form given by

$$\varepsilon_{frozen} - \varepsilon_r = \int \frac{d\omega}{2\pi} \omega B(\omega) \tag{16}$$

where $B(\omega)$ is a function called the "satellite generator". The important feature, implicit in Eq. /15/ and explicit in Eq. /16/ is that the relaxation shift can be expressed as an integral over the excitation spectrum. This is an interesting relation which shows that a static property like the relaxation shift, which is associated with the redistribution of the electrons around the hole is related to the actual spectrum seen in a photoelectron experiment.

The interpretation of the second sum rule in Eq. (15) or (16) may become more transparent if we consider just second order perturbation theory to calculate the self-energy. The dynamics of the system will then be expressed in terms of the density-density response functions $R(\omega)$. In this approximation the second sum rule can be written

$$\varepsilon_{frozen} - \varepsilon_r \sim \int \frac{d\omega}{\omega} \mathrm{Im}R(\omega) \tag{17}$$

The imaginary part of $R(\omega)$ is the spectral function for density excitations and the formula shows that the relaxation shift is related to the density excitations created when a core electron is suddenly removed.

Using the Kramer-Kronig formulas one finds that the integral in the right hand part is proportional to the real part of $R(\omega)$ at zero frequency, i.e. Re $R(0)$, which is the static density response function to an external perturbation (in this case the hole). Thus, the relaxation shift of the core level is directly related to the induced charge distribution around the hole, which occurs after the electrons have had time to adjust to the creation of the hole. On the other hand Eq. (17) shows that the relaxation shift is related to the spectrum of excitations given by $R(\omega)$. Thus the static and dynamic aspects are intimately related. For simplicity, we choose lowest order perturbation theory to illustrate the connection between the two aspects. It is evident from the structure of the theory that a close relation between the static relaxation around the hole and the dynamical aspect of the excitation spectrum in $A(\omega)$ will exist also in the general case.

Practically all approximate calculations which have been published so far refer to the situation where the hole state is well separated from other one-electron states and in this case the hole will effectively only give rise to an external potential which acts as a perturbation on the other electrons. This is what we shall refer to as the normal situation and it is then clear that this is a situation for which a mean field description is appropriate. In this case one could find the relaxed energy e.g. by comparing two self-consistent calculations, one with and one without the hole or using the approximate prescriptions such as the $X\alpha$ method or similar techniques. However, there seems to be a wide class of phenomena where strong fluctuations occur and where dynamic correlations are very strong. In these cases one has to go beyond mean-field approximations to understand the physics.

Let us approach the more general situation step by step and consider the terms to lowest order in the self-energy. This is illustrated schematically in the diagrams shown in Fig. 1.

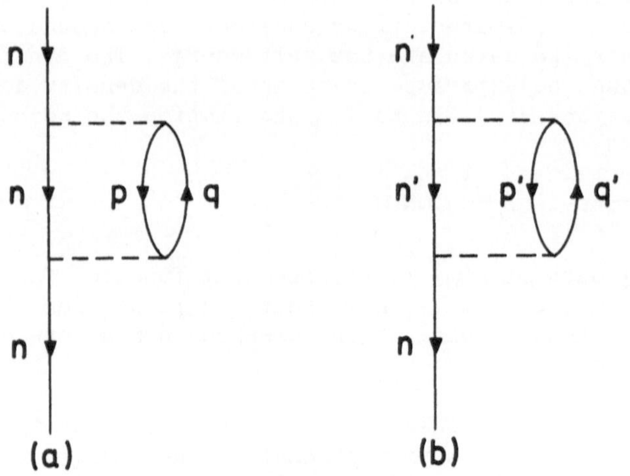

Fig. 1. Self-energy contributions corresponding to virtual excitation of an electron-hole pair.

The diagram (a) gives the normal type of contribution. The electron-hole excitations pq give rise to the relaxation shift of the level n for virtual excitations pq and for an energy such that we can have an energy-conserving transition across the central part of the diagram we have a shake-up line with two holes n and p and an excited electron q. A process where the hole line n runs through the diagrams of all orders means that the hole essentially acts as an external potential, so that diagrams corresponding to the example given in (a) could essentially be approximated by some mean field approach such as the method developed by Hedin and Johansson /17/.

Diagrams of which (b) is the simplest example, correspond to fluctuations in which the hole makes a virtual transition to n' together with the virtual excitation of an electron hole pair p'q'. Diagrams of this type will give additional contributions to the relaxation shift to those accounted for by mean field approximations and they will also give rise to additional structure in the shake-up spectrum. The new situation corresponds to the competition between the configuration where we have one hole n in the ground state configuration and that where we have two holes n' and p' and one excited electron in the state q'.

There is important difference between processes of type (a) and (b). In diagrams of type (a) there is no change in the angular momentum along the hole line and this implies that the intermediate excitation pq has no angular momentum. In processes of type (b), where the hole changes its state from n to n', one may have a change in angular momentum. The conservation rules for the Coulomb matrix element imply that the same change appears in the intermediate excitation p'q'. This means that the hole may now couple to particle-hole excitations of a different symmetry than a monopole, and we shall have e.g. strong dipole transitions in the intermediate state if the change in the transition $n \rightarrow n'$ corresponds to $\Delta l = 1$. The overall symmetry of the intermediate state must, however, correspond to the symmetry of the hole state n.

The processes of type (b) correspond to fluctuations of the hole and the symmetry is determined by the change in angular momentum in the virtual transition $n \rightarrow n'$. If there is no change in angular momentum we shall have a monopole fluctuation. If the transition corresponds to $\Delta l = 1$ we have a dipolar fluctuation around the hole n, going together with a dipolar fluctuation in the other electrons corresponding to virtual excitations p'q'. These two dipolar fluctuations may couple very strongly via the Coulomb interaction and give rise to a large additional relaxation shift. Further changes in l will result in quadrupole fluctuations and so on.

The importance of diagrams of type (b) is determined by the energies of the configuration with one hole n in the ground state and the competing configuration with two holes n' and p' and one excited electron q'. Different situation may occur:

(a) The states n'p'q' correspond to higher excitation energies than that corresponding to the state with one hole n. The virtual transitions will now give an additional relaxation shift corresponding to the fluctuations of the hole. In the satellite region one would see additional structure due to the various excitations n'p'q'. The magnitude of the dynamical shift depends on the energy separation and the strength of the matrix elements and may be as large as 10-20 eV in cases of strong coupling. In this case, the hole corresponds to the ground state of the system and is seen as a strong line in the photoelectron spectrum. Looked upon as a quasi-particle it is literally becoming somewhat shaky depending on the magnitude of the fluctuations.

(b) The states n'p'q' of interest are discrete but fall in the very region of the hole state n. The coupling is then very strong. It may even be that the state n is no longer the lowest state but that one or several of states predominantly of the type n'p'q' may be lower. This indicates that the system is no longer stable in the usual sense when we create a hole n, but a state in the configuration n'p'q' has lower energy. There is no simple quasi-particle picture left and one would have to examine each individual case in detail.

(c) The states n'p'q' of interest may fall in a continuum in the region of the core hole n. This implies that the self-energy has now a finite imaginary part in an energy interval around the unperturbed energy n. This may lead to a complete smearing of the line over an appreciable energy interval. In such a case we see no line at all corresponding to n, and there is simply no quasi-particle state of the N-1 electron system which corresponds to the creation of the hole n. The one electron state n and its orbital has then only a formal significance in the theory but does not correspond to a physical excitation. The fact that some lines were missing in the photoelectron spectra was observed already in the early ESCA studies /18/. Recent studies by Gelius /19/ and by Shirley et.al./20/ give the systematic trends for the 4p and 4s spectra from Z = 42 to Z = 73.

Recent calculations by Wendin and Ohno in our group give a complete verification of the picture just presented. An account of some preliminary results for Xe was given in Ref. /21/. These results by Wendin and Ohno which will soon be published give an almost complete theoretical description for the experimental results by Gelius and Shirley et. al. The high resolution spectra by Gelius shows a fine structure superimposed on the continuous background which is due to spectroscopic states of the competing configuration with two holes and one excited electron. It would be rather awkward to couple up to these spectroscopic states in a full many-body calculation like the one performed by Wendin and Ohno and it is more reasonable to treat the fine structure as a separate problem, assuming the position of the main line to be given. Such a calculation has recently been done by McGuire /22/ and accounts well for the observed fine structure.

The effects seen in the 4p and 4s spectra for the elements Z = 42 to Z = 73 are thus fully explained as examples of very strong dynamical correlations, involving large fluctuations of the hole and even the complete disappearance of the quasi-hole states. There is no doubt that similar strong effects will exist in transition metals and the actinides and would certainly also be of importance in some molecules.

REFERENCES

1. Wigner, E. and Seitz, F., Phys. Rev. 43 (1933) 804; Phys.Rev. 46 (1934) 509; Wigner, E., Phys. Rev. 46 (1934) 1002.
2. Löwdin, P.O., Phil. Mag. Suppl. 5, No. 17 (1956) 1.
3. Hohenberg, P. and Kohn, W., Phys. Rev. 136B (1964) 864; Kohn, W. and Sham, L.J., Phys. Rev. 140A (1965) 1133.
4. Slater, J.C., Phys. Rev. 81 (1951) 385.
5. Hedin, L., Lundqvist, B.I. and Lundqvist, S., Sol. State Comm. 9 (1971) 537; Hedin, L. and Lundqvist, B.I., J. Phys. C. 4 (1971) 2064.
6. Von Barth, U. and Hedin, L., J. Phys. C. 5 (1972) 1629.
7. Gunnarsson, O. and Lundqvist, B.I., Phys. Rev. (in press).
8. Gunnarsson, O., Johansson, P., Lundqvist, S. and Lundqvist, B.I., Int. J. Quant. Chem. (in press).
   Gunnarsson, O. and Johansson, P., Int. J. Quant. Chem. (in press).
9. Gunnarsson, O., Lundqvist, B.I. and Wilkins, J.W. Phys. Rev. B10 (1974) 1319.
10. Gunnarsson, O., J. Phys. F (in press).
11. Gunnarsson, O., Jonson M. and Lundqvist, B.I., (to be published).
12. Wendin, G., J. Phys. B5 (1972), J. Phys. B6 (1973) 42, Phys. Lett. 46A (1973) 101, Phys. Lett. 46A (1973) 119, in Vacuum Ultraviolet Radiation Physics, Pergamon and Vieweg (1974).

13. Amusia, M. Ya., in Vacuum Ultraviolet Radiation Physics, Pergamon and Vieweg (1974).

14. Lundqvist, B.I., Phys. Kondens. Materic, 9 (1969) 236.

15. Langreth, D.C., Phys. Rev. B, 1 (1970) 471.

16. Doniach, S., in Computational Methods in Band Theory, Plenum Press, New York (1971).

17. Hedin, L. and Johansson, A., J. Phys. B 2 (1969) 1336.

18. Siegbahn, K. et.al., ESCA, Atomic Molecular and Solid State Structure Studied by Means of Electron Spectroscopy, Almqvist and Wiksell, Uppsala (1967).

19. Gelius, U., Journal of Electron Spectroscopy, 5 (1974) 985.

20. Shirley, D.A., Martin, R.L., McFeely, F.R., Kowalczyk, S.P. and Ley, Feraday discussions of the Chemical Society, Vancouver, B.C. (1974).

21. Lundqvist, S. and Wendin, G., Journal of Electron Spectroscopy, 5 (1974) 513.

22. McGuire, E.J., Phys. Rev. A, 9 (1974) 1840.

POWER SERIES METHODS FOR CELLULAR CALCULATIONS ON ATOMS, MOLECULES AND SOLIDS

J.C. Slater*

Quantum Theory Project, University of Florida

Gainesville, Florida 32611, U.S.A.

1. CELLULAR FORMULATION OF THE Xα METHOD

Evidence is accumulating [1-5] that a calculation of the total energy of a molecule or crystal by the Xα method, as a function of nuclear positions, gives a surprisingly good approximation to the correct value. Unfortunately, most existing calculations of this type use either the muffin-tin or the overlapping-sphere method, neither of which is based on sufficiently sound foundations to give one complete confidence in the results. It is therefore highly desirable to look into as accurate methods as possible to make this calculation. The cellular formulation seems to be the first step in this direction.

The total energy, according to the Xα method, is

$$\begin{aligned}\langle EX\alpha\rangle = {} & \Sigma(i) n_i \int u_i^*(1)(-\nabla_1^2) u_i(1)\, dv_1 \\ & + \int \rho_e(1)\Sigma(p) \frac{2Z_p}{r_{1p}}\, dv_1 \\ & + \tfrac{1}{2}\int \rho_e(1)\rho_e(2) \frac{2}{r_{12}}\, dv_1 dv_2 \\ & - \frac{9}{2}\alpha \left(\frac{3}{4\pi}\right)^{1/3} \int\{(-\rho\uparrow(1))^{4/3} + (-\rho\downarrow(1))^{4/3}\} dv_1 \\ & + \Sigma(\text{pairs } p,q,\ p\neq q) \frac{2Z_p Z_q}{r_{pq}} \qquad (1)\end{aligned}$$

*Assisted by the National Science Foundation, through Grant GH32006A.

where

$$\rho_e(1) = \rho\uparrow(1) + \rho\downarrow(1) = -\Sigma(i)n_i u_i^*(1)u_i(1) \qquad (2)$$

In Eq. 2 we write the electronic charge density (a negative quantity) in terms of the spin-orbitals u_i, of which some correspond to spin up, some to spin down, denoted by arrows. The first term of Eq. 1 is the kinetic energy, the second the potential energy of the electronic cloud in the field of the nuclei of atomic numbers Z_p, the third the potential energy of interaction of the electronic charge cloud with itself, the fourth the $X\alpha$ exchange-correlation interaction, and the last term is the nuclear interaction energy. Atomic units are used, energies being measured in terms of rydbergs. The second and third terms, representing the coulomb interactions of electrons with themselves and with the nuclei, can be written in terms of the electrostatic potentials derived from the total charge density, electronic and nuclear, which in turn can be derived from Poisson's equation, so that the double volume integral in the third term can be reduced to a single integral over volume, of the product of electronic charge density and the electrostatic potential. Hence each term of Eq. 1 can be written in terms of a volume integral over all space.

To carry out these volume integrals, the cellular method subdivides space into a set of cells, with no more than one nucleus in a cell. As an example, we shall take the case of the diamond crystal. It is well known that this crystal can be described in terms of a body-centered cubic lattice of points. Half of these points are occupied by carbon atoms, half are empty. One can use the well-known Wigner Seitz[6] cells for the body-centered cubic structure, each cell being a truncated cube. Each of the eight corners of the cube is cut off to leave a hexagonal face. The vectors from the origin to half of these eight hexagonal faces point to four atoms, the remaining four vectors point to empty cells. The cells are of two sorts, filled cells (those with a carbon atom at the center) and empty cells, containing only electronic charge. The problem of volume integration reduces to that over one filled and one empty cell.

We must then find a convenient mathematical method to write the charge density, potential, and wave function in a cell of each type. The approximate spherical symmetry of the problem within a cell suggests using spherical coordinates within each cell, with separate expansions inside each, and with boundary conditions of continuity from one cell to another. Leite, Bennett, and Herman[7] have recently given an elementary discussion of how to use these boundary conditions in the determination of energy bands. They have not, however, faced the problem of the calculation of the total energy. It is in setting up this mathematical formulation that the problem of power series expansions comes up.

If we make an expansion in polynomials in x, y, and z, we are at once led to the theory of spherical or solid harmonics. If we consider polynomials of the nth degree, we see in Table 1 the independent polynomials of each n up to 3. An extension of this table would show that the number of such polynomials is (n+1)(n+2)/2, giving respectively 1, 3, 6, and 10 for the cases tabulated. Each of these polynomials can be written as the product of r^n and a function of the angles. Each of these angular functions in turn can be written as a linear combination of spherical harmonics. There are, however, more polynomials of order n than there are spherical harmonics of order $\ell = n$. There are $2\ell+1$ independent spherical harmonics of order ℓ, which can be written as polynomials of the ℓth order, divided by r^ℓ. For all values of ℓ greater than unity, $2\ell+1$ is less than $(\ell+1)(\ell+2)/2$. For $\ell = 1$, there are three spherical harmonics, which are constants times $(x^2-y^2)/r^2$, $(2z^2-x^2-y^2)/r^2$, xy/r^2, yz/r^2, zx/r^2, whereas Table 1 shows six independent functions. The remaining function is $(x^2+y^2+z^2)/r^2 = 1$, a spherical harmonic corresponding to $\ell = 0$. Similarly for n = 3, in addition to the seven spherical harmonics corresponding to $\ell = 3$, we have the three functions $x(x^2+y^2+z^2)/r^3 = x/r$, y/r, and z/r, corresponding to $\ell = 1$. In fact, quite generally, the (n+1)(n+2)/2 independent polynomials corresponding to a given n in Table 1 can be written in terms of r^n times the 2n+1 spherical harmonics corresponding to $\ell = n$, r^n times the 2n-1 corresponding to $\ell = n-2$, r^n times the 2n-3 corresponding to $\ell = n-4$, and so on.

We can state this result in a different form. If we examine all polynomials whose angular dependence is that of a particular spherical harmoic of order ℓ, we shall find that the radial dependence can be r^ℓ, $r^{\ell+2}$, $r^{\ell+4}$, ... In other words, a general expansion of a function in polynomials in x, y, and z, can be written as a sum over all spherical harmonics, each one multiplied by a function $r^\ell(a_o + a_2r^2 + a_4r^4 + \ldots)$, where the a's are arbitrary constants.

The reader will probably at once think of what seems a counter-example. The wave function of the 1s orbital in hydrogen is $\exp(-r) = 1 - r + r^2/2 \ldots$ The corresponding spherical harmonic, for $\ell = 0$, is a constant. Hence here we have a function whose power series expansion in r contains odd as well as even powers of r. The reason for the apparent discrepancy is that the hydrogen problem contains a singularity at the origin, arising from the nucleus. The wave function itself is continuous at the origin, but its derivatives are discontinuous there, as we see from the cusp of the wave function, which decreases linearly in r in all directions from the nucleus. Thus we do not have the continuity of the function and all its derivatives which is required to have a power series expansion about the origin. We shall examine this question further

Table 1. Independent polynomials of the nth degree, for n up to 3.

n = 0:	1
n = 1:	x, y, z
n = 2:	x^2, y^2, z^2, xy, yz, zx
n = 3:	x^3, y^3, z^3, x^2y, x^2z, y^2x, y^2z, z^2x, z2y, xyz

Table 2. Comparison of charge density as function of radius along 100 and 111 directions, in empty cell of diamond. "Correct", superposition of atomic charge densities. "Expansion", sum of eq. 3, using only terms specified in text. Radii 1, 2, and 3, correspond to values of r/μ, where r is measured in atomic units, and $\mu = (9\pi^2/128Z)^{1/3}$ = 0.4872221379 for carbon. Thus they correspond to r = 0.487..., 2(0.487...), and 3(0.487...) atomic units respectively. On this scale, the radius of the Wigner-Seitz sphere is (3.389...)(0.487...) units.

Radius	Correct 100	Expansion 100	Correct 111	Expansion 111
3	0.06719	0.06919	0.10332	0.10583
2	0.04528	0.04719	0.05581	0.05737
1	0.03532	0.03591	0.03681	0.03715
0	0.03226	0.03226	0.03226	0.03226
-1	0.03532	0.03591	0.03433	0.03467
-2	0.04528	0.04719	0.03769	0.03755
-3	0.06719	0.06919	0.03868	0.03903

in Sec. 5, where we shall see that for a wave function surrounding a nucleus, the expansion must be in terms of spherical harmonics of order ℓ, multiplied by power series $r^\ell(a_o + a_1 r + a_2 r^2 + a_3 r^3 + \ldots)$, containing odd as well as even powers of r. Such an expansion of course holds only for those functions which are finite at the origin. We shall not have to consider the other case, of functions which go infinite at the origin.

An example of a case where we have only even powers of r is found in the empty cells of the diamond structure. We shall take up this case in the next section, considering the charge density, and the coulomb potential which can be found from it by solving Poisson's equation. We shall find that when we have found the power series expansion of the charge density, we can at once find the potential in a similar series expansion, without further calculation.

2. POISSON'S EQUATION AND THE COULOMB POTENTIAL

Suppose we have a function of the form we have just been discussing,

$$f(r,\Theta,\phi) = \Sigma(\ell n) a_{n\ell} r^n F_\ell(\Theta,\phi) \qquad (3)$$

where $F_\ell(\Theta,\phi)$ is a spherical harmonic of order ℓ, or a linear combination of such spherical harmonics with a single ℓ value, and where we allow all values of n, equal to and greater than ℓ. Let us then find the Laplacian of this function. For any function $R(r)F_\ell(\theta,\phi)$ we have

$$\nabla^2 R(r)F_\ell(\theta,\phi) = \left(\frac{1}{r^2} \frac{d}{dr} \left(r^2 \frac{dR}{dr}\right) - \frac{\ell(\ell+1)}{r^2} R\right) F_\ell(\theta,\phi) \qquad (4)$$

where we have used the differential equation satisfied by the spherical harmonics. If we substitute the function of Eq. 3 into Eq. 4, we have

$$\nabla^2 f(r,\Theta,\phi) = \Sigma(\ell n) \left(n(n+1)-\ell(\ell+1)\right) a_{n\ell} r^{n-2} F_\ell(\Theta,\phi) \qquad (5)$$

Thus we find that though $r^\ell F_\ell(\Theta,\phi)$ satisfies Laplace's equation, since this corresponds to the term $n = \ell$ in Eq. 5, any other power of r multiplied by F_ℓ will not satisfy Laplace's equation.

From Eq. 5 we can immediately write down the solution of Poisson's equation

$$\nabla^2 V = -8\pi\rho \qquad (6)$$

where ρ is the charge density (in units of number of positive charges numerically equal to the electronic charge, per atomic unit of length cubed,) and V is the coulomb potential arising from this charge, in rydbergs. Let the function $f(r,\Omega,\phi)$ be the charge density which then will be given by the series of Eq. 3. Then we can derive the relation

$$V = \Sigma(\ell n) \frac{-8\pi}{n(n+1)-\ell(\ell+1)} a_{n-2,\ell} r^n F_\ell(\Omega,\phi) \tag{7}$$

We note that the denominator goes to zero for the case $n = \ell$. Thus for V to be finite, the numerator $a_{\ell-2,\ell}$ must be zero, which we have already assumed, but the ratio is indeterminate, and the coefficient is arbitrary. In other words, it is these terms with $n = \ell$ which give the general solution of Laplace's equation which must be added to a particular solution of Poisson's equation to get the general value of the potential. The terms of Eq. 7 for $n>\ell$ are determined from the charge density, which is assumed known.

As a test of these relations, we have made calculations for the diamond crystal, using a charge density which is the sum of atomic charges determined by the Xα method. We start with an Xα calculation for the carbon atom in its valence state, $1s^2 2s 2p^3$, using $\alpha = 0.75$, which Profs. J.W.D. Connolly and J.R. Sabin were good enough to carry out using the modified form of the Herman-Skillman programs in routine use at this laboratory. The author is greatly indebted to Profs. Connolly and Sabin for making this calculation. Its results have already been used by the writer[2] in a comparison of the Xα and Thomas-Fermi-Dirac methods of calculation.

From the results of the computer calculation, we know the charge density of a carbon atom at any distance from the nucleus. These densities were superposed to give a density in the empty cell, using approximately 30 neighboring atoms. It should be noticed that the charge density falls off much more rapidly than the potential with the distance, so that it does not require nearly as many neighbors to get an accurate superposed density as a superposed potential. (Leite, Bennett, and Herman in Ref. 7 used six shells of neighboring atoms to get an accurate potential by superposition). The charge density was calculated for a number of radii along the 100 and 111 directions in the crystal, and for 25 directions in the circle of intersection of the plane $x = y$ and the sphere whose volume is the same as that of the cell (the so-called Wigner-Seitz sphere). The resulting densities were then fitted to a function like that of Eq. 3, using a minimum number of spherical harmonics and powers of r to get fairly adequate agreement. This part of the calculation, like everything except for the original Herman-Skillman calculation, was made with a pocket calculator (HP 35), and to avoid unnecessary labor, the fitting was not carried to perfect agreement.

The symmetry of the charge density inside either an empty or a filled cell in diamond has T_d symmetry. Hence instead of the spherical harmonics, we use solid or cubic harmonics, linear combinations of spherical harmonics of a given ℓ having symmetry properties of basis functions for irreducible representations of the point group of the crystal. The charge density and potential will have the symmetry of the totally symmetric representation of this group. It was found that in addition to the spherical component, corresponding to $\ell = 0$, and adequate representation of the charge density was obtained by using only two additional solid harmonics: for $\ell = 3$, the function xyz/r^3, and for $\ell = 6$ the function $\left(x^2y^2z^2 + (x^4+y^4+z^4-3/5r^4)r^2/22 - r^6/105\right)/r^6$. It was satisfactory to use only the constant, r^2, and r^4 terms in the radial function for $\ell = 0$, and only the terms $n = \ell$ in the other two functions. A comparison of the charge density found from Eq. 3 and that found by direct summation of the atomic potentials is given in Tables 2 and 3. It will be seen that the comparison is quite satisfactory. It should be pointed out, however, that an attempt to represent the function using only terms in $\ell = 3$ and $\ell = 4$ was very unsatisfactory. The term $\ell = 4$ made only a small and inappropriate contribution, whereas that in $\ell = 6$ was very important.

It is worth while to understand what is going on along the 100 and 111 directions. The 100 direction does not point to a nearest neighboring cell, but rather to the second nearest, and symmetry demands that the charge density be an even function of r along this direction. It is the 111 which points to the nearest neighbor, and the -1-1-1 points to the nearest empty cell. The large charge density found for radius 3, in the 111 direction, shows the piling up of charge between the atoms, in the direction of the bond. Clearly as the distance from the neighboring atom increases, as we go from radius 3 to radius 0 and continue to radius -2, we are seeing the decrease of the effect of the overlap charge from the neighboring atom to a low value. In Table 3, the 100 direction corresponds to $\cos\Theta = 1$ or -1, the 111 approximately to $\cos\Theta = 7/12$, the -1-1-1 approximately to $-7/12$, and the 110 direction to $\cos\Theta = 0$. The charge density in Table 3 rises to a maximum at 7/12, and falls to a minimum at -7/12, illustrating the same points brought out in Table 2. Since Table 3 corresponds to a value of r of 3.389, it is natural that the density (0.13749) along the 111 direction at this distance is even greater than the value (0.10583) found at radius 3 in Table 2.

The discussion we have given is for an empty cell. In a filled cell, we have not only the contributions from the tails of distant atoms, as in the empty cell, but also the charge arising from the atom inside the cell. Similar calculations have been made for this case. Here it was found that the $\ell = 4$ term, $(x^4+y^4+z^4-3r^4/5)/r^4$, which was not needed for the empty cell, is rather important, but

Table 3. Comparison of charge density as function of cos Ω, at radius of Wigner-Seitz sphere, in direction x = y, or ∅ = 45°, in empty cell of diamond. "Correct" and "Expansion" have same significance as in Table 2.

cos Ω	Correct	Expansion
1	0.08174	0.08226
11/12	0.09625	0.09354
10/12	0.10348	0.10928
9/12	0.12049	0.12387
8/12	0.13171	0.13379
7/12	0.13575	0.13749
6/12	0.13304	0.13486
5/12	0.12157	0.12677
4/12	0.11364	0.11471
3/12	0.10131	0.10039
2/12	0.08902	0.08555
1/12	0.07789	0.07167
0	0.06836	0.06993
-1/12	0.05782	0.05099
-2/12	0.05355	0.04503
-3/12	0.04862	0.04179
-4/12	0.04440	0.04063
-5/12	0.04142	0.04069
-6/12	0.03957	0.04110
-7/12	0.03889	0.04127
-8/12	0.03987	0.04119
-9/12	0.04317	0.04183
-10/12	0.04628	0.04562
-11/12	0.06075	0.05694
-1	0.08174	0.08266

otherwise the expansion was similar to that for the empty cell. When the values of charge density around the boundaries of the cells were computed from the power series expansion, it was found that there was a satisfactory continuity from one cell to the next, the errors being comparable to those found in Table 2 and 3. If one goes along the 111 direction through the crystal, one finds two filled cells, two empty cells, and so on. The charge density as computed from our expansions gives a very good description of the behavior not only in the bonding region, between the two filled cells, but also in the region of the two empty cells.

Chaney, Lin, and Lafon[8] in their discussion of the tight binding method applied to the energy bands of diamond, have made a calculation of this charge density throughout the cell, computed as we have done it by superposing atomic charges. In Fig. 5 of their paper they plot this charge density along the 111 direction, and their curve is practically identical with the one we have found. They find that this gives only a rough approximation to the charge in the crystal found from their tight-binding calculation. The latter leads to a considerably larger charge density in the bond region than is found from superposition of the spherical atomic charges, on account of the creation of bonding orbitals. But the qualitative nature will not be different, and as a result of the sort of analysis used in setting up Tables 2 and 3, it is believed that the power series expansion would be capable of describing properly the exact charge density in the empty cells, and the contribution of the neighboring atoms to the charge density in the filled cells. We put off the discussion of the charge density in the filled cell arising from the atom in that cell until Sec. 4.

Now that we have discussed the expansion of the charge density in both types of cells, we can use Eq. 7 to set up the potential arising from all atoms except that in a filled cell. The latter potential is given by the computer analysis of the isolated atom. As we have pointed out earlier, the terms of Eq. 7 must be supplemented in each cell by terms satisfying Laplace's equation; that is a constant for $\ell = 0$, terms in xyz for $\ell = 3$, in $x^4+y^4+z^4-3r^4/5$ for $\ell = 4$, and in $x^2y^2z^2 + (x^4+y^4+z^4-3r^4/5)r^2/22 - r^6/105$ for $\ell = 6$. Such terms, in both cells, must be chosen to satisfy the condition of continuity between adjacent cells. They are, in fact, the leading terms in the non-spherical part of the potential, and the constant values for $\ell = 0$ are what bring in Madelung potentials. It was found easy to choose these constants to get continuity to a satisfactory approximation. When this was done, the result was a potential as a function of position which behaved as is to be expected. No effort was made to set up a potential by adding atomic potentials, for comparison, since as has been stated earlier one must use a very large number of neighboring atoms to get satisfactory results.

In the discussion which we have given, it is assumed that a power series expansion of the charge density is given. However, in actual cases this charge density will be found by adding the separate charge densities of the occupied orbitals of the system, and it is not likely that these orbitals will always be expressed in power series. Furthermore, to square a power series and express the product as another power series is not always convenient. How, we may ask, do we get a power series representation of a function such as a charge density, which is given only as a table of values, such as we get out of the computer program used in the present case? Even more difficult, how would we get a power series representation of the potential in which an individual electron moves, which includes not only a coulomb potential but also the exchange-correlation term in the one-third power of the charge density? To answer such questions, we have found a polynomial expansion in a very high-order polynomial to be very convenient. We take up this problem in the next section, preparatory to discussing the charge density and potential in a filled cell.

3. EXPANSION IN A 10TH DEGREE POLYNOMIAL

Most methods of numerical computation are based on an expansion of a function in terms of polynomials. Thus Simpson's rule for integration, and the simple methods of integration of differential equations, are based on fitting a parabola to three successive entries in a table, or a fourth order polynomial to five successive entries. Such a function is not good over any large range of the variable. But we may well ask, how good a fit would we get if we used, for instance, a 10th degree polynomial, with coefficients chosen so that the polynomial is an exact fit at eleven equally spaced points? The answer proves to be that we can get a fit of a very usable accuracy, in the cases we are interested in. Let us then ask how this is to be carried out.

Let us assume a polynomial expansion,

$$f(u) = \sum_{n=0}^{n=10} a_n u^n \tag{8}$$

which reduces to a predetermined set of values $f(5)$, $f(4)$, ... $f(-4)$,$f(-5)$ for $u = 5,4,\ldots -5$. How do we determine the coefficients $a_o \ldots\ldots a_{10}$? It is a straightforward, though tedious, matter to compute the values of the coefficients as functions of the f's. The writer has carried through this derivation, and the results are the following.

One first sets up quantities analogous to first and second

differences, by the definitions

$$d^1(u) = f(u) - f(-u)$$

$$d^2(u) = f(u) - 2f(0) + f(-u) \qquad (9)$$

which are to be computed for u = 1, 2, 3, 4, and 5. Then the coefficients are given as follows:

$$a_o = f(0)$$

$$a_1 = (10!)^{-1} \big(3024000\ d^1(1) - 864000\ d^1(2) + 216000\ d^1(3) - 36000\ d^1(4) + 2880\ d^1(5)\big)$$

$$a_2 = (10!)^{-1} \big(3024000\ d^2(1) - 432000\ d^2(2) + 72000\ d^2(3) - 9000\ d^2(4) + 576\ d^2(5)\big)$$

$$a_3 = (10!)^{-1} \big(-1401960\ d^1(1) + 1048560\ d^1(2) - 292140\ d^1(3) + 50440\ d^1(4) - 4100\ d^1(5)\big)$$

$$a_4 = (10!)^{-1} \big(-1401960\ d^2(1) + 524280\ d^2(2) - 97380\ d^2(3) + 12610\ d^2(4) - 820\ d^2(5)\big)$$

$$a_5 = (10!)^{-1} \big(203490\ d^1(1) - 196560\ d^1(2) + 82215\ d^1(3) - 15960\ d^1(4) + 1365\ d^1(5)\big)$$

$$a_6 = (10!)^{-1} \big(203490\ d^2(1) - 98280\ d^2(2) + 27405\ d^2(3) - 3990\ d^2(4) + 273\ d^2(5)\big)$$

$$a_7 = (10!)^{-1} \big(-11340\ d^1(1) + 12240\ d^1(2) - 6210\ d^1(3) + 1560\ d^1(4) - 150\ d^1(5)\big)$$

$$a_8 = (10!)^{-1} \big(-11340\ d^2(1) + 6120\ d^2(2) - 2070\ d^2(3) + 390\ d^2(4) - 30\ d^2(5)\big)$$

$$a_9 = (10!)^{-1} \big(210\ d^1(1) - 240\ d^1(2) + 135\ d^1(3) - 40\ d^1(4) + 5\ d^1(5)\big)$$

$$a_{10} = (10!)^{-1} \big(210\ d^2(1) - 120\ d^2(2) + 45\ d^2(3) - 10\ d^2(4) + d^2(5)\big) \qquad (10)$$

In computing these a's, there often is extensive cancellation between positive and negative contributions. As a practical matter, to minimize roundoff errors, it is important to compute first the sum of the five terms in d(1) ... d(5), before dividing by the common denominator 10! = 3628800.

When one wishes to have the eleven entries spaced at intervals h about a central entry at x_o, one uses the same series given in Eq. 8, replacing u by $(x-x_o)/h$. Then one has a power series in $x-x_o$, the coefficients of this power series being a_n/h^n. Sometimes one wishes to find the expansion, not about the central entry of the eleven, but about some other point u_1. Then one can prove that the coefficients can be found by the following simple rule. One finds the value of the polynomial of Eq. 8, and of its derivatives up to the 10th derivative (since it is a tenth order polynomial there will be no more non-vanishing derivatives) at the point u_1. From these values we simply construct a Taylor series of the function expanded in terms of $u-u_1$. The results are identical with what would be found by replacing u by $u_1 + (u-u_1)$ in the original polynomial. The result proves to be

$$
\begin{aligned}
f(u) = {} & (a_o+a_1u_1+a_2u_1{}^2+a_3u_1{}^3+a_4u_1{}^4+a_5u_1{}^5+a_6u_1{}^6+a_7u_1{}^7+a_8u_1{}^8 \\
& + a_9u_1{}^9+a_{10}u_1{}^{10}) \\
& +(u-u_1)(a_1+2a_2u_1+3a_3u_1{}^2+4a_4u_1{}^3+5a_5u_1{}^4+6a_6u_1{}^5+7a_7u_1{}^6 \\
& + 8a_8u_1{}^7+9a_9u_1{}^8+10a_{10}u_1{}^9) \\
& + (u-u_1)^2(a_2+3a_3u_1+6a_4u_1{}^2+10a_5u_1{}^3+15a_6u_1{}^4+21a_7u_1{}^5 \\
& + 28a_8u_1{}^6+36a_9u_1{}^7+45a_{10}u_1{}^8) \\
& +(u-u_1)^3(a_3+4a_4u_1+10a_5u_1{}^2+20a_6u_1{}^3+35a_7u_1{}^4+56a_8u_1{}^5 \\
& + 84a_9u_1{}^6+120a_{10}u_1{}^7) \\
& +(u-u_1)^4(a_4+5a_5u_1+15a_6u_1{}^2+35a_7u_1{}^3+70a_8u_1{}^4+126a_9u_1^5 \\
& + 210a_{10}u_1{}^6) \\
& + (u-u_1)^5(a_5+6a_6u_1+21a_7u_1{}^2+56a_8u_1{}^3+126a_9u_1{}^4+252a_{10}u_1{}^5) \\
& +(u-u_1)^6(a_6+7a_7u_1+28a_8u_1{}^2+84a_9u_1{}^3+210a_{10}u_1{}^4) \\
& + (u-u_1)^7(a_7+8a_8u_1+36a_9u_1{}^2+120a_{10}u_1{}^3)
\end{aligned}
$$

$$+(u-u_1)^8(a_8+9a_9u_1+45a_{10}u_1{}^2)$$
$$+(u-u_1)^9(a_9+10a_{10}u_1)$$
$$+(u-u_1)^{10}a_{10} \qquad (11)$$

4. POTENTIAL IN THE ATOMIC PROBLEM

In this section we shall take up the problem of describing in polynomial form the potential in which an electron moves, in an isolated atom. This gives the essential information needed for discussing the filled cell in such a problem as diamond. This potential is the sum of the coulomb potential, that arising from all electrons as well as the nucleus, and the $\rho^{1/3}$ term arising from the exchange-correlation term in the energy of Eq. 1. The sum of these two terms is printed out as part of the computer output in the calculation of the isolated carbon atom. Let us then describe the methods which can be used to evaluate this complete potential in polynomial form. For the purpose of notation, let us write it as $V_N + V_e + V_X$, where the first is the potential arising from the nucleus, V_e that arising from all electrons of the atom, and V_X the exchange-correlation term. The term V_N is positive, so that the energy of an electron in this potential is $-V_N-V_e-V_X$.

For practical purposes it is important to know that the Herman-Skillman calculations[9] are made in terms of the variable x, defined in Table 2, and that calculations are made at lattice points of the so-called Herman-Skillman mesh. These points have an interval in x of 0.0025 from 0 to 0.1, of 0.005 from 0.1 to 0.3, of 0.01 from 0.3 to 0.7, of 0.02 from 0.7 to 1.5, of 0.04 from 1.5 to 3.1, of 0.08 from 3.1 to 6.3, of 0.16 from 6.3 to 12.7, and so on. The whole mesh contains 441 points, but the intervals enumerated include all which we shall be concerned with. Since calculated values are available only at these points, it is more convenient to use these points for our tables, rather than having to interpolate.

We shall now consider practical ways of using the theorems of Sec. 3 to get a polynomial expansion of this potential. We regard this as a test of the practicability of the method, as well as a first step which could be made in carrying out a self-consistent expansion for the crystal. The writer approached the problem from the point of view of getting as accurate an expansion as possible. One could make a separate expansion in each of the ranges of x involved in the Herman-Skillman mesh. Thus, one could expand by fitting the function at the eleven points 0, 0.01, ... 0.10, regarding these as u = -5, -4, ... 5 of Sec. 3. One could make another expansion in the range 0.1 to 0.3, and so on. The tenth degree polynomial method was set up with this problem in mind. When this is done, the fit is practically perfect at each point of the mesh, but the coefficients increase in going from a_o to a_{10}, rather than decreasing as we would hope to find. The polynomials provide a very perfect interpolation scheme, but they are rather valueless for any other purposes.

Instead, it is found that one gets the best convergence, the most rapid decrease of the a_n's with n, if the spacing between points is chosen as large as possible. Consequently we have found that the most useful set of points for fitting with a polynomial are spaced with intervals of 0.64 in x. One has to look at the details of the mesh to find how to pick out ten intervals of 0.64, for which the mesh points all represent computed values, and the most useful set of points proves to be those corresponding to x = 0.06, 0.70, 1.34, 1.98, 2.62, 3.26, 3.90, 4.54, 5.18, 6.46. We take the first and last points to correspond to u = -5 and 5 respectively, with 3.26 corresponding to the center of the interval. We take the computed entries and fit the polynomial to these points. Before doing this, however, we must note the fact that the nuclear potential V_N goes infinite at the origin. This means that the potential at x = 0.06 is very large, the function decreasing rapidly as x increases. As a first step, we found an expansion, not of the whole potential, but only of the non-nuclear part, $V_e + V_X$, which goes smoothly through the origin. Having found this expansion, we then proceeded as follows.

First, an expansion was set up in terms of powers of x. This was done by expanding V_e+V_X in powers of x-3.26, as just described, and then transforming this to a series in powers of x, by use of Eq. 11. Then the nuclear term, proportional to 1/x, was added, as a term in x^{-1}. This nuclear term then falls into the power series, and as we shall see in the next section, it leads to a very convenient method of solving Schrödinger's equation. However, the resulting expansion proves not to be very rapidly convergent; it works well only out to about x = 1.

As a second expansion, the original set of points was modified by omitting the first one, 0.06, and adding one at 7.10, so that the center of the interval came at 3.90. In this case, the nuclear potential V_N was included as part of the potential, since it was to be expanded in powers of x-3.90. In Table 4 we give computed values of both polynomials, at a variety of lattice points for which calculations are available, and for comparison the values from the Herman-Skillman program. We see from this table that the expansion in powers of x agrees quite well with the correct value at distances up to x = 6.30, while that in x-3.90 is almost as satisfactory for x greater than 0.70. It cannot be expected to work for smaller x, since it is not fitted to smaller x's for which the nuclear potential gets very large.

The values of Table 4 illustrate several properties of the polynomial expansions. First, at the points where the exact values are matched, indicated by asterisks in the table, we see that the 10-place accuracy of the HP35 calculator is enough to reproduce these numbers to an accuracy of a few units in the 8th place, roundoff errors coming in the 9th and 10th places. (The calculations are made using the full 10 places of the calculator, but in Table 4 we have given only the number of significant figures given in the computer "Correct" printout.) Second, outside the range of the 11 points which are fitted exactly, the expansion very rapidly loses accuracy. This is shown by the entries for x = 6.94, 7.58, and 8.22, which we have included to illustrate this point. Third, we can see the type of discrepancy which is observed between the points at which the polynomial is fitted. There tends to be a sort of sinusoidal oscilation, the computed functions being too large in one interval, too small in the next. Thus, the series in x is numerically too large between x = 0.06 and 0.70, too small between 0.70 and 1.34, and so on. Fourth, these deviations will not necessarily be the same for different polynomial expansions of the same function. They are in fact of opposite sign for the two expansions illustrated in the table.

Even though these points indicate that the polynomial expansions are of limited accuracy, still one can hardly fail to be impressed by the general success of the polynomials in reproducing the correct values. The errors are in almost every case less than one part in a thousand, quite remarkable when we remember that what we are dealing with is really an elevenpoint interpolation formula holding over a range of x which is about twice the radius of the Wigner-Seitz sphere. In other words, the potential can be represented with the accuracy of this table not only throughout the cell in which the nucleus is located, but well into the neighboring cells as well.

In spite of this accuracy, however, we shall see in the next section that the errors can really be of significance. The reason is that, though they are a very small percentage of the total potential, still their absolute value, in the region of small x, can be quite considerable. Thus the maximum radial density of the 1s orbital comes about at x = 0.34. We see that in this neighborhood, the series in x is some 0.6 ry too large. This, as we shall see in the next section, results in a modified eigenvalue for the 1s which is too low, as compared with the exact 1s, by the order of a half rydberg.

Table 4. Potentials $V_N+V_e+V_X$ for carbon atom as functions of x, for carbon atom. "Correct" refers to output of Herman-Skillman program, "Series in x" to the series expansion around the nucleus, and "Series in 3.90" to the expansion about 3.90. Energies are in rydbergs. Asterisks indicate where series are fitted to correct values.

x	Correct	Series in x	Series in x-3.90
0.01	2444.6397	2443.7301	
0.02	1213.0481	1212.3214	
0.04	597.03631	596.71371	
0.06	391.60004	391.60003*	
0.08	288.85877	289.11226	
0.10	227.22789	227.67651	
0.12	186.17327	186.76799	
0.14	156.89038	157.59019	
0.16	134.97389	135.74472	
0.18	117.97418	118.78795	
0.20	104.41985	105.25358	
0.22	93.373332	94.208453	
0.24	84.208703	85.030398	
0.26	76.491971	77.288621	
0.28	69.912664	70.650779	
0.30	64.242757	64.964898	
0.34	54.986204	55.614808	
0.38	47.770618	48.298567	
0.42	42.005772	42.434051	
0.46	37.306476	37.641347	
0.50	33.410988	33.661908	
0.54	30.135536	30.313643	
0.58	27.347555	27.464577	
0.62	24.949333	25.116717	
0.66	22.866936	22.895767	
0.70	21.044381	21.044380*	21.044381*

0.78	18.012843	17.979665	18.164007
0.86	15.603234	15.560678	15.782515
0.94	13.652921	13.614875	13.809187
1.02	12.051820	12.024199	12.168802
1.10	10.722073	10.705377	10.799303
1.18	9.6063164	9.5981993	9.64976637
1.26	8.6609198	8.6582257	8.6786081
1.34	7.8520900	7.8520897*	7.8520900*
1.42	7.1535036	7.1543703	7.1429583
1.50	6.5446725	6.5454544	6.5293282
1.66	5.5360810	5.5360611	5.5222055
1.82	4.7362502	4.7359946	4.7297482
1.98	4.0887647	4.0887646*	4.0887647*
2.14	3.5565517	3.5568296	3.5597008
2.30	3.1139874	3.1143351	3.1172344
2.46	2.7425724	2.7428406	2.7442968
2.62	2.4284842	2.4284841*	2.4284842*
2.78	2.1611231	2.1609726	2.1600634
2.94	1.9321991	1.9320146	1.9309753
3.10	1.7351318	1.7350180	1.7344093
3.42	1.4164303	1.4165338*	1.4169811*
3.74	1.1735145	1.1737544	1.1739761
4.06	0.98515797	0.98508633*	0.98473163*
4.38	0.83674469	0.83667558	0.83631108
4.70	0.71803227	0.71822457*	0.71851832*
5.02	0.62173315	0.62199826	0.62233527
5.34	0.54258987	0.54246730*	0.54176161*
5.66	0.47675977	0.4764838	0.47548507
5.98	0.42139886	0.4228435*	0.42361914*
6.30	0.37437647	0.3779866	0.37883187
6.94	0.29926170	0.1954056*	0.24677116*
7.58	0.24244270	-1.054662	1.48449913*
8.22	0.19841688	-7.384645	18.1396821

5. SCHRÖDINGER'S EQUATION FOR THE SPHERICAL PROBLEM

We have seen in the preceding section how the potential in which an electron moves can be expressed to a quite good accuracy by a tenth order polynomial in r. Let us next ask how Schrödinger's equation can be solved to get the orbitals within a cell. We continue to consider the spherically symmetrical problem which was treated in Sec. 4. There are two distinct cases, depending on whether we are expanding around $r = 0$, or around a different value $r = r_o$, since in the first case the nuclear potential energy is included in the expansion of the potential as a term in r^{-1}, while in the second case it must be approximated by terms in powers of $r-r_o$. We take up the case of expansion around the nucleus first.

If u is the wave function, Schrödinger's equation is

$$\nabla^2 u = -(\varepsilon + V_N + V_e + V_X)u \tag{12}$$

We assume

$$u = \Sigma(n) u_n r^n F_\ell(\theta,\phi)$$
$$-(\varepsilon + V_N + V_e + V_X) = \Sigma(m) v_m r^m \tag{13}$$

where for a spherical problem only one value of ℓ will appear. By Eq. 5, we then have

$$\nabla^2 u = \Sigma(n)\big(n(n+1)-\ell(\ell+1)\big)u_n r^{n-2} F_\ell(\theta,\phi) \tag{14}$$

This must equal the product of the two summations given in Eq. 13. We equate the coefficients of r^{n-2} on both sides of the equation, and find

$$\big(n(n+1)-\ell(\ell+1)\big)u_n = \Sigma(m) v_m u_{n-m-2} \tag{15}$$

In Eq. 15 we have a recursion formula for the u's, which can be evaluated once we know the v's.

Let us write out this recursion formula more explicitly. For the polynomial expansion of the potential, we know that v_m is non-vanishing only for $m = -1$ (the nuclear coulomb term), and for $m = 0, 1, \ldots 10$. Thus we have

$$u_n = \big(n(n+1)-\ell(\ell+1)\big)^{-1} (v_{-1}u_{n-1} + v_o u_{n-2} + \ldots. + v_{10}u_{n-12}) \tag{16}$$

To get the solution which is regular at the origin, we must have a series starting with r^ℓ, so that we assume that the u_n's are zero for $n < \ell$. Then we can rewrite Eq. 16 in the explicit form

$$u_{\ell+1} = \left[1(2\ell+2)\right]^{-1} v_{-1} u_\ell$$

$$u_{\ell+2} = \left[2(2\ell+3)\right]^{-1} (v_{-1} u_{\ell+1} + v_0 u_\ell)$$

$$u_{\ell+3} = \left[3(2\ell+4)\right]^{-1} (v_{-1} u_{\ell+2} + V_0 u_{\ell+1} + v_1 u_\ell)$$

$$u_{\ell+4} = \left[4(2\ell+5)\right]^{-1} (v_{-1} u_{\ell+3} + v_0 u_{\ell+2} + v_1 u_{\ell+1} + v_2 u_\ell)$$

$$\ldots\ldots\ldots \qquad (17)$$

These equations, which are very simple to compute, give the coefficients of the expansion of the wave function, in terms of a single arbitrary constant u_ℓ.

One must realize of course that there is nothing new about these equations. The standard method of solving Schrödinger's equation by numerical integration starts with a power series expansion for very small values of r, to get the initial entries in the tables. The determination of the v's is made by taking the first three entries in the table of potentials, fitting a polynomial to them, as we have set up our tenth order polynomial, and using the first several equations from Eq. 17 to give a power series holding at small r. The only thing new in the present treatment is the idea of applying the same method to a solution of the wave equation over a much greater range of variables.

We can immediately verify from Eq. 17 the correctness of our statement, made in Sec. 1, that it is the presence of the nuclear charge which leads to an expansion of the radial function in the form r^ℓ $(a_0+a_1r+a_2r^2+a_3r^3 + \ldots)$, containing coefficients a_n with odd n as well as those with even n. We recall that the term v_{-1} comes from the nuclear charge, and is zero if there is no nucleus. If further the potential is an even function of r, regular at the origin, it can be expanded in even powers of r, so that v_1, v_3,... will be zero. Then we can verify at once from Eq. 17 that in the absence of a nucleus, the only nonvanishing u's are u_ℓ, $u_{\ell+2}$, $u_{\ell+4}$,....

Next we take the other case, where the radial expansion is in powers of $r-r_0$. For this case it is more convenient to rewrite the wave function as $\left[P(r)/r\right] F_\ell(\Theta,\phi)$, instead of as $R(r)F_\ell(\Theta,\phi)$. We use Eq. 4, and can write Schrödinger's equation for P as

$$\frac{d^2P}{dr^2} = -\left(\varepsilon + V_N + V_e + V_X - \frac{\ell(\ell+1)}{r^2}\right) P \tag{18}$$

We then include the term $-\ell(\ell+1)/r^2$ as well as V_N in the quantity to be expanded in powers of $r-r_o$, and write

$$-\left(\varepsilon+V_N+V_e+V_X - \frac{\ell(\ell+1)}{r^2}\right) = \Sigma(m)v_m(r-r_0)^m \tag{19}$$

in place of Eq. 13. Further, we expand P as

$$P = \Sigma(n)u_n(r-r_0)^n \tag{20}$$

Both series start with the constant term.

We then find that u_0 and u_1 are arbitrary, forming the two arbitrary constants expected in a general solution of the second-order Schrödinger equation. If we proceed as in the earlier case, we find

$$u_n = \left(n(n-1)\right)^{-1} \Sigma(m)v_m u_{n-m-2} \tag{21}$$

which leads to the following equations:

$$u_2 = (1.2)^{-1}v_0u_0$$

$$u_3 = (2.3)^{-1}(v_0u_1+v_1u_0)$$

$$u_4 = (3.4)^{-1}(v_0u_2+v_1u_1+v_2u_0)$$

$$\ldots\ldots \tag{22}$$

These equations, similar to Eq. 17 for the other case, are equally simple to compute, and give the general solution in terms of the arbitrary constants u_0 and u_1.

To proceed further, let us consider an actual case. In Table 5 we give four different calculations of the 2s P function for carbon, and for comparison the function determined by the computer program. The first series for P_{2s} is that in x, in which the assumed potential is given by the series in x in Table 4. The eigenvalue is taken as the value from the computer output. The initial slope of the curve for P, given from the initial value of the function R at the origin, is arbitrarily taken to come from setting u_0 in 1 in Eq. 13 (modified to express radii in terms of x rather than of r). The "Correct" entries in Table 5 are those from the computer program, but multiplied by a constant to reduce this function to having the same slope at the origin.

To get good convergence of this "Series in x" it is found that one must take some 30 or 35 terms, going up to terms in r^{30} or r^{35}. For any given number of terms, there will be a truncation error from the omitted terms. There is every indication that the series converges for any value of x, but it is tedious to compute more than about 30 or 35 terms with the pocket calculator though nothing in principle prevents going to much higher powers. We have arbitrarily continued the calculation of the series out to a value of x for which the truncation error was in the sixth significant figure. Entries in the table in which this truncation error is significant are enclosed in parentheses. It is clear that this series in x is only good out to about x = 1.2. We shall come back later to the decidedly non-negligible discrepancies between the "Correct" values and those from the "Series in x."

The other series in x-3.9, x-2, and x-1 will have two arbitrary constants, u_0 and u_1 in Eq. 20, or the value and slope of the curve of P as a function of r, at the point r_0, which we may determine to equal the function and slope of the series in x by differentiation of the series in Eq. 13. Hence it at first seemed necessary to take r_0 small enough so that we had a reliable value for the series in x at this point. For this reason an expansion was made in x-1, and u_0 and u_1 for this calculation were taken to get agreement with the series in x in both function and slope at x = 1. To set up this series, the potential as a series in x-3.90 from Table 4 was converted into a series in x-1, by use of Eq. 11, and the coefficients were found from this potential by Eq. 22. The function P as computed by this method is given in Table 5, as "Series in x-1." In this tabulation a number of entries are included at x smaller than 0.7, enclosed in parentheses. These results are unreliable because they lie outside the region in which the potential is reliable, and it was thought interesting to see how far the values of P remained reasonable outside this range. They are clearly not bad down to about 0.4. On the other hand, above x = 2.30, this series begins to have serious truncation errors, even with some 32 terms.

At this point in the calculation, it seemed desirable to join onto a series in a still larger value of r_0, and the series in x-2 was computed. It is to be noted that the entries for both the series in x-2 and in x-1 agree at x = 1.98, an indication of the accuracy of the fitting (as the series in x-1 and the series in x agree at x = 1). This series in x-2 was reliable out to about x = 3.5, not quite far enough to fit the series in x-3.9 onto it. To perform this fit a calculation (not tabulated) in x-3.5 was made, fitting to that in x-2, and in turn using this to fit to a series in x-3.9. The accuracy of this multiple fitting procedure

Table 5. Different calculations of P_{2s} of carbon. Variable x is same as in Table 4. "Correct" denotes computer output, multiplied by a factor 0.3249 to make its initial slope agree with series calculations. "Series in x", expansion around nucleus, uses "Series in x" potential from Table 4. "Series in 3.9" uses corresponding potential from Table 4. "Series in x-2" and "Series in x-1" also use potential of "Series in 3.9" in Table 4, but transformed into series around other values of x by use of Eq. 11. Entries in parentheses are not reliable, as described in text.

x	Correct	Series in x	Series in x-3.9	Series in x-2	Series in x-1
0.1	0.074122	0.074129			(0.102383)
0.2	0.107989	0.107924			(0.118280)
0.3	0.115011	0.114691			(0.118020)
0.4	0.104536	0.103816			(0.104755)
0.5	0.083050	0.081885		(0.082152)	(0.082162)
0.6	0.055064	0.053492		(0.053616)	(0.053627)
0.7	0.023720	0.021819		0.021983	0.021882
0.8	-0.008809	-0.010949		-0.010904	-0.010890
0.9	-0.041037	-0.043328		-0.043332	-0.043223
1.0	-0.071972	-0.074344	(-0.074364)	-0.074358	-0.074344
1.1	-0.100979	-0.103391	(-0.103381)	-0.103379	-0.103365
1.2	-0.127673	-0.130062	(-0.130017)	-0.130013	-0.130000
1.3	-0.151848	(-0.154172)	-0.154071	-0.154072	-0.154060
1.4	-0.173426		-0.175491	-0.175490	-0.175480
1.5	-0.192412		-0.194291	-0.194290	-0.194281
1.66	-0.217585		-0.219132	-0.219131	-0.219125
1.82	-0.236789		-0.237978	-0.237977	-0.237973
1.98	-0.250639		-0.251452	-0.251451	-0.251451
2.14	-0.259804		-0.260226	-0.260225	-0.260228
2.30	-0.264950		-0.264970	-0.264969	(-0.264972)
2.46	-0.266708		-0.266321	-0.266320	
2.62	-0.265658		-0.264864	-0.264864	
2.78	-0.262317		-0.261124	-0.261123	

2.94	-0.257141	-0.255558	-0.255557
3.10	-0.250523	-0.248561	-0.248561
3.26	-0.242802	-0.240470	-0.240470
3.42	-0.234263	-0.231569	-0.231572
3.58	-0.225146	-0.222097	(-0.222392)
3.74	-0.215652	-0.212250	
3.90	-0.205944	-0.202193	
4.06	-0.196158	-0.192061	
4.22	-0.186404	-0.181957	
4.38	-0.176769	-0.171972	
4.54	-0.167321	-0.162170	
4.70	-0.158115	-0.152602	
4.86	-0.149192	-0.143306	
5.02	-0.140581	-0.134309	
5.18	-0.132304	-0.125631	
5.34	-0.124374	-0.117283	
5.50	-0.116800	-0.109270	
5.66	-0.109583	-0.101593	
5.82	-0.102723	-0.094246	
5.98	-0.096216	-0.086788	
6.14	-0.090054	-0.080514	
6.30	-0.084229	-0.074103	
6.46	-0.078731	-0.067978	
6.62	-0.073549	-0.062121	
6.78	-0.068670	(-0.056535)	

Table 6. Different calculations of P_{2p} for carbon. Notation same as in Table 5, except that "Correct" entries are as in computer output, and series are multiplied by constants to reduce them to same values as correct values for small x.

x	Correct	Series in x	Series in x-3.9
0.1	0.0134600	0.0134654	
0.2	0.0469094	0.0468895	
0.3	0.0924295	0.0922931	
0.4	0.1446191	0.1442047	
0.5	0.1997606	0.1988887	
0.6	0.2552962	0.2538767	
0.7	0.3094854	0.3074576	
0.8	0.3611749	0.3585554	
0.9	0.4096356	0.4064795	
1.0	0.4544412	0.4508171	(0.450817)
1.1	0.495376	0.491350	0.491350
1.2	0.532370	0.527995	0.528218
1.3	0.565445	(0.560766)	0.561384
1.4	0.594691		0.590887
1.5	0.620241		0.616835
1.66	0.653853		0.651333
1.82	0.679213		0.677874
1.98	0.697175		0.697645
2.14	0.718615		0.710473
2.30	0.714381		0.718210
2.46	0.715175		0.721287
2.62	0.712033		0.720406
2.78	0.705321		0.716199
2.94	0.695729		0.709229
3.10	0.683776		0.699991
3.26	0.669914		0.688917
3.42	0.654536		0.676381
3.58	0.637977		0.662702

3.74	0.620526	0.648150
3.90	0.602428	0.632954
4.06	0.583891	0.617305
4.22	0.565090	0.601947
4.38	0.546173	0.585261
4.54	0.527264	0.569109
4.70	0.508465	0.552997
4.86	0.489862	0.536994
5.02	0.471525	0.521160
5.18	0.453511	0.505537
6.46	0.324861	0.390117

is shown by the closeness with which the entries in the table for the series in x-3.9 match those for the series in x, over the very narrow range (x = 1.0 to 1.3) in which both have an approximate validity. It was clear, once these various series had been computed, that the series in x-3.9 had a far greater range of convenient convergence than the others, and that if the fitting could be secured in some other way, these intermediate series could be eliminated. The way to carry out this simplification was realized at once, and while it was not carried through for the 2s orbital, it was used for 2p, which we shall present next.

This improved method of fitting is the following. We have pointed out earlier that there are two arbitrary constants, u_0 and u_1, in Eq. 22 for determining the coefficients of the expansion in $r-r_0$. One can then set up one solution with $u_0 = 1$, $u_1 = 0$, and another with $u_0=0$, $u_1 = 1$. These in a way are cosine-like and sine-like, so that they can be denoted C and S. If we now compute these two functions, any solution of the problem can be written as $u_0C + u_1S$. We can determine the constants u_0 and u_1 by requiring the function to have a predetermined value and slope at a joining point, such as x = 1.0 in our case. As we have stated, this has not been done for the 2s orbital, but it was used for the 2p, for which the series in x, the series in x-3.9, and the correct value are given in Table 6. It is clear from this table that the two series join together with satisfactory accuracy in the neighborhood of x=1, where the fitting was carried out. Hence, it seems reasonable to think that we can get satisfactory descriptions of the 2s and the 2p wave functions in carbon, over the whole range from x = 0 to about x = 6.3, by using the series in x in the range 0-1, and the series in x-3.9 between 1 and 6.3. When we recall that the radius of the Wigner-Seitz sphere is about 3.4, we see that these expansions of the wave function will not only carry us throughout the filled cell in diamond, but well out into the adjacent cells as well.

Let us now consider the rather serious discrepancies between the results of our series expansions of the wave function, and the "Correct" values determined from the computer program. In the first place, it should be pointed out that our series solutions for P_{2s} in Table 5, or P_{2p} in Table 6, should be exact solutions of the wave function for the potential given in Table 4 (that is, the series in x out to x = 1.0, and the series in x-3.9 for larger x). All three series, in x-3.9, x-2, and x-1, are determined from the potential series in x-3.9. In this connection, it is striking that over all the range where the series in x-3.9, x-2, and x-1 are giving reliable values, the three agree with each other remarkably well, the differences between them being orders of magnitude smaller than the discrepancy between these series values and the "Correct" results. This should give us confidence in thinking that our series solutions give us a very accurate solution of the potential expressed in the polynomial expansions.

If we examine the nature of the disrepancies between the correct function and the series expansions, we see that they are largely of the type to be expected if the eigenvalue is chosen slightly wrong. In the 2s, the eigenfunction for large r is slightly smaller numerically than the correct function,while in the numerical integration outward of Schrödinger's equation, an eigenfunction which is numerically too small at large r, falling more and more below the correct value, will eventually cross the axis, while one which is numerically too large will eventually go through a minimum without crossing the axis, and will rise again. The correct eigenvalue lies between these values, and can be obtained by fitting the eigenfunction, out on the tail, to another function which is chosen to go to zero at infinity. We have not yet carried out such a calculation for the 2s or 2p functions, but our treatment of the 1s function illustrates the same point, and is given in Table 7. The 1s function has the advantage that the tail of the function lies in the range covered by the expansions in x-3.9, while to get the tail of the 2s and 2p we should have to use another expansion, valid at larger values of x than given in Tables 5 and 6.

In Table 7, we evaluate the series in x for three separate energy values. The first is -20.187 rydbergs, the value found from the computer printout (actually -20.18709128 ry). It is clear that the function is falling too rapidly, so that it would be bound to cross the axis at a finite x. This is a sign that the energy is too high. Therefore a calculation was made with $\varepsilon = -21.187$ ry, one rydberg lower. Even a few points were enough to show that this energy was too low (numerically too large), and the table suggested that a value about half way between, -20.687, would be about right. Hence a series in x, and a series in x-3.90, were computed with this value of energy.

The tail comes in the range between x = 1 and x = 6.30, where the series in x-3.90 is being used. It is very interesting to see the form of the functions C and S in this case. For the 2p function, these functions throughout the range of x were not far from C = 1 and S = x-3.9, which would correspond to their first term. This is a result of the rather small values of the higher derivatives of the function. But for the 1s, x = 3.9 corresponds to a range where the kinetic energy is large and negative, so that C and S resemble the hyperbolic cosine and sine respectively of $k(x-x_0)$, where k is the square root of the absolute value of the kinetic energy, and both functions are very large for x-3.9 in the neighborhood of 3. Thus for positive values of $x-x_0$, the functions both

Table 7. Different calculations for P_{1s} of carbon. Notation as in Table 6. Calculations are made for three different values of ε, -20.187 (value from computer output), -21.187, and -20.687. Calculations of series in x-3.90 made only for ε = -20.687.

x	Correct	Series in x ε=-20.187	Series in x ε=-21.187	Series in x ε=-20.687	Series in x-3.90
0.1	1.005351	1.005445	1.005885	1.005665	
0.2	1.508262	1.507362	1.510302	1.508832	
0.3	1.703859	1.699342	1.707724	1.703529	
0.4	1.717522	1.707008		1.715480	
0.5	1.628696	1.610843		1.625130	
0.6	1.487199	1.461560		1.483142	
0.7	1.323816	1.290410		1.320738	
0.8	1.157100	1.115965		1.156559	
0.9	0.997743	0.948689		1.001249	
1.0	0.851413	0.794297		0.860220	0.860220
1.1	0.720616			(0.736018)	0.725737
1.2	0.605918			(0.629162)	0.608236
1.3	0.506752				0.507053
1.4	0.421933				0.420830
1.5	0.349998				0.347999
1.66	0.257856				0.255131
1.82	0.188742				0.185837
1.98	0.137431				0.134702
2.14	0.099640				0.097214
2.30	0.071985				0.069911
2.46	0.051850				0.050128
2.62	0.037253				0.035853
2.78	0.026708				0.025589
2.94	0.019113				0.018230
3.10	0.013656				0.012967

3.26	0.009744	0.009211
3.42	0.006944	0.006533
3.58	0.004944	0.004633
3.74	0.003516	0.003281
3.90	0.002499	0.002322
4.06	0.001775	0.001642
4.22	0.001260	0.001161
4.38	0.000894	0.000820
4.54	0.000634	0.000580
4.70	0.000449	0.000410
4.86	0.000318	0.000291
5.02	0.000225	0.000208
5.18	0.000159	0.000150
5.34	0.000113	0.000110

have the same sign, and approach proportionality to each other at large x, while for negative $x-x_0$ they have opposite sign. It is very simple to find the ratio of coefficients u_0 and u_1 which will make the sum $u_0C + u_1S$ go asymptotically to zero at large x. In other words, this is the way to describe the decreasing tail of the wave function, using the present method of calculation.

This was done for the case being considered, and the other constant was determined to make the two series continuous at x = 1. In this way we set up the composite description of the 1s function in Table 7. The function does not have an exact continuity of slope at x = 1, as we can see from the fact that the two series do not exactly agree for x = 1.1 and 1.2. But the overall behavior of the series solutions is excellent. It gives values slightly smaller than the correct values from x = 0 to 0.8, slightly higher from 0.9 to 1.4, and slightly smaller from 1.5 to infinity.

We can now understand the reasons for the slight discrepancies in eigenfunction, and the 0.5 ry depression of the eigenvalue, by considering the discrepancies between the correct potential and the tenth order polynomials, which we have noted in connection with Table 4. We have pointed out that the series is too low by about 0.6 ry in the region around 0.34, where the 1s function has its maximum. In fact, if we find the value of the discrepancy, averaged over the charge density of the 1s orbital, we find 0.499 ry, so that we should expect the depression of energy, by first-order perturbation theory, to be almost precisely what we were led to to get the fit of the wave function shown in Table 7. Such corrections to the eigenvalues, necessary to get a proper energy using the polynomial representation of the potential, can thus be made to very good accuracy by using first order perturbation methods.This has not yet been checked for the 2s and 2p orbitals, but the correction to the eigenvalue will be very much smaller than for the 1s. The reason is that by far the larger part of the charge in the 2s and 2p is at values of r for which the errors in the polynomial representations of the potential are almost negligible.

6. DISCUSSION

We have seen in the preceding sections how to deal with charge densities, potentials, Poisson's equation connecting them, and Schrödinger's equation determining the wave functions, all in terms of expansions in powers of r and spherical harmonics of the angles. In a short survey such as this, it is not possible to take up all the problems which come up in a complete self-consistent field treatment of a molecular or crystalline problem by the cellular method; but the results seem promising enough to justify further development of the approach. The fundamental fact is that we must deal with rather complicated three-dimensional functions in an accurate manner, and the expansions in spherical harmonics and powers of r are inherently an efficient way to make such expansions.

In this they contrast to the more popular methods which have been habitually used in molecular calculations, such as the expansions in exponential functions or gaussians. In practically all ab initio calculations using such methods, the limiting feature has been the number of basis functions to be used, and their proper choice. In contrast, present computers, and even present pocket calculators, have the capacity, through their operation x^y, to handle large numbers of powers of r with great ease. One can foresee the possibility of describing a complete molecular orbital by means of a relatively small number of coefficients representing its expansion in each of a number of adjacent cells.

One should point out a few methods of technique which will probably occur to the reader, and which will almost certainly find their uses when the methods advocated here are formulated in terms of computer programs. For instance, it is in principle very simple to multiply two power series in r. This is just what is being done in Eqs. 15 and 21, in the summations $\Sigma(m)v_m u_{n-m-2}$. If we wish to multiply two series of the form of Eq. 3, involving spherical harmonics as well as powers of r, we need the formulas for the products of two spherical harmonics, or two solid or cubic harmonics. The formulas for products of two spherical harmoics are well known, the coefficients being expressible in terms of the Gaunt coefficients, and from these we can easily derive those for products of two solid or cubic harmonics. As another example, we have described methods of joining functions together with continuous value and slope, in the one-dimensional problem of the radial wave functions for an atom. But one often wants to make such joining in three dimensions, from one cell to another. There is a good deal of information in the literature about how to do this, such as the work of von der Lage and Bethe[10] and the recent work of Leite et al.[7] But this could be carried much further.

One problem which is constantly met in the cellular method, as it is met in the muffin-tin and overlapping-sphere methods, is that of writing a function, such as a wave function or charge density, which is expressed in terms of spherical harmonics and power series in r about one center, in a similar form about another center. For functions of r alone, we have given the method in our discussion of Eq. 11. For the three-dimensional case, involving the spherical harmonics, various theorems are familiar, if the radial functions are expressed in terms of spherical Bessel and Neumann functions. But here the answer, when we are using power series for the wave functions of r, is given in that gold mine of information, Per-Olov Löwdin's monumental work on Quantum Theory of Cohesive

Properties of Solids,[11] Sec. 6.1. His formulas are particularly well adapted to the case where the wave functions are given as power series in r.

It is not obvious that the use of power series expansions for radial wave functions has any great advantage over the numerical integration methods of Herman and Skillman, in the inner part of atoms. Thus we have found our expansions in x, holding out only to about x = 1, have a fairly narrow region of convergence even for a light atom like carbon, and they might not hold nearly as well for a heavy atom. But in the outer regions of the atom, from x = 1 or 1.5 out to the boundary of the Wigner-Seitz cell, an expansion such as the one in x-3.9 which we have used may prove to be very valuable. It may well be desirable to combine such an analytic treatment for the outer part of the atom, with a Herman-Skillman type of numerical integration for the inner region.

We have seen that the discrepancies between the results of our power series and the Herman-Skillman results for the carbon atom arise largely from discrepancies of the potential in this inner region. It would not be hard to fit a power series solution for x greater than 1.5, for example, onto a numerical solution for smaller x. A composite scheme of this sort may well turn out to be the most practical for actual molecular and solid-state calculations. It has an obvious resemblance to the pseudopotential method, in that we should be using different descriptions of the core of the atom and of the outer part. But it should be capable of great numerical accuracy, if all calculations are carried out rigorously.

One point should be mentioned in connection with molecular calculations. We have used the diamond crystal as an example, and have considered its filled and empty cells. What, however, would we do about the outside of a finite molecule, or the region outside a surface of a solid? It does not seem likely that the outer sphere of Johnson and Smith would be the ideal solution. The writer is inclined to think that at least in some problems, it would be useful to pack the filled and interstitial empty cells of a finite molecule inside a sort of blanket of empty cells, extending far enough out to enclose essentially all the charge arising from the tails of the surface atoms. One could build up solutions of the Schrödinger equation in these blanketing cells, of the type which we used in Sec. 5 to treat the tail of the carbon ls orbital. These solutions could be set up so as to reduce to zero on the outer faces of the cells, but to join smoothly onto the occupied cells inside the molecule. In this way, no outer sphere would be required.

These are only a few of the many general points which could be brought up here. The present paper should be regarded as a progress report, a pilot exploration of methods which might perphaps be used to set up computer programs for a more accurate treatment of molecules and solids. One may hope in the future for practical programs of this sort, and one may hope for useful results for real and complicated molecules and crystals, arising from them.

REFERENCES

1. See for instance J.C. Slater, Int.J. Quant. Chem. 8S, 81 (1974).
2. J.C. Slater, Int. J. Quant. Chem. 9S(1975).
3. J.B. Danese and J.W.D. Connolly, Int. J. Quant. Chem. 7S 279 (1973); J. Chem. Phys. 61, 3063 (1974).
   J.B. Danese, J.Chem.Phys. 61, 3071 (1974).
4. J.R. Sabin, J.P. Worth, and S.B. Trickey, Phys. Rev. B, 11, 3658 (1975, and references contained there).
5. J.W.D. Connolly, chapter on The Xα Method, to be published in Modern Theoretical Chemistry: Vol. IV, Approximate Methods, G.A. Segal, ed., Plenum Press (in preparation).
6. E. Wigner and F. Seitz, Phys. Rev. 43, 804 (1933); 46, 509 (1934).
7. J.R. Leite, B.I. Bennett, and F. Herman, Phys. Rev. B (in press).
8. R.C. Chaney, C.C. Lin, and E.E. Lafon, Phys. Rev. B, 3 459 (1971).
9. F. Herman and S. Skillman, "Atomic Structure Calculations," Prentice-Hall, Inc., Englewood Cliffs, N.J., 1963.
10. F. von der Lage and H. Bethe, Phys. Rev. 71, 612 (1947).
11. P.-O. Löwdin, Phil.Mag. Supplement (Advances in Physics) 5, 1 (1956).

TEST OF THE CONVENTIONAL QUANTUM CHEMISTRY METHODS ON THE HYDROGEN ATOM†

T.L. Bailey+ and J.L. Kinsey[x]

Quantum Chemistry Group For Research in Atomic, Molecular and Solid-State Theory, Uppsala University, Uppsala, Sweden

INTRODUCTION

This paper is a numerical study of the reliability and accuracy of some of the conventional methods of quantum chemistry by their application to the problem of the hydrogen atom, a case where the quantities of interest are known exactly. The variation principle, $\delta\langle H\rangle = 0$, is used to approximate energy eigenvalues and the corresponding eigenfunctions. As basis functions, two simple sets are chosen: a) Gaussians, $\exp(-k\eta r^2)$; $k = 1, 2,\ldots N$, and b) exponentials, $\exp(-k\eta r)$; $k = 1, 2,\ldots N$. N is the order of the basis set and η is the scale factor. Scaled functions $\bar{\phi}_k$ are obtained from the unscaled ones ϕ_k in the usual way /1/:

$$\bar{\phi}_k = \eta^{3/2}\,\phi_k\,(\eta r)$$

+John Simon Guggenheim Foundation Fellow (1959-60). Permanent address: Department of Physics, University of Florida, Gainesville, Florida.

[x]National Science Foundation Postdoctoral Fellow (1959-1960). Permanent address: Department of Chemistry, Massachusetts Institute of Technology, Cambridge, Massachusetts.

†Sponsored by the Aeronautical Research Laboratories, Wright Air Development Division, Air Research and Development Command, U.S. Air Force; and supported in part by the U.S. Office of Naval Research

Both sets become complete in the interval $0\leq r\leq\infty$ when $N\to\infty$ and, for both, all the necessary integrals are elementary.

The method of calculation, though standard, will be summarized here for completeness and definition of the symbols. The wave function Φ for the system is first approximated in terms of the basis set

$$\Phi = \sum_j c_j\phi_j \tag{1}$$

Application of the variation principle, $\delta\langle H\rangle = 0$, then gives the linear system

$$\sum_j (H_{kj}-E\Delta_{kj})c_j = 0 \tag{2}$$

and the secular equation

$$\det\{H_{k\ell} - E\Delta_{k\ell}\} = 0 \tag{3}$$

The energy matrix $\underset{\sim}{H}$ is defined by $H_{k\ell} = \int\phi_k^* H\phi_k d\tau$, and the overlap matrix $\underset{\sim}{\Delta}$ by $\Delta_{k\ell} = \int\phi_k^*\phi_\ell d\tau$. An eigenstate corresponding to a given eigenvalue is represented by $\underset{\sim}{c}$, the column vector of the coefficients c_k. Thus, the expectation value $\langle F\rangle$ for an arbitrary observable F is given in matrix notation by

$$\langle F\rangle = \underset{\sim}{c}^\dagger \underset{\sim}{F}\, \underset{\sim}{c}/\underset{\sim}{c}^\dagger \underset{\sim}{\Delta}\, \underset{\sim}{c} \tag{4}$$

where $\underset{\sim}{F}$ is defined by $F_{k\ell} = \int\phi_k^* F\phi_\ell d\tau$. A set-scale procedure will be used rather than overall scaling. That is, equations (3) and (2) are to be solved at each value of the scale factor η.

The primary features to be investigated are the convergence of the energies and various other expectation values toward the correct values as the sets are extended, the dependence of the convergency on scaling, the validity and usefulness of various lower bound riteria, and the limitations on numerical accuracy arising from approximate linear dependence. The states of interest are the ground state and the first few excited S states.

Upper bounds for the energies are taken to be the roots of the secular equation and expectation values of several dynamical variables are calculated from equation (4) using the corresponding eigenvectors. Three different lower bound criteria are to be used:

$$\lambda_W^j = \langle H\rangle_j - (\Delta H)_j \qquad \text{(Weinstein /2,3/)} \qquad (5)$$

$$\lambda_T^j = \langle H\rangle_j - \frac{(\Delta H)_j^2}{E_{j+1} - \langle H\rangle_j} \qquad \text{(Temple /4,5/)} \qquad (6)$$

$$\lambda_S^j = \alpha_j - \{(\Delta H)_j^2 - (\langle H\rangle_j - \alpha_j)^2\}^{\frac{1}{2}} \qquad \text{(Stevenson /5,6/)} \qquad (7)$$

$\langle H\rangle_j$ is the expectation value of the energy in the state Φ_j, the jth solution of the (finite) secular equation. $(\Delta H)_j$ is the width of the Hamiltonian operator in the same state:

$$(\Delta H)_j^2 = \int |(H - \langle H\rangle_j)\Phi_j|^2 d\tau \qquad (8)$$

E_j is the exact j^{th} eigenvalue, and α_j is a free parameter. The Weinstein criterion is a special case of the Stevenson bound, obtained when the choice $\alpha_j = \langle H\rangle_j$ is made.

None of these criteria is valid except under certain restrictive conditions which require some assumptions about the exact eigenvalues or eigenfunctions. λ_W^j is a lower bound for the jth eigenvalue only if $\langle H\rangle_j$ is closer to E_j than to any other eigenvalue: $E_{j-1}+E_j \leq 2\langle H\rangle_j \leq E_j+E_{j+1}$. Similarly, λ_S^j gives a valid lower bound for the same state provided that $E_{j-1}+E_j < 2\alpha_j < E_j+E_{j+1}$. Only the second inequality is of practical importance in either case. Violation of the first inequality does not mean that the bound ceases to apply to the state of interest but that it becomes valid for a lower state as well. The Temple criterion gives a lower bound for E_j under a less stringent condition, namely that $E_{j+1} > \langle H\rangle_j$. As $\langle H\rangle_j^j$ increases to E_{j+1}, λ_T^j goes to $-\infty$ /5/.

Loss of numerical accuracy because of approximate linear dependence /7,8/ is an unpleasant feature peculiar to most of the standard non-orthonormal sets of functions. This is because equation (3) is in general only a necessary and not a sufficient condition for solubility for a minimum energy in accordance with the variation principle. Sufficiency is implied only with the additional requirement that

$$\det \underset{\sim}{\Delta} \neq 0 \qquad (9)$$

For non-orthonormal sets this is a more severe restriction than might be supposed from purely mathematical considerations. Since only a finite number of figures can be maintained in a computation, it may easily happen that a set of functions that are not linearly dependent in a strict mathematical sense will have an overlap matrix that is effectively singular numerically. Extension of such a set, rather than permitting a better approximation to the wave function as theoretically expected, may actually lead to a very damaging deterioration of the accuracy and reliability of the results by causing such a near singularity. A convenient measurement of the limitation on numerical results due to this kind of ill behavior is given by the ratio $\delta_{max}/\delta_{min}$ of the two extreme eigenvalues of the overlap matrix. With a value of 10^9, for example, for $\delta_{max}/\delta_{min}$ one can expect to have to carry about nine significant figures throughout the calculation in order to retain even one figure of significance in the final results. For both the sets of functions employed in this work $\log_{10}(\delta_{max}/\delta_{min})_N \simeq 1.5\ (N-1)$.

This effect of approximate linear dependence restricted the range of the order of the basis set N over which the calculations gave meaningful results. These ranges were found to be: $2 \leq N < 7$ for the Gaussian sets and $2 \leq N \leq 5$ for the exponentials. Those for $N = 2$ were done by desk calculator; for the higher orders, they were done with the Quantum Chemistry Group's Alwac III computer, using special programs for finding the eigenvalues (by iteration), and for computing minima in the $\langle H \rangle$ vs. η curves and the η-values at which these occurred.

RESULTS AND DISCUSSION

In this section the numerical results for both sets are presented and discussed /9/. All quantities are quoted in reduced units /10/ reduced Hartree units (H_R) for energy, reduced Bohr radii (b_R) for distances, and units of $\hbar$ for momenta.

The two sets will be discussed separately, but comparisons between them will be made in the discussion of the second set, the exponentials. We note that this set is a very special one for this particular problem since the exact ground state is obtained whenever $k\eta = 1$.

Gaussian Sets

The curves of upper bounds $\langle H \rangle$ versus scale factor η show the expected behavior as the order of the basis set is increased. Representative samples of the $\langle H \rangle$ versus η curves are shown in Figures 1 and 2. As the set is extended, these become flatter, and the minima decrease toward the appropriate exact energies, in

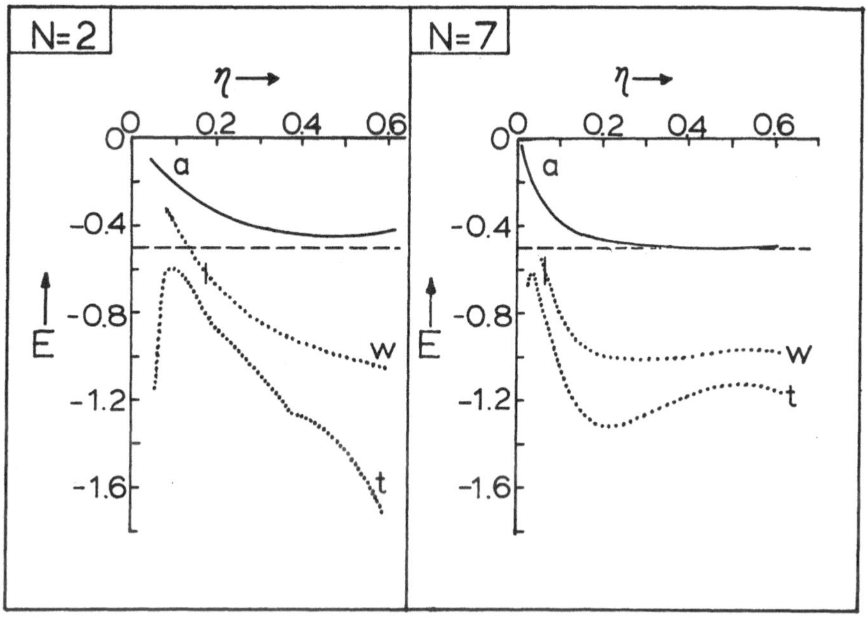

Figure 1. Upper and lower bounds versus scale factor η, calculated for the ground state and for Gaussian sets of orders 2 and 7. Solid curve a: upper bounds $\langle H \rangle_0$; dotted curve w: Weinstein lower bounds; dotted curve t: Temple lower bounds. The dashed horizontal lines give the exact energy, $E_0 = -0.500$. The short vertical lines through the curves w give the minimum η-values for which the Weinstein LB criterion is valid.

accordance with the variation principle and the separation theorem. The optimum upper bounds, i.e., $\langle H \rangle$ calculated at $\partial \langle H \rangle / \partial \eta = 0$, are given in Table I. The upper bounds are fairly poor, even for order 7. At order 7, the ratio $\delta_{max}/\delta_{min} = 0.3835 \times 10^8$, and serious violations of the separation theorem appear in the eigenvalues for the higher excited states. Therefore results calculated for orders higher than 7 would be unreliable, and the order 7 results are of doubtful accuracy. The table shows also that the optimum $\langle H \rangle$ values decrease irregularly, rather than smoothly, so that no reliable extrapolation of the sequence of upper bounds can be made.

If the upper bounds for the energy are poor, the Temple and Weinstein lower bounds, λ_T and λ_W are much worse, as Figures 1 and 2 show clearly. A striking feature of the Temple bounds for the ground state is that these all have their best values at scale factors below $\eta = 0.1$, where rather sharp minima appear. Since the

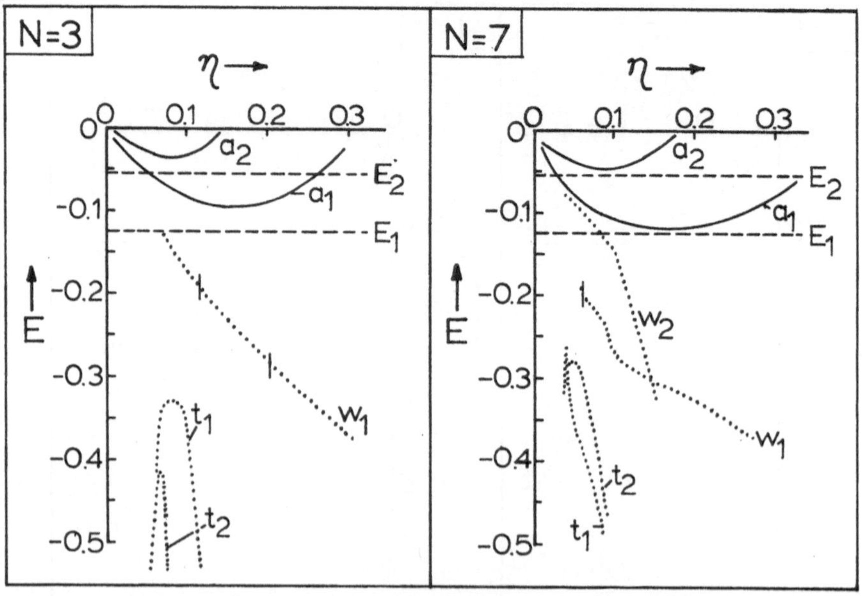

Figure 2. Upper and lower bounds versus η for the first and second excited states; Gaussian sets of orders 3 and 7. The subscripts 1, 2 denote the first and second excited states, respectively. Curves a_1, a_2 are the upper bounds; w_1, w_2 and t_1, t_2 are the Weinstein and Temple LB's; and E_1 and E_2 are the exact energies. The short vertical lines give the ranges of validity of the Weinstein LB criterion.

wave functions are correctly scaled (i.e. $\partial\langle H\rangle/\partial\eta = 0$) in the neigborhood of $\eta = 0.45$, the wave functions are far out of scale at these low $-\eta$ maxima in $\lambda_T^{(0)}$ and thus are bad approximations to the exact wave functions. Nevertheless, these maxima fall in the range of validity of the Temple criterion, and are valid lower bounds. These low-η maxima in $\lambda_T^{(0)}$ show no tendency to converge to the exact energy as the set is extended. No quantitative explanation for their shape is evident, because of the complications introduced by the set-scaling procedure. Qualitatively, it can be seen from the definition $\lambda_T^{(0)} = \langle H\rangle_0 - (\Delta H)^2/(E-\langle H\rangle_0)$ that if $(\Delta H)^2$ decreases with decreasing η in the low -η range more rapidly than does E $E_1-\langle H\rangle_0$, a strong maximum in $\lambda_T^{(0)}$ can occur. At order 5, a second maximum in $\lambda_T^{(0)}$ appears in the η-region where the wave function is approximately correctly scaled; this maximum becomes successively more pronounced for orders 6 and 7. In contrast with the low-η maximum, $\lambda_W^{(0)}$ at the second maximum seems to be converging toward

Table I

Results of calculations for Gaussian sets

Order N		Best scale parameter η_n (n=1,2,3)	$\langle H\rangle_n$ at η_n (n=1,2,3)	$\langle r\rangle_n$ at η_n (n=1,2)	$\langle r^2\rangle_n$ at η_n (n=1,2)	$\langle p^2\rangle_n$ at η_n (n=1,2)
3	n=1	.49007	-.479 198	1.465	2.725	.9580
	2	.16272	-.095 9098	6.819	53.74	.2124
	3	.08145	-.036 4327	-	-	-
4	n=1	.4546	-.481 1965	1.462	2.722	.9670
	2	.18367	-.110 276	6.171	44.04	.2316
	3	.0781	-.038 372	-	-	-
5	n=1	.4614	-.490 742	1.475	2.819	.9824
	2	.1743	-.111 998	6.042	41.34	.2294
	3	.092	-.045 2013	-	-	-
6	n=1	.43814	-.491 894	1.476	2.825	.9845
	2	.182455	-.117 272	5.976	40.71	.2395
	3	.088636	-.046 1166	-	-	-
7	n=1	.4411	-.495 067	-	-	-
	2	.1765	-.117 968	-	-	-
	3	.0939	-.048 5901	-	-	-
Exact	-	-	n=1 -.500 000	1.500	3.000	1.000
			2 -.125 000	6.000	42.000	.2500
			3 -.055 556			

the exact energy as the set is extended. The Weinstein lower bounds for the ground state, $\lambda_W^{(0)}$ are best at the low-η limit of validity of the Weinstein criterion, where the wave functions are far out of scale. The best $\lambda_W^{(0)}$'s are seen to be about the same as the best $\lambda_T^{(0)}$'s, and like the $\lambda_W^{(0)}$'s show no tendency to converge to the exact energy. The $\lambda_W^{(0)}$'s show shallow maxima in the region of correct scaling, but these are less pronounced than the corresponding maxima in $\lambda_T^{(0)}$ and seem to be converging to the exact energy even more slowly. The high-η maxima in both $\lambda_T^{(0)}$ and $\lambda_W^{(0)}$ would be expected to approach the exact energy at set orders much higher than 7. Unfortunately, this process cannot be followed by the computational scheme used here, since the numerical accuracy is being lost so rapidly.

The curves of λ_T versus η for the excited states do not show the interesting structure of $\lambda_T^{(0)}$ versus η, having only single maxima in the η-range of interest. The Weinstein bounds λ_W are considerably better than the Temple bounds, but both are very poor, as seen in Figure 2. This is not surprising; if the ground state wave functions are poor approximations, it is to be expected that the wave functions for the excited states would be worse, with relatively larger operator widths ΔH and relatively deeper lying lower bounds. The irregular shapes of $\lambda_T^{(1)}$ and $\lambda_T^{(1)}$ and $\lambda_W^{(1)}$ for order 7 are probably due to numerical inaccuracy, which is expected to appear in the results for orders 7 and higher.

The expectation values $\langle r \rangle_0$ and $\langle r \rangle_1$ for the ground state and first excited state are shown for orders 3-6 in Figure 3. The shapes of the $\langle r^2 \rangle$ versus η curves were found to be almost identical to those for $\langle r \rangle$ versus η. Values of $\langle r \rangle$, $\langle r^2 \rangle$ versus η for the ground state flatten out in the region where $\partial\langle H\rangle/\partial\eta \approx 0$ as the set is extended, so that $\langle r \rangle$ and $\langle r^2 \rangle$ tend to become almost constant over a fairly wide range of η.

In all cases, the point at which $\langle H \rangle$ is a minimum gives more accurate values for the expectation values than the other points that might be chosen, e.g., λ_W=max., λ_T=max., (ΔH)=min.

Exponential Sets

On the whole, the results for the exponential sets are considerably superior to those obtained for the Gaussians. This is not an unexpected result, because of the possibility of approximating the ground state exactly within the set. Moreover, these functions are less sharply peaked around $r = 0$ than are the Gaussians, and thus

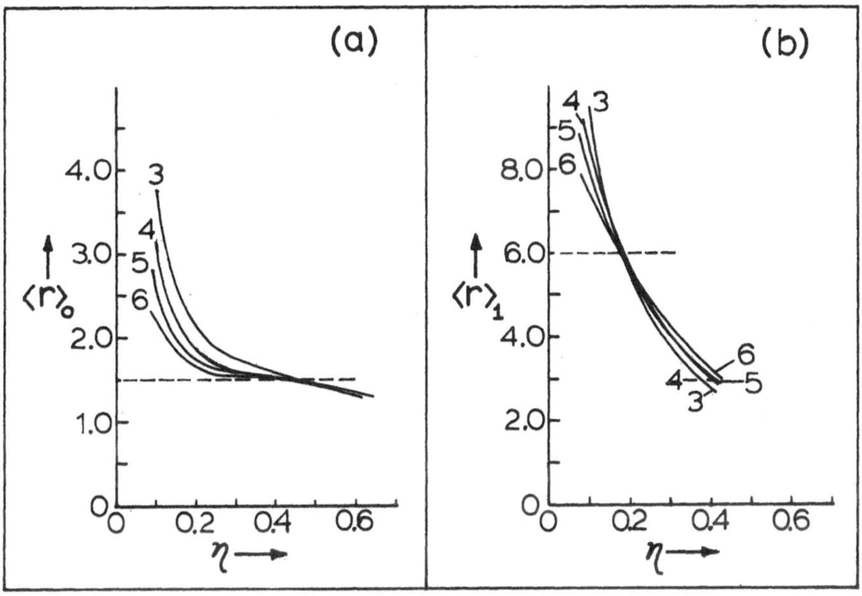

Figure 3. Expectation values of r versus η for ground state (a) and first excited state (b). Gaussian sets, orders 3-6. The dashed horizontal lines give the exact <r>-values.

they are probably better suited for the excited states as well.

Optimum upper bounds $\langle H\rangle_j$ for the first, second, and third excited states (j=1,2,3) are given in Table II. The number of significant figures given for each value in this Table is indicative of the reproducibility of the iterative process used in finding the eigenvalues. These optimum $\langle H\rangle_j$'s appear to converge toward the corresponding exact values as N is increased, although no extrapolation procedure is feasible because of loss in numerical accuracy. The first excited state is already essentially exact to within the available numerical accuracy at N = 5. The energies of the next two higher states show no improvement past that stage because the decreasing number of significant figures balances the apparent lowering of the roots.

A detailed examination of the N = 6 and N = 7 results over the full range of η-values reveals how seriously the numerical accuracy has dropped. There are several violations of both the variation and separation theorems, for example, and some negative values for $(\Delta H)_j^2$ in the order six results. At N = 7, at least one of these phenomena occurs for every value of η examined. Thus, for calculations

Table II

Optimum upper bounds, exponential sets. The numbers in parentheses are the corresponding η-values.

N	First Excited State	Second Excited State	Third Excited State
2	-0.124 568 (0.353)		
3	-0.124 983 53 (0.200 87)	-0.054 890 8 (0.182 057)	
4	-0.124 998 4 (0.207 93)	-0.055 337 5 (0.196 62)	-0.030 509 (0.111 31)
5	-0.124 999 8 (0.212)	-0.055 456 (0.205 67)	-0.030 96 (0.121 205)
6	-0.124 999 9 (0.214)	-0.055 50 (0.212)	-0.031 1 (0.127)
7	-0.124 999 (0.216)	-0.055 50 (0.212)	-0.031 1 (0.126)
Exact	-0.125 000 000	-0.055 555 555	-0.031 250 000

performed using the methods employed here, the maximum accuracy and reliability must be considered to be reached at order five.

The lower bound curves $\lambda^{(j)}$ versus η are, as to be expected, much less flat than those for $\langle H \rangle_j$ versus η, exhibiting in most cases one or more pronounced maxima. Examples are shown in Figure 4 and 5. In spite of their complicated structure, the $\lambda_W^{(j)}$ and $\lambda_T^{(j)}$ versus η curves give fairly good limits at their maxima, in contrast to the results obtained for the Gaussian sets. These maxima are somewhat displaced from the η-values at which the minima $\partial\langle H \rangle_j/\partial\eta = 0$ occur. The displacements are small, however, especially when compared to those which occur for the Gaussian sets, and the tendency seems to be toward smaller displacements as the order of the set is extended.

The Temple bounds are uniformly higher than the Weinstein ones in the neighborhood of their maxima. The Weinstein bounds can, however, be used to obtain criteria for the choice of an α in the Stevenson criterion and some very good lower bounds can be obtained without using any knowledge about the exact location of higher states. This is accomplished in the following way: $\lambda_W^{(j+1)}$ and $\lambda_W^{(j)}$ are assumed to be valid lower bounds for the states j+1 and

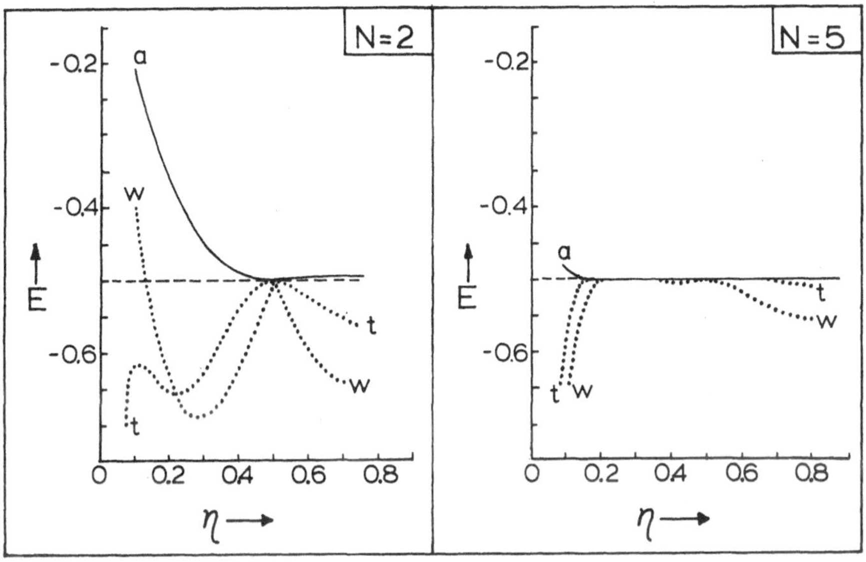

Figure 4. Upper and lower bounds versus scale factor η, for exponential sets of orders 2 and 5. The identifying symbols have the same meanings as those of Figure 1.

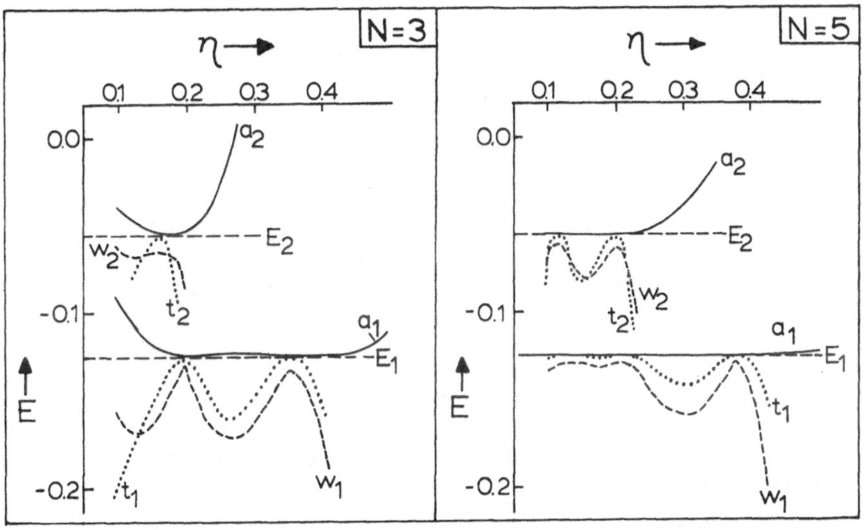

Figure 5. Upper and lower bounds versus η; exponential sets of orders 3 and 7; first and second excited states. The meanings of the identifying symbols are the same as in Figure 2.

j respectively. From them α_j is calculated,

$$\alpha_j = \left(\lambda_w^{(j)} + \lambda_w^{(j+1)}\right)/2$$

and this value used in the Stevenson formula to give $\lambda_s^{(j)}$. This value, if better than $\lambda_w^{(j)}$, replaces it in the formula to give a new α_j and a new $\lambda_s^{(j)}$. The process is repeated until no further improvement results. It is best to begin with the highest state for which $\lambda_w^{(j+1)}$ is believed to be valid and then work down, taking the best Stevenson bound instead of corresponding Weinstein bound where possible:

$$\alpha_{j-1} = \left(\lambda_w^{(j-1)} + \lambda_s^{(j)}\right)/2$$

This process is probably not very useful in general. The Stevenson formula will give a better result than the Weinstein only when $\alpha_j < \langle H\rangle_j$ or, if α is chosen as above, only when $\left(\lambda^{(j)}+\lambda^{(j+1)}\right)/2 < \langle H\rangle_j$ a condition which requires both the lower bounds $\lambda^{(j)}$ and $\lambda^{(j+1)}$ to be fairly good. This was not the case, for example, with the results from the Gaussian sets. It was practicable, however, for the exponentials, and the results are summarized in Table III. These calculations were made only at specific η-values and the best values were taken from among that set. None of these points was at the true minimum of $(\Delta H)_j$, and hence the apparent decrease in $\lambda_s^{(1)}$ between orders four and five.

The dependence of $\langle r\rangle$ on η, for the ground and first excited states and for N = 3-5, is shown in Figure 6. The shapes of the $\langle r^2\rangle$ versus η curves closely resemble those for $\langle r\rangle$ versus η, as they did in the Gaussian case. The general behavior of these expectation values with η is roughly analogous to that of the energy, and the accuracy is quite acceptable when the energy is good.

Table III

Stevenson lower bounds for the first and second excited states, exponential sets.

Order	$\lambda_s^{(1)}$	$\lambda_s^{(2)}$
3	-0.125 052 14	-0.062 944
4	-0.125 006 228	-0.057 883 32
5	-0.125 077 80	-0.057 787 93

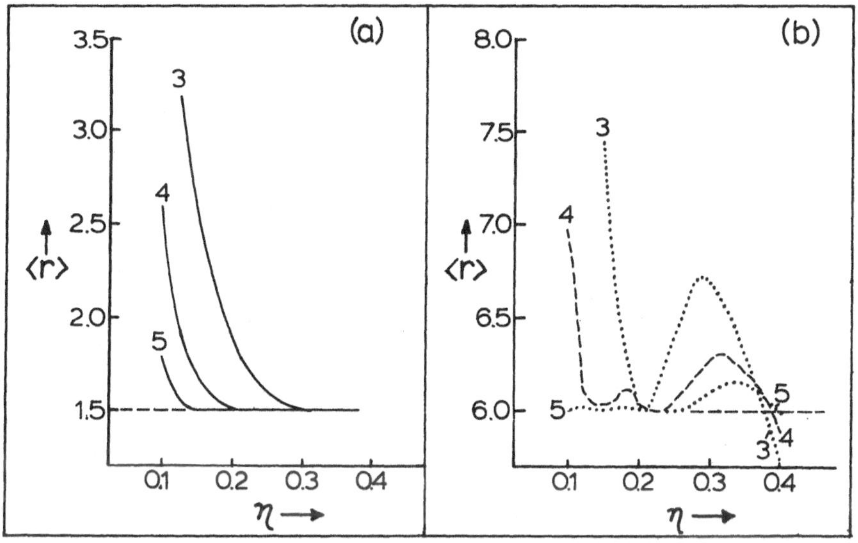

Figure 6. Expectation values of r versus η for the ground state (a) and first excited state (b). Exponential sets, orders 3 - 5. Exact values of <r> are given by the dashed horizontal lines.

The calculated values appear to be converging toward the exact values as N is increased. In contrast to the case of the Gaussian sets, there appears to be no clear-cut basis for choosing the point of minimum <H> over the point of minimum (ΔH), or vice versa, for obtaining the best expectation values.

CONCLUSIONS

For the most part the results speak for themselves and little additional comment is called for. None of the features was entirely unexpected. Rather, perhaps, their value lies in the finality with which they may be judged in this case because of the knowledge of the exact solutions.

The approximate linear dependence, developing in an almost identical manner in the two sets, is clearly a very great disadvantage of them and one which may very well outweigh the gain which derives from the ease with which integrals involving these functions can be computed.

Of the various lower bound criteria, the Temple formula appears to be somewhat more useful than the other two. Not only does it give more accurate limits for the eigenvalues than the Weinstein bound, but it generally has a less restricted range of validity. The requirement of knowledge of the exact excited levels is, of course, to its detriment, but all the criteria must rely on <u>some</u> knowledge of the exact spectrum. Stevenson's criterion, which is also capable of giving rather good bounds, especially when the energy is fairly accurate, may be preferable to Temple's in some cases. Here too, the necessary knowledge prerequisite to confidence that the bound is valid is the unfortunate feature.

No evidence for preferring minimization of the width of the Hamiltonian operator over the minimization of the energy to give functions having good expectation values was forthcoming from this investigation. On the contrary, for the Gaussians the second point was uniformly the more desirable. For the exponentials no such fixed rule was found but the two points were in any case never greatly separated.

HISTORICAL NOTE

The calculations on which this work is based were done at the Quantum Chemistry Group, Uppsala, in 1959-60. The paper itself was circulated, in essentially the form given here, as Technical Note No. 45, Q.C.G., Uppsala, but (although it has been referred to a number of occasions, and has led to further work on lower bounds and related topics) has not yet been published in the open literature. It seems appropriate to us to submit it to this Festschrift, and especially so since the work was undertaken at the suggestion of Professor Löwdin, and was carried out with his guidance.

ACKNOWLEDGEMENTS

We would like to express our thanks to Professor Per-Olov Löwdin for his suggestion of this problem and for his warm hospitality during our stay in Uppsala. We are also grateful to F.M. Jan Olov Nordling for programming and carrying out the Alwac III computations.

REFERENCES

1. P.O. Löwdin, Molecular Spectroscopy 3, 46 (1959).

2. D.H. Weinstein, Proc. Nat. Acad. Sci. 20, 59 (1934).

3. J.K.L. MacDonald, Phys. Rev. 46, 828 (1934).

4. G. Temple, Proc. Roy. Soc. (London) A119, 276 (1928)

5. P.O. Löwdin, Adv. Chem. Phys. 2, 207 (Interscience, New York, 1959).

6. A.F. Stevenson, Phys. Rev. 53, 199 (1938).

7. P.O. Löwdin, Adv. Phys. 5, 1 (1956), particularly p. 49.

8. J.C. Nordling, Lösning av egenvärdesproblem med tillhjälp av bassystem uppvisande dolt approximativt lineärt beroende, Proceedings from the Scandinavian Symposium on the Use of Electronic Computers, Karlskrona, Sweden, NordSAM, Lund 1959.

9. Because of space limitations, only a portion of the results are included here. A complete tabulation of the calculated quantities is given in Tech. Note 45, Quantum Chemistry Group, Uppsala (1970), by T.L. Bailey and J.L. Kinsey.

10. H. Shull and G.G. Hall, Nature 184, 1559 (1959).

NUMERICAL ASPECTS OF WEYL'S THEORY

Michael Hehenberger

Quantum Theory Project, Williamson Hall

University of Florida, Gainesville, Fla. 32611

THE WEYL-TITCHMARSH-RICCATI TECHNIQUE

Requiring

$$\chi(\lambda;x) = \varphi(\lambda;x) + m\psi(\lambda;x), \tag{1}$$

constructed from two linearly independent initial solutions φ and ψ of the second order differential equation

$$\{-\frac{d^2}{dx^2} + q(x)\}\, u(\lambda;x) = \lambda u(\lambda;x) \tag{2}$$

to satisfy a real boundary condition at the right endpoint $x = b$ of the interval $[a,b]$, Weyl /1/ found that m had to lie on a circle in the complex plane. For the singular case, $b \to \infty$, Weyl's circle shrinks either to a limit circle or to a limit point $m_\infty(\lambda)$. For most physical examples the differential operator is of limit point case, implying the existence of a unique square-integrable solution for any complex $\lambda = E + iE$. As demonstrated by Titchmarsh /2/, Im $m_\infty(\lambda)$ is proportional to the spectral density. In fact, as shown in detail by Chaudhuri and Everitt /3/, the spectrum of singular second order differential operators is completely characterized by the analytic properties of the m-function on the real axis.

Clearly, all applications /4-7/ of Weyl's theory carried out at the Quantum Chemistry Group in Uppsala during the past 3-4 years have been centered about the determination of $m_\infty(\lambda)$. The basic computer program, SECO /8/, was extended to solve eigenvalue problems and to determine resonances in the continuous spectrum. In the latter case it was found that, if $m_\infty(\lambda)$ exists on the real axis, it is possible to use real arithmetic in the numerical integration of the 2x2 solution $\underline{Z}(\lambda;x)$, constituted by the linearly independent solutions φ and ψ and their derivatives.

Whereas the treatment of the discrete spectrum can be based on the formula

$$m_\infty(\lambda) = \lim_{x\to\infty} \frac{\varphi(\lambda;x)}{\psi(\lambda;x)} \tag{3}$$

the evaluation of the m-function in the continuum is conveniently carried out using the "generalized Titchmarsh formula" /4/

$$m_\infty(\lambda) = - \left.\frac{\varphi\chi'/\chi - \varphi'}{\psi\chi'/\chi - \psi'}\right|_{(\lambda;x)} \tag{4}$$

Being essentially the quotient of two Wronskians, (4) is valid at any x and has very good stability properties. The important difference between equations (3) and (4) is that the latter requires knowledge of Weyl's solution χ. However, since χ is unique and only its logarithmic derivative is needed, the task of calculating m_∞ can be accomplished in the following way:

(i) If the interval considered is $(0,\infty)$ and the "potential" $q(x)$ is singular at the origin, it is necessary to divide the interval into two subintervals, say $(0,a]$ and $[a,\infty)$. The origin may be treated by any of the standard methods (cf. /4,7/) for giving an accurate value for the logarithmic derivative of the regular solution at the matching point $x = a$. If there is no singularity at the left endpoint, say $x = a$, just use the given left boundary condition. In both cases one has to construct an initial solution matrix

$$\underline{Z}_\alpha(\lambda;a) = \begin{bmatrix} \sin\alpha & \cos\alpha \\ -\cos\alpha & \sin\alpha \end{bmatrix} \tag{5}$$

which defines the regular solution ψ (second column) and the independent solution φ.

(ii) The asymptotic information required to use eq. (4) can be provided by means of analytic solutions for a given asymptotic form of the potential or by a much more powerful method, brought to our attention by Löwdin /9/.

The second order equation (2) can be transformed to a nonlinear first order equation

$$v'(\lambda;x) = P^2(\lambda;x) - v^2(\lambda;x), \tag{6}$$

where

$$P^2(\lambda;x) = q(x) - \lambda, \tag{7}$$

by substituting

$$v(\lambda;x) = i\,\frac{u'(\lambda;x)}{u\,(\lambda;x)}\ . \tag{8}$$

Expanding v(x) in powers of (-i) and substituting back into (6), one arrives at

$$v(\lambda;x) = \sum_{k=0}^{\infty}(-i)^k\, v_k(\lambda;x) \tag{9}$$

where

$$\begin{aligned} v_0 &= P \\ v_1 &= \tfrac{1}{4}\,P^{-2}\,q' \\ v_2 &= -\tfrac{5}{32}\,P^{-5}\,q'^2 - \tfrac{1}{8}\,P^{-3}\,q'' \\ v_3 &= \tfrac{15}{16}\,P^{-8}\,q'^3 + \tfrac{9}{32}\,P^{-6}\,q'q'' + \tfrac{1}{16}\,P^{-4}\,q''' \end{aligned} \tag{10}$$

If applied in the proper region, this asymptotic expansion converges very rapidly and the smallness of the highest-order term v_n is a suitable accuracy criterion.

In the previous applications /6,7/ we used Runge-Kutta integration of the 2x2 solution matrix $\underline{Z}(\lambda;x)$ from x = a until, say, x = b, at which point the expansion (9) was accurate enough to yield the asymptotic logarithmic derivative to be inserted into our generalized Titchmarsh formula (4).

The first main purpose of the present note is to point out inherent inefficiencies in both steps (i) and (ii).

The necessary minimal result of step (i) is to yield the logarithmic derivative tan α of the regular solution at the matching point x = a. For this purpose it is sufficient to use the real form Riccati's equation (6) based on the substitution

$$v'(\lambda;x) = \frac{u'(\lambda;x)}{u\,(\lambda;x)} \tag{11}$$

However, some caution must be exercised to avoid regions in which the regular solution has nodes which may cause $v(\lambda;x)$ in (11) to blow up. A way to avoid this problem is to use instead Pruefer's /10/ transformation

$$w(\lambda;x) = \arctan \frac{u'(\lambda;x)}{u(\lambda;x)} \tag{12}$$

$$r(\lambda;x) = \{u^2(\lambda;x) + u'^2(\lambda;x)\}^{\frac{1}{2}} \tag{13}$$

which, dropping dependence on λ, leads to

$$w'(x) = \{q(x) - \lambda\} \cos^2 w(x) - \sin^2 w(x) \tag{14}$$

and

$$r(x) = r(x_0) \exp\left\{\int_{x_0}^{x} (1 + q(t)\lambda) \sin w(t) \cos w(t) dt\right\} \tag{15}$$

To get $\tan\alpha(a)$, one simply integrates (14) until $x = a$.

Turning to the treatment of the subinterval $[a,b]$, the most time-consuming part is the numerical integration of $\underline{Z}(\lambda;x)$, even if λ is chosen to be purely real. The proposed improvement uses the fact that eq. (4) is valid at any x, hence also at $x = a$. The Riccati expansion (9) certainly diverges in this region, but it may be used in the asymptotic part to provide the initial value for a numerical inward integration from $x = b$ until $x = a$. The arising complication is that χ'/χ necessarily must be complex in the continuous part of the spectrum. To deal with that we separate eq. (6) into real and imaginary parts and get, using the notation $z = z_1 + iz_2$,

$$-v_2' = (q - E) + v_1^2 - v_2^2 \tag{16a}$$

$$v_1' = -\varepsilon + 2v_1 v_2 \tag{16b}$$

In the absence of $\varepsilon = \text{Im}\,\lambda$ eq. (16b) simply reduces to

$$v_1(x) = v_1(x_0) \exp\left\{2\int_{x_0}^{x} v_2(t)\, dt\right\} \tag{16b'}$$

However, it is one of the most appealing features of (16) that the use of a complex λ and even of a complex potential $q(x) = q_1(x) + iq_2(x)$ do not complicate the system of equations in any considerable way. In the latter case, the only changes are to replace $q(x)$ by $q_1(x)$ in (16a) and to add $q_2(x)$ on the right-hand side of (16b).

LÖWDIN'S METHOD FOR NONLINEAR FIRST ORDER EQUATIONS

Both changes discussed in the foregoing section result in the replacement of a second order differential equation by a nonlinear first order equation. The claim is that it is more efficient to deal with the non-redundant information about the logarithmic derivative instead of computing both function and derivative.

Before implementing a new numerical method it is perhaps appropriate to ask for the present "state-of-the-art". The answer to this question is formulated in a recent review article by Shampine et al. /11/ as follows:

"The best methods being used today to solve initial value problems for non-stiff ordinary differential equations are generally agreed to be the Runge-Kutta, Adams and extrapolation methods."

The relevant references of prominent authors representing (in order) these above-mentioned methods are Fehlberg /12/, Krogh /13/ and Gear /14/, and Bulirsch and Stoer /15/. The common philosophy is that an efficient code has to select stepsize (and order) and that it has to meet an error criterion provided by the user. The result presented in /11/ is that for "cheap" function evaluations Runge-Kutta and extrapolation methods are best, whereas the Adams-PECE (=predict-evaluate-correct-evaluate) method is most efficient for expensive function evaluations and coupled systems of equations.

In view of these recent developments it might be worthwhile to reexamine some of the standard methods used in popular quantum-chemical programs. These are mostly based on fixed stepsizes (at least within large blocks) and typically never contain any error parameters provided by the user.

Of course, it may be argued that Numerov's /16/ or de Vogelaere's /17/ methods are the most efficient ones for a potential given in a table of equidistant points and that accuracy and reliability are well-known because of decades of experience with these methods.

However, if higher order methods were used the stepsize could be enlarged and storage requirements for the potential arrays diminished. It is further quite well-known that the efficiency of the Numerov method has its limitations, if coupled systems of equations are treated.

After these more general side remarks let us turn to the second main issue of this note, namely the discussion of Löwdin's method /18/ for treating first order equations.

Löwdin's paper /18/ contains a review of symbolic finite difference calculus with applications to

$$y'(x) = F(x,y(x)) \tag{17}$$

In the process of arriving at a new mixed backward difference - central difference formula he rederived both formulas of Gauss and of Adams-Bashforth type. The latter are the ones which have recently so successfully been implemented for electronic computation /13,14/, introducing the features of variable stepsize and variable order (up to fourteen in /13/).

Löwdin's method is based on

$$h^{-1}y - \frac{1}{3}F = (\mu\delta)^{-1}\left\{F - \frac{1}{3}\nabla F\right\} - \gamma \tag{18}$$

where h is the stepsize, γ is a small correction given by the central difference expression

$$\gamma = \left\{-\frac{1}{180}\mu\delta^3 + \frac{31}{15120}\mu\delta^5 - \frac{557}{907200}\mu\delta^7 \pm \ldots\right\} F \tag{19}$$

and the symbolic operations μ, δ, ∇ are defined by

$$\mu = \frac{1}{2}(E^{\frac{1}{2}} + E^{-\frac{1}{2}}) \tag{20}$$

$$\delta = E^{\frac{1}{2}} - E^{-\frac{1}{2}} \tag{21}$$

$$\nabla = 1 - E^{-\frac{1}{2}} \tag{22}$$

where

$$E f(x) = f(x + h). \tag{23}$$

Löwdin has also developed an iterative method for the start, the details of which are omitted here. In our numerical tests we always found it easier to provide the first nine starting points by other means. Having y_1, y_2, $\ldots y_9$ and hence F_1, F_2, $\ldots F_9$ and the differences ∇F, $\nabla^2 F$, $\ldots \nabla^7 F$ available, one proceeds as follows. For carrying out a step in the "marching process", it is convenient to define the auxiliary quantity

$$M = (\mu\delta)^{-1}\left\{F - \frac{1}{3}\nabla F\right\} \tag{24}$$

which satisfies the recursion equation

$$M_{n+1} = M_{n-1} + \frac{4}{3}F_n + \frac{2}{3}F_{n-1}. \tag{25}$$

The first two values are given by

$$M_1 = h^{-1} y_1 - \frac{1}{3} F_1 - \gamma_1 \tag{26}$$

$$M_2 = h^{-1} y_1 + \frac{2}{3} F_1 + \frac{h}{6} F_1' + \frac{3}{2}\gamma_1 + \frac{1}{21} \delta^2 \gamma_1 \dots \tag{27}$$

and M_3, M_4, ...M_9 are then obtained using (25). However, before doing so, we have to provide an approximation to the central difference formula (19) by means of the extrapolation (cf. 18)

$$\gamma_{n+1} = - \frac{1}{907200} \Big\{ 5040 \nabla^3 + 12600 \nabla^4 + 20820 \nabla^5 +$$

$$28770 \nabla^6 + 36077 \nabla^7 + .. \Big\} F_n \tag{28}$$

By reversing the sign of h, the backward difference formula (28) is easily turned into a forward difference formula and can be used to calculate γ_1 and γ_2 at the start.

Having established a complete scheme of nine values of F's together with values for M_8 and M_9, we are ready to go on with the marching process. First M_{n+1} is calculated from (25), then the extrapolation formula (28) is used to get γ_{n+1}. The remaining task is to solve the "algebraic equation" (cf. eq. 18)

$$y_{n+1} = h(M_{n+1} + \gamma_{n+1}) + \frac{h}{3} F_{n+1} \tag{29}$$

Whereas its solution for linear problems is obvious, nonlinear equations have to be treated in an iterative way. One starts with the zero-order approximation

$$F_{n+1}^{(0)} = E^{-1} F_n$$

$$= \{1 + \nabla + \nabla^2 + \nabla^3 + \dots\} F_n, \tag{30}$$

calculates $y_{n+1}^{(0)}$, gets a new $F_{n+1}^{(1)}$ which leads to $y_{n+1}^{(1)}$ and so on. To improve the order of this iteration process it is possible to use one of the standard procedures, like Aitken's γ^2-process. For details see Löwdin's paper /18/.

Accuracy checks of the resulting solution are conveniently carried out by comparing the central difference relation (19) for the correction with the extrapolation formula (28).

Löwdin /18/ also mentions a way to "aftercorrect" the solution over the entire computed interval by recalculating all γ's by means of (19). In our tests, however, we found that the "preliminary" result never was changed significantly and decided to omit this complication which requires storage of the full γ-array for all grid points.

In modern evaluations (e.g. /11/) of codes for the numerical solution of differential equations an important distinction is made between the time spent for the actual function evaluations and the rest, often called "overhead".

Extreme cases with respect to this distinction are the Runge-Kutta methods and the Adams-PECE method. Runge-Kutta methods have very low overhead but need many function evaluations. Adams-PECE methods spend a lot of overhead to make their decisions about varying order and stepsize and to calculate corresponding weights, but achieve then a given accuracy with a much smaller number of function (derivative) evaluations. The number of function evaluations per step in Löwdin's method depends on the type of the equation. If it is linear, only one evaluation is required, for nonlinear equations it is equal to the number of iterations necessary to solve the "algebraic equation" (29). In praxis, the zero-order approximation (30) for F_{n+1} is usually so accurate that it is realistic to assume only 2-3 function evaluations per step. As to the "overhead" of the method, it must be considered as being very small, at least in the present "standard" version. The only algebraic manipulations necessary to advance one step are calculation of γ_{n+1}, M_{n+1}, $F_{n+1}^{(0)}$ once and y_{n+1} according to (29) 2-3 times.

With the above considerations and Löwdin's /18/ demonstration of the accurateness of his method in mind, we expected the code to perform very well in comparison to other programs available, namely Runge-Kutta (R-K) codes of 4th and 6th /19/ order and the Adams-PECE code DE developed by Shampine & Gordon /20/.

PRELIMINARY NUMERICAL RESULTS

We tested Airy's differential equation

$$y'' = xy \tag{31}$$

and its transformations to Riccati's

$$y' = x - y^2 \tag{32}$$

and to Pruefer's equation

$$y' = x \cos^2 y - \sin^2 y \tag{33}$$

on the intervals [0.0, ± 4.0]. The initial conditions were chosen to give Airy's solution Ai(x) whose logarithmic derivative at x = 0 has the value -0.722 350 252 354 06. Although equations of a very simple type, (31-33) allowed to carry out preliminary tests of all propositions made in the first chapter of this article. The accuracies could be checked against Gordon's /21/ well-known algorithm which is capable of yielding 14-15 figures on the IBM 370/165 in double precision. The results can be summarized as follows. As could be judged from using identical Runge-Kutta methods both for the integration of the second-order equation (31) and the equations (32) and (33), it is more economical to use the first-order equations. Due to the necessity of evaluating trigonometric functions in (33), Pruefer's equation is slightly less efficient than Riccati's equation, but one has to pay this price for the smoother behavior of the solution due to the bounded coefficients. The integration intervals chosen correspond to an exponentially decreasing solution for positive x and to an oscillating behavior for negative values of x. The mixing-in of the independent, exponentially increasing, solution Bi(x) for x > 0 provided a good check for the reliability of the different methods. It was found that the 6th order R-K method, both for (31) and (32-33), gave much better accuracy than the 4th order method for equal number of function evaluations. De Vogelaere's method /17/ was tested for (31) and the well-known fact that it is about as accurate as 4th order R-K, while being about 2-2.5 times faster, could be confirmed. However, in this specific example of Airy's equation, it was still inferior in efficiency to the 6th order R-K method for all accuracies better than 4 significant figures.

It remains to compare the performances of the methods applicable to the nonlinear first order problems (32-33). As expected for such an extremely "cheap" derivative, the 6th order R-K method was more efficient thant Adams-PECE. The 4th order R-K method was not competitive at all, particularly with respect to accuracy. To give an idea, the 6th order R-K method kept 7 figures after 80 steps (= 640 function evaluations), when integrating (32) from x = 0.0 to x = 4.0. The 4th order R-K method, after 320 steps (= 960 function evaluations) only gave 4-5 figures.

The code DE /20/, on the other hand, used 80 steps, but only 160 function evaluations to achieve the same accuracy of 7 significant figures. Still, due to its large overhead, it was slower than 6th order R-K.

Finally, Löwdin's method used 80 steps, corresponding to ca. 160 function evaluations, to obtain the same result as 6th order R-K and Adams-PECE. However, it was 40-50% faster than the Runge-Kutta program.

This behavior was typical of all test calculations. The maximal accuracy achieved by all 3 methods for (32) on [0.0,4.0] was 10 significant figures. For (33) on [0.0,-4.0] the full machine accuracy of 14-15 figures was obtained. In the latter case of a trigonometric-type solution the 6th order R-K routine performed relatively best, compared to the finite-difference methods. Löwdin's method broke down when the stepsize chosen was too big, but was the most efficient again otherwise.

This example showed the value of a code which automatically adjusts its grid. The choice of the proper order in eq. (28) and (30) is simply accomplished by computing the higher differences $\nabla^k F_{n+1}$ and stopping when they drop below a given threshold. Change of stepsize, however, is a less trivial problem whose solution is presently attempted by means of interpolation of the F's and recomputation of M_n and M_{n+1} using eqs. (26-27). A suitable criterion for triggering such a change in stepsize sould be the magnitude of γ_{n+1}.

Since it computes first in each step approximations to F_{n+1} and γ_{n+1} by extrapolation, Löwdin's method would also be well-suited for treating systems of coupled equations.

It is hoped that developments in this direction will contribute to our planned applications of Weyl's theory to problems in more than one dimension.

ACKNOWLEDGEMENTS

This paper is dedicated to my teacher, Professor P.-O. Löwdin, on the occasion of his 60th birthday. His encouragement and constant moral support during the past years have been invaluable.

Support by a grant from the National Science Foundation, NSF GP-42477, is grafefully acknowledged.

REFERENCES

1. H. Weyl, Math. Ann. 68 (1910) 220.

2. E.C. Titchmarsh, "Eigenfunction Expansions Associated with Second Order Differential Equations", Vol. I, Clarendon Press, Oxford (1946) (2nd revised edition: 1962).

3. J. Chandhuri and W.N. Everitt, Proc. Roy. Soc. (Edinburgh) A68 (1968) 95.

4. M. Hehenberger, H.V. McIntosh and E. Brändas, Phys. Rev. A10 (1974) 1494.

5. J. Sitte, Dissertation, Uppsala (1974), (unpublished).

6. E. Brändas and M. Hehenberger, Lecture Notes in Mathematics, Vol. 415 (ed. A. Dold and B. Eckmann), Springer-Verlag, Berlin (1974).

7. M. Hehenberger, B. Laskowski and E. Brändas, "Weyl's theory applied to predissociation by rotation: Mercury hydride", submitted to J. Phys. B.

8. H.V. McIntosh, Seminar Notes, Quantum Chemistry Group Uppsala (1971/72) (unpublished).

9. P.-O. Löwdin, Lecture Notes on Scattering Theory, Uppsala (1949/50) (unpublished).

10. H. Pruefer, Math. Ann. 95 (1926) 499.

11. L.F. Shampine, H.A. Watts and S.M. Davenport, Sandia Laboratories SAND-75-0182, March 1975, to be published in SIAM Review.

12. E.Fehlberg, NASA Technical Report, NASA TR R-287 (Oct. 1968)

13. F.T. Krogh, J. ACM 4 (1973) 545.

14. C.W. Gear, Comm. ACM 14 (1971) 176.

15. R. Bulirsch and J. Stoer, Numer. Math. 8 (1966) 1.

16. B. Numerov, Publ. de l'Observ. Astrophys. Central Russie 2 (1933) 188.

17. R. de Vogelaere, J. Research Natl. Bur. Std. 54 (1955) 119.

18. P.-O. Löwdin, Quarterly of Applied Mathematics X (1952) 97.

19. D. Sarafyan, J. Math. Anal. Appl. 40 (1972) 436.

20. L.F. Shampine and M.K. Gordon, "Computer Solution of ordinary differential equations. The Initial Value Problem", W.H. Freeman & Co., San Francisco (1975)

21. R.G. Gordon, J. Chem. Phys. 51 (1969) 14.

QUANTIZATION AND A GREEN'S FUNCTION FOR SYSTEMS OF LINEAR ORDINARY DIFFERENTIAL EQUATIONS

Harold V. McIntosh

Escuela Superior de Física y Matemáticas, Instituto
Politécnico Nacional, Mexico 14, D.F. México
Instituto Nacional de Energia Nuclear, Avenida Insurgentes
Sur 1079, Mexico 18, D.F. México

One of the traditional puzzles for students of quantum mechanics is the reconciliation of the quantification principle that wave functions must be square integrable with the reality that continuum wave functions do not respect this requirement. By protesting that neither are they quantized the problem can be sidestepped, although some ingenuity may still be required to find suitable boundary conditions. Further difficulties await later on when particular systems are studied in more detail. Sometimes all the solutions, and not just some of them, are square integrable. Then it is necessary to resort to another principle, such as continuity or finiteness of the wave function, to achieve quantization. Examples where such steps have to be taken can be found both in the Schroedinger equation and the Dirac equation. The ground state of the hydrogen atom, the hydrogen atom in Minkowski space, the theta component of angular momentum, all pose problems for the Schroedinger equation. Finiteness of wavefunction alone is not a reliable principle because it fails in the radial equation of the Dirac hydrogen atom. Thus the quantizing conditions which have been invoked for one potential or another seem to be quite varied.

Singular potentials, the criteria for which vary somewhat between the Schroedinger equation, the Dirac equation, and the Klein-Gordon equation, originate another variant of the quantization problem. Continuum wave functions forming the positive energy states of a potential which vanishes at large distances may not be square integrable, but they are at least irredundant in the sense of linear independence in a vector space. Singular potentials exhibit another type of continuum, wherein every energy possesses square-integrable solutions to the wave equation, but they are all linearly dependent on a discrete subset. It is often felt that such potentials are "unphysical" but the appelation is neither true nor particularly admissible as a pretext for not understanding the mathematical properties of such solutions. Physical occurrences of such potentials include the Dirac equation for superheavy nuclei, the ground states of magnetic monopoles, and higher multipole approximations to ordinary potentials. Even if magnetic monopoles do not exist, the same difficulty is still to be found in all the higher multipoles; for instance in electric dipoles.

One wonders why the requirement of square-integrability as the condition of quantization should be so prevalent? No doubt a substantial part of the reason is pedagogical. Square integrability is a rather dramatic characteristic of the solutions of a wave equation, fairly easy to explain in terms of probability, and quite in accordance with the historical development of the philosophy of quantum mechanics. Once students are convinced of the importance of square integrability, the foundations of the course have been laid, and the applications can begin. To dwell further on the foundations, particularly if much mathematics is required, disturbs the balance between speculation and results, and so the matter is usually not pursued further. By the time that the really important discrepancies begin to occur, the simple statement of the quantization principle has become so ingrained as an inviolable axiom, much as happens to the "no crossing" rule, that it is hard to return to basic principles.

If square integrability is not an adequate principle, what really _is_ the principle? For an understanding of this point it seems desirable first to separate quantum dynamics from quantum observability.

Observations are described with the aid of wave functions, in the sense that the square of the amplitude of a wave function is taken to be the probability density for finding a particle, or perhaps an assemblage of particles, in a certain place at a certain time. Properties of the system, such as its energy, its momentum, or some other physical characteristic can be calculated from the wave function by means of a bilinear form and appropriate operators. For such calculations to be possible, the wave functions have to satisfy certain restrictions; for example square integrability is the requirement that there be a unit probability of finding the particle somewhere, anywhere. Consequently the requirement that wave functions should belong to Hilbert space has been widely taken as a basic principle of quantum mechanics. A whole theory of probability is then overlaid on this Hilbert space: the elementary probabilities are taken from bilinear operators on the Hilbert space, following which their origin is of no great concern.

Dynamics, on the other hand, is described by a wave equation, which follows out the temporal development of the wave functions. Even if the system is static, its stasis is described in a special way, by a time-independent wave equation. There are two styles which can be used for this description: operators or functional analysis. On the one hand there is matrix mechanics or the "Heisenberg picture", while on the other there are partial differential equations comprising wave mechanics and the "Schroedinger picture". The choice between these is a matter of taste, although somethings are better expressed one way and others in the other. Differential equation theory tends to be more concrete and computational, giving specific examples and information, while functional analysis allows the information to be summarized from an algebraic viewpoint, dealing with dimensions of solution spaces, mappings between them, and so on.

Having established dynamics and interpretation as two distinct phases of quantum mechanics, the discrepancies between the two have to be taken into account. At the same time it is possible to locate the "quantization principle" and to identify the source of confusion with respect to its application. It is essential to observe that dynamics and interpretation work with two different structures having quite different characteristics. Dynamics is a local theory, dealing with differential equations and derivatives, boundary values and initial conditions, and such like. Interpretation is a global theory, dealing with integrals and integrability, functions and bases, probabilities and statistics. In turn, dynamics can be taken as an expression of the automorphisms of the function spaces of interpretation theory.

The quantization principle is that solutions of the dynamical equations must form a basis for the Hilbert space upon which the interpretation is performed.

Most of the confusion to which we have alluded arises from supposing that a basis for Hilbert space either must or ought to belong to Hilbert space.

The selection of a basis for Hilbert space is an old story. Physicists distinguish "wave functions" and "wave packets" for just this reason, although the way the motivation is usually stated is that they wish to localize their particle. A plane wave may satisfy the Schroedinger equation, but cannot represent a particle because it has a constant amplitude everywhere, even at very remote distances. A Fourier synthesis of plane waves may localize the particle, evidenced by the square integrability of the represented function.

Even though a Hilbert space basis of wave packets could be built up, it is still conceptually simpler to think in terms of plane waves.

Probably the reason that this distinction was lost pedagogically was to avoid burdening an elementary exposition with all the machinery of Fourier analysis, especially in those applications where bound states predominate, and a single solution is at once square integrable and stationary. Mathematical authors of books on Hilbert space have not particularly aided the cause, because of their proclivity for self-contained axiomatic systems. Hilbert space bases of Hilbert space are given a careful analysis, bounded operators receive a preferential treatment; but even such fundamental operators as the derivative wreak havoc with such a restricted theory. Such theories were not precisely what was needed.

Historically the different approaches of Dirac and von Neumann illustrate the contrast between a formalistic but readily usable approach and a mathematically accurate but somewhat complicated exposition. The growth of Schwartz' distribution theory has reconciled many of the technical discrepancies between those two extremes of approach, but the real difficulty has always been more philosophical or conceptual. From the beginning there has been fairly adequate mathematical machinery available once it was clear on what to use it, notably in the form of the theory of the Stieltjes integral.

At least for situations in which the dynamical equation can be written as a set of ordinary linear differential equations there is a very interesting explicit construction connecting the interpretation space of square integrable functions and the solution space for a set of differential equations. The construction originated with Weyl's dissertation /1/ of 1910, played an important role in Schroedinger's formulation of his explanation of quantization, and eventually received an extended application when Titchmarsh and a series of his students began to generalize these concepts at mid-century. These trends were analyzed in Loebl's third volume /2/ of "Group Theory and its Applications" which should be consulted as a predecessor to the present article.

In the Loebl article a single second order differential equation was chosen as an example, because it sufficed to explain how Green's formula establishes a mapping between function space and the space of boundary values for a differential equation. Such a low dimension simplifies the exposition, but at the same time there occur some obscuring simplifications. The fact that Green's function is a scalar is one of them, the fact that some aspects are elegantly formulated in terms of analytic functions of a single complex variable is another.

To present a discussion of some of the details of the higher order systems for Per-Olov Löwdin's anniversary volume somehow seems appropriate, because they involve group theory, projection operators, matrix partitioning, and long forgotten papers which somehow contain all that a contemporary young mathematician could desire.

Birkhoff and Langer /3/ showed how to apply Sturm Liouville theory to systems of first order ordinary differential equations, with the objective to expand functions defined over an interval in terms of the eigenfunctions of the system over the same interval. For purposes of discussion, as well as for numerical integration, the system could be written in <u>standard</u> form

$$\frac{dZ}{dx} = MZ + F \tag{1}$$

Z would be a vector if the intention were to treat a system of equations, but it is better to make Z a square matrix so that all the linearly independent solutions arising from an arbitrary initial condition can be obtained at once. Series solutions of such an equation, in terms of the matrizant, have been known since the last century.

To obtain Green's formula of the Lagrange identity, an adjoint equation is needed, and will generally require a matrix coefficient for the derivative term. For that reason the <u>canonical</u> form of the equation can be introduced:

$$\alpha \frac{dZ}{dx} + \left(\beta + \tfrac{1}{2} \frac{d\alpha}{dx}\right) Z = \lambda\gamma Z + R \tag{2}$$

When F and R are zero, the equation is called homogeneous; otherwise inhomogeneous. If the homogeneous equation 2) is written in operator form

$$L(Z) = \lambda\gamma Z$$

there is an adjoint equation

$$-\alpha^T \frac{dW}{dx} + \left(\beta^T - \tfrac{1}{2} \frac{d\alpha^T}{dx}\right) W = \lambda\gamma^T W \tag{3}$$

whose operator form is

$$M(Z) = \lambda\gamma^T Z$$

Obviously, if

$$\alpha = -\alpha^T$$

$$\beta = \beta^T$$

$$\gamma = \gamma^T$$

the adjoint equation will be the same as the original equation and deserve to be called self-adjoint. The requirement that γ be symmetric is not particularly important, but will usually be imposed to get a symmetric inner product in function space. Likewise it is not necessary to insist that α be nonsingular (and hence that the system be of even order) but the contrary assumption would effectively lower the order of the system, at least at the point of singularity, thereby complicating the analysis.

By direct substitution it can be established that

$$Z^A = (\alpha Z)^{-1\,T},$$

which we could call the adjoint of Z, satisfies the adjoint equation. Therefore, given a self-adjoint differential equation, Z and Z^A can differ at most by a constant multiplicative factor which would have to be the discrepancy in their initial values. From this much alone we can see that the solution of a self-adjoint system of equations is going to have some special properties.

Once the adjoint of an operator has been properly defined, Green's formula follows from a straightforward calculation with two vector solutions ϕ and ψ.

$$\int_b^a \{\phi^T L(\psi) - (M(\phi))^T \psi\}dx$$

$$= \phi^T \alpha\psi|_b - \phi^T \alpha\psi|_a \qquad (4)$$

The formula can be given a real or complex form according to whether the symmetric or the hermitean transpose are used. Moreover, if these vectors are eigenfunctions,

$$L\,(\psi) = \lambda\gamma\psi$$

$$M\,(\phi) = \mu\gamma^T\phi$$

Then

$$(\lambda - \mu)\int_a^b \phi^T \gamma\psi dx = \phi^T \alpha\psi\Big|_a^b \qquad (5)$$

This formula relates inner products in function space to inner products in boundary value space. If ϕ and ψ are taken to be matrices rather than vectors, the result is a relation between Gram matrices.

For real solutions belonging to equal eigenvalues this formula expresses a conservation equation. For a single second order equation it expresses the constancy of the Wronskian of two solutions, whereas in general it states the constancy of a bilinear version of the Wronskian of two solutions, from which other multilinear invariants, including the Wronskian, can be deduced. It is more interesting that when ϕ and ψ are matrices, the result states that the solution matrices conserve a certain bilinear form much as orthogonal matrices conserve distance. The important difference is that the conserved bilinear form is antisymmetric rather than positive definite, so that the solution matrix is required to be symplectic. This condition is particularly evident when the system is written in canonical form and α is the constant unit antisymmetric matrix.

Symplectic matrices have characteristic properties, just as do unitary and orthogonal matrices. Their eigenvalues occur in reciprocal pairs with equal multiplicities. Their eigenvectors are orthogonal with respect to the metric matrix α. These results are not as dramatic as for unitary or orthogonal matrices because the eigenvalues do not need to have absolute value 1, so that a set of symplectic matrices would not usually generate a compact group. Nevertheless they are necessary and sufficient conditions for a complete characterization, and they do have some further consequences. For example, one conclusion is that of 2p linearly independent solutions over a semiinfinite interval, at least p of them must be square integrable, a result which is important for developing the theory further.

Green's formula is the link through which many properties of the solution of a system of differetial equations may be established by allowing passage between function space and boundary value space, quite aside from its use in establishing the symplectic nature of the solution matrix for the system. Perhaps the next interesting result after Green's formula is the derivation of Green's function, which relates the solutions of an inhomogeneous equation to the inhomogeneous term and the solutions of the corresponding homogeneous equation. Green's functions have been determined for an extremely wide variety of differential equations. For this, it seems strange that the explicit form for a first order matrix equation is not more readily accessible, but it does not seem to be found in several of the more widely used textbooks.

It is well known that if

$$\frac{dZ}{dx} = MZ + F$$

and if

$$\frac{\partial G(x,x_0)}{\partial x} = M(x)G(x,x_0)$$

$$G(x_0,x_0) = \underline{1}$$

is the solution of the homogeneous equation from unit initial conditions, then

$$Z(x) = G(x,x_0)Z(x_0) + \int_{x_0}^{x} G(x,\sigma)F(\sigma)d\sigma \qquad (6)$$

is the solution of the inhomogeneous equation.

If the canonical form were used instead, we would have to write

$$M = \alpha^{-1}(-\beta + \lambda\gamma)$$

$$F = \alpha^{-1}R$$

and then return to the standard formulation. But this would only mean that

$$\alpha \frac{\partial G}{\partial x} + \beta G = \lambda\gamma G$$

$$G(x_0,x_0) = \underline{1}$$

so that in either event, G, would be a solution of the homogeneous equation from the unit matrix as initial condition. As a consequence,

$$Z(x) = G(x,x_0)Z(x_0) + \int_{x_0}^{x} G(x,\sigma)\alpha^{-1}(\sigma)R(\sigma)d\sigma \qquad (7)$$

The only trace of the canonical form lies in the presence of the factor α^{-1} under the integral. Usually it is said that G is the kernel of an integral operator, or that it is Green's function for a Volterra, which is to say an initial value, type of equation.

To solve a Sturm-Liouville, boundary value problem, we would begin by referring the value of the solution at an arbitrary point to the boundary values at the points a and b:

$$Z(a) = G(a,x)Z(x) + \int_x^a G(a,\sigma)\alpha^{-1}(\sigma)R(\sigma)d\sigma$$

$$Z(b) = G(b,x)Z(x) + \int_x^b G(b,\sigma)\alpha^{-1}(\sigma)R(\sigma)d\sigma$$

However, boundary <u>conditions</u> are probably more interesting than boundary <u>values</u>, but they can be accommodated by using the matric matrix α and suitable vectors $\underline{a}$ and $\underline{b}$. We require

$$\underline{a}^T\alpha Z(a) = 0$$

$$\underline{b}^T\alpha Z(b) = 0$$

which eventually results in two matrix equations

$$0 = \underline{a}^T\alpha(a)G(a,x)Z(x) + \int_x^a \underline{a}^T\alpha(a)G(a,\sigma)\alpha^{-1}(\sigma)R(\sigma)d\sigma$$

$$0 = \underline{b}^T\alpha(b)G(b,x)Z(x) + \int_x^b \underline{b}^T\alpha(b)G(b,\sigma)\alpha^{-1}(\sigma)R(\sigma)d\sigma$$

They can be combined into a single matrix equation by using the partitioning technique and writing explicit submatrices:

$$0 = \begin{pmatrix} \underline{a}^T\alpha(a)G(a,x) \\ \\ \underline{b}^T\alpha(b)G(b,x) \end{pmatrix} Z(x) + \begin{pmatrix} \int_x^a \underline{a}^T\alpha(a)G(a\sigma)\alpha^{-1}(\sigma)R(\sigma)d\sigma \\ \int_x^b \underline{b}^T\alpha(b)G(b,\sigma)\alpha^{-1}(\sigma)R(\sigma)d\sigma \end{pmatrix}$$

Fortunately, enough of an explicit form for the inverse of the coefficient of Z(x) can be written to be useful. Say that $\underline{a}^T\alpha(a)$ has r rows, and that the system contains 2p equations. Then there are 2p-r columns forming a matrix A for which

$$\underline{a}^T\alpha(a)A = 0$$

as well as r columns forming a matrix B for which

$$\underline{b}^T\alpha(b)B = 0$$

The subsequent results are not affected by the fact that these two matrices are not unique.

After a bit of study we arrive at the result

$$\begin{pmatrix} \underline{a}^T\alpha(a)G(a,x) \\ \\ \underline{b}^T\alpha(b)G(b,x) \end{pmatrix}^{-1} = \begin{pmatrix} G(x,b)B & G(x,a)A \end{pmatrix} \begin{pmatrix} \left(a^T\alpha(a)G(a,b)B\right)^{-1} & 0 \\ \\ 0 & \left(b^T\alpha(b)G(b,a)A\right)^{-1} \end{pmatrix}$$

In order to simplify several subsequent formulas it is convenient to introduce the definitions

$$\Delta_{11} = \underline{a}^T\alpha(a)G(a,b)B$$

$$\Delta_{22} = \underline{b}^T\alpha(b)G(b,a)A.$$

With their help we can begin by writing a fairly explicit form for Z

$$Z(x) = -\int_x^a G(x,b)B\Delta_{11}^{-1}\underline{a}^T\alpha(a)G(a\sigma)\alpha^{-1}(\sigma)R(\sigma)d\sigma$$

$$+ \int_x^b G(x,a)A\Delta_{22}^{-1}\underline{b}^T\alpha(b)G(b,\sigma)\alpha^{-1}(\sigma)R(\sigma)d\sigma$$

which can be made into a single integral

$$Z(x) = \int_a^b \Gamma(x,\sigma)R(\sigma)d\sigma \tag{8}$$

by defining

$$\Gamma(x,\sigma) = \begin{cases} G(x,b)B\Delta_{11}^{-1}\underline{a}^T\alpha(a)G(a,\sigma)\alpha^{-1}(\sigma) & a\leq\sigma\leq x \\ \\ -G(x,a)A\Delta_{22}^{-1}\underline{b}^T\alpha(b)G(b,0)\alpha^{-1}(\sigma) & x\leq\sigma\leq b \end{cases}$$

$\Gamma(x,\sigma)$ is a matrix version of Green's function, which can serve as the kernel in a Fredholm type of equation. Some interesting observations follow from defining

$$P_> = G(\sigma,b)B\Delta_{11}^{-1}\underline{a}^T\alpha(a)G(a,\sigma)\alpha^{-1}(\sigma)$$

$$P_< = G(\sigma,a)A\Delta_{22}^{-1}\underline{b}^T\alpha(b)G(b,\sigma)\alpha^{-1}(\sigma)$$

First, we get the multiplication table

$$P_>\alpha(\sigma)P_> = P_> \qquad P_<\alpha(\sigma)P_> = 0$$

$$P_>\alpha(\sigma)P_< = 0 \qquad P_<\alpha(\sigma)P_< = P_<$$

taken with respect to the metric matrix α. From this we conclude that the P's are orthogonal idempotents. Then again, we find

$$\Gamma(x,\sigma) = \begin{cases} G(x,\sigma)P_>(\sigma) & x\geq\sigma \\ -G(x,\sigma)P_>(\sigma) & x\leq\sigma \end{cases}$$

which makes the Fredholm Green's function a projection of the Volterra Green's function. It can even be written as a left hand projection.

$$\Gamma(x,\sigma) = \begin{cases} P_>(x)\alpha(x)G(x,\sigma)\alpha^{-1}(\sigma) & x\geq\sigma \\ -P_<(x)\alpha(x)G(x,\sigma)\alpha^{-1}(\sigma) & x\leq\sigma. \end{cases}$$

Having two orthogonal idempotents makes us curious about their sum. Fortunately the sum is readily obtainable by observing that

$$\begin{pmatrix} G(\sigma,b)B\Delta_{11}^{-1} G(\sigma,a)A\Delta_{22}^{-1} \end{pmatrix} \begin{pmatrix} \underline{a}^T\alpha(a)G(a,\sigma) \\ \\ \underline{b}^T\alpha(b)G(b,\sigma) \end{pmatrix} = 1$$

so that

$$G(\sigma,b)B\Delta_{11}^{-1}\underline{a}^T\alpha(a)G(a,\sigma) + G(\sigma,a)A\Delta_{22}^{-1}\underline{b}^T\alpha(b)G(b,\sigma) = \underline{1}$$

and finally

$$\alpha^{-1}(\sigma) = P_>(\sigma) + P_<(\sigma)$$

or better

$$\underline{1} = P_>(\sigma)\alpha(\sigma) + P_<(\sigma)\alpha(\sigma)$$

This result is necessary to verify explicitly that equation (7) gives a solution to the inhomogeneous equation, and it is also worthwhile to note that it is just the discontinuity of Green's matrix for coincident arguments. Such an irregularity is well known for scalar Green's functions; they are continuous up to a certain order, but then have a deltafunction discontinuity in their last derivative.

The derivation just given is valid for any system of equations, self-adjoint or not, and for any assignment of boundary conditions to one or the other of the two endpoints. Even so, the formulas have a fairly plausible interpretation. Denominators such as Δ_{11}^{-1} refer to the projection of a solution starting from one of the boundary points on the boundary condition, calculated from the adjoint equation, taken from the other endpoint. According to Green's formula this inner product will be the same no matter the interior point at which it is calculated, and it will vanish only when some initial value at one end gives a solution which meets the boundary condition at the other end. This is in accord with the dichotomy, that whenever the Sturm-Liouville problem has a solution, the inhomogeneous equation does not, and conversely.

When the system of equations is self-adjoint, the boundary values and the boundary conditions satisfy the same differential equation, even though they evolve from possible different initial conditions and hence would not coincide. Anyway, if more conditions were specified at one end than at the other, matrices of two different dimensionalities would be involved. Thus there is a further configuration of high symmetry, wherein half of the boundary conditions are specified at one end and half at the other. At each endpoint the antisymmetric matrix α is a metric matrix for a symplectic geometry, and it may happen that the boundary conditions lie in a maximal isotropic subspace for such a metric. In that case, boundary values would simultaneously serve as boundary conditions. The very highest symmetry would then arise when the two kinds of solutions were considered to be identical, with no distinction between initial value and initial condition.

Indeed, this is the situation most familiar to persons experienced with Green's functions, which are often visualized as products of functions meeting the respective boundary conditions and normalized to have a unit irregularity at their point of crossing. To further compare the derivation just given with a familiar situation, it might be noticed that the self-adjoint form of a second order differential operator

$$-(pu')' + zu = \lambda u$$

is

$$\begin{pmatrix} 0 & -1 \\ 1 & 0 \end{pmatrix} \frac{d}{dx} \begin{pmatrix} u \\ pu' \end{pmatrix} + \begin{pmatrix} -z & 0 \\ 0 & 1/p \end{pmatrix} \begin{pmatrix} u \\ pu' \end{pmatrix} = \lambda \begin{pmatrix} 1 & 0 \\ 0 & 1 \end{pmatrix} \begin{pmatrix} u \\ pu' \end{pmatrix} ,$$

while the self-adjoint form of the one-dimensional Dirac equation:

$$\frac{d}{dx} \begin{pmatrix} \phi_1 \\ \phi_2 \end{pmatrix} = \begin{pmatrix} 0 & m+(E-V) \\ m-(E-V) & 0 \end{pmatrix} \begin{pmatrix} \phi_1 \\ \phi_2 \end{pmatrix}$$

would be:

$$\begin{pmatrix} 0 & -1 \\ 1 & 0 \end{pmatrix} \frac{d}{dx} \begin{pmatrix} \phi_1 \\ \phi_2 \end{pmatrix} + \begin{pmatrix} -V-m & 0 \\ 0 & -V+m \end{pmatrix} \begin{pmatrix} \phi_1 \\ \phi_2 \end{pmatrix} = E \begin{pmatrix} 1 & 0 \\ 0 & 1 \end{pmatrix} \begin{pmatrix} \phi_1 \\ \phi_2 \end{pmatrix} .$$

For the Schroedinger equation the coefficient matrix γ is degenerate, although still symmetric. The bilinear form in function space therefore depends only on the wave function and not on its derivative. For the Dirac equation, we have to use the sum of the squares of the two components, likewise a familiar result.

A self adjoint system of differential equations with canonical boundary conditions is particularly well suited to a discussion of the theory of singular differential equations, because the canonical boundary conditions are already compatible with the definition of the Weyl circle, or its generalization to a higher order system as a maximal isotropic subspace. The details of this generalization may be found in any of the standard references, or in Loebl's third volume /2/. Nevertheless, there is a further detail, the application of this theory to a doubly infinite interval, which has an interesting connection with Green's function, and which is well to bring out this time.

To overcome difficulties of normalizing functions which have a nonzero amplitude over all of an interval which tends to infinity, it is helpful to describe eigenfunction expansions in terms of a Stieltjes integral. For all finite intervals the distribution function of this integral is a step function, whose limiting behaviour is pertinent to the spectral classification of the differential system over an infinite interval. A system of 2p differential equations will allow an eigenfunction expansion of not only scalar functions defined over the solution interval, but even vector-valued functions up to dimension 2p. An expansion formula would be expected to have the form

$$f(x) = \sum_{k=1}^{\infty} c_k\, u_k(x)$$

$$c_k = (f,\, u_k) = \int_a^b f^T(x)\gamma(x)u_k(x)dx.$$

In such a formula, the functions $u_k(x)$ would be the Sturm-Liouville eigenfunctions, c_k their expansion coefficients, and f the arbitrary function to be expanded. However, the Sturm-Liouville functions are supposed to be orthonormal over the interval of representation, and thus of unit square integral. But for purposes of solving an initial value problem, we are more interested in using solutions normalized to a unit initial value. Probably we would use the unit antisymmetric matrix as an initial value if the system were self-adjoint and we wanted a canonical basis. Suppose that the vectors $\{\xi_{i1}\}$ form a maximal isotropic subspace at an initial point and that $\{\xi_{i2}\}$ are their canonical conjugates. Therefore we must have

$$u_k = \sum r_{ki1}\, \xi_{i1} + r_{ki2}\, \xi_{i2}$$

and by substitution

$$f = \sum_{k=1}^{\infty} \sum_{r=1}^{2} \sum_{s=1}^{2} (f, \xi_{kr})\xi_{ks}$$

This formula is to be written as a Stieltjes integral

$$f = \sum_{r=1}^{2} \sum_{s=1}^{2} \int_{-\infty}^{\infty} (f, \xi_r)\xi_s d\rho_{rs}(\mu)$$

whose distribution function is a matrix. Its elements have discontinuities at the eigenvalues of magnitude $r_{kr}\, r_{ks}$, namely products of pairs of expansion coefficients.

One way of determining the spectral matrix is to expand some known function with the hope of isolating its coefficients. In the process, Green's formula may be used to reduce the integral over function space to a sum over the boundary values. Moreover, if the function chosen for expansion belongs to the Weyl surface, the terms which belong to the endpoints will drop out, freeing the formula from an explicit dependence on the endpoints. An implicit dependence remains, because the Weyl surface determines the Sturm-Liouville boundary conditions to be used, but even these vestiges disappear when real boundary conditions are invoked.

It is rare that the same function will belong to the Weyl surfaces at both ends of the two-sided interval, so that the expansion formula would best be applied to a composite function which solves the system of equations in each of two subintervals and has a joining discontinuity at an internal initial point. All integrals would be written as a sum of two parts, for each of which Green's formula would be individually valid.

Finally, it is slightly simpler to apply Green's formula to the Parseval equality rather than to the expansion formula. Taking all this into account, we begin by writing

$$\int_a^b f^H \gamma f dx = \sum_{k=1}^{\infty} |(f^H \gamma u_k)|^2.$$

From the left side:

$$\int_a^b f^H \gamma f dx = \int_a^0 f^H \gamma f dx + \int_0^b f^H \gamma f dx$$

$$= \frac{f^H \alpha f(0) - f^H \alpha f(a) + f^H \alpha f(b) - f^H \alpha f(0)}{\lambda - \lambda^*}$$

Similarly,

$$\sum_{k=1}^{\infty} |(f^H \gamma u_k)|^2 = \sum_{k=1}^{\infty} \frac{|f^H \alpha u_k(0) - f^H \alpha u_k(a) + f^H \alpha u_k(b) - f^H \alpha u_k(0)|^2}{|\lambda - \lambda_k|^2}$$

As noted, a sequence of choices will simplify this equation. By supposing that f belongs to the Weyl surface at each endpoint, the terms $f^H \alpha f(b)$ and $f^H \alpha f(a)$ are eliminated. By using the f's to determine Sturm-Liouville boundary conditions are the endpoints, terms of the type $f^H \alpha u_k(b)$ or $f^H \alpha u_k(a)$ are eliminated. The resulting formula

$$\frac{f^H \alpha f(0^-) - f^H \alpha f(0^+)}{\lambda - \lambda^*} = \sum_{k=1}^{\infty} \frac{|f^H \alpha u_k(0^-) - f^H \alpha u_k(0^+)|^2}{|\lambda - \lambda_k|^2}$$

thus depends on the discontinuity in f at the origin, which is the point at which it does not satisfy the differential equation.

Since the spectral density matrix is divided naturally into quadrants, some algebraic maneuvering and a careful choice of f's is required to obtain the separate quadrants. If we take

$$f_1 = \phi + \psi M_a \qquad x \le 0$$
$$= \phi + \chi M_b \qquad x \ge 0$$

we obtain the result

$$\frac{\operatorname{Im}(M_a - M_b)^{-1}}{\operatorname{Im}(\lambda)} = \int_{-\infty}^{\infty} \frac{d\rho_{11}(\mu)}{|\lambda-\mu|^2} \tag{9}$$

The choice of

$$f_2 = \phi M_a^{-1} + \psi \qquad x \le 0$$
$$\phi M_0^{-1} + \psi \qquad x \ge 0$$

leads to

$$\frac{\operatorname{Im}(M_a(M_b - M_a)^{-1} M_b)}{\operatorname{Im}(\lambda)} = \int_{-\infty}^{\infty} \frac{d\rho_{22}}{|\lambda-\mu|^2} \ . \tag{10}$$

The off-diagonal block can be gotten from a consideration of $(f_1, \gamma f_2)$, which follows a similar expansion and leads to

$$\frac{\operatorname{Im}\left((M_b - M_a)^{-1} M_a\right)}{\operatorname{Im}(\lambda)} = \int_{-\infty}^{\infty} \frac{d\rho_{12}}{|\lambda-\mu|^2} \tag{11}$$

Finally, $\rho_{21} = \rho_{12}^H$

It is an interesting result that these expressions are just the discontinuities in Green's matrix along its diagonal. Therefore we can say that the imaginary part of the discontinuity in the complex Green's matrix is the spectral density function, while the discontinuity in the real Green's matrix is merely $\underline{1}$. This explains why the complex poles of Green's function and the poles of the spectral density are the same; both depend on the same denominator $(M_a - M_b)^{-1}$.

By and large the generalization for a system of equations of Weyl's spectral theory provides the link which is needed between Hilbert space theory and differential equation theory. By the use of such devices as hyperspherical harmonics or separation of variables many quantum mechanical systems can be reduced directly to this form. At the same time, it might be expected that a fairly explicit theory could be developed directly for partial differential equations. In this respect, the use of functional analysis offers a good idea of what to expect. Nevertheless, the principal advantage of a concrete theory such as Weyl's would seem to be the lessons which it teaches us about the diversity of bases for function spaces which can occur in practice, and the danger of supposing that just one type of basis function - the bound state wave function - is typical of them all.

If Weyl's theory is capable of clarifying our philosophical understanding of quantum mechanics, we might go on to ask whether it has any merit as a numerical procedure? By studying the behavior of $|M_a-M_b|$ we have a single scalar function whose zeroes locate the eigenvalues over a finite interval, and whose limiting behaviour will give us some idea of the nature of the spectrum. It is of some advantage that M_a and M_b can be obtained as solutions of a Ricatti equation, but equally disadvantageous if they have to be obtained from complex eigenvalues, because of the fourfold increase in real multiplications involved.

In summary, we have called attention to the vexing problem of explaining just what is the quantization condition for quantum mechanics, and indicated that the extension of Weyl's second order theory of differential equations to systems of equations can be given a particularly elegant formulation which does not seem to be mentioned in any of the common differential equations textbooks. There still remains the explicit demonstration of the correctness of this interpretation through the exhibition of the resolution of a variety of typical examples, which may have to be put off until a subsequent birthday.

REFERENCES

1. H. Weyl, "Über gewöhnliche Differentialgleichungen mit Singularitäten und die zugehörige Entwicklungen willkürlicher Funktionen," Mathematische Annalen 68, 220-269 (1910).

2. H.V. McIntosh, Quantization as an Eigenvalue Problem, in Group Theory and its Applications (E.M. Loebl, ed.), Academic Press, New York, 1975. pp 333-368.

3. George D. Birkhoff and Rudolph E. Langer, "The Boundary

Problems and Developments Associated with a System of Ordinary Linear Differential Equations of the First Order," Proceedings of the American Academy of Arts and Sciences 58, 49-128 (1923). Reprinted in Collected Mathematical Papers of George David Birkhoff, Dover Publications Inc., New York, 1968. pp.345-424.

4. F.R. Gantmacher, The Theory of Matrices, Chelsea Publishing Company, New York, 1959. Volume II, pp. 125-131.

ON RESONANT POTENTIAL SCATTERING

M. Berrondo* and G. García-Calderón

Instituto de Física, University of México, Apdo. 20-364

México 20, D.F., México

It is indeed with pleasure that we dedicate the present contribution to Professor Per Olov Löwdin in the sixtieth anniversary of his birth. From a glance at his bibliography, we could say that his interest in scattering theory has been dormant since his early days as a docent in Uppsala, time at which he lectured on the subject /1/, and led him to study Hulthén's variational principle for the phase shifts /2/. Dormant however, does not mean inexistent: it is enough to recall the number of times he raises the question:"what about the continuum?" during seminars and conferences. In a paper with H. Shull /3/, they actually included the continuum spectrum in a configuration interaction calculation for the helium atom. This was achieved by means of a "discretization" of the continuum, with excellent results for the ground state of He.

The question which we pose in the present article - as well as a partial answer - regards the "discretization" of the continuum for scattering states. In this case, we feel forced to move into the complex energy plane. The so-called poles on the second sheet unveil the structure of the spectrum, as well as explaining the appearance of sharp peaks in the cross sections. Rather than regarding them as poles, we want to solve Schrödinger's equation or complex energies, thus defining a set of functions which can be further utilized in calculations and approximations.

As usual, Löwdin has an alternative answer to the problem, this time in terms of Fourier-Plancherel transforms /4/ and bracketing functions. Perhaps this means that the continuum has awoken....

*Consultant at the Instituto Mexicano del Petróleo, México.

1. BOUNDARY CONDITIONS

In contrast with other fields in Physics, the resonances which we find in non-relativistic quantum mechanics appear as an intrinsic property of the potential, independent of any kind of external field. Indeed, resonant phenomena in mechanical systems, for instance, are readily understood in terms of the response of the system to an external perturbation - usually frequency dependent - acting as a source in the equations of motion. There appears a resonance peak whenever the driving force has a frequency very close to the natural frequency of the system. The responding amplitude varies rapidly in the neighbourhood of the resonant frequency.

The Schrödinger equation instead is a homogeneous equation for the wave function: the behaviour of a particle urged by the potential is solely determined by the boundary conditions we impose on the wave function, since there is no source term to which the system responds.

Let us limit our discussion to central potentials $V(r)$. Defining the energy $E = k^2$ and the effective potential $V_\ell(r) = V(r) + \ell(\ell + 1)/r^2$ for angular momentum ℓ, the radial equation is:

$$\psi''(r) + \left(k^2 - V_\ell(r)\right)\psi(r) = 0 \tag{1.1}$$

Since there is no sink or source at the origin, the radial wave function $\psi(r)$ must vanish at $r = 0$. For non-singular potentials, this condition determines uniquely the solution /5/, up to a multiplicative constant. We define the regular solution $\phi_\ell(r)$ as the one which behaves as:

$$\phi_\ell(r) \underset{r \to 0}{\to} r^{\ell+1}/(2\ell+1)!! \tag{1.2}$$

In favourable cases - namely when $\int_c^\infty |V(r)| dr < \infty$ - we can express this regular solution in terms of the Jost solutions $f_\ell^\pm(k,r)$ defined by their asymptotic behaviour:

$$f_\ell^\pm(k,r) \underset{r \to \infty}{\to} h_\ell^\pm(kr) \to e^{\pm i(kr - \ell\pi/2)}, \tag{1.3}$$

where $h_\ell(kr)$ are the Riccati-Hankel functions /7/. The regular solution is the linear combination:

$$\phi_\ell(k,r) = (2ik^{\ell+1})^{-1}\left(f_\ell^-(k)f_\ell^+(k,r) - f_\ell^+(k)f_\ell^-(k,r)\right), \tag{1.4}$$

with the Jost functions f(k) defined in terms of the Wronskians:

$$f_\ell^\pm(k) = k^\ell W\left(f_\ell^\pm(k,r),\ \phi_\ell(k,r)\right) \equiv k^\ell W_\ell^\pm(k), \tag{1.5}$$

and are chosen so that $f_\ell(k) = 1$ in the absence of interaction.

The asymptotic form of Eqn. (1.4) exhibits the regular solution as a mixture of incoming and outgoing waves. The S-function is hence defined as:

$$S_\ell(k) = \frac{f_\ell^-(k)}{f_\ell^+(k)} = \frac{W\left(f_\ell^-,\ \phi_\ell\right)}{W\left(f_\ell^+,\ \phi_\ell\right)}\ . \tag{1.6}$$

We can also notice from Eqn. (1.4) that, when $f_\ell^+(k)$ vanishes, the function $\phi_\ell(r)$ has a purely outgoing component. Since the flux must be conserved, this happens only for complex values of k (for physical angular momenta ℓ). As is well known /6,8/, for short range potentials - such that $\int_0^\infty e^{\alpha r}|V(r)|r dr<\infty$ for all $\alpha>0$ - $f_\ell^\pm(k)$ are both entire functions in the complex k-plane. In this case, the S-matrix, Eqn. (1.6), becomes a meromorphic function of k, with poles at the vanishing values of $W_\ell^+(k)$. Zeroes of $W_\ell^+(k)$ lying on the imaginary axis correspond to bound states. A complex pole k_n lying on the lower half-plane may also have a physical meaning: The energy associated with this pole is $k_n^2 = \varepsilon_n - i\Gamma_n/2$, and we want to compare it with neighbouring poles $k_{n\pm1}$. If the condition $|k_n^2 - k_{n\pm1}^2| >> \Gamma_n$ is fulfilled, then the pole k_n leads to a sharp peak in the partial cross section $\sigma_\ell(E)$ around $E\simeq\varepsilon_n$. This situation is interpreted as a resonance. The corresponding phase shift undergoes a sudden variation approximately equal to π in the vicinity of ε_n, passing through $\pi/2$ (mod π). When the background contribution is negligible, $\sigma_\ell(E)$ shows a Breit-Wigner shape. It is evident that this does not define the resonance as exactly as the bound states /7/. Nevertheless, the important point is the fact that we can characterize S(k) in terms of its singularities in the k-plane, especially for short-range potentials.

Let us first study the case in which the potential is cut-off, i.e. it actually vanishes beyond a certain value of the radius r = a. The Riccati-Hankel functions become now exact solutions for $r\geq a$, but neither of them vanishes at the origin for real values of k. Instead, we define the Gamow states $\{u_{n\ell}(r)\}$ as the solutions of the complex Schrödinger equation /9/ which vanish at the origin, and fulfill the "outgoing" boundary conditions:*

$$u_{n\ell}''(r) + \left(k_{n\ell}^2 - V_\ell(r)\right)u_{n\ell}(r) = 0 \tag{1.7a}$$

*We shall sometimes omit the subindex ℓ in $u_{n\ell}(r)$ and $k_{n\ell}$

$$u_{n\ell}(0) = 0;\; u'_{n\ell}(a)/u_{n\ell}(a) = h_\ell^{+'}(k_n a)/h_\ell^{+}(k_n a) \equiv b_\ell(k_n a). \quad (1.7b)$$

Our interest in these functions becomes apparent when we compare the boundary condition at $r = a$ with the definition of the S-matrix, Eqn. (1.6): the values $\{k_{n\ell}\}$ in Eqns. (1.7) correspond to the poles of $S_\ell(k)$. For cut-off potentials and fixed ℓ, there is an infinite number of simple poles in the lower half-plane, symmetrically distributed with respect to the imaginary axis. In the upper half-plane there is only a finite number of simple poles, all situated on the imaginary axis, thus implying that the number of bound states is finite. A detailed discussion of these properties is found in Ref. /8/.

It is evident that the boundary condition, Eqn. (1.7b), depends on the state n, and is complex, so it is neither homogeneous nor self-adjoint. Thus orthogonality and completeness cannot be taken for granted. This, together with the fact that the functions associated with poles in the lower half-plane increase exponentially for $r>a$, has limited the use of Gamow states.

Orthogonality in the case of homogenous boundary conditions follows from the use of Green's formula.

For Gamow states, the latter reads:

$$\left(u'_n(r)u_m(r) - u_n(r)u'_m(r)\right)_0^a = (k_m^2 - k_n^2)\int_0^a u_n(r)u_m(r)dr. \quad (1.8)$$

Substituting the boundary conditions Eqn. (1.7b), we see that

$$\int_0^a u_n(r)u_m(r)dr + \frac{b_\ell(k_n a)-b_\ell(k_m a)}{k_n^2 - k_m^2}\, u_n(a)u_m(a) = 0 \text{ for } k_n \neq k_m. \quad (1.9a)$$

The contribution from the surface $r = a$ is a direct consequence of the peculiar choice of boundary conditions. It reflects the fact that the flux associated to $u_n(r)$ is not conserved, so $k_{n\ell}$ must be complex to compensate for this. For $n = m$, we shall <u>choose</u> the condition /10/:

$$\int_0^a u_n(r)u_m(r)dr + \left.\frac{db_\ell(ka)}{d(k^2)}\right|_{k = k_n} u_n^2(a) = 1 \quad (1.9b)$$

to normalize the Gamow states. For bound states, the surface contribution in Eqns. (1.9) is equal to the integral $\int_a^\infty u_n(r)u_m(r)dr$, so the <u>usual</u> <u>orthonormalization</u> is recovered. For states associated to poles on the lower half-plane, the integrals from 0 to ∞ diverge, while Eqns. (1.9) are finite and meaningful results. Furthermore, they are independent of the value of a. This is readily seen by taking the partial derivative of the LHS's of Eqns. (1.9) with respect to a, and using the boundary conditions, Eqns. (1.7b).

We would finally like to mention briefly two different alternatives for the boundary conditions at r = a:

i) in Wigner's R-matrix theory /12/, we find the functions $w_n(r)$ where

$$w_n'(a)/w_n(a) = b^W \text{ - real constant.} \tag{1.10}$$

This is a homogeneous, self-adjoint boundary condition, so use of Green's formula between w_m^* and w_n yields

$$\int_0^a w_m^*(r)w_n(r)dr = \delta_{nm}. \tag{1.11}$$

The eigenvalues associated to these states are all real, and they form a complete set in the interior region $r \leq a$ /5/.

ii) Kapur & Peierls' choice /13/ corresponds to a homogeneous, but non-self-adjoint boundary condition:

$$v_{n\ell}'(a)/v_{n\ell}(a) = b_\ell^S(ka) \equiv h_\ell^{+'}(ka)/h_\ell^+(ka), \tag{1.12}$$

and leads to the orthonormality relation between v_n and the adjoint of v_m:

$$\int_0^a (v_m^\dagger)^* v_n dr \equiv \int_0^a v_m(r)v_n(r)dr = \delta_{nm}. \tag{1.13}$$

These two choices present obvious mathematical conveniences and have been extensively used in scattering problems. They suffer the drawback of depending on the choice of a and, in Kapur-Peierls case, also on the incident energy k^2. In addition, resonances are introduced in an indirect and complicated fashion. Poles may even exist in the absence of a potential. In this sense, Gamow states constitute the most natural set of functions, since their properties do not depend neither on the cut-off nor on the incident energy. They are intrinsically related to the potential, and include the bound states as a subset as well.

2. REDUNDANT SINGULARITIES AND WEYL'S m-FUNCTION

The results obtained above can be painlessly extended to short range potentials - such that $\int_0^\infty e^{\alpha r}|V(r)|rdr<\infty$ for all $\alpha>0$ - since the boundary condition (1.3) defines the Jost solutions uniquely in this case for any complex value of k. These functions $f_\ell^\pm(k,r)$ are entire functions of k /6,8/, and the only singularities of $S_\ell(k)$ are the ones given by:

$$W_\ell^+(k) = 0. \tag{2.1}$$

Let us recall that the Gamow states are defined by the boundary condition at a, Eqn. (1.7b). Actually this condition corresponds to the vanishing of the Wronskian evaluated at r = a. Since its value is independent of r, in the general case of short range potentials we can evaluate it asymptotically, and define the Gamow states as the solutions of:

$$u''_{n\ell}(r) + \left(k_{n\ell}^2 - V_\ell(r)\right)u_{n\ell}(r) = 0, \tag{2.2a}$$

$$u_{n\ell}(0) = 0,\ u'_{n\ell}(r)/u_{n\ell}(r) \underset{r\to\infty}{\to} h_\ell^{+'}(k_n r)/h_\ell^+(k_n r) \to ik_{n\ell}. \tag{2.2b}$$

Furthermore, the equations (1.9) replacing the orthonormality relations are also independent of the value of a, so, we are able to rewrite them in terms of the asymptotic forms of the Gamow states:

$$\int_0^\infty \left(u_n(r)u_m(r) - \tilde{u}_n(r)\tilde{u}_m(r)\right)dr + i\tilde{u}_n(0)\tilde{u}_m(0)/(k_n+k_m) = \delta_{nm} \tag{2.3}$$

Here

$$u_{n\ell}(r) \underset{r\to\infty}{\to} \tilde{u}_{n\ell}(r) \tag{2.4}$$

defines the asymptotic Gamow state. The Green function $G_\ell^+(r,r';k)$, which is defined by the asymptotic outgoing boundary condition, allows us to write the S-matrix as:

$$S_\ell(k) = \lim_{r\to\infty} (-)^\ell\, e^{-2ikr} \left(1 - 2ik\, G_\ell^+(r,r;k)\right). \tag{2.5}$$

For potentials

$$V(r) \text{ such that } \int_c^\infty |V(r)|dr < \infty, \tag{2.6}$$

the Jost solution $f_\ell^+(k,r)$ can be analytically extended into the upper half-plane. In the lower half-plane however, it might have singularities, as is actually the case for the exponential potential /6,14/, where it presents poles on the negative imaginary axis, or for the Yukawa potentials, with logarithmic divergences also on the negative imaginary axis /6-8/. The singularities of the S-matrix have now two different sources: they are zeroes of $W_\ell^+(k)$ or poles of $W_\ell^-(k)$. The latter might be situated on the positive imaginary axis of the k-plane, without corresponding to bound states; it is for this reason that they were christened redundant singularities /14/. It is interesting to notice that these do not appear if we start with the potential V(r), cut it off at a certain distance r = a, and then take the limit $a\to\infty$. The reason being that $f_\ell^+(k,r)$ is entire in k no matter how large a is. This implies that making the analytical continuation of $W_\ell^+(k)$ does not commute with the limit $a\to\infty$. Physically, both ways of proceeding should be equivalent; this is actually the case, since the "physical region" for scattering is for real positive k. The question which arises is whether or not the analytic continuation is stable /8/, and this depends on the asymptotic behaviour of the potential.

The poles of $S_\ell(k)$ and the Green function $G_\ell^+(k)$ coincide for short range potentials. This is not true any more for long range potentials fulfilling (2.6). It is due to the fact that the Green function involves functions with outgoing boundary conditions both in the numerator and the denominator; in contrast, $S_\ell(k)$ compares outgoing with incoming waves. In other words, all the poles of $G_\ell^+(k)$ are still given by Eqn. (2.1), while the redundant singularities "cancel out". This is readily seen from Eqn. (1.4): From a well known theorem by Poincaré /15/, the regular solution $\phi_\ell(k,r)$ is entire in k, since its boundary condition does not depend on k, Eqn. (1.2). Hence if $f_\ell^+(k,r)$ is singular at $k = k_0$, $W_\ell^+(k_0)$ is also singular, but the quotient $f_\ell^+(k,r)/W_\ell^+(k)$ must be an analytic function of k /16/, except for the values defined by Eqn. (2.1), in which k_0 is not included. As a result, the function

$$G_\ell^+(r,r';k) = -\phi_\ell(k,r_<)\cdot f_\ell^+(k,r_>)/W_\ell^+(k) \tag{2.7}$$

has no redundant singularities.

Let us look more closely at the function

$$\xi_\ell^+(k,r) = f_\ell^+(k,r)/W_\ell^+(k), \tag{2.8}$$

which does not have redundant singularities. It is a particular case of Weyl's solution /5,17/ for non-singular potentials (2.6) /18/. Weyl's solution is defined as the (unique) square integrable solution /5,17/ for Imk>0. We shall assume that it can be analytically continued into the lower half-plane. It is a linear combination of the regular and irregular solutions $\phi_\ell(k,r)$ and $\theta_\ell(k,r)$ respectively /19/. The irregular solution is defined from the boundary condition at the origin:

$$\theta_\ell(h,r) \underset{r \to 0}{\to} (2\ell-1)!!/r^\ell \tag{2.9}$$

Hence $\theta_\ell(k,r)$ is an entire function in E and the Wronskian

$$W(\theta_\ell,\phi_\ell) = 1. \tag{2.10}$$

The solution (2.8) is given as:

$$\xi_\ell^+(k,r) = \theta_\ell(k^2,r) + m_\ell(k)\phi_\ell(k^2,r), \tag{2.11}$$

where Weyl's m-function is:

$$m_\ell(k) = -W(\xi_\ell^+,\theta_\ell) = -\frac{W(f_\ell^+,\ \theta_\ell)}{W(f_\ell^+,\ \phi_\ell)} \quad . \tag{2.12}$$

From the above discussion, we see that $m_\ell(k)$ has no redundant singularities, and its poles are given by (2.1), as well as the Green function's.

For short range potentials, $S_\ell(k)$ and $m_\ell(k)$ have the same poles. For long range potentials (2.6), $m_\ell(k)$ has no redundant singularities, so all the poles on the upper half-plane are situated on the imaginary axis and correspond to bound states.

The analytic properties of $m_\ell(k)$ and $G_\ell^+(k)$ are simpler than those of $S_\ell(k)$. In particular, we can write /16/:

$$\oint \frac{G_\ell^+(r,r';k)\cdot 2kdk}{k^2 - \kappa^2} = 0 \text{ for real } \kappa, \tag{2.13}$$

where the contour is taken just above the real k-axis, closed with a large semicircle in the upper half-plane, excluding the bound state poles. For truncated potentials, this is readily transformed into a dispersion relation for $\left(h_\ell^+(ka)\right)^2\cdot S_\ell(k)$ /16/. So we can interpret Eqn. (2.13) as a generalized dispersion relation for fixed ℓ.

We would finally like to mention that $m_\ell(k)$ can be also expressed in terms of $G_\ell^+(k)$, in analogy with Eqn. (2.5), this time around the origin:

$$m_\ell(k) = \lim_{r \to 0} \frac{(2\ell+1)!!(2\ell-1)!!}{r^{2\ell+1}} \left((2\ell+1)G_\ell^+(r,r;k)/r - 1\right). \qquad (2.14)$$

3. CONCLUSIONS

In spite of the mathematical disadvantage of using non-homogenous boundary conditions to define the Gamow states, physically they seem to be more appealing, since they only depend on the particular form of the potential. For potentials (2,6), they are associated to the poles of the Green function $G^+(k)$; for the particular case of short range potentials, these coincide with the poles of the analytic continuation of S(k). Since we can obtain the S-matrix for real k - and hence the physical cross section - from the Green function for potentials (2.6), we should study the analytical structure of the Green function - or equivalently of m (k) - instead of the S-matrix or the partial scattering amplitude. The Green function can be expanded in terms of the Gamow states.

Studying the resonances by looking at the poles of the Green function would make us think that they appear as a response to a (unit) source. But we should also remember the fact that these poles appear already in Weyl's solution (2.8), which fulfills the homogenous equation. In fact, locating the source of Green's function away from the origin produces an unwanted geometrical effect due to the redundant singularities /16/.

It is our conclusion in this paper that the states associated to the poles of the Green function - either outgoing or standing wave - should be included in eigenfunction expansions used in approximate methods as perturbation theories, or the variational method. We hope that the study of elastic scattering as considered in this paper, is already a further step in that direction.

REFERENCES

1. P.O. Löwdin, "Spridningsteori", unpublished notes.

2. P.O. Löwdin & A. Sjölander, Ark. Fys. 3, 155 (1951).

3. P.O. Löwdin & H. Shull, J. Chem. Phys. 23, 1362 (1955).

4. P.O. Löwdin, to be published.

5. E.C. Titchmarsh, "Eigenfunction Expansions Associated with Second-Order Differential Equations", Part I (Oxford Univ.

Press, 1946), Chaps. I & III.

6. See e.g. V. de Alfaro & T. Regge, "Potential Scattering", (North-Holland Publ. Co., Amsterdam, 1965).

7. J.R. Taylor, "Scattering Theory", (J. Wiley, New York, 1972).

8. H.M. Nussenzveig, "Causality and Dispersion Relations", (Academic Press, New York, 1974). Chaps. 11 - 12.

9. G. Gamow, Z. Phys. 51, 204 (1928); A.J.F. Siegert, Phys. Rev. 56, 750 (1939); J. Humblet & L. Rosenfeld, Nucl. Phys. 26, 529 (1961).

10. N. Hokkyo, Prog.Theor.Phys. 33, 1116 (1965). For other normalization procedures, see also Y.B. Zel'dovich JETP 12, 542 (1961) and T. Berggren, Nucl. Phys. A109, 265 (1968).

11. R.E.Peierls & G. García-Calderón, to be published.

12. E.P. Wigner, Phys. Rev. 70, 15 (1946); L. Eisenbud & E.P. Wigner, ibid. 72, 29 (1947); A.M. Lane & R.G. Thomas, Rev. Mod. Phys. 30, 257 (1958).

13. P.L. Kapur & R.E. Peierls, Proc. Roy. Soc. A166, 277 (1938); R.E. Peierls, ibid. A253, 16 (1959).

14. S.T. Ma, Phys. Rev. 69, 668 (1946).

15. H. Poincaré, Acta Math. 4. 213 (1884).

16. J.M. Lozano & M. Moshinsky, Nuovo Cimento 20, 59 (1961).

17. H. Weyl, Math. Ann. 68, 220 (1910).

18. E. Brändas, M. Hehenberger & H.V. McIntosh, Int. J. Quantum Chem. IX, 103 (1975).

19. K. Dodaira, Am. J. Math. 72, 502 (1950).

LAGUERRE POLYNOMIALS,

REMINISCENCES FROM UPPSALA

L. B. Rédei

Department of Theoretical Physics, Umeå

University, S-901 87 Umeå, Sweden

I joined Professor Löwdin's group in Uppsala as a young chemistry student from Hungary in the summer of 1957. The first couple of months I spent doing odd jobs in the office like pasting Physics Abstract cuttings on sheets of white paper and helping the secretaries with the binding of Technical Reports. In the meantime I was busy reading Löwdin's articles trying to learn and understand his approach to quantum chemistry. After a few months Professor Löwdin decided that I might have made enough progress to be given a problem. It was on the ground state of the Helium atom, where extensive use was to be made of Laguerre polynomials. At this time Professor Löwdin had put strong emphasis on the importance of using a complete set of base functions like Laguerre polynomials rather than e.g. bound state hydrogen-like wave functions which are not complete unless the continuum states are also included /1/. I was supposed to use the variational method. This was my first research problem and I started to work on it with great enthusiasm. After calculating matrix elements for three months I got to the stage of putting it on the computer. To my horror and surprise the result of the first run was far below the experimental value of the ground state energy. Inexperienced as I was, even I had realized that something must have gone wrong. I did not say anything to Professor Löwdin but went home quietly and spent a torturous week trying to locate the mistake. Finally I found it. It turned out that I had taken a formula for Laguerre polynomials from a book which used a somewhat unorthodox sign convention, different from ours. When later on I told Professor Löwdin about this, he said: "This will be a useful lesson for you. You see, you should never take a formula from a book without checking it yourself". It is my pleasure to be able to show my teacher that I have indeed learnt from this lesson. Last year in the course of my work on the isospin properties of many-particle systems I came upon a new representation for Laguerre polynomials which inspite of its simple and perhaps useful form is not quoted in the standard texts. It is the following:

$$L_n(x) = (-1)^n \frac{e^x}{n!} \left(x \frac{d^2}{dx^2} + \frac{d}{dx}\right)^n e^{-x} , \qquad (1)$$

where $L_n(x)$ are the Laguerre polynomials with the sign conventions of the Bateman Manuscript Project /2/. This formula is the analogue of the well-known representation of the Hermite polynomials:

$$H_n(x) = (-1)^n e^{x^2} \left(\frac{d}{dx}\right)^n e^{-x^2} . \qquad (2)$$

For the proof of equation (1) you can do it yourself or you may look. it up in my paper /3/. Whichever is quicker.

REFERENCES

1. P.O. Löwdin & H. Shull, Phys.Rev. 101, 1730 (1956).
2. Bateman Manuscript Project, Higher Transcendental Functions, Vol. II. (New York, 1953).
3. L.B. Rédei, Acta Scientiarum Mathematicarum (Szeged), Vol. 37, 115 (1975).

PARTITIONING TECHNIQUE FOR DETERMINANTAL EQUATIONS

Frank Weinhold

Department of Chemistry, Stanford University

Stanford, California 94305

I. INTRODUCTION

The partitioning of large arrays into block components which could themselves be manipulated as algebraic entities is a technique which has been used with particular effectiveness by P.-O. Löwdin and his school. This "partitioning technique" underlies the Löwdin analysis /1/ of the relationship between perturbation and variational treatments of Schrödinger's equation, and leads to the resolvent algebra, inner projections, and other important formal developments. In this brief note we describe how this technique can also simplify the treatment of determinantal equations, which permit a unified approach to certain problems of numerical approximation, interpolation, and quadratures which arise frequently in quantum chemistry, as well as to the determination of rigorous error bounds for the quality of approximate wavefunctions and the associated quantum-mechanical properties /2/.

II. SOLUTIONS OF DETERMINANTAL EQUATIONS

One may frequently be interested in what occurs as a determinantal function passes through one of its roots. A well-known example is the ordinary Rayleigh-Ritz secular determinant, whose zeros mark the Hylleraas-Undheim-MacDonald /3/ upper bounds to the true energy levels. In general one may consider a determinantal function $D = D(z) = \det |d_{ij}(z)|$, some or all of whose elements depend on an unknown quantity z, and where it is desired to find solutions of $D(z) = 0$. Although determinantal functions can be quite clumsy to manipulate, expressions for their roots are sometimes surprisingly simple and convenient.

Let us suppose that the vanishing determinant of interest can be block-partitioned into the form

$$D = \det|\, d_{ij}(z) \,| = \det\begin{vmatrix} \underline{G} & \underline{f} \\ \underline{g}^{\dagger} & \underline{A} \end{vmatrix} = 0, \qquad (1)$$

where $\underline{G}$ and $\underline{A}$ are square matrices of order N and M, respectively,

$$(\underline{G})_{ij} = d_{ij}\,, \qquad i,j = 1,2,\ldots,N$$
$$(\underline{A})_{ij} = d_{N+i,N+j}\,, \qquad i,j = 1,2,\ldots,M \qquad (2)$$

and where the entire dependence on the unknown quantity z is contained in the block $\underline{A} = \underline{A}(z)$ of order M. This latter assumption excludes the interesting case of the secular determinant, but by so confining dependence on z to a single block one obtains closed-form solutions rather than the iterative or perturbation-type solutions which are familiar when z appears in very determinantal row.

The condition D = 0 is equivalent to the existence of a null eigenvector for the matrix of the determinant in Eq. (1). When this eigenvector is written in partitioned form (with elements $\underline{\eta}_N$ and $\underline{\eta}_M$), the eigenvalue problem is

$$\begin{pmatrix} \underline{G} & \underline{f} \\ \underline{g}^{\dagger} & \underline{A} \end{pmatrix}\begin{pmatrix} \underline{\eta}_N \\ \underline{\eta}_M \end{pmatrix} = \begin{pmatrix} \underline{G}\underline{\eta}_N + \underline{f}\underline{\eta}_M \\ \underline{g}^{\dagger}\underline{\eta}_N + \underline{A}\underline{\eta}_M \end{pmatrix} = 0. \qquad (3)$$

If $\underline{G}$ is non-singular, the first of these equations permits the component $\underline{\eta}_N$ to be written as

$$\underline{\eta}_N = -\,\underline{G}^{-1}\underline{f}\underline{\eta}_M\,, \qquad (4)$$

while the second equation becomes, in view of (4),

$$(\underline{A} - \underline{g}^{\dagger}\underline{G}^{-1}\underline{f})\underline{\eta}_M = 0. \qquad (5)$$

Eq. (5) is again a null eigenvector problem (of order M rather than N+M), and can be expressed as the equivalent determinantal equation

$$\det|\, \underline{A} - \underline{g}^{\dagger}\,\underline{G}^{-1}\underline{f} \,| = 0. \qquad (6)$$

One can therefore write the overall identity

$$\det \begin{vmatrix} \underline{G} & \underline{f} \\ \underline{g} & \underline{A} \end{vmatrix} = \det|\underline{A} - \underline{g}^{\dagger}\underline{G}^{-1}\underline{f}| = 0, \tag{7}$$

which is apparently a generalization of the usual formula for a determinant of scalar elements.

Eq. (7) shows that the key numerical step in determinantal reduction is the formation of the product $\underline{g}^{\dagger}\underline{G}^{-1}\underline{f}$. Of course, it is usually undesirable to literally compute the inverse matrix $\underline{G}^{-1}$, but the product $\underline{G}^{-1}\underline{f}$ is readily obtained by the Gaussian elimination procedure from the associated set(s) of inhomogeneous linear equations. Thus, the product $\underline{g}^{\dagger}\underline{G}^{-1}\underline{f}$ can be evaluated with fast, stable algorithms even when the order N og $\underline{G}$ is large, or when $\underline{G}$ (as may often be the case) is non-symmetric.

For small M, Eq. (6) is treated by ordinary algebraic manipulations. For M = 1, the desired solution for $\underline{A} = a_{11} = a_{11}(z)$ is simply

$$a_{11} = \underline{g}^{\dagger}\underline{G}^{-1}\underline{f}, \quad (M = 1), \tag{8}$$

while for M = 2, we may denote

$$t_{ij} = (\underline{g}^{\dagger}\underline{G}^{-1}\underline{f})_{ij} \tag{9}$$

so that Eq. (6) is written as

$$\det \begin{vmatrix} a_{11}-t_{11} & a_{12}-t_{12} \\ a_{21}-t_{21} & a_{22}-t_{22} \end{vmatrix} = 0. \tag{10}$$

In one case of frequent interest, both $a_{12} = a_{21}$ and $t_{12} = t_{21}$, and the unknown appears only in the off-diagonal element a_{12}; in this case

$$a_{12} = t_{12} \pm [(a_{11}-t_{11})(a_{22}-t_{22})]^{\frac{1}{2}}. \tag{11}$$

Equations such as (8) and (11) were previously obtained by ordinary determinantal manipulations [cf. Eqs. (32) and (44) of Ref. 2], but their derivation is noticably simplified by the simple formula (7) obtained from the partitioning technique.

III. APPLICATION TO METHODS FOR NUMERICAL APPROXIMATION

Determinantal equations such as Eq. (1) are sometimes a convenient way to formulate various numerical "fitting" procedures, such as interpolation, least-squares or moment-matching approximations, and the associated quadratures. Suppose for example that a function f(x) is to be approximated by a linear combination of N chosen functions $g_i(x)$,

$$f(x) \simeq \tilde{f}(x) = \sum_{i=1}^{N} c_i g_i(x) \ , \tag{12}$$

where the interpolant $\tilde{f}$ is to agree with known values f_k of the function at the N specified points x_k,

$$f_k = f(x_k) = \sum_{i=1}^{N} c_i g_i(x_k) \ , \ k = 1,2,\ldots,N. \tag{13}$$

The condition that there exist some coefficients c_i satisfying (12) and (13) can be expressed by the single determinantal equation

$$\det \begin{vmatrix} g_1(x_1) & g_2(x_2) & \cdots & g_N(x_1) & f_1 \\ g_1(x_2) & g_2(x_2) & \cdots & g_N(x_2) & f_2 \\ \vdots & \vdots & & \vdots & \vdots \\ g_1(x_N) & g_2(x_N) & \cdots & g_N(x_N) & f_N \\ g_1(x) & g_2(x) & \cdots & g_N(x) & \tilde{f}(x) \end{vmatrix} = 0. \tag{14}$$

That Eq. (14) is true only if (12) and (13) are satisfied can be seen by multiplying the first column of the determiannt by c_1, the second by c_2, and so forth up to c_N, then observing that the sum of the results is identical to the final column, a well-known condition for the vanishing of a determinant. Eq. (14) is equivalent to Eq. (1) if the partitioned elements are identified as

$$\begin{aligned} \underline{A} &= \tilde{f}(x) \ , \ (\underline{f})_k = f_k \ , \\ (\underline{g})_k &= g_k(x) \ , \ (\underline{G})_{jk} = g_k(x_j) \ . \end{aligned} \tag{15}$$

the solution of Eq. (12) is therefore given immediately by Eq. (8),

$$\tilde{f}(x) = \underline{g}^{+}\underline{G}^{-1}f = \sum_{i=1}^{N} g_i(x)(\underline{G}^{-1}\underline{f})_i \ , \tag{16}$$

e.g., by $c_i = (\underline{G}^{-1}\underline{f})_i$. When the g_i's are menbers of the ordinary power series $1, x, x^2, \ldots$, Eq. (16) is equivalent to the ordinary Lagrange interpolation formula, but with coefficients expressed in a form which avoids rations of determinants. Other choices of the g_i's lead to other standard interpolation schemes.

One can obtain a more general point of view by letting $[f(x)]_k$ denote the kth "property" of the function f(x) , with $[g_i(x)]_k$ the corresponding property of the ith basis function $g_i(x)$. The requirement that the approximant $\tilde{f}(x)$ should have N such properties in common with the true function f(x) can then be expressed as

$$[f(x)]_k = \sum_{i=1}^{J} c_i [g_i(x)]_k \ . \qquad (17)$$

For example, $[f(x)]_k$ might represent the value of the function at the kth point (as was previously discussed),

$$[f(x)]_k = f(x_k), \qquad (18a)$$

or the kth moment of the function,

$$[f(x)]_k = \int x^k f(x) dx, \qquad (18b)$$

or the convolution of f with some other function $h_k(x)$,

$$[f(x)]_k = \int h_k(x) f(x) dx, \qquad (18c)$$

or the corresponding finite sum over some J points x_j,

$$[f(x)]_k = \sum_{j=1}^{J} h_k(x_j) f(x_j) \ , \qquad (18d)$$

or the kth derivative of the function at some specified point x_0,

$$[f(x)]_k = \left. \frac{d^k f(x)}{dx^k} \right|_{x_0} \ , \qquad (18e)$$

and so forth; the corresponding $[g_i(x)]_k$ brackets are then defined in an obvious analogous manner. In eqs. (18c) and (18d), the special choices

$$h_k(x) = \delta(x - x_k), \qquad (19a)$$

$$h_k(x) = x^k, \qquad (19b)$$

correspond to the interpolation and moment-fitting procedures of (18a) and (18b), respectively, while the choice

$$h_k(x) = g_k(x) \tag{20}$$

gives the least-squares approximant, as one easily confirms.

The condition that an approximant (12) should satisfy the properties (17) is once again equivalent to the determinantal equation (1) if one identifies

$$\underline{A} = \tilde{f}(x), \quad (\underline{f})_k = [f(x)]_k,$$
$$(g)_k = g_k(x), \quad (\underline{G})_{jk} = [g_k(x)]_j, \tag{21}$$

for any choice of the definition of $[\ldots]_k$. The desired approximant $\tilde{f}(x)$ is accordingly given by the explicit solution (16) in terms of known quantities $\underline{f}$ and $\underline{G}$. One therefore obtains a simple but practical prescription for calculating interpolants of rather general character by means of the partitioning technique.

IV. APPLICATION TO ERROR BOUNDS

Determinantal equations of the form of Eq. (1) arise frequently in the determination of upper and lower bounds to quantum-mechanical properties /2/ and the overlap of an approximate wavefunction with the (unknown) true wavefunction ψ /4/. Such bounds often arise from the known positivity of certain Gram determinants, whose elements are scalar products among a chosen set of quantum-mechanical vectors $|\chi_i\rangle$,

$$d_{ij} = \langle\chi_i|\chi_j\rangle, \tag{22}$$

which may include, or depend upon, the unknown ψ. The positivity of the resulting determinant,

$$D = \det |d_{ij}(\psi)| \geq 0, \tag{23}$$

is a consequence of very general geometric properties of the quantum mechanical Hilbert space, so that elements $d_{ij}(\psi)$ whose precise value are otherwise unknown are constrained to occupy the range for which (23) is satisfied. In this manner, rigorous bounds to the permissible values of the elements d_{ij} are found at the limit $D = 0$ of inequality (23). The determination of error bounds therefore becomes the problem of determining the roots of a (Gram) determinantal function, as in Eq. (1), and for this purpose the general expressions such as (8) or (11) are quite useful.

The general approach to error bounds based on Gramian determinantal inequalities has been further pursued by Wang /5/, Spruch and coworkers /6/, Cohen and coworkers /7/, and others /8/. General reviews of the formal methods are given in Refs. 2 and 4, and a variety of numerical applications havesubsequently been made by Weinhold and coworkers /9/, Sims and coworkers /10/, and others /11/. Related determinantal mehtods have been employed in determining rigorous bounds for quantum-mechanical sum rules /12/ and van der Waals force constants /13/. In conjunction with a determinantal statement of the Hylleraas-Undheim-MacDonald interleaving theorem /3/, Eq. (8) also permits a particularly simple derivation /2/ of the <u>operator inequalities</u> /14/, as properly generalized to the case of indefinite operators.

ACKNOWLEDGEMENTS

It is a pleasure to join his many friends around the world in a sincere expression of respect and appreciation to Professor Per-Olov Löwdin for his many contributions to quantum chemistry. The results described in Section III arose at Uppsala in descussions with Erkki Brändas and Harold McIntosh during the latter's seminars on "Surfaces, Defects, and Far Neighbors" in 1972; special thanks are due these workers as well as Professor Löwdin and other members of the group for the stimulating and pleasant atomsphere of the Quantum Chemistry Group.

REFERENCES

1. P.-O. Löwdin, in, C.H. Wilcox (ed.), <u>Perturbation Theroy and Its Applications in Quantum Mechanics</u> (John Wiley, New York, 1966) pp. 255-294, and references therein.

2. F. Weinhold, Advan. Quantum Chem. <u>6</u>, 299 (1972).

3. E.A. Hylleraas and B. Undheim, Z. Phys. <u>65</u>, 759 (1930); J.K.L. MacDonald, Phys. Rev. <u>43</u>, 830 (1933).

4. F. Weinhold, J. Math. Phys. <u>11</u>, 2127 (1970).

5. P.S.C. Wang, J. Chem. Phys. <u>52</u>, 4464 (1970); Chem. Phys. Letters <u>11</u>, 318 (1971).

6. R. Blau, A.R.P. Rau, L. Spruch, Phys. Rev. A <u>8</u>, 119 (1973), and subsequent papers.

7. M. Cohen and T. Feldmann, Can. J. Phys. <u>48</u>, 1681 (1970); M. Cohen, T. Feldmann, and R.P. McEachran, J. Phys. B <u>5</u>, 193 (1972); J.G. Leopold, J. Katriel, and M. Cohen, Chem. Phys. <u>3</u>, J.G. Leopold, M. Cohen, and J. Katriel, J. Phys. B <u>8</u>, 513 (1975).

8. See, e.g., P. Bonelli and G.F. Majorino, Nuovo Cim. B 69, 209 (1970); R.R. Merkel, J. Chem. Phys. 62, 3198 (1975).

9. F. Weinhold. Proc. Roy. Soc. (London) A327, 209 (1972); J. Chem. Phys. 59, 355 (1973); D.P. Shong and F. Weinhold, Can. J. Chem. 51, 260 (1973); M.T. Anderson and F. Weinhold, Phys. Rev. A 9, 118 (1974); Phys. Rev. A 11, 442 (1975).

10. J.S. Sims and R.C. Whitten, Phys. Rev. A 8, 2220 (1973); J.S. Sims and J.R. Rumble, Jr., Phys. Rev. A 8, 2231 (1973); J.S. Sims, S.A. Hagstrom, and J.R. Rumble, Jr. (to be published).

11. See, e.g., E.N. Svendsen, Chem. Phys. Letters 13, 425 (1972), and Refs. 5-8.

12. F. Weinhold. J. Phys. A 1, 655 (1968).

13. F. Weinhold. J. Phys. B 2, 517 (1969); J Chem. Phys. 50, 4136 (1969).

14. P.-O. Löwdin, Phys. Rev. 139, A357 (1965); J. Chem. Phys. 43, S175 (1965).

LOWER BOUNDS TO ENERGY EIGENVALUES

Charles E. Reid

Quantum Theory Project, University of Florida, Gainesville, Florida 32611, U.S.A.

Since the publication of Schrödinger's equation in 1926, a vast amount of human effort and, more recently, of computer time, has been expended on the calculation of approximate energy levels of atoms and molecules. Much, perhaps most, of this work has utilized methods based on the variation theorem. This includes the treatment of the helium atom by Hylleraas and that of the hydrogen molecules by James and Coolidge, as well as all calculations by the Hartree-Fock method and its many variants, or by configuration interaction. It is well known that these methods yield upper bounds to the true eigenvalues. From the early days of quantum mechanics there has been some interest in the much more difficult problem of calculating lower bounds also, so that the eigenvalues could be rigorously delimited within a known range. Temple's /1/ formula was derived in a non-quantum mechanical context, but it and the related formula of D.H. Weinstein /2/ were soon applied to energy levels. These methods were improved by Stevenson and Crawford /3,4/ and, despite severe limitations, are still occasionally used. Starting about 1960, there was an upsurge of interest in lower bounds, as a result of the adaptation of A. Weinstein's /5/ intermediate problem method to quantum mechanics by Bazley and Fox /6/ and later of the bracketing function by Löwdin /7/. A flurry of activity during the next few years showed that these methods, at least as usually understood, were restricted to rather trivial systems such as the helium atom, and the amount of attention given to lower bounds has again dwindled. Whether recent breakthroughs /7,8/ will revive interest still remains to be seen.

There may be several motives for lower bound calculations. For some reasons of logic or even esthetics may suffice - that without both upper and lower bounds the agreement of quantum mechanics with experiment has not been rigorously tested. For others there is the hope that such calculations will be developed to a sufficient degree of versatility that practically useful calculations can be made with them. For a species that is difficult to treat experimentally, such as an unstable radical postulated as an intermediate in a chemical reaction, upper bound calculations are practical but may leave serious uncertainty as to their accuracy, and it would be desirable to be able to place an upper limit on this uncertainty. This use appears to be a rather distant goal for lower bounds research; present methods are far from being able to accomplish it.

This paper is a review of the most widely useful methods of lower bound calculation, intended to show the relations among them, their successes, difficulties, and limitations, and to suggest what problems need to be solved if further progress is to be made. It will be assumed that the reader is acquainted with the usual methods of approximate calculation of energy eigenvalues, as presented in standard textbooks such as that of Messiah /9/. Ideas from the theory of Hilbert space, such as the spectral resolution of operators, will be presented and briefly described where needed. No attempt will be made at mathematical rigor in the presentation of these topics, but many textbooks are available for those who want a rigorous treatment (for example, von Neumann /10/; Halmos /11/; Helmberg /12/.

Throughout this article all energies are reported in Hartree units (1 Hartree = 4.3594 attojoules = twice the ionization energy of the hydrogen atom).

1. METHODS BASED ON THE SQUARE OF THE HAMILTONIAN

All of the early methods for lower bound calculation were based on integrals of the form $\langle H\phi|H\phi\rangle$ or, equivalently $\langle\phi|H^2|\phi\rangle$, where H is the Hamiltonian and ϕ a trial function obeying boundary conditions appropriate to the system. These are most readily derived by a method suggested by MacDonald /13/ and Stevenson /3/, which consists of applying the variation theorem to the operator $(H - \alpha)^{2}$*, where α is a constant. If the eigenvalues of H are E_1 E_2,... (in non-decreasing order) then those of $(H - \alpha)^2$ are $(E_1 - \alpha)^2$, $(E_2-\alpha)^2$,... Now if α lies in a discrete part of the spectrum of H, then one of the E's - call it E_m - must lie closer to α than any other (or at least as close, if α lies midway between two E's). Then $(E_m-\alpha)^2$ is the lowest of the eigenvalues of $(H - \alpha)^2$, and for a normalized

*Wherever a number (such as α in this case) is used as an operator, it is to be interpreted as a multiplicative operator and might more rigorously be written $\alpha \cdot I$, where I is the identity operator.

trial function ϕ the variation theorem gives

$$<\phi|(H-\alpha)^2|\phi> \geq (E_m-\alpha)^2 \tag{1}$$

Calling the expression on the left Δ^2, we have

$$\Delta^2 = <H^2>-2\alpha<H>+\alpha^2 \geq (E_m-\alpha)^2$$

where $<H^2>$ and $<H>$ have been written for $<\phi|H^2|\phi>$ and $<\phi|H|\phi>$ respectively. Thus

$$\Delta \geq E_m - \alpha \geq -\Delta$$

or

$$\alpha + \Delta \geq E_m \geq \alpha-\Delta \tag{2}$$

and an eigenvalue must lie between these limits. Nothing in this theorem tells us which eigenvalue lies in this range, nor whether there is only one.

Various special methods result from particular choices of α. Temple's formula (originally derived differently) follows from the choice

$$\alpha = \tfrac{1}{2}(E_1 + E_2)$$

Substitution of this value into Eq. (1) leads after some algebraic manipulation to

$$E_1(E_2 - <H>) \geq E_2<H> - <H^2>$$

Now we can divide both sides of this inequality by $(E_2 - <H>)$ without changing the sign of the inequality, if this quantity is positive; that is, if $<H>$ approximates E_1 well enough that it is at least smaller than E_2. This gives Temple's formula:

$$E_1 \geq <H>-\left(<H^2> - <H>^2\right)/\left(E_2-<H>\right)$$

The obvious difficulty with this formula is that it requires knowledge of E_2, the first excited state energy. Moreover, if an approximate value of E_2 is to be used, it must be a lower bound to E_2 and so cannot be calculated by the variation method. Since it is at least as difficult to calculate an exact value or lower bound for

E_2 as for E_1, Temple's method has little use in purely theoretical calculations. However, if the introduction of some experimental data is acceptable, measured values of E_2 may be used. In this manner it has been applied to simple systems such as the helium atom /14,15,16/ and the hydrogen molecule /17/.

In the procedure of Stevenson and Crawford /4/ α is fixed at the highest permissible value, which is $\frac{1}{2}(E_1 + E_2)$ if the ground state is to be bounded. This is obtained from experimental values if possible; otherwise the highest "safe" estimate (whatever that may mean) is used. The trial function ϕ is then varied, so as to minimize Δ, thus maximizing the lower bound. If ϕ is expressed as a linear combination of basis functions with variable coefficients, the minimization becomes a simple matrix eigenvalue problem; the ϕ so found will generally be different from that which minimizes the upper bound.

In the method of D.H. Weinstein /2/ the value $\langle H\rangle$ is assigned to α. This leads to

$$\Delta = (\langle H^2\rangle - \langle H\rangle^2)^{\frac{1}{2}}$$

that is, Δ becomes the standard deviation of H with respect to the wave function ϕ. Thus, we have as the bounds

$$\langle H\rangle + \Delta \geq E_m \geq \langle H\rangle - \Delta$$

As with other methods of this type, Weinstein's criterion assures us that an eigenvalue lies in the indicated range, but it does not tell us which one. It is usually assumed that if ϕ is intended as an approximation to the ground-state wave function, $\langle H\rangle$ will lie between E_1 and $\frac{1}{2}(E_1 + E_2)$, and the groundstate eigenvalue is then bounded by $\langle H\rangle - \Delta$ and $\langle H\rangle$; but usually there is no rigorous assurance of this.

Goodisman and Secrest/17/ devised a method in which, for a given value of α, they chose ϕ to minimize Δ, and from this ϕ calculated $\langle H\rangle$, which they used as a new value of α; this was continued iteratively until no further change occurred. There appears to be no assurance that this procedure will converge, nor that its limit will be lower bound if it does. However, the numerical results of Goodisman and Secrest in their treatment of H_2 indicate that convergence to a lower bound did occur in this example. Goodisman /18/ applied this method to excited states, choosing values of α such that after convergence the interval $(\langle H\rangle \pm \Delta)$ contained the desired eigenvalue. Obviously this requires some previous knowledge of the eigenvalues, and also requires that their separation exceeds Δ.

Stevenson's derivation can be extended to provide some information on the optimization of lower bounds. If the lower bound is designated by L, then Eq. (2) becomes

$$L = \alpha - (\langle H^2\rangle - 2\alpha\langle H\rangle + \alpha^2)^{\frac{1}{2}}$$

A little rearrangement gives

$$L^2 - 2\alpha L = \langle H^2\rangle - 2\alpha\langle H\rangle$$

and differentiation shows that

$$\partial L/\partial\alpha = (\langle H\rangle - L)/(\alpha - L)$$

This is necessarily positive, and so the lower bound improves with increasing value of α, up to the value $\frac{1}{2}(E_1 + E_2)$, above which L is not a lower bound. Therefore the Temple formula, which depends on this value of α, gives the best bound possible for a given trial function ϕ; the method of Stevenson and Crawford improves this by introducing some flexibility into ϕ. The Weinstein method involves a usually poorer choice of α. It has the apparent advantage of not requiring knowledge of E_2, but this is largely illusory, since as pointed out above there is no assurance of the validity of the method unless it is known that $\langle H\rangle$ is less than $\frac{1}{2}(E_1 + E_2)$.

An idea of the potentiality and limitations of these methods can be gleaned from the typical results given in Table 1. It appears that for a given number of basis functions even the Temple formula provides a lower bound much further from the eigenvalue than the upper bound furnished by the variation method - about 70 times as far in Kinoshita's 39-function calculation. Even when Kinoshita optimized the lower bound for the same basis (thereby getting a poorer upper bound), the former was still 15 times as far from the eigenvalue as the latter. The Weinstein method leads to much poorer bounds - 10 times as far from the eigenvalue as the Temple bound in the calculation of Goodisman and Secrest - and is only slightly improved by the Goodisman - Secrest iteration technique. The very discouraging results of Keaveny and Christofferson /19/ could be improved if Temple's formula could be used, but no values of E_2 are available.

An additional disadvantage of these methods is the difficulty in evaluating the integrals involved in $\langle H^2\rangle$. The integrands include terms with a factor of $1/r_{12}^2$, where r_{12} is the distance between the electrons. For atoms analytical evaluation is possible but cumbersome (see, for example, Öhrn and Nordling /20/, but for molecules the problem is much more serious. Goodisman and Secrest canceled out the singularity of $r_{12} = 0$ by choosing what they called "cusp-corrected basis functions." This permitted analytical evaluation of a two-dimensional factor of the six-dimensional integral but

Table 1. Typical Results of Methods of the H^2 Type

Reference	System	Basis Functions	Estimated Eigenvalue	Upper Bound	Lower Bound	Error in Upper Bound	Error in Lower Bound	Method[a]
Kinoshita /15/	He atom (ground state)	39 Hylleraas type	-2.9037244[b]	-2.9037225	-2.9038737	19×10^{-7}	1493×10^{-7}	T
Kinoshita/53/	He atom (ground state)	same		-2.9037200	-2.9027906	44×10^{-7}	662×10^{-7}	SC
		80 Hylleraas type		-2.9037237	-2.9037467	7×10^{-7}	223×10^{-7}	SC
Wilets & Cherry /14/	He atom (ground state)	10 Hylleraas type		-2.903603	-2.9136	1.2×10^{-4}	99×10^{-4}	T
		18 Hylleraas type		-2.9037063[c]	-2.90549	1.8×10^{-5}	177×10^{-5}	T
Goodisman/17/ & Secrest	H_2 molecule (ground state)	20 "cusp-corrected"	-1.17445[d]	-1.17421	-1.1786	2.4×10^{-4}	41×10^{-4}	T
					-1.221		47×10^{-3}	W
					-1.216		42×10^{-3}	GS
Goodisman[e]/18/	H_2 molecule (various $^1\Sigma_g^+$ states)	6 "cusp-corrected"	-0.68126[d]	-0.2487	-0.9783	0.4326	0.2970	GS
			-	+0.2357	-0.2523			
Keaveny & Christofferson /19/	H_3^+ ion	9 Gaussians	-1.3376[f]	-1.2737	-2.1513	0.0639	0.8137	W

cont.

Table 1. continued

[a]T Temple
SC Stevenson and Crawford
W Weinstien
GS Goodisman and Secrest

[b]From the 1078-term expansion of Pekeris /16/.

[c]Corrected for an apparent misprint in the original article.

[d]Kolos and Roothaan /54/.

[e]The nuclear repulsion energy, not included in Goodisman's article, has been added.

[f]Schwartz and Schaad /55/.

still left a four-dimensional integral to be evaluated numerically. Moreover, the choice that gives even this modest degree of convenience is not necessarily suitable for obtaining close bounds.

2. INTERMEDIATE HAMILTONIAN METHODS

The second group of methods to be discussed is dependent on the construction and solution of what is called an "intermediate problem." The basic idea is due to Alexander Weinstein (not to be confused with D.H. Weinstein, whose contribution to this field was discussed in Section 1); it was further developed by Aronszajn /21/ and was first applied to quantum mechanical problems by Bazley /22/ and Bazley and Fox /6/. A thorough treatment of this early work, together with an extensive set of references, can be found in the book by Weinstein and Stenger /23/. Löwdin /7/ recast this idea in much more perspicuous form by the use of his concept of "inner projections", and it is on this form that the following treatment is based.

Let the hamiltonian be expressible in the form $H^o + V$, where H^o (the "unperturbed hamiltonian") has known eigenfunctions and eigenvalues, and V (the "perturbation") is positive definite. Then it is possible to construct a set of <u>intermediate hamiltonians</u> $H^{(k)}$ which are exactly soluble and are related to H^o and H by the inequalities

$$H^o \leq H^{(1)} \leq \ldots \leq H^{(k)} \leq H^{(k+1)} \leq \ldots \leq H \quad (4)$$

It follows from the ordering theorem (proved later) that the eigenvalues follow the same inequality; that is, if the eigenvalues of each of the hamiltonians $H^o, H^{(k)}(k=1,2...)$, and H are in non-decreasing order, then the nth eigenvalues (for example) are related by

$$E_n^o \leq E_n^{(1)} \leq \ldots \leq E_n^{(k)} \leq E_n^{(k+1)} \leq \ldots \leq E_n$$

Each of the intermediate eigenvalues is thus a lower bound for the corresponding eigenvalue of H. The ordering theorem results in some severe limitations on the usefulness of these methods, as will be shown later.

A few theorems on operators are needed before the intermediate hamiltonians are described.

(a) Operator Inequalities

If A and B are hermitian operators, then the statement

$$A \leq B$$

means that for every vector u in their common domain

$$\langle u|A|u\rangle \leq \langle u|B|u\rangle$$

Weyl /24/ showed that if $A \leq B$ and their eigenvalues are arranged in nondecreasing order, multiple eigenvalues being included according to their multiplicity, then the nth eigenvalue of A is less than or equal to the nth eigenvalue of B. To prove it let the eigenvalues and orthonormal eigenvectors of A be λ_k and u_k, those of B, μ_k and v_k. Construct a vector v:

$$v = \sum_{k=1}^{n} \alpha_k v_k$$

with the α's determined so that

$$\langle v|u_k\rangle = 0 (k = 1,2,\ldots,n-1) \tag{3}$$

and

$$\langle v|v\rangle = 1$$

Then on the one hand

$$\langle v|B|v\rangle = \sum_{k=1}^{n} |\alpha_k|^2 \mu_k \leq \mu_n$$

while on the other

$$<V|B|V> \geq <V|A|V> \geq \lambda_n$$

the last inequality following from the variation principle and Eq. (3). This is called the "ordering theorem" by Löwdin, since it indicates that ordering of the eigenvalues must be accomplished (and therefore must be possible) before the comparison can be made; Weinstein and Stenger /23/ call it the "monotonicity principle."

(b) Properties of Projectors

If M is a subspace and $M^{\perp}$ its orthogonal complement, it can be shown that every vector u can be expressed uniquely in the form

$$u = v + w \tag{5}$$

where $v \epsilon M$ and $w \epsilon M^{\perp}$. This generates a mapping of the u's onto the v's, which consitutes a linear operator called the (orthogonal) projector or projection operator, onto the subspace M. It is easily shown from this definition that if P is an orthogonal projector, then

$$P^2 = P = P^{\dagger}$$

Moreover, P is positive semidefinite, since

$$<u|P|u> = <u|P^2|u> = <Pu|Pu> \geq 0$$

The operator 1 - P is also a projector; it projects onto $M^{\perp}$. It must therefore also be positive semidefinite, and so

$$<u|1 - P|u> \geq 0$$

or, with a little rearrangement,

$$<u|P|u> \leq <u|u> \tag{6}$$

This relation is essential in the construction of inner projections.

(c) Inner Projections

If V is a positive definite (or semidefinite) hermitian operator, then it has a positive hermitian square root; that is, a positive hermitian operator $V^{\frac{1}{2}}$ such that $(V^{\frac{1}{2}})^2 = V$. Now by applying Eq. (6) to the vector $V^{\frac{1}{2}}u$ we find

$$\langle u|V^{\frac{1}{2}}PV^{\frac{1}{2}}|u\rangle = \langle V^{\frac{1}{2}}u|P|V^{\frac{1}{2}}u\rangle \leq \langle V^{\frac{1}{2}}u|V^{\frac{1}{2}}u\rangle = \langle u|V|u\rangle$$

Expressed as an operator inequality, this says that

$$V^{\frac{1}{2}}PV^{\frac{1}{2}} \leq V$$

The form on the left was introduced in this context by Löwdin /7/, who called it an "inner projection."

(d) Projector onto the Span of a Given Set of Vectors

Finally, we need an explicit expression for the orthogonal projector onto the span M of a given set of vectors $f_1, f_2, \ldots, f_n$, which must be linearly independent but not necessarily orthogonal. The derivation can be simplified by using $|\underset{\sim}{f}\rangle$ to represent a row vector of the ket forms of the f's:

$$|\underset{\sim}{f}\rangle = (|f_1\rangle, |f_2\rangle, \ldots, |f_n\rangle)$$

while $\langle\underset{\sim}{f}|$ represents a column vector made up of the bra forms; $\langle\underset{\sim}{f}|\underset{\sim}{f}\rangle$ is then a square matrix whose i, j-element is $\langle f_i|f_j\rangle$. With this notation the orthogonal projector P_m onto M can be written

$$P_m = |\underset{\sim}{f}\rangle \langle\underset{\sim}{f}|\underset{\sim}{f}\rangle^{-1}\langle\underset{\sim}{f}|$$

To show this we note that for any vector u, P_m u is a linear combination of the f's and so lies in M; if we can show that $u - P_m u \varepsilon M^{\perp}$, it will follow from the uniqueness of the representation (5) that P_m is the desired projector. Now

$$\langle\underset{\sim}{f}|P_m u\rangle = \langle\underset{\sim}{f}|\underset{\sim}{f}\rangle\langle\underset{\sim}{f}|f\rangle^{-1}\langle\underset{\sim}{f}|u\rangle = \langle\underset{\sim}{f}|u\rangle$$

and it follows that

$$\langle\underset{\sim}{f}|u-P_m u\rangle = 0$$

Thus $u - P_m u \varepsilon M^{\perp}$.

We now use this projector to construct an inner projection of the perturbation operator V. We find immediately that the inner projection $V^{(n)}$ is given by

$$V^{(n)} = V^{\frac{1}{2}}|\underset{\sim}{f}\rangle\underset{\sim}{\Delta}^{-1}\langle\underset{\sim}{f}|V^{\frac{1}{2}} \qquad (7)$$

where $\Delta \equiv \langle\underset{\sim}{f}|\underset{\sim}{f}\rangle$. This form of the inner projection is occasionally usable, as for example in the treatment of a harmonic oscillator perturbed by a fourth-power term in the potential energy; it has also been applied to the quartic oscillator /25/. More often, however the operator $V^{\frac{1}{2}}$ is not practical to deal with, and following a suggestion of Aronszajn /21/, we get rid of it by replacing the set of functions $|\underset{\sim}{f}\rangle$ by a new set $|\underset{\sim}{g}\rangle$, defined by

$$|f_i\rangle = V^{\frac{1}{2}}|g_i\rangle \text{ or } |g_i\rangle = V^{-\frac{1}{2}}|f_i\rangle$$

Then

$$\Delta_{ij} = \langle g_i|V|g_j\rangle$$

and

$$V^{(n)} = V|\underset{\sim}{g}\rangle\underset{\sim}{\Delta}^{-1}\langle\underset{\sim}{g}|V = V|\underset{\sim}{g}\rangle\langle\underset{\sim}{g}|V|\underset{\sim}{g}\rangle^{-1}\langle\underset{\sim}{g}|V \tag{8}$$

This eliminates $V^{\frac{1}{2}}$, but it has the disadvantage that the result of operating on a function by this form of the projector is a linear combination not of the chosen functions $|\underset{\sim}{g}\rangle$, but of the functions $V|\underset{\sim}{g}\rangle$. However, another choice, proposed by Bazley /22/, is also possible; we replace $|\underset{\sim}{f}\rangle$ by $|\underset{\sim}{h}\rangle$ defined by

$$|\underset{\sim}{f}\rangle = V^{-\frac{1}{2}}|\underset{\sim}{h}\rangle \quad \text{or} \quad |\underset{\sim}{h}\rangle = V^{\frac{1}{2}}|\underset{\sim}{f}\rangle.$$

Then

$$\underset{\sim}{\Delta} = \langle\underset{\sim}{h}|V^{-1}|\underset{\sim}{h}\rangle$$

and

$$V^{(n)} = |\underset{\sim}{h}\rangle\underset{\sim}{\Delta}^{-1}\langle\underset{\sim}{h}| = |\underset{\sim}{h}\rangle\langle\underset{\sim}{h}|V^{-1}|\underset{\sim}{h}\rangle^{-1}\langle\underset{\sim}{h}| \tag{9}$$

With this form of projection, $V^{(n)}|u\rangle$, for any given function $|u\rangle$, is a linear combination of the functions $|\underset{\sim}{h}\rangle$. In accordance with a suggestion of Löwdin, the projections given by Eqs. (7), (8), and (9) will be called the standard, Aronszajn, and Bazley projections respectively. It should be noted that the procedure indicated here of first choosing the f's and then replacing them with the g's or h's is only for the purpose of derivation; in actual use of the Aronszajn or Bazley projection the g's or h's are chosen directly. Further generalizations, including $V^{(n)} = V^{\alpha}|\underset{\sim}{j}\rangle\langle\underset{\sim}{j}|V^{2\alpha-1}|\underset{\sim}{j}\rangle^{-1}\langle\underset{\sim}{j}|V^{\alpha}$, were given by Sack /26/.

We can now return to the construction of the intermediate hamiltonians. The intermediate hamiltonian $H^{(n)}$ of Bazley and Fox is given by

$$H^{(n)} = H^{o} + V^{(n)}$$

These have the properties required by Eq. (4); moreover, if the functions used for the projection are chosen from a complete set, then $H^{(n)} \to H$ as $n \to \infty$, in the sense that for any suitable vector u, $\langle u|H^{(n)}|u\rangle \to \langle u|H|u\rangle$. It is also true that the individual eigenvalues $E_k^{(n)} \to E_k$, provided the essential spectrum of H^o lies entirely above E_k.

Several procedures are available for finding the eigenvalues of the intermediate hamiltonians. Bazley and Fox /6/ give the following method. The eigenvalue equation to be solved is

$$H^{o}|u\rangle + V^{(n)}|u\rangle = \varepsilon|u\rangle$$

or

$$H^{o}|u\rangle + V|\underset{\sim}{g}\rangle\underset{\sim}{\Delta}^{-1}\langle\underset{\sim}{g}|V|u\rangle = \varepsilon|u\rangle$$

Now the quantity $\Delta^{-1}\langle\underset{\sim}{g}|V|u\rangle$ is a column vector; calling it $\underset{\sim}{\alpha}$, we can rewrite this equation as

$$H^{o}|u\rangle + V|\underset{\sim}{g}\rangle\underset{\sim}{\alpha} = \varepsilon|u\rangle$$

from which

$$(H^{o}-\varepsilon)|u\rangle = -V|\underset{\sim}{g}\rangle\underset{\sim}{\alpha}$$

If ε is not equal to any of the eigenvalues of H^o, then $H^o - \varepsilon$ has an inverse, and this equation has the solution

$$|u\rangle = -(H^{o}-\varepsilon)^{-1}V|\underset{\sim}{g}\rangle\underset{\sim}{\alpha}$$

Multiplying this from the left by $\underset{\sim}{\Delta}^{-1}\langle\underset{\sim}{g}|V$, we convert the left side to $\underset{\sim}{\alpha}$, and so

$$\underset{\sim}{\alpha} = -\underset{\sim}{\Delta}^{-1}\langle\underset{\sim}{g}|V(H^{o}-\varepsilon)^{-1}V|\underset{\sim}{g}\rangle\underset{\sim}{\alpha}$$

With a little algebraic manipulation this becomes

$$\left(\underset{\sim}{\Delta} + \langle\underset{\sim}{g}|V(H^{o}-\varepsilon)^{-1}V|\underset{\sim}{g}\rangle\right)\underset{\sim}{\alpha} = 0$$

A nontrivial solution requires that

$$\text{detn}\ \{\underset{\sim}{\Delta} + \langle \underset{\sim}{g}|V(H^{o}-\varepsilon)^{-1}V|\underset{\sim}{g}\rangle\} = 0 \tag{10}$$

and the solutions of this equation are the eigenvalues. It may be of interest to compare this to related developments in other areas, such as Löwdin's /27/ "localized perturbation theory" or the use of separable potentials in scattering theory /28/.

Now if the eigenfunctions of H^{o} are ψ_1^{o}, ψ_2^{o},..., then the inverse of $H^{o}-\varepsilon$ is

$$(H^{o}-\varepsilon)^{-1} = \sum_k \frac{|\psi_k^{o}\rangle\langle\psi_k^{o}|}{E_k^{o} - \varepsilon}$$

where the summation is intended to include integration over the continuous spectrum, if any. Thus the elements of the determinant in Eq. (10) are sums of the form

$$\langle g_i|V|g_j\rangle + \sum_k \frac{\langle g_i|V|\psi_k^{o}\rangle\langle\psi_k^{o}|V|g_j\rangle}{E_k^{o} - \varepsilon} \tag{11}$$

This expression cannot be evaluated unless something is done about the infinite sum in each of these elements. Bazley and Fox /6/ suggested two procedures for this. One of these, the method of special choice of elements, is applicable whenever the g's can be chosen in such a way that each of the functions Vg_i is a linear combination of only a finite number of the eigenfunctions of H^{o}. Then only finitely many of the quantities $\langle g_i|V|\psi_j\rangle$ differ from zero, all others vanishing because of the orthogonality of the ψ's. The trouble with infinite sums then does not arise.

The second procedure for eliminating the infinite sum consists of truncating H^{o}. If the eigenvalues of H^{o} are arranged in non-decreasing order, then we can see from the spectral resolution

$$H^{o} = \sum_i |\psi_i^{o}\rangle E_i^{o}\langle\psi_i^{o}|$$

that the operator $H^{(j,0)}$, defined by

$$H^{(j,0)} \equiv \sum_{i=1}^{j} |\psi_i^o\rangle E_i^o \langle\psi_i^o| + E_{j+1}^o \sum_{i=j+1}^{\infty} |\psi_i^o\rangle\langle\psi_i^o|$$

$$= \sum_{i=1}^{i} |\psi_i^o\rangle E_i^o \langle\psi_i^o| + E_{j+1}^o \left(1 - \sum_{i=1}^{i} |\psi_i^o\rangle\langle\psi_i^o|\right)$$

$$= \sum_{i=1}^{j} |\psi_i^o\rangle \left(E_i^o - E_{j+1}^o\right) \langle\psi_i^o| + E_{j+1}^o \qquad (12)$$

satisfies the inequality $H^{(j,0)} \leq H^o$. Then if

$$H^{(j,k)} = H^{(j,0)} + V^{(k)}$$

we have the inequality

$$H^{(j,k)} \leq H^{(k)} \leq H$$

Now to apply Eq. (10) to the determination of the eigenvectors of this operator, we need the spectral resolution of $(H^{(j,0)} - \varepsilon)^{-1}$, which is

$$(H^{(j,0)} - \varepsilon)^{-1} = \frac{1}{E_{j+1}^o - \varepsilon} + \sum_{i=1}^{j} |\psi_i^o\rangle \left\{ \frac{1}{E_i^o - \varepsilon} - \frac{1}{E_{j+1}^o - \varepsilon} \right\} \langle\psi_i^o|$$

Only a finite number of terms is involved in evaluating integrals over this operator, and so again the difficulty of the infinite sum is avoided. A simpler method of getting eigenvalues of $H^{(j,k)}$ is ascribed by Bazley and Fox /6/ to W. Börsch-Supan. If the Bazley projection is used in generating the operator $V^{(k)}$, then the space spanned by the combined sets of function $h_1, h_2, \ldots h_k$ and $\psi_1^o, \psi_2^o, \ldots \psi_j^o$ is an invariant space for $H^{(j,k)}$, which therefore has eigenvectors that lie entirely within this space, and the corresponding eigenvalues can be determined from a matrix whose order is no more than j + k, and less if these functions are not entirely independent.

Bazley and Fox applied their procedures to three illustrative problems - the helium atom, the oscillator with a fourth-power perturbation, and a hydrogen-like atom with a noncoulombic potential. Using the modified hamiltonian $H^{(2,2)}$, they found a lower bound for the ground state of the helium atom of -3.0008 hartrees, as compared to a known eigenvalue of -2.9037 hartrees.

Unless the Börsch-Supan method is used, a major difficulty in the Bazley-Fox procedure is the complicated form of the determinantal equation which has to be solved; each element is of the form given by Eq. (10). In the calculation on the helium atom, Bazley and Fox used a 2x2 determinant, with three terms in the summation part of each element. This kept the difficulties manageable, but they would be more serious for a large basis set. Probably the most practical method of solution in this case would be to have a computer generate a table of values of the determinant for various values of ε, and then find the root by inverse interpolation.

Several modifications have been proposed in the Bazley-Fox method; some of these will be described here. In Gay's /29/ method (the essential feature of which was first suggested by Löwdin) the operator $(H^o-\varepsilon)^{-1}$ and the difficult determinantal equation (Eq. (10)) associated with are eliminated by using a new manifold $|\tilde{j}>$ defined by

$$|\tilde{g}> = V^{-1}(H^o-\varepsilon)|\tilde{j}> \tag{13}$$

Substitution into the expression for $\tilde{\Delta}$ and Eq. (10) gives

$$\text{detn } \{<\tilde{j}|(H^o-\varepsilon)V^{-1}(H^o-\varepsilon) + (H^o-\varepsilon)|\tilde{j}>\} = 0 \tag{14}$$

This equation has several important advantages over Eq. (10). With $(H^o-\varepsilon)^{-1}$ out of the way, there is no need either to choose the basis functions from some special set (as in Bazley and Fox's first procedure) or truncate the operator $(H^o-\varepsilon)^{-1}$ (as in the second). Both these procedures may be undesirable; in the method of special choice, the set needed for evaluation of the matrix elements is not necessarily one that provides good lower bounds, while the truncation of an operator, though certainly leading to valid lower bounds, often gives rather poor ones. Another advantage of Gay's equation is that the removal of the unknown from the denominator of the elements of the determinant greatly simplifies the solution. Although it is more difficult than the ordinary linear matrix eigenvalue problem, Gay was able to solve it by a straightforward and practical iterative scheme. Against these advantages are some less desirable features. The definition (13) makes the projection manifold, and therefore the intermediate hamiltonian itself, dependent on ε.This means that we never get more than one root of any intermediate hamiltonian, even if we find all the roots of Eq. (14). If, for example, ε_1 and ε_2 are roots of this equation, then ε_1 is an eigenvalue -- the only one we know -- of the intermediate hamiltonian obtained by using ε_1 for ε in Eq. (13). ε_2 is not an eigenvalue of this hamiltonian at all but of another one, generated by using ε_2 for ε in (13). Since we have only one eigenvalue of any one hamiltonian, we have no direct way of knowing whether it is the lowest eigenvalue or some other, and so there is no way of applying the ordering theorem to determine for which state of H we have a lower bound. Gay was able to show that his root of this equation was a lower bound of the ground state by appealing to the fact that the kth eigenvalue of the intermediate hamiltonian is not only below the kth eigenvalue of H but also above the kth eigenvalue of H^o. For the system he studied (the helium atom) the two lowest eigenvalues of H^o are -4.00 and -2.50, while the best known upper bound for the ground state of H is -2.9037... It follows that any root lying between -4 and -2.9037 must be a lower bound to the ground state, and this includes the root -2.9059 found by Gay. This difficulty becomes more serious if H^o has more than one eigenvalue below the ground state of H.

The second modification of the Bazley-Fox method that we will take up was proposed by Miller/30/. He observed that if a truncation is performed on H^o, and V is a multiplicative operator (as it usually is in applications), then it is not necessary to perform an inner projection on V; it is possible to get the eigenvalues of the operator $H^{(k,0)}+ V$. The method is almost identical with that used in deriving Eq. (10). If ε and u are the eigenvalue and eigenvector of $H^{(k,0)} + V$, so that

$$(H^{(k,0)} + V)u = \varepsilon u$$

then by using the third form of $H^{(k,0)}$ given in Eq. (12) we find

$$\sum_{i=1}^{k} |\psi_i^o>(E_i^o-E_{k+1}^o)<\psi_i^o|u>+E_{k+1}^o|u>+V|u> = \varepsilon|u>$$

Setting $\alpha_i = (E_i^o - E_{k+1}^o)\psi_i^o|u>$ and solving formally for $|u>$ gives

$$|u> = (\varepsilon - E_{k+1}^o-V)^{-1}\sum_{i=1}^{k} |\psi_i^o>\alpha_i$$

Multiplying by $(E_j^o - E_{k+1}^o)<\psi_j^o|$ leads to

$$\alpha_j = (E_j^o-E_{k+1}^o)\sum_{i=1}^{k} <\psi_j^o|(\varepsilon-E_{k+1}^o-V)^{-1}|\psi_i^o>\alpha_i$$

This has a nontrivial solution only if

$$\text{detn}\ \{(E^o_{k+1}-E^o_j)^{-1}\delta_{ij}-\langle\psi^o_j|(E^o_{k+1}+V-\varepsilon)^{-1}|\psi^o_i\rangle\} = 0 \qquad (15)$$

and this is the equation that must be solved to find the eigenvalues of $H^{(k,0)}+V$.

Miller's method has two important advantages: it does not require V to be positive definite, and it is not necessarily made useless by the presence of many eigenvalues of H^o - even infinitely many, plus a continuum - below the ground state of H. These are such important advantages that they might make this the leading method for lower bounds calculation, if they were not counterbalanced by two formidable disadvantages. The first of these is the difficulty of evaluating integrals over the operator $(E^o_{k+1}+V-\varepsilon)^{-1}$. Even for the helium atom, with $V = 1/r_{12}$, Miller used numerical integration for these integrals, though they could be evaluated analytically by an adaption of method of Öhrn and Nordling /20/. For an N-electron atom, the most practical choice for V is the sum of the N(N-1)/2 electron repulsion terms. For N > 2 analytical evaluation appears hopeless, and the rapidly increasing number of dimensions in the integral will quickly make numerical evaluation tax the capacity of any computer. The other difficulty lies in the solution of equations such as Eq. (15). There seems to be no better way of accomplishing this than to select two or more values of ε near a suspected root, evaluate all of the integrals and then the determinant for each of them, locate an approximate root by inverse interpolation, and continue this procedure iteratively.

The intermediate hamiltonian method as described above fails when an attempt is made to apply it to a problem in which any part of the essential spectrum of H^o lies below the eigenvalue of H that is to be bounded. The reason is that the inner projections are of finite rank and so are compact operators. It follows from a theorem of Weyl /31/ that when such an operator is added to an operator with an essential spectrum, the essential spectrum is unchanged. As an example, consider the lithium atom. If H^o is chosen to be

$$\sum_{i=1}^{3} (-\tfrac{1}{2}\nabla_i^2 - \frac{3}{r_i})$$

(which is the only practical choice), then the essential spectrum of H^o consists of a continuum extending from -9 to infinity. Since this continuum will persist on the addition of an inner projection of the electron repulsion terms, the ground state of the intermediate hamiltonian will never exceed -9. This is, of course, a lower bound to the ground state of H (about -7.4781), but it is too low to be of any interest.

A modification to avoid this problem has been developed independently by Fox /7/ and by the present author /8/.* This method depends on the fact that although the electron repulsion operator is a three-electron operator, it is the sum of several two-electron operators with two-electron basis functions, the resulting operators are compact in the subspace of two-electron functions, but not in the full space of three-electron functions.

To illustrate this consider again the lithium atom, and let ψ_1 represent the ground-state hydrogenic function for atomic number 3, and ψ_k some other eigenfunction of the same operator. For the Bazley projection basis use only the single two-electron function $\psi_1(1)\psi_1(2)$, and similarly for $1/r_{13}$ and $1/r_{23}$. It can then be shown that the effect of V' on the Slater determinant $A\psi_1(1)\bar{\psi}_1(2)\psi_k(3)$ (where the bar indicates a β spin function) is given by

$$V'A\psi_1(1)\bar{\psi}_1(2)\psi_k(3) = \langle\psi_1(1)\psi_1(2)|r_{12}|\psi_1(1)\psi_1(2)\rangle^{-1}A\psi_1(1)\bar{\psi}(2)\psi_2(3)$$

The value of the integral is 35/48, so this function is an eigenfunction of $H^O + V'$ with the eigenvalue

$$-9(1 + 1/2k^{-2}) + 48/35$$

The ground state eigenvalue (with k=2) is found to be -8.75357; the beginning of the continuum (obtained by letting $k \to \infty$) is at -7.62857. Thus even with this simple projection basis the onset of the continuum has been raised by more than 1.37 Hartrees.

With a basis of ten hydrogenic s-functions, a lower bound of -7.79528 was found /32/; extrapolation to the entire set of discrete hydrogenic s-functions indicates a value of -7.79438, and inclusion of all 2p functions raises it by about 0.014 more. The beginning of the continuum for ten s-functions is at -7.55867; extrapolation indicates a value of -7.57767 if all discrete s-functions could be used, and again the 2p's raise this by about 0.014. These results suggest that no basis of discrete hydrogenic functions can raise the

---

* Since there has been some discussion of priority in this development, it may be worthwhile to present the facts here. Fox's work can be found in a report of the Applied Physics Laboratory, Johns Hopkins University, dated January, 1969. This report was not published at that time but was distributed to a private mailing list. In 1969 and 1970 fragmentary hints of the method appeared in abstracts of meetings of the American Mathematical Society and the Society of Industrial and Applied Mathematics. It was finally published in November, 1972. The present author's work was done independently in 1971 and published in July, 1972.

continuum above the ground state of the lithium atom hamiltonian itself.

There is nothing in this formalism, however, that prevents the use of a nonhydrogenic basis, though this necessitates the truncation of H^o. It is rather obvious that the practical way to truncate H^o is to apply the standard truncation formula (Eq. (12)) to each of its one-electron parts. In fact, this is not only convenient but necessary to obtain reasonably good bounds, since an attempt to apply the truncation procedure directly to H^o would lead to an infinitely degenerate eigenvalue lower than -9 Hartrees, at which the continuum of H^o starts. Now if $H^{o\prime}$ is the truncated approximation to H^o, then $H^o - H^{o\prime}$ is a positive semidefinite operator, and if it is written

$$(H^o - H^{o\prime} + \eta) - \eta$$

where η is an arbitrary positive constant, then the operator in parentheses is positive definite, and an inner projection on it should recover part of the loss sustained in truncating H^o. To perform a Bazley projection on this operator we need to calculate elements such as $\langle\phi_1|(H^o - H^{o\prime} + \eta)^{-1}|\psi_2\rangle$. The spectral resolution of this operator is

$$(H^o - H^{o\prime} + \eta)^{-1} = \sum_{k=1}^{n} |\psi_k^o\rangle \eta^{-1} \langle\psi_k^o| + \sum_{k=n+1}^{\infty} |\psi_k^o\rangle (E_k^o - E_{n+1}^o + \eta)^{-1} \langle\psi_k^o|$$

where the second sum includes integration over the continuum. It follows that the integrals can be trivially found if either of the ϕ's is one of the first n eigenfunctions of H^o (which make up the truncation manifold), while if both are orthogonal to this manifold, the inverse required is simply the resolvent of the hydrogenic hamiltonian. A practical computational algorithm for integrals of Slater-type orbitals over the hydrogenic resolvent was devised for this application /33/. The method has not been fully tested, but it has yielded a previously unpublished result of -7.63638 which appears to be the best lower bound for the ground state of the lithium atom now available. In this treatment the lowest point in the essential spectrum consists of an infinitely degenerate eigenvalue at -7.63500.

3. METHODS DEPENDENT ON A BRACKETING FUNCTION

Suppose that the Schrödinger equation can be so transformed that the eigenvalue is to be found by solving an equation of the form

$$\varepsilon = f(\varepsilon) \tag{16}$$

If the solution cannot be found directly, it may be desirable to try an iterative solution. Start with an approximate value ε_o, one calculates $\varepsilon_1 = f(\varepsilon_o)$, then $\varepsilon_2 = f(\varepsilon_1)$, and so on. Under proper conditions such a procedure may converge to a root. What is of more interest here, however, is that if the derivative $f'(\varepsilon)$ exists and is negative throughout the closed interval from the root to the initial approximant ε_o, then ε_o and ε_1 will lie on opposite sides of the root. This is easily shown by using the theorem of mean value. If E is a root, then $f(E) = E$, and

$$\varepsilon_1 = f(\varepsilon_o) = f(E) + (\varepsilon_o - E)f'(\xi) = E + (\varepsilon_o - E)f'(\xi)$$

where ξ lies between E and ε_o. Thus

$$\varepsilon_1 - E = (\varepsilon_o - E)f'(\xi)$$

Because the last factor is negative, ε_1 and ε_o differ from E in opposite directions. Since in quantum mechanical applications it is easy to find an upper bound, this would give a lower bound. Because the values ε and $f(\varepsilon)$ bracket a root, a function such as $f(\varepsilon)$ is called a <u>bracketing function</u>.

The formalism for applying this idea to lower bounds for energy levels in quantum mechanics was discovered and extensively developed by Löwdin /7,34/ though somewhat similar ideas had been used independently in scattering theory by Hahn, O'Malley, and Spruch /35/. The derivation involves the concept of partitioning. First we select a reference manifold, which for now will be taken as a single normalized function ϕ. The projection operator $|\phi\rangle\langle\phi|$ will be called O, and the complementary projector 1 - O will be called P. This procedure is called partitioning because it has the effect of partitioning the complete Hilbert space into two mutually orthogonal subspaces - the subspaces onto which O and P project. As mutually orthogonal projectors O and P have the properties that

$$O^2 = O = O^\dagger$$
$$P^2 = P = P^\dagger$$

and

$$O\,P = P\,O = o$$

O and P do not generally commute with H, though they will do so if ϕ is an eigenfunction of H; of course, they commute with constants such as ε. With the help of these properties it is easy to show that

$$P(\varepsilon-PHP) = P(\varepsilon-PHP)P = (\varepsilon-PHP)P = P(\varepsilon-H)P \tag{17}$$

Now since $O + P = 1$, the Schrödinger equation

$$H\psi = E\psi$$

can be written

$$(O+P)H(O+P)\psi = E(O+P)\psi$$

Multiplying both sides by O on the left gives (since $OP = 0$)

$$OHO\psi + OHP\psi = EO\psi \tag{18}$$

Similarly, multiplication by P gives

$$PHO\psi + PHP\psi = EP\psi$$

The last equation can be formally solved for $P\psi$:

$$(E-PHP)P\psi = PHO\psi \quad \text{or} \quad P\psi = (E-PHP)^{-1}\, PHO\psi$$

The operator $(E - PHP)^{-1}P$ occurs frequently in this theory; it is commonly designated by T. Thus

$$P\psi = THO\psi \tag{19}$$

Several properties of T and other forms* of it can be derived by manipulations such as those of Eq. (17). Thus since P commutes with $(E - PHP)$, it also commutes with $(E - PHP)^{-1}$. From this it is easily shown that $PT = T$, and so $OT = o$.

Another often used property is

$$P(E-H)T = P$$

This follows from

$$P(E-H)T = P(E-H)PT = P(E-PHP)\ (E-PHP)^{-1}P = P^2 = P$$

Finally, it is to be noted that T exists whenever E is not an eigenvalue of PHP. Going back to the solution (19) for $P\psi$, we can substitute this into Eq. (18) to get

---

* In earlier publications on the bracketing method T was defined as $P[\alpha + P(E - H)P]^{-1}P$, where α is an arbitrary constant other than zero. Since the equivalent of the two forms (which can be seen by assigning α the value E) was noticed, this less elegant form has been largely dropped.

$$O H O \psi + O H T H O \psi = E O \psi$$

or, since $O = |\phi\rangle\langle\phi|$,

$$|\phi\rangle\langle\phi|H|\phi\rangle\langle\phi|\psi\rangle + |\phi\rangle\langle\phi|HTH|\phi\rangle\langle\phi|\psi\rangle = E|\phi\rangle\langle\phi|\psi\rangle$$

Since $|\phi\rangle$ is not identically zero, we can divide it out of every term of this equation. If $\langle\phi|\psi\rangle \neq 0$ (that is, if ϕ is not orthogonal to the true wave function), we can take this factor out also. Thus we find

$$E = \langle\phi|H + HTH|\phi\rangle \tag{20}$$

Now this is in the form of Eq. (16), since the right side depends on E through the dependence of T on E. We must next show that the expression on the right side is a bracketing function; that its derivative with respect to E is negative. It is first necessary to generalize the definition of T to

$$T(\varepsilon) = P(\varepsilon - PHP)^{-1} P$$

for any ε; in place of Eq. (20) we can write

$$\varepsilon_1 = \langle\phi|H + HT(\varepsilon)H|\phi\rangle \tag{21}$$

Then since the derivative of the inverse of an operator is given by

$$\frac{dQ^{-1}}{d\varepsilon} = -Q^{-1} \frac{dQ}{d\varepsilon} Q^{-1}$$

and

$$\frac{d}{d\varepsilon} (\varepsilon - PHP) = 1$$

we find

$$\frac{dT}{d\varepsilon} = -P(\varepsilon-PHP)^{-1} \cdot 1 \cdot (\varepsilon-PHP)^{-1} P = -T^2$$

Using this in Eq. (21) leads to

$$\frac{d\varepsilon_1}{d\varepsilon} = -\langle\phi|HT^2H|\phi\rangle$$

This is certainly negative, since apart from the minus sign it is the square of the norm of the vector $TH\phi$; moreover, it shows that the derivative of ε_1 exists throughout every interval that does not contain an eigenvalue of PHP. It follows that the function

$$\varepsilon_1 = f(\varepsilon) = \langle\phi|H + HT(\varepsilon)\, H|\phi\rangle \qquad (22)$$

is a bracketing function, and if it is evaluated for an upper bound ε, ε_1 will be a lower bound, if no eigenvalue of PHP lies between E and ε.

The difficulty with this procedure is that, although we have a formal definition of $T(\varepsilon)$, we do not have any practical representation of it that can be used in numerical work. Löwdin /7/ attacked this problem by constructing a modified operator which, when used instead of T in Eq. (22), will give a number - call it ε_1' - that is known to be less than or at least no larger than ε_1. This will then be a lower bound, and if the modified operator is suitably chosen it may be practical to calculate ε_1' but not ε_1.

Our procedure for doing this is to replace H in the definition of T by an intermediate hamiltonian H' (which may be of the type $H^{(k)}$ or $H^{(j,k)}$ described in Section 2) to give a modified T, which we can call T'; the other H's in Eq. (22) are left unchanged. If ε is lower than the lowest eigenvalues of PHP and of PH'P, then

$$\varepsilon - PHP < \varepsilon - PH'P < 0$$

Now if A and B are operators and $A > B > 0$, then $B^{-1} > A^{-1} > 0$; this follows from the identity

$$B^{-1} - A^{-1} = A^{-1}(A-B)A^{-1} + A^{-1}(A-B)B^{-1}(A-B)A^{-1}$$

which is easily verified by expanding the expression on the right. Applying this (with the signs changed) to Eq. (23), we find

$$T' < T < 0$$

Now the quantity $\langle\phi|HTH|\phi\rangle$ is of the form $\langle u|T|u\rangle$, where $u = H\phi$, and so this shows that

$$\langle\phi|HT'H|\phi\rangle < \langle\phi|HTH|\phi\rangle$$

whence $\varepsilon_1' < \varepsilon_1$. The advantage is that because of the truncation of H, the spectral resolution of T', in contrast to that of T, contains only finitely many terms and so may be practical to calculate. A disadvantage is the need for integrals over the operator H^2. This approach was proposed by Wang /36/; after deriving the inequalities given above, she applied the method to the hydrogen molecule ion, H_2^+, and obtained the following results:

Upper bound	-1.10117
Exact eigenvalue	-1.10263
Lower bound	
from a 2-function truncation	-1.10529
- - 4-- -	-1.10438

Although these bounds look encouraging, it should be recognized that they are obtained on a very simple system, for which approximate methods are unnecessary, since an exact solution is available. There appear to be no other published calculations by Wang's method.

Another method, or rather group of methods, starts by treating hamiltonians of the form

$$H = H^o + V$$

where V is positive, and H^o has known eigenfunctions ψ_1^o, ψ_2^o, ... with corresponding eigenvalues E_1^o, E_2^o, ... For the reference function ϕ the ground state eigenfunction of H^o, ψ_1^o, is chosen. This immediately leads to several simplifications; thus

$$OH^o = H^oO = E_1^oO$$

$$PH^o = H^oP$$

and

$$OH^oT = 0$$

Similarly integrals such as $\langle\phi|H^oT|...\rangle$ all vanish. Substituting into Eq. (22) then leads to

$$\begin{aligned} \varepsilon_1 &= \langle\phi|H^o + V + (H^o + V)\,T\,(H^o + V)|\phi\rangle = E_1^o + \langle\phi|V + VTV|\phi\rangle \\ &= E_1^o + \langle\phi|t|\phi\rangle \end{aligned} \tag{24}$$

where

$$t = V + VTV$$

To get rid of the intractable operator T, we can make use of the identity

$$(A - B)^{-1} = A^{-1} + A^{-1}B(A - B)^{-1}$$

where A and B are any operators for which the indicated inverses exist. Applying this to $\varepsilon - PHP$ gives

$$(\varepsilon\text{-PHP})^{-1} = \left((\varepsilon\text{-PH}^{o}\text{P})\text{-PVP}\right)^{-1} = (\varepsilon\text{-PH}^{o}\text{P})^{-1}+(\varepsilon\text{-PH}^{o}\text{P})^{-1}\text{PVP}(\varepsilon\text{-PHP})^{-1}$$

If we now multiply by P and define a new operator T_o by

$$T_o = (\varepsilon\text{-PH}^{o}\text{P})^{-1}\text{P} = \text{P}(\varepsilon\text{-PH}^{o}\text{P})^{-1} = \text{P}(\varepsilon\text{-PH}^{o}\text{P})\text{P}$$

we find

$$T = T_o + T_oVT$$

We are still not entirely rid of T. However, multiplication by V as a right factor gives

$$TV = T_o(V + VTV) = T_ot$$

Then by the definition of t

$$t = V + VTV = V + VT_ot$$

and so

$$V^{-1}t = 1 + T_ot$$

which easily leads to

$$V^{-1} - T_o = t^{-1}$$

In general we cannot find any practical means of inverting this operator to get t itself, but if t is positive definite, this form can be used directly to construct a lower bound t' to t by a Bazley projection (Eq. (9)). Using this instead of t in Eq. (24), we find

$$\varepsilon_1' = E_1^o + \langle\phi|t'|\phi\rangle < E_1^o + \langle\phi|t|\phi\rangle = \varepsilon_1$$

ε_1' is therefore a lower bound to the true eigenvalue also - a poorer lower bound than ε_1 itself, except for the great advantage that it is practical in some cases to calculate ε_1' but not ε_1.

This argument depends on the positive definiteness of t, and so it is necessary to examine the conditions under which this holds. Since V is positive definite, it is clearly sufficient that $-T_o$ should be positive semidefinite. If ψ_k^o is an eigenfunction of H^o other than ϕ, then

$$(H^{o}-\varepsilon)\psi_k^o = (E_k^o-\varepsilon)\psi_k^o$$

and

$$P\psi_k^o = \psi_k^o$$

These two equation led to

$$(PH^{o}P-\varepsilon)\psi_k^o = (E_k^o-\varepsilon)P\psi_k^o$$

from which we find

$$-T_o\psi_k^o = (E_k^o-\varepsilon)^{-1}\psi_k^o$$

The eigenfunctions of $-T_o$ are, then, the same as those of H^o; the corresponding eigenvalue for $\phi(=\psi_1^o)$ is zero, while for other eigenfunctions ψ_k^o the eigenvalue is $(E_k^o-\varepsilon)^{-1}$. Thus we can immediately write the spectral resolution:

$$-T_o = \sum_{k=2}^{\infty} \frac{|\psi_k^o\rangle\langle\psi_k^o|}{(E_k^o-\varepsilon)} \tag{25}$$

From this it is easy to see that $-T_o$ is positive semidefinite if all the denominators are positive; that is, if $\varepsilon<E_2^o$. For some systems this condition is attainable. Thus for the helium atom the ground state energy is -2.9037; if the electron repulsion term is chosen for V, then $E_2^o = -2.5$, and it is easy to find a variational upper bound lying between these values. By contrast, for the hydride ion H^- $E_2^o = -0.625$, which is lower than the true ground-state energy $E_1 \approx -0.528$. Then since we must have $\varepsilon>E_1$, the requirement $\varepsilon<E_2^o$ cannot be met, and this method fails. For the other ions isoelectronic to the helium atom (Li^+, Be^{++},...)$E_2^o>E_1$, and no difficulty arises.

In all these cases V^{-1} is essentially the interelectronic distance r_{12}, and standard procedures /37/ are available for calculating integrals over it; the integrals over $-T_o$ offer more of a problem. The simplest procedure is to choose the Bazley projection manifold $|\underset{\sim}{h}\rangle$ from the set of eigenfunctions of H^o. Then by the spectral resolution of $-T_o$ (Eq. (25)) all nondiagonal terms vanish, and the diagonal elements are given by

$$\langle\psi_k^o|-T_o|\psi_k^o\rangle = \begin{cases} o & k = 1 \\ \dfrac{1}{E_k^o - \varepsilon} & k \neq 1 \end{cases}$$

The objection to this procedure is that this set of functions is not complete unless the continuum functions are included. It is not practical to include these, and without them the inner projection cannot give a good approximation to the operator t.

Another possibility is to truncate $-T_o$, as in Eq. (12), giving an operator $-T_o(j)$ (if the j lowest functions are retained). It is not hard to show that in this case truncation gives an upper bound; that is $-T_o(j) > -T_o$. But since the Bazley projection formula requires inverting the matrix

$$\langle \tilde{h}|V^{-1}-T_o|\tilde{h}\rangle$$

and an upper bound before inversion yields a lower bound after inversion, this is exactly what is needed, and the infinite sums encountered in using Eq. (25) directly are avoided. Several examples of applications of these methods to simple systems are available /38,39,40/.

A third procedure, first suggested by Löwdin, is to avoid the operator $(\varepsilon-PH^o)^{-1}$, which occurs in T_o, by replacing the manifold $|\tilde{h}\rangle$ with a manifold $|\tilde{j}\rangle$ related to $|\tilde{h}\rangle$ by

$$|\tilde{h}\rangle = (\varepsilon-H^o)|\tilde{j}\rangle \tag{26}$$

Then we find

$$\langle \tilde{h}|T_o|\tilde{h}\rangle = \langle \tilde{j}|(\varepsilon-H^o)T_o(\varepsilon-H^o)|\tilde{j}\rangle$$

Now since $PT_o = T_o$ and P commutes with H^o, this can be transformed into

$$\langle \tilde{j}|(\varepsilon-H^o)PT_o(\varepsilon-H^o)|\tilde{j}\rangle = \langle \tilde{j}|P(\varepsilon-H^o)T_o(\varepsilon-H^o)|\tilde{j}\rangle = \langle \tilde{j}|P(\varepsilon-H^o)|\tilde{j}\rangle$$

since $P(\varepsilon - H^o)T_o = P$. The other integrals must, of course, be transformed to the j-manifold also, but this is straightforward and causes no difficulties. In this method the reference manifold itself depends on ε, and as a result the slope $d\varepsilon_1'/d\varepsilon$ is not necessarily negative, and ε_1', regarded as a function of ε, is not a bracketing function. However, ε_1' is a lower bound, since $\varepsilon_1' < \varepsilon_1$ and ε_1, which is calculated from a bracketing function, is a lower bound.

Choi and Smith /41/ used this ε-dependent projection to calculate lower bounds for the ground state of the helium isoelectronic series. The differences between upper and lower bounds ranged from 0.0002 hartrees from helium to 0.005 hartrees for Ne^{8+}.

For the ground state of H^-, and for excited singlet states of other members of this isoelectronic series, the condition $\varepsilon < E_2^o$ cannot be realized. One modification to permit treating such systems is the use of a multi-dimensional reference manifold. The eigenfunctions corresponding to the g lowest eigenvalues of H^o (where g is a suitably chosen integer) are used, and the projector O becomes

$$O = \sum_{i=1}^{g} |\psi_i^o\rangle\langle\psi_i^o|$$

With O defined in this manner, the spectral resolution of $-T_o$ becomes

$$-T_o = \sum_{i=g+1}^{\infty} \frac{|\psi_i^o\rangle\langle\psi_i^o|}{E_i^o - \varepsilon}$$

and the requirement for $-T_o$ to be positive definite is that $\varepsilon < E_{g+1}^o$. If only a finite number of states of H^o have eigenvalues below the desired eigenvalue of H, then g can be chosen to meet this condition. The operator T_o offers the same difficulties as in the preceding case, and the same methods are available for handling them.

When O is a multi-dimensional projector, ε_1 is no longer a single number, but a square matrix. What value is then to be considered a lower bound? Choi and Smith /41/ showed that the lowest eigenvalue of the matrix $\underset{\sim}{\varepsilon}_1$ is a lower bound to the ground state. More generally, it was shown by T.M. Wilson /42/ that each of the eigenvalues of $\underset{\sim}{\varepsilon}_1$, arranged in nondecreasing order, is a lower bound to the corresponding state of H (counting only those with the same symmetry as the functions ψ_i^o), provided that ε is an upper bound to that state and is less than E_{g+1}^o.

The use of these methods can be illustrated by the calculations of T.M. Wilson and the present author /43/ on the triplet states of the He atom and the Li^+ ion. For the helium atom the ordering of the triplet states is shown in the following table:

Triplet states of H^o	Triplet states of H
-2.5	
-2.22	-2.175229
-2.125	
-2.080	-2.0687
-2.0555	

Since two states of H^o lie below the lowest state (2^3S) of H, we must choose $g \geq 2$ to find a lower bound to the lowest triplet state; similarly $g \geq 4$ for the next state (3^3S). The best bounds were found by using the Löwdin projection (Eq. (26)) with the j-manifold consisting of 40 functions of the type introduced by Hylleraas /44/. These consist of an exponential term of the form $e^{-\alpha(r_1 + r_2)}$ multiplied by suitably chosen nonnegative integral powers of $(r_1 + r_2)$, $(r_1 - r_2)$, and r_{12}. For the 2^3S state of the helium atom, Pekeris's /16,45/ upper bound is -2.175229 hartree; using $\varepsilon = -2.1752$ hartrees gives a lower bound of -2.17527 hartrees, only 0.00003 less than the upper bound. For the 3^3S state the difference between upper and lower bounds was 0.018 hartree. Similar results were obtained for the Li^+ ion.

T.M. Wilson /46/ made an extensive study of the behaviour of the eigenvalues of the $\tilde{\varepsilon}_1$ matrix as ε is varied. This gave, among other results, an answer to the question mentioned in the description of Gay's method - that is, for what eigenvalue of the original hamiltonian is a given root the lower bound? For each root of Gay's equations, one of the branches of the function ε_1 (plotted against ε) crosses the $\varepsilon = \varepsilon_1$ line. It is necessary only to determine which branch crosses at any given root; the root is then a lower bound to the corresponding eigenvalue of H.

4. MISCELLANEOUS METHODS AND COMBINATIONS

This survey has covered the three main types of approaches to lower bound calculations, without trying to be exhaustive. A few other methods of more limited applicability are briefly mentioned here.

It is well known that the solutions of the differential equation

$$(H - \varepsilon)\psi = 0$$

diverge to ±∞ as the variables approach infinity for general values of ε, converging to zero only for eigenvalues. Rosenthal and Wilson /47/ took advantage of this by looking for two closely spaced values of ε for which the divergence is in opposite directions; an eigenvalue then lies between. They applied this method successfully to the quartic osciallator and, with much greater difficulty, to the hydrogen molecule ion and the helium atom. Whether it will be useful for more complex systems is doubtful.

The eigenvalues of an oscillator with potential energy $x^2+\lambda x^4$ can be expressed in terms of a continued fraction /48/ or, equivalently, a sequence of Padé approximants. The approximants to the continued fraction converge upward to the eigenvalue and so provide lower bounds. Similar results were found by Rousseau /49/, and a proof was given by Loeffel et. al./50/.

Weltin /51/ has given a method which works if a basis can be found in which at least part of the matrix representing the hamiltonian is tridiagonal. Unfortunately, for most problems of significant interest this is not practical.

E.B. Wilson /52/ modified Gay's equation by writing H - V for H^o, obtaining

$$\text{detn}\{\langle \tilde{j}|(H-\varepsilon)V^{-1}(H-\varepsilon)-(H-\varepsilon)|\tilde{j}\rangle\} = o$$

In this modification V is largely arbitrary, though it must be positive definite. Wilson showed that a suitably chosen constant for V leads to Temple's formula, and that improved bounds for the helium atom could be found by choosing $V = c/r_{12}$ with c>1, the largest permissible value of c being set by the need for establishing, as in Gay's method, which eigenvalue is bounded by a given root.

Combinations of methods may be of value. Bazley /22/ determined a lower bound for E_2 for the helium atom by intermediate hamiltonians and then used this (with Kinoshita's 39-term wave function) in Temple's formula to calculate a lower bound of -2.9037474, differing by only 3.7×10^{-6} from Kinoshita's upper bound. This eliminates the semiempirical feature of Temple's method while getting a far closer bound than is obtainable from intermediate hamiltonians alone.

Experience with the bracketing function method indicates that when it can be applied, very close bounds can be found. With present practically usable formalisms this can be done only if we can find a soluble H^o<H with only finitely many eigenvalues below the eigenvalue of H which is to be bounded. For the lithium atom the usual

division of H into H^o and V does not meet this requirement. Accordingly one of the goals of the author's research on intermediate hamiltonians is to find an H' with essential spectrum beginning above the ground state energy of H. Then the choice $H^o = H'$, $V = H-H'$, will permit further improvements to be made by the bracketing function method.

5. CONCLUSIONS

Over the years since Temple's formula appeared interest in lower bounds has varied, but it has never been the subject of major effort by many quantum theorists. Probably this is due to the feeling that nonrigorous estimates of the accuracy of upper bounds does not justify the difficulty of calculating them. Whether this situation will continue indefinitely remains to be seen. Lower bound calculations are now in a rudimentary state comparable to that of upper bound calculations while Hylleraas was working on the helium atom. At that time the extensive work now routinely carried out on upper bounds for atoms and molecules was not foreseeable; will lower bounds undergo a similar development and be calculated in a similarly routine manner? If so, the basic methods reported here seem sure to be of fundamental importance, and in this volume devoted to the work of Professor Löwdin it is fitting to point out the major significance of the contributions that he has made to this area.

REFERENCES

1. G. Temple, Proc. Roy. Soc. (London) A119, 276 (1928).
2. D.H. Weinstein, Proc. Nat. Acad. Sci. 20, 529 (1934).
3. A.F. Stevenson, Phys. Rev. 53, 199 (1938).
4. A.F. Stevenson, and M.F. Crawford, Phys. Rev. 54, 375 (1938).
5. A. Weinstein, Mem. Sci. Math. 88, (1937).
6. N.W. Bazley, and D.W. Fox, J. Res. Natl. Bur. Std. 65B, 105 (1961); Phys. Rev. 124, 483 (1961).
7. D.W. Fox, SIAM J. Math. Anal. 3, 617 (1972).
8. C.E. Reid, Int. J. Quantum Chem. 6, 793 (1972).
9. A. Messiah, Quantum Mechanics, North Holland Publishing Co., Vol. I, 1961, Vol. II, 1962.
10. J. von Neumann, Mathematical Foundations of Quantum Mechanics, Princeton University Press, 1955 (trans. from German edition, 1932).

11. P.R. Halmos, Introduction to Hilbert Space and the Theory of Spectral Multiplicity, Chelsea (1951).

12. G. Helmberg, Introduction to Spectral Theory in Hilbert Space, North Holland Publishing Co. (1969).

13. J.K.L. MacDonald, Phys. Rev. 46, 828 (1934).

14. L. Wilets, and I.J. Cherry, Phys. Rev. 103, 112 (1956).

15. T. Kinoshita, Phys. Rev. 105, 1490 (1957).

16. C.L. Pekeris, Phys. Rev. 126, 1470 (1962).

17. J. Goodisman, and D. Secrest, J. Chem. Phys. 45, 1515 (1966).

18. J. Goodisman, J. Chem. Phys. 47, 5247 (1967).

19. I. T. Keaveny, and R. E. Christofferson, J. Chem. Phys. 50, 80 (1969).

20. Y. Öhrn, and J. Nordling, Arkiv för Fysik 31, 471 (1966).

21. N. Aronszajn, Proc. Symp. on Spectral Theory and Differential Problems, Stillwater, Oklahoma (1951).

22. N.W. Bazley, Proc. Nat. Acad. Sci. 45, 850 (1959).

23. A. Weinstein, and W. Stenger, Methods of Intermediate Problems for Eigenvalues, Academic Press (1972).

24. H. Weyl, Math. Ann. 71, 441 (1911).

25. C.E. Reid, J. Mol. Spectrosc. 36, 183 (1970).

26. R.A. Sack, Int. J. Quantum Chem. 6, 989 (1972).

27. P.-O. Löwdin, J. Mol. Spectrosc. 14, 119 (1964).

28. R.G. Newton, Scattering Theory of Waves and Particles (McGraw-Hill Book Co., New York, 1966), pp. 247, 274.

29. J.G. Gay, Phys. Rev. 135, A1220 (1964); N.W. Bazley, and D.W. Fox, Phys. Rev. 148, 90 (1966).

30. W.H. Miller, J. Chem. Phys. 42, 4305 (1965).

31. H. Weyl, Rend. Circ. Mat. Palermo 27, 373 (1909).

32. C.E. Reid, Chem. Phys. Letters 26, 243 (1974).

33. C.E. Reid, J. Comp. Phys. 15, 522 (1974).

34. P.-O. Löwdin, Int. J. Quantum Chem. 2, 867 (1969).

35. Y. Hahn, T.F. O'Malley, and L. Spruch, Phys. Rev. 128, 932 (1962); 130, 381 (1963).

36. P.S.C. Wang, J. Chem. Phys. 48, 4131 (1968).

37. J.L. Calais, and P.-O. Löwdin, J. Mol. Spectrosc. 8, 203 (1962).

38. C.F. Bunge, and A. Bunge, J. Chem. Phys. 43, S194 (1965).

39. J.H. Choi, and D.W. Smith, J. Chem. Phys. 43, S189 (1965).

40. C.E. Reid, J. Chem. Phys. 43, S186 (1965).

41. J.H. Choi, and D.W. Smith, J. Chem. Phys. 45, 4425 (1966).

42. T.M. Wilson, J. Chem. Phys. 47, 3912 (1967).

43. T.M. Wilson, and C.E. Reid, J. Chem. Phys. 47, 3920 (1967).

44. E. Hylleraas, Z. für Physik 48, 469 (1928); 54, 347 (1929); 65, 209 (1930).

45. C.L. Pekeris, Phys. Rev. 112, 1649 (1958); 115, 1216 (1959); 126, 143 (1962).

46. T.M. Wilson, J. Chem. Phys. 47, 4706 (1967).

47. C.M. Rosenthal, and E.B. Wilson, Int. J. Quantum Chem. IIs, 175 (1968).

48. C.E. Reid, Int. J. Quantum Chem. 1, 521 (1967).

49. C.C. Rousseau Dissertation Texas A & M UNiversity (1968).

50. J.J. Loeffel, A. Martin, B. Simon, and A.S. Wightman, Phys. Letters 30B, 656 (1969).

51. E. Weltin, Int. J. Quantum Chem. 4, 257 (1970).

52. E.B. Wilson, J. Chem. Phys. 43, S172 (1965).

53. T. Kinoshita, Phys. Rev. 115, 366 (1959)

54. W. Kolos and C.C.J. Roothann, Rev.Mod. Phys. 32 32, 219 (1960).

55. M.E. Schwartz and L.J. Schaad, J. Chem. Phys. 47, 5325 (1967).

BOUNDS TO THE SUM-RULE FUNCTION FROM INNER-PROJECTIONS

Ragnar Ahlberg

Department of Quantum Chemistry, Box 518

751 20 Uppsala, Sweden

1. INTRODUCTION

Many physical properties are evaluated from the average of an operator in quantum mechanics. Such averages can almost never be exactly computed.

In variation and/or perturbation theory one frequently tries to approximate the wavefunction, but one can also try to approximate the operator.

It is often desirable that the approximations to the operators are chosen in some optimal sense, so that, e.g., the exact average is a possible limit. Inner projections fulfil such requirements (Löwdin /1,2/ and references therein, Lindner and Löwdin /3/, and Ahlberg /4/). It has been shown that these optimal properties are the same as those provided by the Hylleraas variation principle /5/, or by the Padé approximants /6/.

In this note inner projections are used to obtain bounds to the sum-rule function, S(k). The bounds are used to check interpolation schemes. One, two and three dimensional examples are given, which exhibit increasing accuracy and consistency with previous results. It is found that the ratio between an S(k) and its inner projection approaches 1 as $k\to-\infty$.

2. DEFINITIONS

Inner projections were introduced as a means to approximate definite operators, i.e., operators whose eigenvalues are all of the same sign. Three forms have been used:

1) The primitive form of the projection of A:

$$A^1 = A^{\frac{1}{2}} O A^{\frac{1}{2}} = A^{\frac{1}{2}} |\underline{f}\rangle\langle\underline{f}|\underline{f}\rangle^{-1}\langle\underline{f}| A^{\frac{1}{2}} \tag{1}$$

2) The Aronszajn projection:

$$A^1 = A|\underline{g}\rangle\langle\underline{g}|A\underline{g}\rangle^{-1}\langle\underline{g}|A \tag{1'}$$

3) The Bazley projection:

$$A^1 = |\underline{h}\rangle\langle\underline{h}|A^{-1}\underline{h}\rangle^{-1}\langle\underline{h}| \tag{1"}$$

The elements of $\underline{f}$, $\underline{g}$ and $\underline{h}$ are suitable functions.

Analogies to the two latter forms were given by Aronszajn /7/, Bazley /8/and by Bazley and Fox /9/. Löwdin /10/ showed that inner projections to definite operators yield bounds, and Ahlberg /4/ showed that they exhibit stationary properties also for a certain class of normal operators.

3. APPLICATION TO THE SUM RULE FUNCTION

The sum rule function for v is defined from

$$S(k) = 2\langle\psi_o|v(H-E_o)^{k+1}\ Pv\psi_o\rangle \tag{2}$$

where H is the Hamiltonian, ψ_o the ground state wavefunction, E_o the ground state energy of the system, $P = 1 - |\psi_o\rangle\langle\psi_o|$, and v a suitable perturbation (usually the dipole operator). Extensive discussions are given by Barnsley /11/ and Hirschfelder et. al. /12/.

Some of the S(k)'s can be computed and/or determined from refraction index data. Other moments may be interpolated from

$$S(k) = S(0)\left(a_o + a_1/(2.5-k) + a_2/(2.5-k)^2 + \ldots\right. \tag{3}$$

or /14/

$$\ell n\ S(k)/S(0) = k\left(a_o + a_1/(2.5-k) + a_2/(2.5-k)^2 + \ldots\right. \tag{3'}$$

Dalgarno and Kingston /13/ used the theorem

$$\lim_{k\to\infty} S(-k)/S(-k-p) = (E_1-E_o)^{-p} \qquad (4)$$

(E_1 is the energy of the first excited state) and obtained

$$a_o = E_1-E_o \qquad (5)$$

Bell's formula /14/ gives

$$a_o = \ln (E_1 - E_o) \qquad (5')$$

The number of terms kept in the right-hand side of (3) and (3') of course depends on the number of known S(k)'s, which are used to determine a_1, a_2... . One is then able to interpolate other moments. Pack /15/and Whisnant and Byers Brown /16/ found the accuracy of these procedures to be satisfactory. Inner projections provide further insight in these matters, and can give estimates when no other procedure is available.

The operator in Eq. (2) is positive. Then, under the assumptions specified below, one may write

$$S(k) \geq S^1(k) = 2\langle\psi_o|v\underline{h}\rangle\langle\underline{h}|(H-E_o)^{-k-1}\underline{h}\rangle^{-1}\langle\underline{h}|v\psi_o\rangle \qquad (6)$$

or if

$$\{h_i\} = \{(H-E_o)^{m_i} Pv\psi_o\} \qquad (7)$$

$$S(k) \geq \underline{a}^+\underline{A}^{-1}\underline{a} = S^1(k) \qquad (6')$$

where

$$a_i = S(m_i-1) \qquad (8)$$

and

$$A_{ij} = S(m_i + m_j - k - 2) \qquad (9)$$

Also Goscinski /17/, and Langhoff and Karplus /18/ have considered special cases of (6').

We will now give the conditions under which $\underline{A}^{-1}$ exists. For a matrix problem of order n, we have to require that $\{h_i\}$ consists of n linearly independent functions orthogonal to ψ_o. Since the rank of $\underline{A}$ is n if and only if this condition is fulfilled, it implies that $\underline{A}^{-1}$ exists. E.g. when $v\psi_o$ has nonzero components of at least n nondegenerate excited states, eq. (7) can be used.*

*One case in which eq. (7) cannot be used is the harmonic oscillator (Linderberg, private communication).

The following theorem is important for the test of the sum rule function:

$$S^1(k)/S(k)\to 1 \qquad \text{as } k\to-\infty$$

if the set (7) contains at least one function with m_i close to k in the limiting procedure, i.e. $(m_i - k) = p$ where p is finite.

Proof: If the theorem holds in the one-dimensional case, it holds always because one can never do worse by introducing further variational freedom (Lindner and Löwdin [3]). The relation (6') now becomes

$$S(k)\geq S^2(m-1)/S(2m-k-1) = S^1(k) \tag{10}$$

One has

$$\lim_{\ell\to-\infty} S(\ell)/f_1(E_1 - E_o)^{\ell} = 1 \tag{11}$$

and

$$\lim_{k\to-\infty} \frac{S^1(k)}{S(k)} = \lim_{k\to-\infty} \frac{S^2(m-1)}{S(2m-k-2)\cdot S(k)} = 1. \tag{12}$$

The last step follows since $m\to-\infty$ as $k\to-\infty$ and since we make repeated use of eq. (11). Q.E.D.

4. COMPUTATIONS

One-Dimensional Inner Projections:

The relation (10) is used in this case. For each k in the table (first entry) $m_i = m$ is chosen to be k + 2. The frequency moments are taken from Pack /15/(who used the Dalgarno - Kingston interpolation formula) in all the examples. Though scarce, the results for k = -1.5 are of interest, because those moments are very effective in computation of van der Waals constants (Ahlberg and Goscinski /19/ and Pack /15/). It is seen that the relative errors range from 50 to 0%, and decrease with k according to the previous theorem.

Two-Dimensional Inner Projections:

The range $-3.5\geq k\geq -6$ was examined. In each case $m_1 = k + 3$ and $m_2 = k + 2$. The new bounds were improved as expected. The trends from the one-dimensional case are conserved.

Table. The first, second and third entries display one-, two- and three-dimensional approximations respectively. The fourth entry gives the raw data from Pack /15/, which are given for comparison.

k-values in S(k)

System	-1.5	-2.0	-2.5	-3.0	-3.5	-4.0	-4.5	-5.0	-5.5	-6.0
	2.48	4.0	6.37	10.1	16.1	25.7	41.0	65.7	106.	170.
					16.6	26.4	42.0	67.0	107.	172.
										172.
	2.9704	4.5	6.9172	10.75	16.85	26.58	42.17	67.17	107.4	172.2
He	1.07	1.13	1.20	1.27	1.36	1.45	1.57	1.69	1.83	1.99
					1.43	1.51	1.62	1.74	1.88	2.03
										2.03
	1.4186	1.3849	1.3877	1.4182	1.472	1.546	1.640	1.754	1.887	2.042
Ne		1.43	1.67	1.88	2.12	2.43	2.82	3.32	3.95	4.74
						2.64	3.05	3.57	4.21	5.02
										5.10
		2.667	2.532	2.544	2.671	2.904	3.247	3.711	4.318	5.098
Ar		4.56	8.91	13.6	18.6	25.0	34.0	46.9	65.9	93.7
						25.6	35.1	49.1	69.7	99.4
										99.8
		11.09	13.38	16.66	21.32	27.94	37.46	51.17	71.77	100.5
Kr		4.81	11.5	21.4	33.9	50.1	72.5	105.	154.	227.
						51.5	74.2	108.	158.	235.
										236.
		16.78	21.68	28.99	39.64	55.18	78.02	111.8	162.1	237.3
Xe		5.89	17.9	41.1	76.1	126.	199.	314.	499.	801.
						129.	202.	318.	507.	822.
										828.
		27.07	39.14	58.30	88.25	135.3	209.8	329.1	520.7	829.6

Three-Dimensional Inner Projections:

Due to the lack of data only $k = -6$ could be treated. One then has to add the function with $m_3 = k + 4$ to the "basis"-set. Again the new bounds were improved.

DISCUSSION

It is seen that the results from the Dalgarno - Kingston interpolation scheme in no case (essentially) violate the inequalities obtained from the inner projections and that these yield increasingly accurate bounds as the dimensionality is increased. The bounds get tighter as k decreases and this reflects the fact that $S'(k)/S(k) \to 1$ as $k \to -\infty$. It is expected that the extrapolations from (6') will be useful in computations when S(0), S(-1), S(-2), and S(-3) are available /19, 20, 21/.

ACKNOWLEDGEMENTS

The author is indebted to Professor Löwdin for his encouragement of and interest in these matters. Discussions with Professor Linderberg and Docent Goscinski have been of great value.

REFERENCES

1. Löwdin, P.O., Int. J. Quantum Chem. 2, 867 (1968).
2. Löwdin, P.O., Int. J. Quantum Chem. 4, 231 (1971).
3. Lindner, P. and Löwdin, P.O., Int. J. Quantum Chem. S2, 161 (1968).
4. Ahlberg, R. Int. J. Quantum Chem. S8, 363 (1974).
5. Brändas, E. and Goscinski, O., Phys. Rev. A1, 552 (1970).
6. Goscinski, O. and Brändas, E., Int. J. Quantum Chem. 5, 131 (1968).
7. Aronszajn, N. Proceedings of the symposium spectral theory and differential problems, Stillwater, Oklahoma (unpublished), 1951.
8. Bazley, N., Phys. Rev. 120, 164 (1960).
9. Bazley, N. and Fox D.W, J. Res. Natl. Bur. Std. (U.S.) 65B, 105 (1961).
10. Löwdin, P.O., Phys. Rev. 139, 357 (1965).
11. Barnsley, M.F. Thesis: Best possible bounds for some atomic

and molecular properties, University of Wisconsin, Madison. (1972).

12. Hirschfelder, J. ., Byers Brown, W. and Epstein, S.T., Adv. Quantum Chem. 1, 255 (1969).

13. Dalgarno, A. and Kingston, A.E., Proc. Roy. Soc. A259, 663 (1960).

14. Bell, R.J. Proc. Phys. Soc. 86, 17 (1965).

15. Pack, R.T., Chem. Phys. Lett. 13, 205 (1972).

16. Whisnant, D.M. and Byers Brown, W., Mol. Phys. 26, 1105 (1973).

17. Goscinski, O., Int. J. Quantum Chem. 2, 761 (1968).

18. Langhoff, P.W. and Karplus, M., in Baker B.A. Jr. and Gammel, J.L., Eds., The Padé Approximants in Theoretical Physics (Acad. Press New York), 41 (1970).

19. Ahlberg, R. and Goscinski, O., J. Phys. (Atom. Molec. Phys.), B7, 1194 (1974).

20. Arai S. and Kaneko S., J. Phys. Soc. 26, 1562 (1969).

21. Dehmer J.L., Inokuti M. and Saxon R.P., Phys. Rev. A 12, 102 (1975).

INVESTIGATIONS INTO THE PROPERTIES OF PROJECTED SPIN FUNCTIONS

Ruben Pauncz

Department of Chemistry, Technion, Israel Institute of Technology, Haifa, Israel

INTRODUCTION

In most molecular calculations one uses a spin free Hamiltonian, so that spin is a good quantum number and the wavefunction should be eigenfunction of S^2. When dealing with closed shell systems we can easily set up a wavefunction which belongs to a resultant spin 0 by using doubly occupied orbitals. On the other hand, if we try to improve the wavefunction using different orbitals for different spins, then we immediately have the problem of constructing many electron spin eigenfunctions. Löwdin /1/ solved the problem in a very effective way be means of the projection operator formalism. In this basic paper he showed how to obtain the spin eigenfunction starting from a primitive spin function and how to calculate the energy associated with the total wavefunction. He also showed that by a suitable orthogonalization procedure one can construct from the projected spinfunctions another set which can be obtained by the so-called genealogical scheme (branching diagram functions). In this paper we shall investigate some special properties of the projected wavefunctions and we shall consider in some detail the relation between the two sets (projected functions and branching diagram functions).

RELATION BETWEEN THE PROJECTED AND BRANCHING DIAGRAM FUNCTIONS

Consider an N electron problem and the spin quantum number S and define μ and ν by the relation

$$\mu = \tfrac{1}{2}N + S \qquad\qquad \nu = \tfrac{1}{2}N - S \tag{1}$$

Let us denote the primitive spinfunctions which have μ α's and ν β's by $\theta_i(\mu,\nu)$. Each can be characterized by a pathdiagram (Löwdin /17) where an α corresponds to an arrow pointing in the direction of 45° degrees and a β to an arrow pointing in the direction of -45° degrees. Let us select those primitive spinfunctions whose pathdiagrams lie entirely above the axis and denote them by $\theta_1, \theta_2, \ldots, \theta_f$ ($f=f(N,S)$ is the number of independent spin states belonging to N and S). We can order them either by the lexicographic order, as in Löwdin's treatment, or by the last letter sequence. This latter will be more convenient for some purposes. We can associate with each primitive function a Yamanouchi symbol, where the α is replaced by 1 and the β is replaced by 2. The last letter sequence will mean that those functions whose last letter is 2 precede those where the last letter is 1. If the last letter is the same then we consider the last but one letter, and so on.

To each Yamanouchi symbol of this kind we can associate a Young tableau, in which the first row has μ boxes and the second row has ν boxes. A standard tableau is obtained if we write the numbers into the boxes from 1 to N; we write the number in the first row if we have 1 in the Yamanouchi symbol, and in the second row if we have 2. These standard tableaux will be arranged according to the same ordering as the corresponding Yamanouchi symbols, we shall denote them by $T_1, T_2, \ldots, T_f$. Finally we shall construct the branching diagram functions, where we shall use again the Yamanouchi symbol, but now 1 means addition and 2 means subtraction of the next spin. In such a way we have established a one-to-one correspondence between primitive spin functions, Young tableaux and branching diagram functions. This correspondence and the same ordering for each set will be maintained during the whole treatment. The branching diagram functions will be denoted by $X_1, X_2, \ldots, X_f$; they are spin eigenfunctions belonging to the quantum number $S=S_z=\frac{1}{2}(\mu-\nu)$. In the following treatment we shall always deal with this quantum number, so we shall not indicate this fact in the notation.

The first branching diagram function X_1 which corresponds to the Yamanouchi symbol $Y_1=(111\ldots122\ldots2)$ can easily be written down. Here we couple together the first μ spins to a resultant $\frac{1}{2}\mu$ and the last ν spins to a resultant $\frac{1}{2}\nu$. Subsequently the two subsystems are coupled together to a resultant $\frac{1}{2}(\mu-\nu)$ by the use of the appropriate vector coupling (Clebsch-Gordan) coefficients.

$$X_1 = \sqrt{\left(\frac{2S+1}{\mu+1}\right)} \sum_{p=0}^{\nu} (-)^p \binom{\mu}{p}^{-1} \left(\alpha^{\mu-p}\beta^p\right)\left(\alpha^p\beta^{\nu-p}\right) \tag{2}$$

Here the square brackets are defined as follows: $\left(\alpha^k\beta^\ell\right)$ is the sum of all primitive spinfunctions with k α's and ℓ β's. In the formula the first bracket refers to the first μ electrons and the second one to the last ν electrons.

We would like to establish the following relation between the primitive spinfunctions and the branching diagram functions

$$\begin{aligned}
\theta_1 &= X_1 T_{11} && +\cdots \\
\theta_2 &= X_1 T_{12} + X_2 T_{22} && +\cdots \qquad (3) \\
&\cdots \\
\theta_f &= X_1 T_{1f} + X_2 T_{2f} + \cdots + X_f T_{ff} && +\cdots
\end{aligned}$$

where $+\cdots$ means contributions from spin eigenfunctions which belong to S larger than $S_z=\frac{1}{2}(\mu-\nu)$. In order to establish this relation we shall prove that

$$\langle\theta_i|X_j\rangle = 0 \quad \text{when } i<j \qquad (4)$$

Let us consider the orthogonal representation generated by the branching diagram functions.

$$P\,X_i = \sum_{j=1}^{f} X_j\, U_{ji}(P) \qquad (5)$$

It has been shown (Pauncz /2/) that if we use the one-to-one correspondence between Young tableaux and branching diagram functions then this representation is identical with Young's orthogonal representation. This fact is important because in the latter we can write down explicitly the matrices corresponding to the transpositions (k, k+1) from the tableaux. The knowledge of the representation matrices enables us to construct the basic units of the symmetric group algebra

$$e_{ij} = (f/N!)\sum_P U_{ji}(P^{-1})\,P \qquad (6)$$

These basic units has been used by Löwdin and Goscinski /3/ in an interesting analysis of the spin degeneracy problem and the construction of the wavefunction. Because of the fact that we have orthogonal matrices we can write these units in the following form

$$e_{ij} = (f/N!)\sum_P U_{ij}(P)\,P \qquad (7)$$

We observe that the Hermitean adjoint of e_{ij} is e_{ji}.

Let us apply e_{ij} to the first primitive spinfunction. This function remains invariant under all permutations which belong to the subgroup $g_a = S_\mu \otimes S_\nu$, where S_μ contains all the permutations of the numbers 1, 2,...,μ and S_ν consists of the permutations of the numbers $\mu+1, \mu+2,\ldots \mu+\nu$. Let us divide the symmetric group S_N into the cosets of the subgroup G_a and denote the coset generators by t_k.

$$e_{ij}\theta_1 = \frac{f}{N!}\sum_{k=1}^{nk}\sum_{P_a\varepsilon G_a} U_{ij}(t_k P_a)\, t_k P_a \theta_1 \tag{8}$$

$$P_a\theta_1 = \theta_1 \tag{9}$$

$$e_{ij}\theta_1 = \frac{f}{N!}\sum_{m=1}^{f}\sum_{k=1}^{nk}\sum_{Pa\varepsilon Ga} U_{im}(t_k)U_{mj}(P_a)t_k\theta_i \tag{10}$$

$$\sum_{Pa\varepsilon Ga} U_{mj}(Pa) = \delta_{1m}\delta_{1j}n_g \tag{11}$$

Equation (11) is obtained using the lemma of Matsen and Cantu /4/ and the fact that the first branching diagram function is invariant under all the permutations of the subgroup G_a. Therefore the matrix representations of $\sum_{Pa\varepsilon Ga} P_a$ is a matrix whose left upper corner element is n_g (the order og G_a) and all the rest is zero.

$$E_{ij}\theta_1 = \delta_{ij}\, n_g \frac{f}{N!}\sum_{k=1}^{nk} U_{i1}(t_k)t_k\theta_1 \tag{12}$$

Following the analysis of Löwdin and Goscinski we see that we obtain the set of wavefunctions $X_1, X_2, \ldots, X_f$ by applying the operators $e_{11}, e_{21}, \ldots, e_{f1}$ to θ_1. Finally we get

$$\langle\theta_1|X_j\rangle = \langle\theta_1|e_{j1}\theta_1\rangle = \langle e_{ij}\theta_1|\theta_1\rangle = 0 \text{ for } j>1. \tag{13}$$

This shows that θ_1 has contribution only from X_1 (and the higher spin functions) and the branching diagram functions $X_2,\ldots X_f$ do not contain the primitive function θ_1.

In order to prove relation (4) for i=2,...,f-1 we use the observation (Pauncz /5/) that using the basic transpositions (k,k+1) one can set up a chain of transpositions which lead from the first tableau to the last one

$$T_i = (k,k+1)\, T_j \quad \text{where } j>i \tag{14}$$

The same relation is true for the primitive functions, while for the branching diagram functions we have the relation

$$(k,k+1)X_i = -\rho_i X_i + X_j\sqrt{(1-\rho^2)} \tag{15}$$

$$\rho_i = 1/d^{i}_{k,k+1} \tag{16}$$

$d^{i}_{k,k+1}$ is the <u>axial distance</u> in tableau T_i between the numbers k and k+1 <u>i.e.</u> the number of steps needed to arrive from k to k+1. (Left and down are counted as positive, up and right, negative). As an illustration consider the first and second tableaux.

T_1

1	2	•	ν	•	•	•	•	μ
μ+1	μ+2	•	μ+ν					

T_2

1	2	•	ν	•	•	•	•	μ+1
μ	μ+2	•	μ+ν					

$\theta_1 = \underbrace{\alpha\alpha\ldots\ldots\alpha}_{\mu}\ \underbrace{\beta\beta\ldots\beta}_{\nu}$ $\qquad$ $\theta_2\ \ \underbrace{\alpha\alpha..\ \beta}_{\mu}\ \ \underbrace{\alpha\beta\ldots\ \beta}_{\nu}$

$\theta_2 = (\mu,\mu+1)\theta_i \qquad\qquad T_2 = (\mu,\mu+1)T_1$

$$d^{1}_{\mu,\mu+1} = \mu$$

$$(\mu,\mu+1)X_1 = -X_1/\mu + X_2\sqrt{(1-\mu^{-2})} \tag{17}$$

Applying the transposition $(\mu,\mu+1)$ to the first equation of (3) the left hand side will be θ_2, and the right hand side a linear combination of X_1 and X_2 as given by (17). Therefore relation (4) is true for i=2. If we apply successively the appropriate transpositions of the form (k,k+1) we can establish the validity of (4) for all values of i. It is important to observe that if a transposition (k,k+1) yields T_i from T_j then the same transposition will never yield a T_m (m>j) from T_l ($l<i$).

After establishing the validity of (3) let us apply the projection operator O_{ss} to the set of equations. We shall have on the left hand side the projected spin functions, on the right hand side $X_1, X_2, \ldots, X_f$ will remain unchanged as they are eigen functions belonging to the quantum number $S=S_z$ while the higher spin functions will be annihilated. Therefore we have the relations:

$$\Theta_1 = X_1 T_{11}$$

$$\Theta_2 = X_1 T_{12} + X_2 T_{22} \tag{18}$$

$$\Theta_f = X_1 T_{1f} + X_2 T_{2f} + \ldots + X_f T_{ff}$$

As the branching diagram functions form an orthonormal set, it is evident from the structure of equations (18) that the projected functions formed in this way (from the primitive functions whose pathdiagrams lie entirely above the axis) form a linearly independent system (Löwdin's theorem /1/). For an alternative proof of the theorem see Gershgorn /6/. Defining the row vectors $\underline{X} = (X_1, X_2, \ldots, X_f)$ and $\underline{\Theta} = (\Theta_1, \ldots \Theta_f)$ (18) can be written as a matrix equation

$$\underline{\Theta} = \underline{X}\,\underline{T} \tag{19}$$

where $\underline{T}$ is an upper triangular matrix. Comparison of X_1 with Löwdin's formula for the projected function Θ_1 yields

$$T_{11} = \sqrt{(2S+1)/(\mu+1)}$$

The rest of the matrix can be obtained by using the chain of transpositions (k,k+1) and the relations (14) and (15). The reasoning used in proving the theorem yields also an algorithm for the construction of the triangular matrix which gives the relation between the projected functions and the branching diagram functions.

Let us form the overlap matrix of the projected functions with the elements $\Delta_{ij} = \langle\Theta_i|\Theta_j\rangle$. From (19) it follows

$$\underline{\Delta} = \tilde{\underline{T}}\,\underline{T} \tag{20}$$

The overlap matrix can easily be evaluated following the rules given by Löwdin /1/ and then relation (20) shows that we can evaluate $\underline{T}$ from $\underline{\Delta}$ by using the well-known Cholesky algorithm /7/. This is an alternative way for the construction of $\underline{T}$.

We can invert $\underline{T}$ and the result is again an upper triangular matrix and we can obtain the branching diagram functions in terms of the projected spin functions.

$$\underline{X} = \underline{\Theta}\,\underline{T}^{-1} \tag{21}$$

REPRESENTATIONS OF THE SYMMETRIC GROUP GENERATED BY THE PROJECTED FUNCTIONS

If we apply a permutation to one of the projected functions Θ_i the result could be expressed as a linear combination of the linearly independent projections. (belonging to the same S).

$$P\,\Theta_i = \sum_{j=1}^{f} \Theta_j A_{ji}(P) \tag{22}$$

One can show that in such a way one obtains a representation of the symmetric group which is irreducible. The representation generated by the projected functions has some interesting properties. For N=3 and 4 this representation is identical with Young's natural representation, the peculiar property of which is that the matrices contain only integers. Actual calculations have shown that up to N=6 the elements are +1,0, or -1. For N=7 we have the first time the appearance of an element 2. The representation has the drawback that it is not unitary, moreover these matrices do not have the immediate importance in the energy expression.

Let us define the following matrices:

$$B_{ij}(P) = \langle \Theta_i | P | \Theta_j \rangle \tag{23}$$

Löwdin /1/ has shown that when the wavefunction is constructed using the projected spin function, then the matrix elements of the Hamiltonian and the identity operator are given in terms of the $B_{ij}(P)$

$$H_{ij} = \sum_P (-1)^p\, H(P)\, B_{ij}(P)$$

$$S_{ij} = \sum_P (-1)^p\, I(P)\, B_{ij}(P) \tag{24}$$

The relation between the $\underline{A}$ and $\underline{B}$ matrices is easily obtained

$$\underline{B}\,(P) = \underline{\Delta}\,\underline{A}\,(P) \tag{25}$$

The $\underline{B}$ matrices do not generate a representation of the symmetric group in the usual sense, they rather give a sandwich representation.

$$\underline{B}(PR) = \underline{\Delta}\,\underline{A}\,(PR) = \underline{\Delta}\,\underline{A}\,(P)\,\underline{A}\,(R) = \underline{\Delta}\,\underline{A}\,(P)\,\underline{\Delta}^{-1}\,\underline{\Delta}\,\underline{A}\,(R)$$

$$= \underline{B}(P)\,\underline{\Delta}^{-1}\,\underline{B}(R) \qquad (26)$$

The inverse of the overlap matrix which appears in this sandwich representation, can easily be obtained from (20)

$$\underline{\Delta}^{-1} = \underline{T}^{-1}\,\underline{\tilde{T}}^{-1} \qquad (27)$$

Actual calculations up to N=10 have shown that this matrix has only integer elements. These are useful relations in view of the importance of the B(P) matrix in the energy expression.

Acknowledgments

The author would like to express his sincere gratitude to Professor Per-Olov Löwdin for all his inspiration and scientific advice and expresses his best wishes for the 60ieth birthday. The author is grateful to Andreas Schmelczer for discussions and calculations of $\underline{\Delta}^{-1}$ matrices. This paper is part of a research project supported by the U.S. - Israel Binational Science Foundation.

References

1. P.O. Löwdin, Calcul des fonctions d'onde moléculaire (Centre National de la Recherche Scientifique, Paris, 1958) p. 23.
2. R. Pauncz, Alternant Molecular Orbital Method (Saunders, Philadephia, 1967) p. 216.
3. P.O. Löwdin and O. Goscinski, Int. J. Quantum Chem. S2, 533 (1970).
4. F.A. Matsen and A.A. Cantu, J. Phys. Chem. 73, 2488 (1969).
5. R. Pauncz, Chem. Phys. Letters 31, 443 (1975).
6. Z. Gershgorn, Int. J. Quantum Chem. 2, 341 (1968)
7. E. Bodewig, Matrix Calculus (North Holland Pub. Co, Amsterdam. 1956) p. 110.

Appendix

$\underline{\Delta}^{-1}$ matrices up to N = 6

N=3 S=½

2	1
1	2

N=4 S=0

4	2
2	4

N=4 S=1

2	1	1
1	2	1
1	1	2

N=5 S=½

6	3	3	3	3
3	4	2	2	1
3	2	4	1	2
3	2	1	4	2
3	1	2	2	4

N=5 S=3/2

2	1	1	1
1	2	1	1
1	1	2	1
1	1	1	2

N=6 S=0

12	6	6	6	6
6	8	4	4	2
6	4	8	2	4
6	4	2	8	4
6	2	4	4	8

N=6 S=1

6	4	3	3	4	3	3	2	2
4	6	3	3	4	2	2	3	3
3	3	4	2	2	2	1	2	1
3	3	2	4	2	1	2	1	2
4	4	2	2	6	3	3	3	3
3	2	2	1	3	4	2	2	1
3	2	1	2	3	2	4	1	2
2	3	2	1	3	2	1	4	2
2	3	1	2	3	1	2	2	4

N=6 S=2

2	1	1	1	1
1	2	1	1	1
1	1	2	1	1
1	1	1	2	1
1	1	1	1	2

MANY-BODY THEORY OF MOLECULAR COLLISIONS+

David A. Micha

Quantum Theory Project Departments of Chemistry and Physics

University of Florida, Gainesville, Fl 32611

1. INTRODUCTION

Fundamental aspects of chemical reactivity have been actively studied in recent years in terms of molecular motions on given potential energy surfaces /1/. These surfaces are obtained after separation of electronic and nuclear motions has been accomplished within the adiabatic approximation. The separation of electronic and nuclear portions of the problem has proved fruitful from the computational point of view, but has prevented a deeper understanding of chemical bonding in a reactive process. The molecular-orbital or valence-bond (VB) methods employed in reactive problems are the same ones used to describe stable chemical species. In this contribution we want to discuss some of the conceptual understanding that may be extracted from a simultaneous treatment of both electronic and nuclear motions in reactive molecular collisions.

The formalism of reactive collisions has also been significantly developed in recent years, with most of the efforts directed towards systems of a few particles without internal structure /2,3/. We have recently adapted these developments to atom-diatomic collisions by treating atoms as particles with spin, in such a way that only diatomic potential energies are required to begin with /4/. It is of interest to provide a more fundamental approach by incorporating details of atomic electronic structure. This requires that atoms be treated as composite particles, and leads to a manybody approach to reactive collisions. In following sections we briefly develop the equations for two coupled rearrangement channels with

+ Supported by NSF Grant MPS-75-01077

an operator approach similar to ohters used for bound-state /5/ and non-reactive scattering problems /6/. We next construct channel N-electron states for the atom-diatomic problem and analyse the critical question of decomposition of the electronic hamiltonian H_{el} in a way appropriate to reactions. Expansion of the coupled equations provides then insight on the electronic structure of breaking of forming chemical bonds. These developments are finally illustrated for $H + H_2$.

The present treatment has some similarities with methods of atoms-in-molecules /7/ and diatomics-in-molecules /8,9/, but differ from these in the way it includes electronic exchange at all stages fo the rearrangement process.

II. EQUATIONS FOR TWO COUPLED REACTION CHANNELS

Concentrating on the process A + BC→AB + C we introduce the channel 1 coordinates ($\underset{\sim}{R}_1$, $\underset{\sim}{r}_1$) for the arrangement A + BC, with $\underset{\sim}{R}_1$ the relative position of A with respect to the center of mass of BC, and $\underset{\sim}{r}_1$ the position of C with respect to B. Channels 2 and 3 are obtained by cyclic perturbation of A,B, and C. Indicating with H the total (nuclear plus electronic) hamiltonian, we introduce

$$H_i = \lim_{R_i \to \infty} H \tag{1a}$$

$$H_0 = \lim_{R_i \to \infty,\ r_i \to \infty} H \tag{1b}$$

and

$$H = H_1 + V^{(1)} = H_3 + V^{(3)} = H_0 + V \tag{2}$$

where H_i is the free-motion hamiltonian in channel i, and $V^{(i)}$ the remaining coupling. We seek scattering solutions of Schrödinger's equation for cases in which H decomposes in more than one way /10/. The reaction above is a particular case in which only channel 1 and 3 are coupled.

Introducing the complex energy z, we have for the total propagator G the expression

$$G = (z - H)^{-1} = G_0 + G_0 VG. \tag{3}$$

Adding and substracting terms,

$$G = G_0 + G_0 (V^{(1)} + V^{(3)})\ G + G_0\ (V - V^{(1)} - V^{(3)})\,G. \tag{4}$$

Partly solving for G this gives

$$G = G' + G' (V^{(1)} + V^{(3)}) G \tag{5a}$$

$$\begin{aligned} G' &= [1 - G_0 (V - V^{(1)} - V^{(3)})]^{-1} G_0 \\ &= (G_0^{-1} - V + V^{(1)} + V^{(3)})^{-1} \\ &= (z - H_1 + V^{(3)})^{-1} \\ &= (z - H_3 + V^{(1)})^{-1} \end{aligned} \tag{5b}$$

We next define, for i=1,3,

$$L_i = G' V^{(i)} G, \tag{6}$$

which on account of Eq. (5a), and solving for L_i, gives (with $i \neq j$)

$$\begin{aligned} L_i &= G' V^{(i)} (G' + L_i L_j) \\ &= (1 - G' V^{(i)})^{-1} G' V^{(i)} (G' + Lj) \\ &= [(G')^{-1} - V^{(i)}]^{-1} V^{(i)} (G' + Lj) \\ &= G_j V^{(i)} (G' + Lj) \end{aligned} \tag{7}$$

where

$$G_j = (z - H_j)^{-1} . \tag{8}$$

Turning next to the scattering wave function $\Psi_{1\alpha}$, obtained from the freemotion wavefunction $\Phi_{1\alpha}$ for state α in channel 1, the formal theory of scattering gives

$$\Psi_{1\alpha} = \Phi_{1\alpha} + GV^{(1)}\Phi_{1\alpha} \tag{9}$$

or, replacing from Eq. (5a),

$$\Psi_{1\alpha} = \Psi_{1\alpha}{}^{(1)} + \Psi_{1\alpha}{}^{(3)}, \tag{10a}$$

$$\Psi_{1\alpha}^{(1)} = \Phi_{1\alpha} + L_3 V^{(1)} \Phi_{1\alpha}, \qquad (10b)$$

$$\Psi_{1\alpha}^{(3)} = (G' + L_1)\, V^{(1)} \Phi_{1\alpha}. \qquad (10c)$$

Making use of Eq. (7) in Eq. (10b) it immediately follows that

$$\Psi_{1\alpha}^{(1)} = \Phi_{1\alpha} + G_1 V^{(3)}\, \Psi_{1\alpha}^{(3)}, \qquad (11a)$$

while Eq. (10c) gives

$$\Psi_{1\alpha}^{(3)} = (1 + G_3 V^{(1)}).\, G' V^{(1)} \Phi_{1\alpha} + G_3 V^{(1)} L_3 V^{(1)} \Phi_{1\alpha}$$

which together with

$$(1 + G_3 V^{(1)})\, G' = G_3\, (G_3^{-1} + V^{(1)})\, G'$$

$$= G_3\, (G')^{-1} G' = G_3$$

leads to

$$\Psi_{1\alpha}^{(3)} = G_3 V^{(1)}\, \Psi_{1\alpha}^{(1)}. \qquad (11b)$$

Equations (11 a,b) are our basic ones. They provide components of the total wavefunction that correspond to the last interaction occurring in channels 1 or 3. They involve only free motion propagators G_j and apply to quite general hamiltonians. In differential form (multiplying by G_i^{-1}) they are

$$(z - H_1)\, \Psi^{(1)} = V^{(3)}\, \Psi^{(3)}, \qquad (12a)$$

$$(z - H_3)\, \Psi^{(3)} = V^{(1)}\, \Psi^{(1)}, \qquad (12b)$$

which when added would reproduce Schrödinger's equation. Coupled wave and transition operators could similarly be obtained. For example, from the definition of the transition operator for the $1 \to i$ process,

$$T_{i1} \Phi_{1\alpha} = V^{(i)} \Psi_{1\alpha}, \qquad (13)$$

it may be proved from the previous equations that

$$T_{11} = V^{(1)} G_3 T_{31} , \quad (14a)$$

$$T_{31} = V^{(3)} + V^{(3)} G_1 T_{11} , \quad (14b)$$

which have been previously proposed as the correct coupled equations for the rearrangement problem /11/.

III. CONSTRUCTION OF CHANNEL ELECTRONIC STATES

We shall construct electronic states appropriate to each reaction channel for the N electrons of the triatomic system, starting with the exact atomic states. In principle we should include all electronic bound states of each atom, while in practice a selected set of states compatible with the asymptotic scattering wavefunctions would be used. The properly normalized and antisymmetrized eigenstates $| A \ell >$ of the electronic hamiltonian H_A of atom A satisfy

$$H_A | A \ell > = E_\ell^A | A \ell > \quad (15a)$$

$$< A\ell | A\ell' > = \delta_{\ell\ell'} , \quad (15b)$$

$$\mathbf{A}_A | A\ell > = | A\ell > , \quad (15c)$$

where the antisymmetrizer $\mathbf{A}_A = \mathbf{A}_A^\dagger = \mathbf{A}_A^2$ for A is

$$\mathbf{A}_A = (N_A!)^{-1} \sum_{P \epsilon S_A} (-1)^p P \quad (16)$$

i.e., a sum over all permutations in the symmetric group S_A of A. Similar expressions hold for atoms B and C.

Beginning with M_0 chosen N-electron states for the zeroth-channel (with A,B, and C free), we collect all M_0 states in the $1 \mathrm{x} M_0$ row matrix

$$| \underset{\sim}{\varphi}_0 > = (N_{\ell,m,n} \mathbf{A}_{A,B,C} | A\ell > | Bm > | Cn >) \quad (17)$$

where N is a normalization constant and $\mathbf{A}_{A,B,C}$ is a supplementary antisymmetrizer satisfying

$$\mathbf{A} = \mathbf{A}_{A,B,C} \mathbf{A}_C \mathbf{A}_B \mathbf{A}_A . \quad (18)$$

We have then

$$|\underset{\sim}{\varphi}_0|\underset{\sim}{\varphi}_0> = \underset{\sim}{\Delta}_{00} \qquad (19)$$

where $\underset{\sim}{\Delta}_{00}$ is an overlap $M_0 \times M_0$ matrix.

Turning next to channel 1, we make use of diatomic states $N_{mn}\mathbf{A}_{B,C}|Bm>|Cn>$ to construct by superposition eigenstates $|BCp>$ of the electronic spin $\underset{\sim}{S}_1$ of diatomic BC, that also diagonalize the electronic hamiltonian H_{BC} of the diatomic, and satisfy

$$S_1{}^2|BC\,p> = s_1(s_1+1)|BC\,p> \qquad (20a)$$

$$S_{1Z}|BC\,p> = m_1|BC\,p> \qquad (20b)$$

$$<BC\,p'|\,H_{BC}|BC\,p> = E_p{}^{BC}\,(r_1)\delta_{pp'}. \qquad (20c)$$

From these, M_1 triatomic states may be constructed and collected in the $1 \times M_1$ row matrix

$$|\underset{\sim}{\varphi}_1> = (N_{\ell,p}\,\mathbf{A}_{A,BC}|A\ell>|BCp>) \qquad (21)$$

which may also be written as $|\underset{\sim}{\varphi}_1> = |\underset{\sim}{\varphi}_0>\underset{\sim}{\Delta}_{01}$ with the $M_0 \times M_1$ matrix $\underset{\sim}{\Delta}_{01}$ dependent on $(\underset{\sim}{R}_1, \underset{\sim}{r}_1)$. Finally, this last collection of states may be used to construct $M^{(1)}$ valence-bond structures for channel 1, which diagonialize the total electronic spin operators S^2 and S_z and are grouped in the $1 \times M^{(1)}$ row matrix $|\underset{\sim}{\varphi}^{(1)}>$ so that

$$S^2|\underset{\sim}{\varphi}^{(1)}> = s(s+1)|\underset{\sim}{\varphi}^{(1)}>, \qquad (22a)$$

$$S_Z|\underset{\sim}{\varphi}^{(1)}> = m|\underset{\sim}{\varphi}^{(1)}>. \qquad (22b)$$

This row of states may be reexpressed as

$$|\underset{\sim}{\varphi}^{(1)}> = |\underset{\sim}{\varphi}_1>\underset{\sim}{\Delta}^{(1)}{}_1 = |\underset{\sim}{\varphi}_0>\underset{\sim}{\Delta}^{(1)}{}_0, \qquad (23)$$

with matrices of coefficient that satisfy

$$<\underset{\sim}{\varphi}_1|\underset{\sim}{\varphi}_1> = \underset{\sim}{\Delta}_{11} = \underset{\sim}{\Delta}_{01}{}^{\dagger}\underset{\sim}{\Delta}_{00}\underset{\sim}{\Delta}_{01}, \qquad (24a)$$

$$<\tilde{\varphi}^{(1)}|\tilde{\varphi}^{(1)}> = \tilde{\Delta}^{(11)} = \tilde{\Delta}^{(1)}{}_1{}^{\dagger}\tilde{\Delta}_{11}\tilde{\Delta}^{(1)}{}_1. \quad (24b)$$

Similar constructions provide us with VB structures in channels 2 and 3. Overlaps between VB structures of different channels may be viewed as reaction recoupling coefficients, given by

$$<\varphi^{(i)}|\varphi^{(j)}> = \tilde{\Delta}^{(ij)} = \tilde{\Delta}^{(i)}{}_0{}^{\dagger}\tilde{\Delta}_{00}\tilde{\Delta}_0{}^{(j)}. \quad (24c)$$

Completeness relations within each of the finite spaces introduced above take the form

$$|\tilde{\varphi}_0>(\tilde{\Delta}_{00})^{-1}<\tilde{\varphi}_0| = \tilde{I}_0 , \quad (25a)$$

$$|\tilde{\varphi}_i>(\tilde{\Delta}_{ii})^{-1}<\varphi_i| = \tilde{I}_i , \quad (25b)$$

$$|\tilde{\varphi}^{(i)}>(\tilde{\Delta}^{(ii)})^{-1}<\varphi^{(i)}| = \tilde{I}^{(i)} . \quad (25c)$$

The step-by-step procedure followed here drastically reduces the number of required basis functions and gives states naturally adapted to rearrangement processes.

IV. DECOMPOSITION OF THE ELECTRONIC HAMILTONIAN

We make use of a non-relativistic electronic hamiltonian H_{el}, and further neglect for simplicity the electron mass compared to nuclear ones. Contributions from spin-orbit and mass polarization effects could be incorporated later on. The decomposition of H_{el} presents special problems when dealing with scattering. Usage of the asymptotic hamiltonians defined by Eqs. (1a,b) can not be avoided because they are brought in by the scattering boundary conditions. Difficulties arise because this asymptotic hamiltonians do not commute with the total antisymmetrizer **A**. For example, matrix elements of the electronic portion $H_{1,\ el}$ of H_1 between two VB structures of channel 1 would contain spurious terms as $R_1 \to \infty$, contributed by the electronic kinetic energy operators. On the other hand, matrix elements of H_{el} behave properly. In effect, for

$$H_{el} = \sum_I H_I + \sum_{JK}' V_{JK} , \quad (26)$$

where the V_{JK} indicate Coulomb interactions between atoms J and K, we find

$$<\tilde{\varphi}^{(1)}|H_{el}|\tilde{\varphi}^{(1)}> = \tilde{\Delta}^{(1)\dagger}_1<\varphi_1|H_{el}|\varphi_1>\tilde{\Delta}^{(1)}{}_1 \quad (27a)$$

$$H_{el}|\underset{\sim}{\varphi}_1> = (N_{1,p}\, \mathbf{A}_{A,BC}\, H_{el}|A1>|BCp>)$$

$$\underset{R_1 \to \infty}{\sim} |\underset{\sim}{\varphi}_1>(\underset{\sim}{E}_1 + \underset{\sim}{e}_1) \tag{27b}$$

where we have introduced the $M_1 \times M_1$ matrices

$$\underset{\sim}{E}_1 = [(E_1^A + E_p^{BC})\delta_{11'}\delta_{pp'}] \tag{28}$$

and $\underset{\sim}{e}_1$, which measures the error introduced by restricting the basis of BC to a finite number. Within this error we then get

$$<\underset{\sim}{\varphi}^{(1)}|H_{el}|\underset{\sim}{\varphi}^{(1)}> \sim \underset{\sim}{\tilde{E}}^{(1)} = \lim_{R_1\to\infty} \underset{\sim}{\Delta}_1^{(1)\dagger} \underset{\sim}{E}_1 \underset{\sim}{\Delta}_1^{(1)} \tag{29}$$

which is the correct limit. But we also find

$$<\underset{\sim}{\varphi}^{(1)}|H_{el}|\underset{\sim}{\varphi}^{(1)}> = <\underset{\sim}{\varphi}^{(1)}|H_{1,el} + V^{(1)}|\varphi^{(1)}>$$

$$\underset{R_1\to\infty}{\sim} \underset{\sim}{\tilde{H}}^{(1)} + \underset{\sim}{\tilde{V}}^{(1)} \tag{30}$$

Hence we can express H_{el} as the sum of the two terms (with $\underset{\sim}{\Gamma}^{(i)} = (\underset{\sim}{\Delta}^{(ii)})^{-1} = \underset{\sim}{\Gamma}^{(i)\dagger}$)

$$\overline{H}_{1,el} = H_{1,el} + |\varphi^{(1)}>\underset{\sim}{\Gamma}^{(1)}\ (\underset{\sim}{\tilde{E}}^{(1)} - \underset{\sim}{\tilde{H}}^{(1)})\Gamma^{(1)}<\underset{\sim}{\varphi}^{(1)}| \tag{31a}$$

$$\overline{V}^{(1)} = V^{(1)} - |\underset{\sim}{\varphi}^{(1)}<\Gamma^{(1)}\ \underset{\sim}{\tilde{V}}^{(1)}\Gamma^{(1)}<\underset{\sim}{\varphi}^{(1)}| \tag{31b}$$

which have the correct limits within the space of $\underset{\sim}{\varphi}^{(1)}$ and satisfy

$$H_{el} = \overline{H}_{1,el} + \overline{V}^{(1)} = H_{1,el} + V^{(1)} \tag{32}$$

Similar procedures can be followed to obtain the proper decompositions in channels 2 and 3.

In what follows we shall need matrix elements of

$$\overline{V}_{JK} = V_{JK} - |\underset{\sim}{\varphi}^{(i)}>\underset{\sim}{\Gamma}^{(i)}\underset{\sim}{\tilde{V}}_{JK}\Gamma^{(i)}<\underset{\sim}{\varphi}^{(i)}| \tag{33}$$

which we can obtain noticing that, e.g. for JK = BC, the interaction V_{BC} appears as the matrix

$$<\underset{\sim}{\varphi}^{(1)}|V_{BC}|\underset{\sim}{\varphi}^{(1)}> \; = <\underset{\sim}{\varphi}^{(1)}|H_{BC} - H_B - H_C|\underset{\sim}{\varphi}^{(1)}>. \qquad (34)$$

Each term in this expression may be obtained in the appropriate basis, which may be transformed into $\underset{\sim}{\varphi}^{(1)}$. For example,

$$<\underset{\sim}{\varphi}_1|H_{BC}|\underset{\sim}{\varphi}_1> \; = \underset{\sim}{E}_1{}^{BC} + \underset{\sim}{e}_1{}^{BC} , \qquad (35a)$$

$$\underset{\sim}{E}_1{}^{BC} = (E_p{}^{BC}\delta_{11'}\delta_{pp'}) , \qquad (35b)$$

while $\underset{\sim}{e}_1{}^{BC}$ includes all asymptotically spurious terms, that cancel at $R_1 \to \infty$ with the corresponding ones in the second term of Eq. (33). Similarly,

$$<\underset{\sim}{\varphi}_0|H_B|\underset{\sim}{\varphi}_0> \; = \underset{\sim}{E}_0{}^{B} + \underset{\sim}{e}_0{}^{B} , \qquad (36a)$$

$$\underset{\sim}{E}_0^{B} = (E_m^{B} \, \delta_{11'}\delta_{mm'}\delta_{nn'}) . \qquad (36b)$$

Going back to the basis $^{(1)}$ the result is

$$<\underset{\sim}{\varphi}^{(1)}|\bar{V}_{BC}|\underset{\sim}{\varphi}^{(1)}> = \underset{\sim}{V}^{(1)}{}_{BC} + \underset{\sim}{v}^{(1)}{}_{BC} , \qquad (37a)$$

$$\underset{\sim}{V}^{(1)}{}_{BC} = \underset{\sim}{\Delta}^{(1)}{}_1 \, \underset{\sim}{E}^{BC}_1 \Delta^{(1)}{}_1 - \underset{\sim}{\Delta}^{(1)\dagger}{}_0 (\underset{\sim}{E}^{B}_0 + \underset{\sim}{E}^{C}_0)\underset{\sim}{\Delta}^{(1)}{}_0 , \qquad (37b)$$

where $\underset{\sim}{v}^{(1)}_{BC}$ goes asymptoically to zero for $R_1 \to \infty$ or $r_1 \to \infty$.

V. EXPANSION OF THE COUPLED REARRANGEMENT EQUATIONS

Recalling that the wavefunction $\Psi^{(i)}$ accounts for a process with the last interaction in channel i, we expand it in the basis specially constructed for this channel. We assume, as usually done, that the basis $\underset{\sim}{\varphi}^{(i)}$ is practically complete by choice, in which case we can use the completeness relation Eq. (25c) to expand states and operators in channel i, to get

$$\Psi^{(i)} = |\underset{\sim}{\varphi}^{(i)}>\underset{\sim}{\psi}^{(i)} \qquad (38)$$

where we have introduced the $M^{(i)}$ x 1 column matrix $\underset{\sim}{\psi}^{(i)}$, with elements dependent on (R_i, r_i). Replacing in Eqs. (12 a,b) and projecting the i-th equation on $\langle\underset{\sim}{\varphi}^{(i)}|$ we find the set of $M^{(1)} + M^{(3)}$ coupled equations

$$(z - \langle\underset{\sim}{\varphi}^{(i)}|\ \sum_J K_J + \overline{H}_{i,el}|\underset{\sim}{\varphi}^{(i)}\rangle)\underset{\sim}{\psi}^{(i)}$$
$$= \langle\underset{\sim}{\varphi}^{(i)}|\ \overline{V}^{(j)}|\underset{\sim}{\varphi}^{(j)}\rangle\underset{\sim}{\psi}^{(j)}\ , \tag{39}$$

where $i \neq j$, and we have used the hamiltonian decomposition of Section IV. Solving this set with scattering boundary conditions we would obtain the solution

$$\Psi_{1\alpha} = |\underset{\sim}{\varphi}^{(1)}\rangle\psi_{1\alpha}^{(1)} + |\underset{\sim}{\varphi}^{(3)}\rangle\psi_{1\alpha}^{(3)}. \tag{40}$$

We are then naturally lead to view the scattering state as a sum of channel contributions each one of which contains all the VB structures (both bonding and non-bonding) for that channel. This may be contrasted with the expansion used for potential surface calculations, in which one would use a basis with a mixture of some VB structure from every channel. The two expansion procedures are of course equivalent, but only the coefficients in the expansion of Eq. (40) have the required physical meaning.

The nuclear kinetic energy operators K_J in Eq. (39) would introduce momentum-dependent coupling terms in each channel i. Making an adiabatic approximation on the left-hand side, that is neglecting derivatives of $|\varphi^{(i)}\rangle$ with respect to neclear variables, and introducing the relation of Eq. (25c) for channel j in the left-hand side we get

$$(z - \sum_J K_J - \langle\underset{\sim}{\varphi}^{(i)}|\overline{H}_{i,el}|\underset{\sim}{\varphi}^{(i)}\rangle)\underset{\sim}{\Psi}^{(i)}$$
$$= \underset{\sim}{\Delta}^{(ij)}\underset{\sim}{\Gamma}^{(j)}\langle\underset{\sim}{\varphi}^{(j)}|\overline{V}^{(j)}|\underset{\sim}{\varphi}^{(j)}\rangle\underset{\sim}{\Psi}^{(j)}\ . \tag{41}$$

This shows that by using the proper basis sets, the rearrangement between channels i and j may be viewed as the result of recoupling of valence-bond structures.

Proceeding further in the spirit of the diatomics-in-molecules approximation, we could take for the need matrix elements

$$\langle\underset{\sim}{\varphi}^{(i)}|\overline{H}_{i,el}|\underset{\sim}{\varphi}^{(i)}\rangle \cong \sum_J \underset{\sim}{E}_J^{(i)} + \underset{\sim}{V}_{JK}^{(i)} \tag{42a}$$

$$\langle \underset{\sim}{\varphi}^{(j)} | \overline{V}^{(j)} | \underset{\sim}{\varphi}^{(j)} \rangle \cong \underset{\sim}{V}_{IJ}^{(j)} + \underset{\sim}{V}_{JK}^{(j)} \tag{42b}$$

that is, we would use the correct asymptotic forms, but with matrices evaluated at the proper nuclear distances. Calculations would then only require information about diatomics. Equations (39) to (42) are our main results, in order of increasing approximation.

VI. EXAMPLE OF $H + H_2$

The previous developments may be applied to the sample case of $H + H_2$ with 1s orbital per atom. Indicating the spin-orbital of atom A for electron spin projection m_A with the short notation $a(1)m_A(1)$, and similarly for B and C, we can construct $M_0 = 8$ states in channel 0 of type

$$\langle 123 | m_A m_B m_C \rangle = N_0 \mathbf{A} a(1) m_A(1)\, b(2) m_B(2)\, c(3) m_C(3). \tag{43}$$

For channel 1 we first construct eigenfunctions of $S_1^{\,2}$ and S_{1z}, which is readily done for convenience in terms of Clebsh-Gordan coefficients /12/,

$$\langle 23 | BCp \rangle = N_p \mathbf{A}_{BC} \sum_{m_B m_C} b(2) m_B(2)\, c(3) m_C(3) \langle s_1 m_1 \tfrac{1}{2}\tfrac{1}{2} | \tfrac{1}{2} m_B \tfrac{1}{2} m_C \rangle \tag{44}$$

The resulting singlet ($s_1 = 0$) and triplet ($s_1 = 1$) Heitler-London states already diagonalize H_{BC}, giving the Heitler-London potenitals ${}^1E_{BC}(r_1)$ and ${}^3E_{BC}(r_1)$ of Σ symmetry. Next we can write $M_1 = 8$ states of tysp φ_1 in the form

$$\langle 123 | s_1 m_1 m_A \rangle = N_1 \tfrac{1}{3}(1 - P_{12} - P_{13})\, a(1) m_A(1) \langle 23 | BCp \rangle \tag{45}$$

with $p = (s_1 m_1)$. The VB structures appropriate to channel 1 and with total electronic spin quantum numbers (s,m) are then

$$| s m s_1 \rangle_1 = \sum_{m_A m_1} | s_1 m_1 m_A \rangle \langle s m \tfrac{1}{2} s_1 | \tfrac{1}{2} m_A s_1 m_1 \rangle, \tag{46}$$

of which we find only $M^{(1)} = 2$ for the doublet states with $s = m = \frac{1}{2}$, corresponding to the lowest potential surface, that we indicate as

$$| \tfrac{1}{2}\tfrac{1}{2} s_1 \rangle_1 = | s_1 \rangle_1 \,. \tag{47}$$

For the components $\Psi^{(i)}$ of the total wavefunction we find

$$|\Psi^{(i)}(\underset{\sim}{R}_i,\underset{\sim}{r}_i)\rangle = |0\rangle_i \; {}^1\psi^{(i)}(\underset{\sim}{R}_i,\underset{\sim}{r}_i) + |1\rangle_i \; {}^3\psi^{(i)}(\underset{\sim}{R}_i,\underset{\sim}{r}_i) \quad (48)$$

with the two terms corresponding to singlet and triplet VB structures in channel i. The standard VB treatment of the potential surface for H_3 would instead use $|0\rangle_1$ and $|0\rangle_3$ as a basis.

An approximation frequently used in the literature, and which leads to the London-Eyring-Polanyi surface, follows by neglecting overlaps. The VB recoupling coefficients may then be obtained from summation properties of Clebsh-Gordan coefficients, in the form

$$\underset{\sim}{\Delta}^{(ij)} \cong \underset{\sim}{\Delta}_0^{(i)\dagger}\underset{\sim}{\Delta}_0^{(j)}$$

$$\cong (\langle sms_I s_i (s_J s_K) | sms_J s_j (s_K s_I)\rangle) \quad (49)$$

which is a matrix of 6-j coefficients from recoupling of electronic spins, with $s_A = s_B = s_C = \frac{1}{2}$. In particular $\underset{\sim}{\Delta}^{(ii)}$ becomes the unity matrix. Energy matrices are given by

$$\underset{\sim}{E}_I^{(i)} = E_{1s}\underset{\sim}{I}^{(i)} , \quad (50a)$$

$$\underset{\sim}{E}_{JK}^{(i)} = \begin{bmatrix} {}^1E_{JK}(r_i) & 0 \\ 0 & {}^3E_{JK}(r_i) \end{bmatrix} . \quad (50b)$$

Further using the approximations of Eqs. (42a,b) in Eq. (41) we end up with four coupled equations for the components ${}^1\psi^{(1)}$, ${}^3\psi^{(1)}$, ${}^1\psi^{(1)}$, and ${}^3\psi^{(3)}$, in six nuclear variables. These equations require only diatomic potential data and may be shown to be equivalent to the ones we have previously derived from spin-dependent potentials and the Faddeev equations /4/.

VII. DISCUSSION

Our wish for integrating the formal theory of rearrangement collisions with approaches to molecular electronic structure has required that we consider several new problems. Perhaps the most challenging one is that we must correctly handle the scattering boundary conditions when dealing with composite particles, in our case atoms. The rearrangement formalism of section II is one of several equivalent treatments that we selected to more conveniently discuss the decomposition of the total hamiltonian. Our procedure for obtaining channel electronic states naturally leads to a VB language. However, computational work need not be carried within the VB formalism but may instead be performed with more developed molecular orbital programs, insofar VB structures may be expressed as superpositions of configurations.

Our discussion has been based for simplicity on a unique distribution of the N electrons among the three nuclei, so that we are not here considering the coupling of rearrangement with electron transfer. This however may be done by extending the formalism. Very little modification is needed to incorporate spin-orbital coupling, which we would include in the atomic hamiltonians H_I. Channel electronic states would then be constructed for total electronic (spin and orbital) angular momenta. The present treatment can however describe electronic transitions in reactions. Electronic transitions may occur in our treatment even when introducing an adiabatic approximation, because here the dynamics is not controlled by potential energy surfaces but instead by electronic channels recoupling.

Finally, an important consequence of the present developments is that advantages are seen to result from considering <u>all</u> (bonding and non-bonding) VB structures in <u>each</u> reaction channel, letting them recouple for different rearrangements in accordance with the dynamics of the problem. This may be contrasted with procedures followed to construct potential energy surfaces, in which use is made of <u>some</u> VB structures of <u>all</u> different bonding arrangement, i.e. different channels.

ACKNOWLEDGEMENTS

The present contribution has been inspired by the work and lectures of Per-Olov Löwdin on operator methods of theoretical chemistry and on molecular electronic structure. The author thanks Erkki Brändas, Yngve Öhrn, and Charles Reid for a helpful discussion.

REFERENCES

1. See e.g. D.A. Micha, Advan. Chem. Phys. <u>30</u>, 7 (1975).

2. L.D. Faddeev Sov. Phys. JETP <u>12</u> 1014 (1961); Sov. Phys. Dokl. <u>6</u> 384 (1961).

3. K.M. Watson and J. Nutall <u>Topics in Several Particle Dynamics</u> (Holden-Day, S. Francisco, 1967).

4. (a) D.A. Micha, J. Chem. Phys. <u>57</u> 2184 (1972); (b) P. McGuire and D.A. Micha Chem. Phys. Letters <u>17</u> 207 (1972); (c) D.A. Micha J. Chem. Phys. <u>60</u> 2480 (1974); (d) D.A. Micha and J.-M. Yuan J. Chem. Phys. <u>63</u> 000 (1975); (e) J.-M. Yuan and D.A. Micha J. Chem. Phys. <u>64</u> (1976).

5. P.O. Löwdin Int. J. Quantum Chem. II, 867 (1968) and previous references therein.

6. B.A. Lippmann and J. Schwinger Phys. Rev. 79 469 (1950).

7. G.G. Balint-Kurti and M. Karplus in Orbital Theories of Molecules and Solids ed. N.H. March (Clarendon Press, Oxford, 1974).

8. F.O. Ellison J. A. C. S. 85, 3540 (1963).

9. J.C. Tully J. Chem. Phys. 58 1396 (1973); 59 5122 (1973).

10. We follow here an approach by Y. Hahn and K.M. Watson Phys. Rev. A5, 1718 (1972).

11. D.J. Kouri and F.S. Levin Phys. Rev. A10, 1616 (1974).

12. A.R. Edmonds Angular Momentum in Quantum Mechanics (Princeton Univ. Press, Princeton, N.J., 1960).

ON THE LÖWDIN BRACKETING FUNCTION

Erkki Brändas

Quantum Theory Project, University of Florida

Gainesville, Florida 32611

I. INTRODUCTION

In his series of publications "studies in perturbation theory", /1-2/ Löwdin introduced the bracketing function

$$f(z) = \langle\phi|H+HT(z)H|\phi\rangle \tag{1}$$

where H is the Hamiltonian, ϕ the reference function belonging to the Hilbert space h and T the reduced resolvent

$$T(z) = P(z-PHP)^{-1}P \tag{2}$$

$$P = 1-O;\ O = |\phi\rangle\langle\phi|;\ \langle\phi|\phi\rangle = 1. \tag{3}$$

In addition to the bracketing theorem, derived from the properties of f, he also demonstrated that perturbation schemes of all kinds were contained in (1) as well as in its multidimensional extensions. An essential point in these discussions were the transformation of the time independent Schrödinger equation

$$H\Psi = E\Psi \tag{4}$$

to the inhomogeneous equation

$$(z-H)\Psi(z) = (z-f(z))\phi \tag{5}$$

In (5) $\Psi(z) = (1+T(z)H)\phi$ denoted the trial wave function and z the energy (complex in general). When $E = f(E)$, (5) reduced to the homogeneous equation (4). At this end it should be noted, that the

the generalization of equation (4) to (5), in addition to giving a deeper understanding of the time independent equation (1), also provided a natural formulation for the time dependent problem. /6/

In this work we will review some of these aspects. For convenience we will emphasize a variational formulation exhibiting the properties of f. A time dependent formalism will furthermore be presented, in which Löwdin's bracketing function (and its derivatives) will be of fundamental importance. As byproducts of this approach we will obtain a direct derivation of Wigner's Theorem /3/, including generalizations and a simple discussion of the Fermi golden rule. /4-5/

II. THE TIME INDEPENDENT SCHRÖDINGER EQUATION

In order to give a simple but general proof of (5) we will use a resolvent identity, which will later also be of use in the time dependent case,

$$O(z-H)^{-1} = O(z-OH(z)O)^{-1}\, O(1+HT(z)), \tag{6}$$

with the effective operator defined as

$$H(z) = H + HT(z)H. \tag{7}$$

The identity (6) can be verified directly by applying (z-H) and (z-OHO) to the left and the right hand side of (6) respectively. For a discussion of these types of transformations in connection with a super operator formulation see reference 7. By using the transpose of (6) together with the definition of $\Psi(z) = (1+T(z)H)\phi$ one gets directly

$$(z-H)\,\Psi(z) = O(z-H(z))O\phi \tag{8}$$

In (8) O may be multidimensional, but for O defined by (3), equations (8) and (5) will be identical. Although we will treat the one dimensional case here extensions are usually straightforward. In the final analysis we will put z = E, where E is a real variable. It is convenient, however, to keep z complex as long as possible in order to include a treatment of the continuous part of spectrum in the formulation from the beginning. The limiting procedures will then be specified, when necessary, and we will generally assume that z is situated in the upper half of the complex energy plane.

From the definition

$$\Phi(z) = T(z)H\phi \tag{9}$$

with the notation

$$\Phi^*(z) = \Phi(\bar{z}), \tag{10}$$

where $\bar{z}$ is the complex conjugate of z, we may easily obtain functionals for f(z), which are stationary for variations around $\Phi = \Phi(z)$ or $\Psi = \Psi(z)$. Note that since ϕ does not depend on z we have $\phi^* = \phi$. The functions Φ and Φ^* belong to the conjugate Hilbert Spaces h and h^*. Now the functional

$$f(\Psi,z) = \langle\phi^*|H|\phi\rangle + \langle\phi^*|H|\Phi\rangle + \langle\Phi^*|H|\phi\rangle - \langle\Phi^*|z-H|\Phi\rangle \tag{11}$$

exhibits the stationary property mentioned above. Replacing Φ by $\tilde{\Phi}$ will then give rise to an error $\delta f(z) = f(\tilde{\Phi},z) - f(\Phi,z)$, which is quadratic in $\delta\Phi = \tilde{\Phi} - \Phi$, i.e.

$$\delta f(z) = \langle\delta\Phi^*|H-z|\delta\Phi\rangle \tag{12}$$

which in certain circumstances may have a definite sign. Since $z = f(\Phi,z)$ is equivalent to

$$z = \frac{\langle\{\phi + \Phi(z)\}^*|H|\phi + \Phi(z)\rangle}{\langle\{\phi + \Phi(z)\}^*|\phi + \Phi(z)\rangle} \tag{13}$$

we will have a direct generalization of an expectation value to the complex case. A discussion of such principles for an arbitrary splitting of H has been considered previously. /8/ For the case $H = H_0 + V$ and $\phi = \phi_0$, where $H_c\phi_0 = E_c\phi_0$, the B-W perturbation expansion of eq. (11) $\{\tilde{\Phi}(z) = \phi_1(z)+\ldots+\phi_n(z);\ \phi_i(z) = (T_0(z)V)^{i-1}\phi_0;$ $\varepsilon_i(z) = \langle\phi_0|V|\phi_{i-1}(z)\rangle;\ T_0(z) = P(z-PH_0P)^{-1}P\}$

$$f(\tilde{\Phi},z) = E_0 + \varepsilon_1 + \varepsilon_2(z)+\ldots+\varepsilon_{2n+1}(z) \tag{14}$$

which for $z = E = f(\tilde{\Phi}, E)$ gives an upper bound to the ground state energy of the perturbed operator H in accordance with the Rayleigh-Ritz principle. This is Wigner's theorem /3/ that states that the B-W series truncated at odd orders can be written as an expectation value. Extensions of this theorem containing Wigner's result was discussed by Löwdin /2/. Löwdin also showed that for $T_0(E)<0$ and $V>0$ to the B-W series truncated to even order will always give a lower bound to the exact ground state energy. The latter result follows directly from considering the Lippmann-Schwinger functional /9/.

$$f(\Psi,z) = E_0 + \langle\phi_0^*|V|\Psi\rangle + \langle\Psi^*|V|\phi_0\rangle - \langle\Psi^*|V-VT_0V|\Psi\rangle \tag{15}$$

which is stationary for variations around $\Psi(z) = \phi_0 + \Phi(z)$. Truncating the B-W expansion of Φ at the nth term (see e.g. (15)) we obtain

$$f(\tilde{\Psi},z) = E_0 + \varepsilon_1 + \varepsilon_2(z)+\ldots+ \varepsilon_{2n+2}(z) \tag{16}$$

with the error given by

$$\delta f(z) = f(\tilde{\Psi},z)-f(\Psi,z) = <\delta\Psi^*|VT_0(z)V-V|\delta\Psi> \tag{17}$$

For $V>0$ and $T_0(E)<0$, $\delta f(E)<0$ and one obtains a lower bound for the solution of $E = f(\tilde{\Psi},E)$. To conclude that (20) yields a rigorous lower bound to the ground state, it is necessary to use the bracketing property of f, i.e. for $E<E_g$, where E_g is the exact ground state energy /8/ fulfilling $E<E_g<f(\Psi,E)$. Assuming that E given by (20) gives an upper bound to E_g introduces a direct contradiction, since $\delta f<0$ or $E = f(\tilde{\Psi},E)<f(\Psi,E)<E_g$.

It should be observed at this moment that expansion (16) could be substantially improved by noting that the functions $(T_0(z)V)^i\phi_0, i = 1,\ldots n-1$, together with ϕ_0 spans an n-dimensional linear space. By performing an optimal variation of our functionals (11) or (19) in this space an improved result occurs which was identified as the (N,NH) and (N,N) Pade approximant of the BW series /10. For other functionals giving rise to other approximants see reference 8.

In this context one may wonder, in what sense $(T_0V)\phi_0$ is a good correction to ϕ_0. It is clear that an explanation based on perturbation theory only, is not sufficient. To answer this question we rewrite (11) in the B-W case as

$$f(\Phi,z) = E_0 + \varepsilon_1 + \varepsilon_2(\Phi,z) + \varepsilon_3(\Phi,z); \tag{18}$$

$$\varepsilon_2(\Phi,z) = <\phi_0^*|V|\Phi> + <\Phi^*|V|\phi_0>-<\Phi^*|z-H|\Phi> \tag{19}$$

$$\varepsilon_3(\Phi,z) = <\Phi^*|V|\Phi>. \tag{20}$$

Neglecting third and higher order terms in (18) we realize that $\varepsilon_2(\Phi,z)$ is a Hylleraas type /11/ functional which is stationary for variations around $\Phi = T_0(z)V\phi_0$. For $z = E<E_1^0$, where E_1^0 is the first excited state of H_0, $\varepsilon_2(\Phi,E)$ assumes its lowest value for $\Phi = T_0(E)V\phi_0$, i.e. for the first order correction to ϕ_0. In analogy with the steepest descent procedure we now ask ourselves: "What is the optimum amount of ϕ_1 to be added to ϕ_0, i.e. the "step length?" Replacing Φ by $\lambda\Phi$ in (18) realizing that the functionals

$\varepsilon_2(\Phi,z)$ and $\varepsilon_3(\Phi,z)$ become the appropriate second and third order corrections to the energy ($z = E$) it is obtained after variation of λ /12-13/

$$\lambda = \varepsilon_2(z)(\varepsilon_2(z)-\varepsilon_3(z))^{-1} \tag{21}$$

Equation (21) corresponds to the simplest Pade approximant or the "geometric approximation". Continuing the search for additional corrections to the wavefunction, subject to largest possible energy descent (for $z = E$ in (18)) neglecting fifth and higher order corrections yield from (11) that the functional to be optimized is given by

$$\varepsilon_4(\chi,z) = \langle\phi_0^*|VT_0(z)V|\chi\rangle+\langle\chi^*|VT_0(z)V|\phi_0\rangle-\langle\chi^*|z-H_0|\chi\rangle, \tag{22}$$

which yields $\chi(z) = (T_0(z)V)^2\phi_0$, i.e. the second order correction. New step lengths, defining the appropriate amounts of $(T_0V)\phi_0$ and $(T_0V)^2\phi_0$ are easily obtained by optimization of $\Phi(z) = \lambda_2(T_0(z)V)\phi_0 + \lambda_3(T_0(z)V)^2\phi_0$ in (11). The $(2,3)$ Pade approximant to the B-W series are then obtained. In this way the $(N, N+1)$ approximants appear. An analogous analysis applied to the R-S case yield the general variation perturbation result /14-15/

$$E = E_0+\varepsilon_1+\sum_{k=2}^{n} \lambda_k\varepsilon_k \tag{23}$$

where the λ_k's are obtained from optimizing the functional (11) in the linear space spanned by all the perturbation corrections. The λ_k's obtained here will not allow for an identification of (23) in terms of Pade approximants, but the latter may still serve as a good approximate summation rule. This gives the direct relationship between perturbation theory and the steepest descent or rather the conjugate gradient methods /16/. For a comparison of the actual convergence properties exhibited by the various expansions touched upon here we refer to (14)-(16).

Before leaving this section it should be noted that the elegant inner projection technique of Löwdin /2,17/ derived from the theory of operator inequalities bypasses any ad hoc variational principle, giving directly the optimal result /18/. Our variational formulation emphasized here, however, complements this approach in that it elucidates the structural aspects of perturbation theory and its explicit relation to stationary principles /8/. Furthermore, the extension of E to complex values leads naturally in the variational formulation to a description in terms of conjugate Hilbert spaces. This extension will as we will see simplify the analysis in the following section.

III. THE TIME DEPENDENT SCHRÖDINGER EQUATION

In the introduction we mentioned that Löwdin's inhomogeneous equation (5) also provided a natural formulation for the time dependent "diffusion equation" (atomic units used, i.e. h = 1)

$$H\Psi(t) = i\frac{\partial}{\partial t}\{\Psi(t)\} \tag{24}$$

Separating out positive and negative times from the beginning we introduce the propagators

$$G^{\pm}(t) = \mp i\theta(\pm t)e^{-iHt} \tag{25}$$

satisfying

$$(i\frac{\partial}{\partial t} - H)G^{\pm}(t) = \delta(t) \tag{26}$$

where $\theta(t)$ is the Heaviside function and $\delta(t)$ is Dirac's δ-function. Evolution of Ψ into positive or negative times then yield $\Psi^{\pm}(t)$ defined by

$$\Psi^{\pm}(t) = \pm iG^{\pm}(t)\Psi^{\pm}(0). \tag{27}$$

We note that equation (36) is invariant under the combined operations $t\to -t$ and $H\to -H$, which is consistent with $G^{\pm}\xrightarrow[H\to -H]{t\to -t} -G^{\mp}$ and $\Psi^{\pm}\xrightarrow[H\to -H]{t\to -t}\Psi^{\mp}$. The transformation $H\to -H$ becomes meaningful if one remembers that reciprocal time energy units follow from the consideration of $\hbar^{-1}H$ in (24) before introducing atomic units. At time t = 0 we will define the initial condition $\Psi^{\pm}(0)$, such that for complete initial information $\Psi^{+}(0) = \Psi^{-}(0) = \Psi(0)$. We then decompose the normalized wavepacket $\Psi(0)$ into its O and P part, i.e.

$$\Psi(0) = O\Psi(0) + P\Psi(0) = \phi(0) + \chi(0) \tag{28}$$

For simplicity we will assume that $\chi(0) = 0$, which means that our initial normalized wave packet serves as a reference function in the partitioning technique formulation, see below. Our problem is now to evaluate the probability amplitude

$$a^{\pm}(t) = \langle\phi(0)|\Psi^{\pm}(t)\rangle \tag{29}$$

Inserting the Fourier-Laplace transform, $G^{\pm}(t)$, see equation (25), of the full resolvent

$$G(z) = (z-H)^{-1} \tag{30}$$

i.e.

$$G^{\pm}(t) = \frac{1}{2\pi} \int_{C^{\pm}} G(z) e^{-izt} dz \tag{31}$$

where $C^{\pm}$ refer to appropriate contours in the upper and the lower half plane of z respectively, into (27) & (29) using the resolvent identity (6) we obtain

$$\Psi^{\pm}(t) = \pm\frac{i}{2\pi} \int_{C^{\pm}} \frac{\Psi(z)}{(z-f(z))} e^{-izt} dz \tag{32}$$

where $\Psi(z)$ satisfies (5). This shows explicitly the relation between the inhomogeneous equation (5) and the Fourier transformation of(24). It is furthermore obtained

$$a^{\pm}(t) = \pm\frac{i}{2\pi} \int_{C^{\pm}} \frac{e^{-izt}}{(z-f(z))} dz = \pm \frac{i}{2\pi} \int_{C^{\pm}} W(z) e^{-izt} dz \tag{33}$$

$$W(z) = (z-f(z))^{-1} \tag{34}$$

Denoting the residues of the Weinstein function W(z) at the poles $z_k^{\pm}$, where $z_k^{+}(z_k^{-})$ is situated in the lower (upper) half plane, by $R_k^{\pm}$ we get from the residue theorem

$$a^{\pm}(t) = \sum_k R_k^{\pm} e^{-iE_k t} e^{\mp\varepsilon_k t} \tag{35}$$

where $z_k^{\pm} = E_k \mp i\varepsilon_k$. By means of eq. (33) we may determine $a^{\pm}(t)$ directly in terms of the first order derivative of f(z) evaluated at $z_k^{\pm}$ (satisfying $z_k = f(z_k)$) by the relation

$$R_k^{\pm} = (1-f'(z_k^{\pm}))^{-1} \tag{36}$$

If a double pole at z_k should occur the residue theorem will give a contribution to the integral (32) containing a linear term in t, i.e.

$$\frac{1}{f''(z_k)^2}\{\frac{2}{3} f'''(z_k) + 2it\, f''(z_k)\}\, e^{-iE_k t} e^{\mp\varepsilon_k t} \tag{37}$$

In general a multiple pole of order n will generate a polynomial of order n-1 in t multiplying the exponential factor. Since $a^{\pm}(0) = 1$ we obtain the sumrule

$$\Sigma R_k^{\pm} = \sum_k (1-f'(z_k^{\pm}))^{-1} = 1 \tag{38}$$

At a time t different from zero our normalization integral (=1 at t = 0) has therefore changed to ($2\varepsilon_k = \Gamma_k$)

$$<\phi^{\pm}(t)|\phi^{\pm}(t)> = |a^{\pm}(t)|^2 = \sum_k |R_k^{\pm}|^2 e^{\mp\Gamma_k t}$$

$$+ \sum_{k\neq l} R_k^{\pm} R_l^{\mp} e^{-i(E_k-E_l)t} e^{\mp(\varepsilon_k+\varepsilon_l)t}. \tag{39}$$

We have here assumed that the poles in the upper and lower halves of the z-plane are mirror images of each other in the real axis (Schwartz reflection principle). The general residue formulas presented here could of course easily be modified to include the situation when this is not the case. We rewrite (39) as below

$$|a^{\pm}(t)|^2 = \sum_k |R_k^{\pm}|^2 e^{\mp\Gamma_k t} + \text{interference terms} \tag{40}$$

where the second term constitutes a generalized interference term. From (39) we realize that stationary states corresponding to real poles in W(z) will contribute at all times, while complex poles corresponding to resonances in (or in general building up) the continuous spectrum are responsible for decaying probabilities and phases. If the whole spectrum of H is continuous or if ϕ is orthogonal to the bound states of H, we obtain $|a^{\pm}(t)|^2 \xrightarrow[t\to\pm\infty]{} 0$, i.e. the wave packet spreads out indefinitely and the probability amplitude becomes zero at infinite time.

A recently discussed trace algebra /19/ may now easily be imposed, by neglecting the generalized interferences in (40). By defining a trace, (see below) we assume implicitly that only probabilities, (and not phases), are available at t = 0. Experimentally one measures of course averages over short time intervals, from which the postulate of random phases are invoked. We will therefore primarily be interested in the time development of the diagonal part of the density matrix i.e.

$$<\phi^{\pm}(t)|\phi^{\pm}(t)> = (\mathrm{Tr}\{\rho^{\pm}(t)\})_d = \sum_k |R_k^{\pm}|^2 e^{\mp\Gamma_k t} \tag{41}$$

The entropy (referring to subspace 0) is then defined as (omitting the ± sign for simplicity)

$$S(t) = -\frac{(\mathrm{Tr}\{\rho(t)\ln\rho(t)\})_d}{(\mathrm{Tr}\{\rho(t)\})_d} = -\frac{\sum_k |R_k|^2 e^{-\Gamma_k t}(\ln|R_k|^2 - \Gamma_k t)}{\sum_k |R_k|^2\, e^{-\Gamma_k t}} \tag{42}$$

In principle (25) and (30)-(32) may hold even if H (and its associated spectrum) depends on t, although (31) then differs from a simple Fourier-Laplace transform. Denoting all the real poles at $t\to\pm\infty$ by k_r we find that

$$\lim_{t\to\pm\infty} S(t) = -\frac{\sum_{k_r} |R_{k_r}|^2 \ln|R_{k_r}|^2}{\sum_{k_r} |R_{k_r}|^2} \tag{43}$$

Before ending the time dependent description we will briefly comment on the time proportional transition probability connected with the Fermi golden rule formula for transitions between a discrete state (here chosen to be $\phi(0)$) and the continuum /4,5,20/. We start by assuming that $\phi(0)$ and $\Psi(E_k)$ are similar in the sense that $|R_k|^2 >> |R_\ell|^2$ $\ell \neq k$, or from (36) $|f'(z_k)| << |f'(z_\ell)|$. For times that are not too large we obtain, with the notation $P_k(t) = |R_k|^2 e^{-\Gamma_k t}$

$$P(t) = |a(t)|^2 \approx P_k(t) = P_k(0)e^{-\Gamma_k t} \tag{44}$$

where Γ_k is the solution to the coupled equations (from $z_k^\pm = E_k \mp i\frac{\Gamma_k}{2} = f(z_k^\pm)$)

$$E_k = \mathrm{Re}\{f(E_k,\Gamma_k)\} \tag{45}$$

$$\Gamma_k = 2\mathrm{Im}\{f(E_k,\Gamma_k)\} \tag{46}$$

Equations (45)-(46) has to be solved (in general) by means of analytic continuation into the unphysical sheet of the z-plane. The starting point for this analysis is the dispersion relation (at $z = E_k, \phi = \phi(0)$)

$$f^{\pm}(E_k,0) = \langle\phi|H|\phi\rangle + \langle\phi|H\mathscr{P}(E_k-PHP)^{-1}PH|\phi\rangle \mp i\pi\langle\phi|HP\delta(E-PHP)PH|\phi\rangle \tag{47}$$

where $\mathscr{P}$ denotes the principal value of the corresponding integral. The following assumption will now lead to the golden rule formula. If Γ_k is "small" compared to E_k, we may decouple the equations (45)-(46) by putting $\Gamma_k = 0$ in the real and imaginary parts of f. By splitting the Hamiltonian into two parts $H = H_0 + V$, with $H_0\phi(0) = E_k\phi(0)$, we then obtain from (47)

$$\Gamma_k = 2\pi\left(\frac{d\rho}{d\omega}\right)_{\omega=E_k} |\langle\phi(0)|V|\Psi(E_k)\rangle|^2 \tag{48}$$

In order to obtain (48) we have used the spectral resolution

$$PHP = \int|\Psi(\omega)\rangle\omega d\rho(\omega)\langle\Psi(\omega)| \tag{49}$$

inserted into (47). Without splitting the Hamiltonian the decoupled result (45)-(46) will appear as

$$\Gamma_k = 2\pi\left(\frac{d\rho}{d\omega}\right)_{\omega=E_k} |\langle\phi|H|\Psi(E_k)\rangle|^2 \tag{50}$$

Various forms of this expression within Feschbach's description of scattering resonances /4/ have been examined by Miller /5/.

CONCLUSION

In this investigation various aspects of the Löwdin bracketing function was discussed. Due to obvious space limitations we had to leave out many of the very important aspects of many-electron theory, which although inherent in the present formulation require special attention.

It is interesting to note, however, that from the transform eq. (13) the inhomogeneous equation (5) assumed complete generality from which the appropriate knowledge could be extracted or inserted and propagated.

One of the weaknesses in this formulation is of course the spectral density concept, assumed here without a critical examination. In simple cases this concept was examined by means of Weyl's complex eigenvalue theory and compared with the more general partitioning technique formulation /21/. Although extensions of this theory to coupled differential equations are straightforward, other avenues based on a direct treatment of the Weinstein function by means of a "double-variance" method have furthermore been tried as a complementary /22/.

It should finally be observed that by rotating the contours (31) 90 degree's counter clockwise a temperature dependent Schrödinger equation is obtained. An analysis similar to the time dependent case studied here, reveals that contrary to the time symmetry with respect to t = 0 (with complete initial knowledge) an assymmetry arises, which is associated with the occurance of positive and negative temperatures as dictated by the spectral properties of H. A formulation of this situation in terms of the Löwdin bracketing function follows naturally, but will be reported elsewhere.

ACKNOWLEDGEMENTS

This article is dedicated to Professor Per-Olov Löwdin on the occasion of his sixtieth birthday. His friendly but determinate criticism united with thought provoking and elucidatory comments, his contageous enthusiasm and encouragement twinned with non-smoking signs, have provided a healthy environment for us in more than one aspect. May this situation reach a stationary level!

This work has been supported by a grant from the National Science Foundation, NSF MPS74-03948.

REFERENCES

1. P.O. Löwdin, J. Chem. Phys. 19, 1396 (1951).
2. P.O. Löwdin, J. Mol. Spectry. 10, 12 (1963); 13, 326 (1964); J. Math. Phys. 3, 969 (1962); 3, 1171 (1962); J. Mol. Spectry. 14, 112 (1964); 14, 119 (1964); 14, 131 (1964); J. Math. Phys. 6, 1341 (1965); Phys. Rev. 139, A357 (1965); J. Chem. Phys. 43, S175 (1965); C.H. Wilcox, Ed., Perturbation Theory and its Application in Quantum Mechanics (Wiley, New York, 1966), page 255; Int. J. Quant. Chem. 2, 867 (1968); 5, 685 (1971).
3. H. Feshbach, Ann. Phys. 5, 357 (1958); 19, 287 (1962).
4. W.H. Miller, Chem. Phys. Letters 4, 627 (1970).
5. E. Brändas, Physica (to be published).
6. A.P. Grecos and I. Prigogine, Physica 59, 77 (1972).
7. D.A. Micha and E. Brändas; J. Chem. Phys. 55, 4792 (1971); J. Math. Phys. 13, 155 (1972).
8. B.A. Lippmann and J. Schwinger, Phys. Rev. 79, 469 (1950).
9. O. Goscinski and E. Brändas Int. J. Quant. Chem. 5, 131 (1971); 7, 133 (1973).

10. E.A. Hylleraas, Z. Physik 65, 209 (1930).

11. P. Goldhammer and E. Feenberg, Phys. Rev. 101, 1233 (1956).

12. O. Goscinski, Int. J. Quant. Chem. 1, 769 (1967).

13. O. Goscinski and E. Brändas; Chem. Phys. Letters 2, 299 (1968).

14. E. Brändas and O. Goscinski, Int. J. Quant. Chem. 3S, 383 (1970); Phys. Rev. 1A, 552 (1970).

15. R.J. Bartlett and E.J. Brändas, J. Chem. Phys. 56, 5467 (1972).

16. P.O. Löwdin, Int. J. Quant. 4S, 231 (1971).

17. P. Lindner and P.O. Löwdin, Int. J. Quant. Chem. 2S, 161 (1968).

18. P.O. Löwdin, Preliminary Research Report, Uppsala Quantum hemistry Group (1975), Uppsala University.

19. P.O. Löwdin, Advan. Quant. Chem. 3, 323 (1967).

20. E. Brändas, M. Hehenberger and H.V. McIntosh, Int. J. Quant. Chem. 9, 103 (1975).

21. P. Froelich and E. Brändas, Phys. Rev. 12A, 1 (1975).

NUMERICAL INFINITE-ORDER PERTURBATION THEORY

Rodney J. Bartlett[†] and David M. Silver[††]

[†]Battelle Memorial Institute, Pacific Northwest Laboratories Richland, Washington 99352, U.S.A.

[††]Applied Physics Laboratory, The Johns Hopkins University Laurel, Maryland 20810, U.S.A.

I. INTRODUCTION

With the emergence of the Hartree-Fock independent particle model in physics and chemistry, the correlation problem, which Löwdin defined /1/, has constituted one of the primary research areas within quantum chemistry. Löwdin's contributions to this topic are enormous, including among them his work in density matrix theory /2/, in extended independent particle methods where he proposed the projected Hartree-Fock theory /3/ and the alternate molecular orbital method /4/, and in infinite-order perturbation theory /5/.

Infinite-order perturbation theory is an outgrowth of the partitioning technique /5a,b/ or the projection operator approach /5d/ to Schrödinger's equation which leads to closed-form expressions for the energy and the wavefunction. These may be easily transformed into various types of perturbation theory by different expansions of an operator inverse.

---

*This work was supported in part by the Battelle Institute and the Energy Research and Development Adminitration under Contract No. AT(44-1) 1830 at Battelle Memorial Institute, Pacific Northwest Laboratories, and in part by the Department of the Navy, Naval Sea Systems Command under Contract No. N00017-72-C4401 at The Johns Hopkins University Applied Physics Laboratory.

Infinite-order perturbation theory /5,6/ is, of course, equivalent to Schrödinger's equation, and, consequently, such a result cannot be exactly realized numerically for more than one electron, but hope exists that sufficiently good approximations to this infinite order result may be found for most properties of interest. In this respect, it is felt that a procedure built upon the "linked-cluster" expansion /7,8,9/ provides much promise for such a solution. If so, however, higher order diagrams in the "linked-cluster" expansion than are normally calculated must sometimes be considered either to justify lower-order approximations or to provide the crucial effects that are needed to obtain high accuracy for some property.In the "linked-cluster" expansion, in particular, this appears to be almost prohibitively difficult, since the increase in the number of diagrams with increase in order tends to be overwhelming.

To resolve this difficulty, one needs to formulate an evaluation of higher-order diagrams that does not suffer from prohibitive increases in computational procedure. This may be done either by including higher-order diagrams through iterative calculations as in Brueckner's theory of nuclear matter /8/, or Löwdin's generalization to an exact SCF theory /5e/; or, a more direct, but recursive, evaluation of higher-order contributions. Doubtlessly, combinations of both approaches should be utilized for optimum results, but the present paper is primarily concerned with the second possibility and its contribution toward the goal of numerical infinite-order perturbation theory.

Section II gives the pertinent formulae of Löwdin's partitioning approach to perturbation theory /5/, while Section III describes the recursive procedure that permits higher-order double-excitation diagrams to any order to be computed from the same code employed in second- and third-order. In the final section, a numerical example is presented for the two-body contribution to the correlation energy of LiF through sixth-order. This provides for the first time a critical evaluation of the convergence properties of classes of higher-order diagrams, which has normally been assumed to be adequately approximated by a geometric series /9,10,11/.

II. INFINITE ORDER PERTURBATION THEORY

Consider a nondegenerate ground state of the Schrödinger equation

$$H\psi = E\psi. \tag{1}$$

To use Löwdin's partitioning technique /5/, we define the projector O, and its complementary projector P, with the properties,

$$O\psi = \omega_o, \; 1 = O + P, \; O^+ = O, \; P^+ = P, \; O^2 = O, \; P^2 = P, OP = PO = 0 \tag{2}$$

Then, Eqn. (1) may be rewritten as

$$H(O + P)\,\psi = E(O + P)\psi. \tag{3}$$

Multiplying Eqn. (3) first by O, and then by P, we obtain the two equations

$$\begin{aligned} OHO\psi + OHP\psi &= EO\psi \\ PHO\psi + PHP\psi &= EP\psi \end{aligned} \tag{4}$$

Solving for $P\psi$, and defining $T = (E - PHP)^{-1}P$, it follows that

$$P\psi = TH\omega_o \tag{5}$$

$$E = \langle\omega_o|H + HTH|\omega_o\rangle. \tag{6}$$

Once a separation of the hamiltonian is made such as

$$H = H_o + V, \tag{7}$$

and where it is convenient to have

$$H_o\omega_o = E_o\omega_o, \tag{8}$$

Eqn. (6) becomes the fundamental formula of perturbation theory /6/

$$E = E_o + \langle\omega_o|V + VTV|\omega_o\rangle = E_o + \langle\omega_o|t|\omega_o\rangle. \tag{9}$$

Brillouin-Wigner (BW), Rayleigh-Schrödinger (RS), and other types of perturbation theory are conveniently derived from Eqn. (9) by different expansions of the previously defined T operator. Furthermore, using E from Eqn. (9) in the T operator and substituting into $t = V + VTV$, the reaction operator may be expressed as

$$t = V + V \frac{P}{(E_o - H_o) - (V - \langle t\rangle_o)} V, \tag{10}$$

Consequently, for the ground state energy, the basic problem of quantum chemistry is the approximate determination of the reaction operator, t.

One very promising way to proceed is to employ the "linked-cluster" expansion of diagrammatic perturbation theory.

III. RECURSIVE METHOD FOR HIGHER-ORDER DIAGRAMS

Most of the ab initio work which has been done either assumes an independent particle model or considers the independent particle solution to define an unperturbed problem which is to be improved with configuration interaction.

With the usual definitions defining the Hartree-Fock model,

$$(h(1) + u(1)\chi_p(1) = \varepsilon_p \chi_p(1), \tag{11}$$

$$u(1) = \sum_j \int \chi_j(2) g_{12}(1 - P_{12}) \chi_j(2) dt_2 \tag{12}$$

we have

$$H_o = \sum_i (h(i) + u(i)) \tag{13}$$

$$H_o \omega_o = E_o \omega_o \tag{14}$$

$$\omega_o = A(\chi_1(1)\chi_2(2)\ldots\chi_n(n)) \tag{15}$$

$$E_o = \sum_{i=i}^{n} \varepsilon_{i=1} \tag{16}$$

$$V = -\sum_i u(i) + \sum_{i<j} g(i,j) \tag{17}$$

Then, from Eqn. (9), and invoking the "linked-cluster" expansion,

$$\Delta E = E - E_o = \langle \omega_o | t | \omega_o \rangle = \sum_{k=0}^{\infty} \langle \omega_o | V\left((E_o - H_o)^{-1} V\right)^k | \omega_o \rangle_L . \tag{18}$$

$$E_{corr} = \sum_{k=1}^{\infty} \langle \omega_o | V\left((E_o - H_o)^{-1} V\right)^k | \omega_o \rangle_L \tag{19}$$

The "linked-cluster" expansion provides an important simplification for the reaction operator. When the standard RS expansion of t is employed (see Löwdin /6,5d-e/) certain terms appear which have an N^2 dependence on the number of particles in the system rather than N, the so-called "unlinked" terms. However, closer investigation demonstrates that all of these terms mutually cancel. Hence, these terms are "unphysical" in the sense that they do not contribute to the correct result. Furthermore, most methods of computation, such as a conventional, variational, configuration interaction calculation, must compute both linked and unlinked terms and is thus ultimately limited by the N^2 dependence of the latter. Also, since the cancellation of unlinked clusters occurs within a given order, but does not necessarily occur within a single type of excitation, a restricted CI employing only double excitations, for example, retains unlinked terms that the inclusion of quadruple excitations would cancel. (See reference 12 for a discussion of the relationship between CI and MBPT.) Consequently, since the complete cancellation of all the "unlinked" terms is already explicitly incorporated into the diagrammatic linked cluster expansion for the energy, this formula provides a distinct simplification of the reaction operator.

To evaluate the reaction operator approximately by perturbation theory, however, raises some justified concern. If one stops at some order, the question arises about the significance of the next order. If it were desired to compute the next order, though, it could require a nearly prohibitive addition to the computer code. To avoid this problem, a recursive scheme has been formulated and described below.

Henceforth, the convention is used that i, j, k, l represent HFSCF spin orbitals occupied in the ground state ("hole states") and a, b, c, d represent excited, "virtual" orbitals ("particle states"), with p, q, r, s representing either. As long as HFSCF orbitals are employed for both the occupied and excited states, the components shown in Figure 1, where -X represents the interaction with the negative of the Hartree-Fock effective potential, mutually cancel.

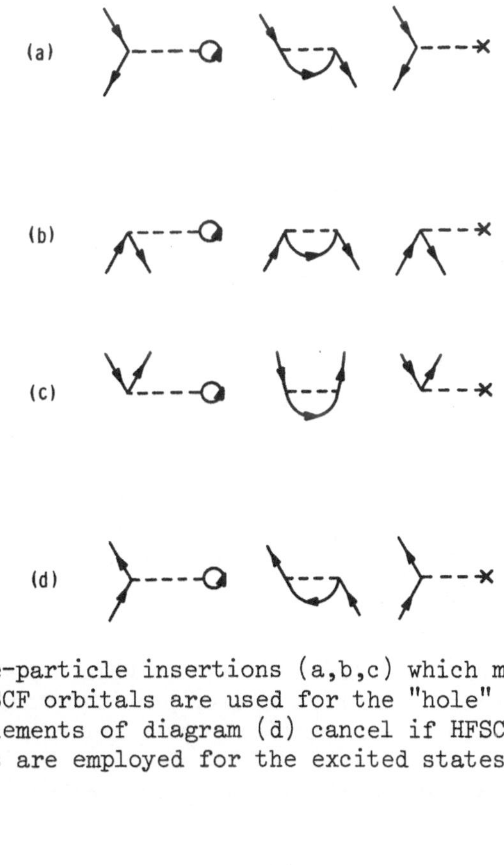

Figure 1. Single-particle insertions (a,b,c) which mutually cancel if HFSCF orbitals are used for the "hole" states, while the elements of diagram (d) cancel if HFSCF "virtual" states are employed for the excited states.

This greatly reduces the number of diagrams which need to be considered in a given order. The case where the "virtual" space is modified, which removes the cancellation in Figure 1d, has been extensively investigated previously /13, 14/, with the result that as long as one goes beyond second order there is little to be gained from such a modification.

Subject to the restriction to HFSCF orbitals, all of the second and third-order antisymmetrized diagrams contributing to the correlation energy are presented in Figure 2. Using the rules listed in reference 12, these diagrams have their algebraic equivalents

$$2(A) = \sum_{i>j} \sum_{a>b} |\langle ij||ab\rangle|^2/D(ijab) \quad (20)$$

$$2(B) = \sum_{i>j} \sum_{a>b} \sum_{c>d} \langle ij||ab\rangle \langle ab||cd\rangle\langle cd||ij\rangle/D(ijab)D(ijcd) \quad (21)$$

$$2(C) = \sum_{i<j} \sum_{a<b} \sum_{k<l} \langle ab||ij\rangle\langle ij||kl\rangle\langle ab||kl\rangle/D(ijab)D(klab) \quad (22)$$

$$2(D) = -\sum_{ijk} \sum_{abc} \langle ab||ij\rangle\langle cj||bk\rangle\langle ik||ac\rangle/D(ijab)D(ikac) \quad (23)$$

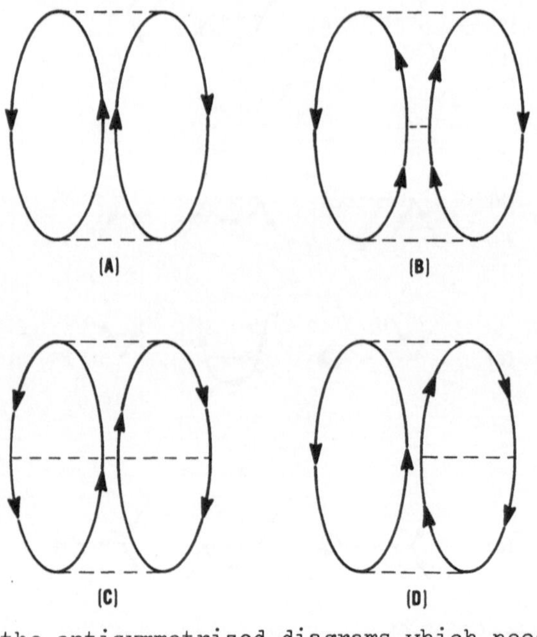

Figure 2. All the antisymmetrized diagrams which need to be considered in second and third-order, provided that HFSCF orbitals compose the single particle basis set.

The integral notation is

$$<pq||rs> = \int_{\tau_2} \int_{\tau_1} d\tau_2 d\tau_1 \chi_p^*(1)\chi_q^*(2) r_{12}^{-1}(1-P_{12})\chi_r(1)\chi_s(2), \quad (24)$$

and the standard denominator is given by

$$D(ijab) = \varepsilon_i + \varepsilon_j - \varepsilon_a - \varepsilon_b \quad (25)$$

Note that diagrams 2(A), 2(B), and 2(C) each represent two Goldstone diagrams, while 2(D) represents eight, including the common ring diagram. As the order increases, the use of antisymmetrized graphs effects a tremendous simplification.

Using the definitions,

$$<pq||rs>_1 = <pq||rs> \quad (26)$$

$$(pq||rs)_n = <pq||rs>_n/D(pqrs), \qquad n\geq 1 \quad (27)$$

diagram 2(D) may be written in ordered form as

$$2(D) = \sum_{i>j} \sum_{a>b} \sum_{ck} (ab||ij)_1 ((ca||ik)_1 <kb||jc>_1 - (cb||ik)_1 <ka||jc>_1$$
$$- (ca||jk)_1 <kb||ic>_1 + (cb||jk)_1 <ka||ic>_1) \quad (28)$$

Generalizing to any $n\geq 1$, a set of intermediates may be defined as

$$X_{n+1}(abij) = \sum_{c>d} (cd||ij)_n <ab||cd>_1 \qquad i>j,\ a>b \quad (29)$$

$$Y_{n+1}(abij) = \sum_{k>l} (ab||kl)_n <ij||kl>_1 \qquad i>j \quad a>b \quad (30)$$

$$Z_{n+1}(abij) = \sum_{k,c} ((ca||ik)_n <kb||jc>_1 - (cb||ik)_n <ka||jc>_1$$
$$- (ca||jk)_n <kb||ic>_1 + (cb||jk)_n <ka||ic>_1), \qquad i>j,\ a>b \quad (31)$$

and, finally, for $n\geq 2$,

$$<ab||ij>_n = X_n(abij) + Y_n(abij) + Z_n(abij) \qquad i>j,\ a>b \quad (32)$$

The pseudo-integral notation for these intermediates is used since from Eqns. (29)-(32) it is easily seen that

$$<ab||ij>_n = -<ab||ji>_n = -<ba||ij>_n = <ba||ji>_n \quad (33)$$

However, these intermediates do not retain all of the symmetry properties of the integrals themselves.

By means of these intermediates, it is obvious that for n = 2,

$$E_3 = \sum_{i>j} \sum_{a>b} (ab||ij)_1 \ <ab||ij>_2. \qquad (34)$$

However, the real advantage of the quantities defined in Eqns. (29)-(32) becomes evident in higher order. In CI language, the complete fourth-order contribution to the reaction operator, subject to HFSCF orbitals, arises due to single, double, triple, and quadruple excitations. Of these, all of the possible double-excitation contributions are drawn in Figure 3. By writing the algebraic equivalent of each of these antisymmetrized graphs, it may be shown that

$$E_4 = \sum_{i>j} \sum_{a>b} |<ij||ab>_2|^2/D(ijab). \qquad (35)$$

In fact, the origin of the diagrams in Figure 3 may be seen to be derived from the products of the X, Y, and Z intermediates. The ladder diagrams, Figure 3(A) and Figure 3(B), correspond to X^2 and Y^2, respectively, while the four hole-particle diagrams Figures 3(I-L) are associated with Z^2. The cross terms between the particle-ladder, hole-ladder, and hole-particle diagrams give rise to Figures 3(C-H). Notice the factor of two is also included by the two distinct "time" orders in each twosome of diagrams.

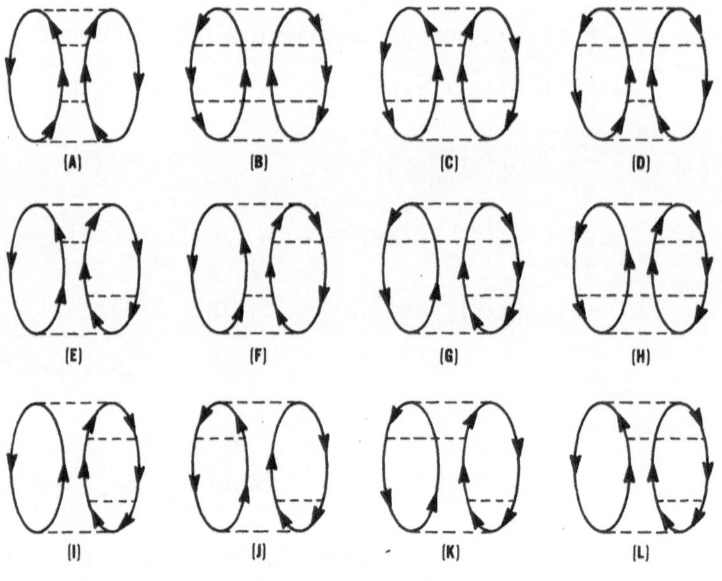

Figure 3. All the antisymmetrized double-excitation diagrams which need to be considered in fourth-order, subject to a HFSCF single particle basis set.

The evaluation of $\langle ab||ij\rangle_2$ in third-order provides the fourth-order energy at negligible additional cost, since E_4 may be evaluated with Eqn. (35) by employing the same code as used for E_2. Furthermore, this scheme is generalizable in a recursive manner. By repeating the procedure with the n = 2 intermediate now divided by the denominator to replace the quantities for n = 1 in Eqns. (29) -(31), $\langle ab||ij\rangle_3$ is obtained, providing the double excitation contribution of E_5 and E_6. Or, in general,

$$E_{2n-1} = \sum_{i>j} \sum_{a>b} (ab||ij)_{n-1} \langle ab||ij\rangle_n \quad (36)$$

$$E_{2n} = \sum_{i>j} \sum_{a>b} \langle ab||ij\rangle_n^2/D(ijab) \quad (37)$$

Thus, by simply repeating the procedure required to evaluate the three diagrams in Figure 2, one obtains the contribution of the 60 antisymmetrized or 1104 Goldstone "double excitation" diagrams in fifth order, and so forth, up to any order, without any addition to the computer code. This result, which initially might seem surprising, may also be viewed as a consequence of ordinary perturbation theory (see Appendix).

This procedure suggests the possibility of evaluating a type of diagram up to an order where the contribution from the highest included order is found to be numerically negligible, and, thereby, to obtain the complete contribution of a particular class of diagrams. In addition to eliminating the uncertainty with respect to higher order contributions in perturbation theory, if this may be achieved for all classes of diagrams that are important for some property, a numerical infinite-order result would be effectively obtainable.

One of the primary advantages of the pictorial description of the terms in the linked cluster expansion is that certain contributions of the diagrams may be summed to all orders by denominator modifications. Obviously, if we replace the standard denominator, D(ijab), by

$$D^S(ijab) = D(ijab)+\Delta(ijab)$$

$$\Delta(ijab) = -\langle ab||ab\rangle - \langle ij||ij\rangle + \langle ai||ai\rangle + \langle bi||bi\rangle + \langle aj||aj\rangle + \langle bj||bj\rangle \quad (38)$$

the diagonally scattered contributions to 2(B), 2(C), and 2(D), and similar components of all higher orders, are included in the computation of 2(A). In addition, once 2(B), 2(C), and 2(D) are evaluated, now respectively subject to the restrictions ab $\neq$ cd, ij $\neq$ kl, and ck ≠ ai, aj, bi, or bj, higher order contributions are included into the third-order calculation as well. Analogously, the simple expedient of restricting the summation indices as above eliminates these terms from the recursive determinations of higher orders, while continuing to sum new parts of even higher-order diagrams to all orders. Kelly has suggested additional denominator modifications involving the computed pair energies themselves /9/, which then requires iterative procedures. Once these are included, contributions from quadruple excitations are introduced. Clearly, denominator modifications, plus recursive calculations of high-order diagrams, can be utilized to provide powerful approximations for the reaction operator.

IV. NUMERICAL APPLICATION

Many-body theory often uses the terminology two-, three-, and four-body, etc., interactions. This refers to the number of different occupied states ("hole lines") that are simultaneously involved in a diagram contributions. (Note: this does not refer to the number of excitations in a CI sense). Diagram 2(A) and 2(B) are exclusively two-body contributions since only two occupied states, i and j, are involved. However, diagrams 2(C) and 2(D) include more than two-body interactions. In 2(D) three "occupied" states are possible, and in 2(C) there may be three or four. When k = i or j in Eqn. (31), the two-body contribution to Z_2 is obtained, while k ≠ i or j gives the three-body component. Similarly in Eqn.(30), k = i and l = j gives the two-body part, while the others are given by either k or l being different than i or j and then both being different from i or j. In obvious notation, we are thus led to

$$\langle ab||ij\rangle_n = II_n(abij)+III_n(abij)+IV_n(abij)+V_n(abij)+\ldots,\ n\geq 2 \tag{39}$$

$$II_n(abij) = X_n(abij)+Y_n(abij;2)+Z_n(abij;2) \tag{40}$$

$$III_n(abij) = Y_n(abij;3)+Z_n(abij;3) \tag{41}$$

$$IV_n(abij) = Y_n(abij;4)+Z_{n(n\geq 3)}(abij;4) \tag{42}$$

.
.
.

Note that even though the calculation through E_3 need only involve two-, three-, and four-body interactions, the recursive procedure outlined automatically introduces the requisite number of occupied states into the higher orders. Thus, an evaluation of E_4 using the n = 2 intermediate includes five- and six-body contributions. Furthermore, an unrestricted application of Eqns. (30) and (31) to give $Y_{n+1}(abij)$ and $Z_{n+1}(abij)$ followed by utilization of Eqns. (36) and (37) continues to introduce the maximum number of occupied states permitted for the double-excitation diagrams in any order of perturbation theory. Alternately, restricting Eqn. (30) to the kl = ij term and restricting Eqn. (31) to the k = i or j terms permits only two-body effects to be included in any order. This is the "pair-restricted" approximation and is the only approximation which the recursive procedure preserves through all orders.

To illustrate the procedure computationally and to investigate some assumptions commonly made about higher-order diagrams, we have carried out a pair-restricted calculation of LiF molecule through sixth-order, using the shifted denominator of Eqn. (38). The basis set, which is essentially that of McLean and Yoshimine /15/ augmented by some correlating functions, is listed in Table I. The HFSCF energy is -106.99158 hartree at R = 2.955bohr. McLean /16/ defines the experimental correlation energy to be -0.445 hartree. Since the two-body contributions to diagram 2(C) and the higher-order diagrams which the hole-ladder interaction generates are summed by the denominator shift, only Eqn. (29) and Eqn. (31), where the latter is restricted to k = i or j, need be evaluated. These numbers are reported in Table II. The total correlation energy of -0.37174 hartree, amounts to 83.5% of the experimental value.

The convergence of the pair approximation is seen to be quite good, with ∿ 94% of the final result arising from E_2^S, another 5% from E_3^S, and ∿ 1%, ∿ 0.2%, and ∿ 0.04% from E_4^S, E_5^S, and E_6^S, respectively. Since chemical accuracy is normally considered to be ∿ 1 kcal/mole or ∿0.0016 hartree, computation beyond E_4^S would not be necessary for this particular example.

Any conclusions drawn are dependent on the choice of basis set, but whithin this limitation, it may be justifiably stated that the value -0.37174 hartree is essentially the independent pair correlation energy limit. Since the three- and four-body interactions should result in a net positive contribution to the correlation energy, any improvement over the independent pair approximation will give a smaller net value for the correlation energy within this basis set.

In previous MBPT calculations, contributions from higher-order diagrams are conventionally estimated by assuming various geometric-type approximations /9,10,11/. In the case of the particle-ladder diagrams a geometric approximation based on the ratio of the third-order to the second-order diagrams would give a particle-ladder

Table 1. Slater orbital basis set. The functions listed without an asterisk are taken from the basis set of McLean and Yoshimine /15/, while the asterisk indicates functions added by the authors, some of which replace the 4f functions in the McLean-Yoshimine basis set.

	Orbital	Exponent
Li	1s	4.6990, 2.4780
	2s	1.7700, 0.8100, 0.64300
	$2p_\tau$	1.4651
	$3d_\tau$	2.500
	$2p_\pi$	0.81417
	$3p_\pi$	1.0056*
	$3d_\pi$	0.8966
	$3d_\tau$	0.8966*
F	1s	14.2010, 7.9380
	2s	3.3320, 2.0570
	3s	9.9620
	$2p_\tau$	9.4350, 4.2490, 2.3560, 1.4340
	$3p_\tau$	3.3000*
	$3d_\tau$	2.5837
	$2p_\pi$	9.4350, 4.2490, 2.3560, 1.4340
	$3p_\pi$	3.0670*
	$3d_\pi$	2.2073
	$3d_\tau$	1.2000*, 2.3955*, 4.8000*

Table II. Two-Body Contributions of Different Classes of Diagrams to the Correlation Energy of LiF (R = 2.955 bohr)[a]. The diagram type is indicated in square brackets by figure number and letter. All numbers include the denominator shift of Eqn. (38).

Order	Particle Ladder	Hole-Particle Ladder	Interaction Terms	Total
Second	-0.3484 [2(A)]			-0.3484
Third[b]	+0.0292 [2(B)]	-0.0480 [2(D)]		-0.0189
Fourth[b]	-0.0034 [3(A)]	-0.0103 [3(I-L)]	+0.0100 [3(E,F)]	-0.0036
Fifth	+0.0004	-0.0025	+0.0014	-0.0007
Sixth	-0.0001	-0.0007	+0.0006	-0.0002
			Total	-0.3717
			Exp.[c]	-0.445

[a]All energies are in hartrees.

[b]The two-body contributions of diagrams [2(C)], and [3(B-D,G,H)] and their higher order analogs which include a "hole-hole" interaction are incorporated into the lower orders in the calculation via the <ij||ij> term in the shifted denominator.

[c]Reference 16.

contribution to $E_4^S + E_5^S + E_6^S$ which is +0.00226hartree compared to the correct result of +0.00299hartree, or a difference of about 24%. For the pair contributions to the hole-particle diagrams, the analogous values are -0.00765 hartree from the geometric approximation and -0.01346 hartree for the correct result, which is a difference of about 43%. The net effect on the total correlation energy between the geometric approximation and the computed total $E_4^S + E_5^S + E_6^S$ is 1.7%, or about 4 kcal/mole. Thus, the normal geometric approximation is not applicable to the individual diagram types and is not extremely accurate for the combined results in this pair-restricted calculation. However, the contributions from E_4^S, E_5^S, and E_6^S are small enough that it makes only little difference in the final answer.

It has been demonstrated that complete classes of higher-order diagrams may be conveniently computed by recursive-type formulae and that rigorous basis set limits for various approximations may be obtained. Once the three-and four-body interactions are included in the recursive calculation the limiting result would be the "linked-cluster" contribution of all double excitations. Although the present study was restricted for simplicity only to HFSCF orbitals, for both occupied and excited states, the diagrams arising from one-particle insertions that need to be considered for more general cases may be similarly included without difficulty. Furthermore, the generalizations needed to incorporate single, triple, and quadruple excitations may also be made. Thus, the hope persists that, essentially, the complete basis set limit numerical approximation for the reaction operator is a realizable goal.

APPENDIX

We will restrict ourselves to Brillouin-Wigner (BW) perturbation theory for simplicity, although the recursive calculation of higher orders may be easily proven in the Rayleigh-Schrödinger case as well.

The wavefunction and energy in BW perturbation theory are defined as

$$\omega_n = (T_o V)^n |\omega_o\rangle$$

$$E_{n+1} = \langle\omega_o|V|\omega_n\rangle$$

$$T_o = P(E - PH_oP)^{-1}P$$

Introducing the representation of the operators as a matrix of n-particle configurations, $|\underline{f}\rangle$, where $P|\underline{f}\rangle = |\underline{h}\rangle$, it is apparent that we can define the column matrices, $\underline{a}_k$ and $\underline{A}_k$, as

$$\underline{a}_1 = \langle\underline{h}|V|\omega_o\rangle$$

$$\underline{a}_2 = \langle\underline{h}|V|\underline{h}\rangle\, \underline{a}_1 = \underline{V}\underline{a}_1$$

$$\vdots$$

$$\underline{a}_n = \underline{V}\underline{a}_{n-1}$$

$$\underline{A}_n = \langle\underline{h}|E - H_o|\underline{h}\rangle^{-1}\underline{a}_n = \underline{T}_o\underline{a}_n$$

Consequently, for the energies,

$$E_2 = \underline{a}_1^+\, \underline{T}_o\, \underline{a}_1 = \underline{A}_1^+\underline{a}_1$$

$$E_3 = \underline{A}_1^+\, \underline{a}_2$$

$$E_4 = \underline{a}_2^+\, \underline{T}_o\, \underline{a}_2 = \underline{A}_2^+\, \underline{a}_2$$

$$\vdots$$

$$E_{2n-1} = \underline{A}_{n-1}^+\underline{a}_n$$

$$E_{2n} = \underline{a}_n^+\, \underline{T}_o\, \underline{a}_n = \underline{A}_n^+\, \underline{a}_n .$$

These may be seen to be the same kind of expressions as occur in Eqns. (36) and (37). By simply replacing $\underline{a}_{n-1}$ by $\underline{a}_n$, higher orders could be similarly generated. The difference, of course, is that in MBPT we have all of the benefits of the formulation of the "linked-cluster" expansion, which eliminates any need for n-particle configurations, and thus removes the "unlinked-cluster" contributions to all orders, in addition to providing a more tractable and flexible computational scheme.

ACKNOWLEDGEMENT

I (RJB) am privileged to contribute to this volume in honor of Prof. Löwdin. As a co-director of my dissertation at the Florida Quantum Theory Project, I was very fortunate to be introduced to the exciting possibilities in quantum theory and in perturbation theory, in particular, by Prof. Löwdin. I have benefited immeasurably from his knowledge, inspiration, and contagious enthusiasm, and I want to sincerely express my appreciation for his invaluable assistance during these several years.

REFERENCES

1. P.O. Löwdin, Adv. Chem. Phys. 2, 207 (1959).
2. P.O. Löwdin, Phys. Rev. 97, 1474, 1490, 1509 (1955).
3. P.O. Löwdin, Revs. Mod. Phys. 32, 328 (1960).
4. P.O. Löwdin, Nikko Symposium, Molecular Physics, edited by M. Kotani (Maruzen Company, Ltd., Tokyo, 1954), p. 13.
5. a. P.O. Löwdin, J.Chem. Phys. 19, 1396 (1951); b. J. Mol. Spectry. 10, 12 (1963); c. J. Mol. Spectry. 13, 326 (1964); d. J. Math. Phys. 3, 969 (1962); e. J. Math. Phys. 3, 1171 (1962); f. J. Mol. Spectry. 14, 112 (1964); g. J. Mol. Spectry. 14, 119 (1964); h. J. Mol. Spectry. 14, 131 (1964); i. J. Math. Phys. 6, 1341 (1965); j. Phys. Rev. 139, A357 (1965); k. J. Chem. Phys. 43, 5175 (1965); l. Perturbation Theory and Its Applications in Quantum Mechanics (ed. C.H. Wilcox, John Wiley and Sons, New York, 1966), p.255; m. Int. J. Quant. Chem. 2, 867 (1968).
6. P.O. Löwdin, Adv. Chem. Phys. 14, 283 (1969).
7. J. Goldstone, Proc. R. Soc. Lond. A239, 267 (1957).
8. K.A. Brueckner, Phys. Rev. 97, 1353 (1955); 100, 36 (1955); K.A. Brueckner, C.A. Levinson, and H.M. Mahmood, Phys. Rev. 95, 217 (1954).
9. H.P. Kelly, Adv. Chem. Phys. 14, 129 (1969).
10. T. Lee, N.C. Dutta, and T.P. Das, Phys. Rev. Letters 25, 204 (1970).
11. U. Kaldor, Phys. Rev. A7, 427 (1973).
12. R.J. Bartlett and D.M. Silver, Int. J. Quant. Chem. 9S, 183 (1975).
13. R.J. Bartlett and D.M. Silver, J. Chem. Phys. 62, 3258 (1975); Int. J. Quant. Chem. 8S 271 (1974); Phys. Rev. A10, 1927 (1974).
14. D.M. Silver and R.J. Bartlett, Phys. Rev., A13, 1 (1976).
15. A.D. McLean and M. Yoshimine, "Tables of Linear Molecule Wave Functions," supplement, IBM Journal of Research and Development November, 1967.
16. A.D. McLean, J. Chem. Phys. 39, 2653 (1963).

CALCULATION OF THE BROMINE NUCLEAR PSEUDOQUADRUPOLE COUPLING CONSTANT IN THE LiBr MOLECULE USING A DENSITY-OF-STATES FUNCTION DEDUCED FROM OVERLAP INTEGRALS*

H.B. Jansen[†] and P. Pyykkö[+]

[†]Scheikundig Laboratorium, Vrije Universiteit, Amsterdam-Buitenveldert, The Netherlands

[+]Institutionen för Fysikalisk Kemi, Åbo Akademi, 20500 Åbo (Turku), Finland

I. INTRODUCTION

The total observed quadrupole-like coupling constant of a nucleus can be divided into several components,

$$(eqQ/h)_{apparent} = b_{tot} = b_q + b_{pq} + \dots \qquad (1)$$

where the genuine quadrupole contribution b_q is proportional to the nuclear quadrupole moment Q and the pseudoquadrupole coupling constant b_{pq} is a magnetic polarizability, proportional to the square of the nuclear magnetic moment $g_n \beta_n I$. Expressions for the various contributions to b_{pq} have been derived by Ramsey /1/. The other terms in the series (1), a review of the earlier work and estimates of the magnitude of b_{pq} have been published both for molecules /2,3/ and metals /4/.

Experimental evidence for nuclear pseudoquadrupole effects have recently been obtained for metallic gallium /5/. Experimental evidence for a nuclear pseudoquadrupole contribution of a few ppm for the Br nucleus in the LiBr molecule has also been presented /6/.

The traditional second-order perturbation theory, using virtual

*As one of the authors (P.P.) received his introduction to Quantum Chemistry in the Uppsala group and both of us have benefitted from the Löwdin summer schools, it is a pleasure to dedicate this article to Professor Per-Olov Löwdin on the occasion of his 60th birthday.

orbitals as intermediate states, did not lead to converging results /7/. Therefore we here use for the calculation of b_{pq} a continuous integration method /8/ based on an approximate density-of-states function deduced from overlap integrals. This method proved remarkably stable in the calculation of internuclear spin-spin coupling constants.

II. THEORY

The second-order interaction energy between a nucleus and the molecular electronic system can be written as

$$E = \underline{I} \cdot \underline{\underline{J}} \cdot \underline{I}. \qquad (2)$$

For an axially symmetric site we can define the parallel component, $J_{\parallel}$, and the transversal component, $J_{\perp}$, of the diagonal tensor, $\underline{\underline{J}}$. In terms of this tensor the pseudoquadrupole coupling constant can be expressed as

$$b_{pq} = \frac{4}{3} I(2 I - 1) (J_{\parallel} - J_{\perp}). \qquad (3)$$

The non-relativistic expressions for the tensor $\underline{\underline{J}}$ have been derived by Ramsey /1/. We consider here only the dipole-dipole Hamiltonian

$$H_2 = g_n g \beta_n \beta \left[\frac{\underline{I} \cdot \underline{s}}{r^3} - \frac{3(\underline{I} \cdot \underline{r})(\underline{s} \cdot \underline{r})}{r^5} \right] \qquad (4)$$

and the Fermi contact Hamiltonian

$$H_3 = - \frac{8\pi}{3} g_n g \beta_n \beta \underline{I} \cdot \underline{s} \delta(\underline{r}) \qquad (5)$$

The contribution from the cross term between these two is believed to be the leading term in the pseudoquadrupole interaction. Including only this term we get

$$J_{\parallel}^{(2,3)} - J_{\perp}^{(2,3)} = 16\pi g_n^2 \beta_n^2 \beta^2 \sum_i^{occ} \sum_j^{unocc} \langle i|(3 z^2 - r^2)r^{-5}|j\rangle \times$$
$$\times \langle j|\delta(\underline{r})|i\rangle/(\varepsilon_i - \varepsilon_j + (ii|jj)), \qquad (6)$$

where, due to the delta function, only σ-type MO:s contribute. If virtual orbitals are used as the unoccupied molecular orbitals j, however, the sum over j does not converge /7/.

Similar difficulties for the internuclear spin-spin coupling can be overcome by transforming the sum over j into a continuous integration /8/. In the present case we introduce a "scanning molecular orbital" (SMO) centered at the bromine nucleus. The ground state wave function in our calculation is computed using a GTO basis. We therefore choose as the SMO:s the following functions

$$\chi_s(\alpha) = (2\,\alpha/\pi)^{3/4} \exp(-\,\alpha\, r^2) \tag{7}$$

$$\chi_p(\alpha) = 2^{7/4}\,\alpha^{5/4}\,\pi^{-3/4}\, x \exp(-\,\alpha\, r^2) \tag{8}$$

$$\chi_d(\alpha) = 2^{11/4}\,(15)^{-\frac{1}{2}}\,\alpha^{7/4}\,\pi^{-3/4}\,(2x^2-y^2-z^2) \exp(-\alpha r^2), \tag{9}$$

x being the molecular axis. The variable of integration is the orbital exponent α of the SMO. This set of functions is non-orthogonal but complete in the subspaces of s-, pσ- and dσ-type /9/.

The sum over the unoccupied MO:s is now replaced by a continuous integration,

$$\sum_{j}^{\text{unocc}} \rightarrow \sum_{\mu=s,p,d} \int_0^\infty D_\mu(\alpha)\, d\alpha, \tag{10}$$

where $D_\mu(\alpha)$ is the density of states function for the symmetry type μ. These functions are deduced from overlap integrals through the following heuristic argument:

Consider two s-type GTO:s $|\alpha\rangle$ and $|\alpha'\rangle$ with the orbital exponent

$$\alpha' = \alpha + \Delta\alpha \tag{11}$$

and the overlap integral

$$\langle\chi_s(\alpha)|\chi_s(\alpha')\rangle = (\alpha\alpha')^{3/4}\,(2/(\alpha+\alpha'))^{3/2}$$

$$\simeq 1 - \frac{3}{16}\left(\frac{\Delta\alpha}{\alpha}\right)^2. \tag{12}$$

Then introduce the projection operator /12/

$$O = |\alpha\rangle\,\langle\alpha| \tag{13}$$

giving the norm of the orthogonal complement

$$N = \langle\alpha'|(1-O)^\dagger\,(1-O)|\alpha'\rangle = 1 - |\langle\alpha|\alpha'\rangle|^2$$

$$= \frac{3}{8} \left(\frac{\Delta\alpha}{\alpha}\right)^2 + O((\Delta\alpha)^4). \tag{14}$$

Then the "amplitude of new state" per unit α is

$$D_s(\alpha) = \frac{dn}{d\alpha} = \lim_{\Delta\alpha\to 0} \frac{N^{\frac{1}{2}}}{\Delta\alpha} = \sqrt{\frac{3}{8}} \frac{1}{\alpha}. \tag{15}$$

For the p- and d-type densities-of-states we find similarly that

$$D_p(\alpha) = \sqrt{\frac{5}{8}} \frac{1}{\alpha} \tag{16}$$

$$D_d(\alpha) = \sqrt{\frac{7}{8}} \frac{1}{\alpha}. \tag{17}$$

In order to remove the part of the Hilbert space, spanned by the occupied MO:s, we orthogonalize the SMO:s against them

$$\phi = \chi_\mu(\alpha) - \sum_i^{occ} \langle\chi_\mu(\alpha)|i\rangle \tag{18}$$

III. NUMERICAL CALCULATIONS

The ground state wave function for the Li Br molecule was computed at the experimental internuclear distance of 4.0655 a.u. using a GTO basis set where the contracted GTO:s (CGTO) were obtained by fitting to the STO basis functions of Matcha /10/ for the same molecule. The 128 primitive GTO:s were contracted to 51 CGTO:s No 4f functions were included. Some ground state properties obtained with our wave function are compared with those of Matcha in Table I.

The orbital energies, ε_j, for the SMO:s are calculated in the field of a completely filled ground state. For this purpose, and for calculating the Coulomb integral $(ii|\phi\phi)$ in the energy denominator, the SMO:s of Eq. (18) are renormalized, whereas for the calculation of the b_{pq} they are not. The integration over α is carried out numerically using the Simpson method with ln α as the variable and $\ln(2^{\frac{1}{2}})$ as the step. The results are given in Table II. The value of the quadrupole anomaly, ${}_{79}\Delta_{81}$, defined by

Table I

Expectation values of the total energy E, and the field gradient q for the most important MO:s.

Quantity	Previous work[a]	This work
E	- 2579.8901	- 2575.59
$q^{(1\sigma)}$	- 0.002435	+ 0.02766
$q^{(2\sigma)}$	- 0.0022	+ 0.0152
$q^{(3\sigma)}$	- 1011.56	- 997.43
$q^{(5\sigma)}$	- 144.56	- 142.92
$q^{(9\sigma)}$	- 8.313	- 8.299

[a] The total energy is given in Ref. 10 and the field gradient contributions were calculated using that wave function.

Table II

Contributions to the Br^{79} nuclear pseudoquadrupole coupling constant, b_{pq}, and the quadrupole anomaly, ${}_{79}\Delta_{81}$, for the two bromine isotopes in LiBr.

Contribution	b_{pq}^{79} (Hz)	${}_{79}\Delta_{81}$
This work	- 674 [d]	$+ 7 \cdot 10^{-6}$
Previous estimates	+ 116 [a] + 312 [b]	 $- 3 \cdot 10^{-6}$
Exp. [c]	--	$+ (5 \pm 5) \cdot 10^{-6}$

[a] Ref. 2.

[b] Ref. 7, including the four first virtual orbitals of Matcha.

[c] Ref. 6. The error limit corresponds to one standard deviation.

[d] Integrated from $\alpha = 0.0039625$ to $\alpha = 4096$.

the equation

$$\frac{b_{tot}(79)}{b_{tot}(81)} = (1 + {}_{79}\Delta_{81}) \frac{Q(79)}{Q(81)} , \tag{19}$$

and resulting from the present pseudoquadrupole contribution,

$${}_{79}\Delta_{81} \cong \frac{b_{pq}(79)}{b_{tot}(79)} - \frac{b_{pq}(81)}{b_{tot}(81)}, \tag{20}$$

is also given.

In addition to the total contribution, the present method also gives an analysis of it in terms of the spectral density function

$$\Phi_\mu(\alpha) = \alpha\, D_\mu(\alpha) \sum_i^{occ} \langle i|(3z^2 - r^2)r^{-5}|\phi\rangle\langle\phi|\delta(\underline{r})|i\rangle /$$

$$(\varepsilon_i - \varepsilon_\phi + (ii|\phi\phi)). \tag{21}$$

This function is depicted in Figure 1.

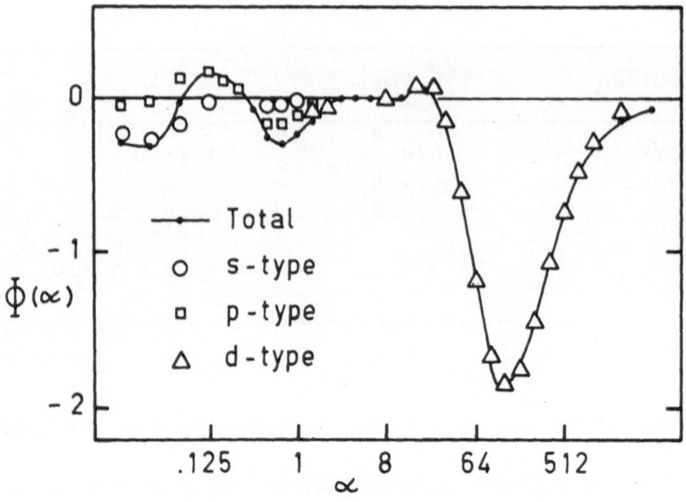

Figure 1. The spectral density function Φ for the Br nuclear pseudoquadrupole coupling in LiBr.

The total contribution is equal to

$$\int_{-\infty}^{\infty} d(\ln\alpha)\sum_{\mu} \Phi_{\mu}(\alpha) \qquad \text{(in a.u.)} \tag{22}$$

IV. DISCUSSION

Despite of its approximate character the present spectral density function has the advantage of a remarkable stability. We here find for the nuclear pseudoquadrupole coupling constant of Br in LiBr a result which agrees both in sign and order-of-magnitude with experiment. Although relativistic effects /11/, correlation effects and the other terms of Ramsey have not been included, they are not expected to change the qualitative picture. We also find that this molecular property is a most unusual static polarisability in that excitations both from 1σ and 9σ, with excitation energies ranging from several tens of keV to a few eV, give contributions of a comparable order of magnitude.

The dominant contribution around α = 100 arises from excitations from the 1σ and 2σ MO:s and is thus determined by the nonspherical character of the bromine 1s and 2s shells.

REFERENCES

1. N.F. Ramsey, Phys Rev. 89,527 (1953).
2. P. Pyykkö and J. Linderberg, Chem. Phys. Lett. 5, 34 (1970).
3. P. Pyykkö, Chem. Phys. Lett. 6, 479 (1970).
4. P. Pyykkö, J. Phys. F: Metal Phys. 1, 102 (1971).
5. S.L. Segel and J.D. Stroud, J. Phys. F: Metal Phys. 00, 0000 (1975).
6. R.C. Hilborn, T.F. Gallagher and N.F. Ramsey, J. Chem. Phys. 56, 855 (1972).
7. P. Pyykkö, Preprint No. 19-70, Research Institute for Theoretical Physics, University of Helsinki (1970).
8. H.B. Jansen, J.A.B. Lohman and P. Pyykkö, Chem. Phys. 0, 000 (0000).
9. S.F. Boys and I. Shavitt, Proc. Roy. Soc. (London) 254, 487 (1960).

10. R.L. Matcha, J. Chem. Phys. 53, 485 (1970).

11. P. Pyykkö, E. Pajanne and M. Inokuti, Intern. J. Quantum Chem. 7, 785 (1973).

12. P.-O. Löwdin, Phys. Rev. 139, A 357 (1965).

ON INVERSION SYMMETRY IN MOMENTUM SPACE

Per Kaijser and Vedene H. Smith, Jr.

Department of Chemistry, Queen's University, Kingston

Ontario, K7L 3N6, Canada

I. INTRODUCTION

The non-commutativity of postition and momenta is one of the fundamental differences between quantum and classical mechanics. As a consequence in quantum theory, it is customary to choose either the position space or the momentum space representation to describe the bound electronic states of atomic and molecular systems. The position space representation is by far the more commonly chosen one due not only to the simpler form of the Hamiltonian operator but also to the fact that it yields a charge density distribution which enables one intuitively to visualize the form of the system. However, the momentum space representation is equally valid although its picture is more difficult for the human mind to grasp.

The Schrödinger equation is not commonly solved in the momentum space representation for finite systems /1,2/. The access to the momentum space representation is normally accomplished via a Dirac-Fourier transformation of the wavefunction in the position space representation /3,4/. In this manner one can obtain the one-particle charge density $\rho(\vec{p})$ in momentum space. Instead of transforming the wavefunction itself, one can transform the one-particle charge density matrix /5,6/ $\gamma(\vec{r}, \vec{r}')$ (note that the whole density matrix and not only its diagonal part, the density, has to be considered /7/ or equivalently (except possibly for a set of measure zero) each of its eigenfunctions separately /8-11/. The Dirac-Fourier transformed orbitals /3,4/

$$\hat{\Psi}(\vec{p}) = (2\pi)^{-3/2}\int e^{-i\vec{p}\cdot\vec{r}}\,\Psi(\vec{r})d^3r \tag{1}$$

are in the following denoted by the symbol "$\hat{}$".

The one-particle charge density matrix in position space may be written in the form

$$\gamma(\vec{r}, \vec{r}\,') = \underline{\Psi}(\vec{r})\underline{n}\ \underline{\Psi}^{\dagger}(\vec{r}\,') \tag{2}$$

where $\underline{\psi}(\vec{r})$ is the row vector of the eigenfunctions of $\gamma(\vec{r},\vec{r}\,')$, the (charge density) natural orbitals $\{\Psi_i(\vec{r})\}$ and $\underline{n}$ stand for the diagonal matrix containing their occupation numbers /5/. The one-particle momentum density matrix then becomes

$$\rho(\vec{p},\vec{p}\,') = \hat{\underline{\Psi}}(\vec{p})\ \underline{n}\ \hat{\underline{\Psi}}^{\dagger}(\vec{p}\,') \tag{3a}$$

where the diagonal part represents the momentum density

$$\rho(\vec{p}) = \rho(\vec{p},\vec{p}) \tag{3b}$$

In the fixed nuclei approximation, which is adopted in the present discussion, the total electronic system has to be at rest, which implies that

$$\int \vec{p}\rho(\vec{p})d^3p = 0 \tag{4}$$

A sufficient, but not necessary condition for eqn. (4) to hold is that $\rho(\vec{p})$ obey inversion symmetry; i.e.

$$\rho(\vec{p}) = \rho(-\vec{p}) \tag{5}$$

As Löwdin /4/ has shown this property is satisfied for those $\rho(\vec{p})$ which are obtained from real wavefunctions. In the case of real self-adjoint Hamiltonians H, it is well-known that both the exact wavefunction Ψ and its complex conjugate Ψ^* are solutions of the Schrödinger equation corresponding to the same eigenvalue E. This degeneracy leaves one with the freedom to choose the real combination $\Psi + \Psi^*$ and the pure imaginary combination $\Psi - \Psi^*$, both of which satisfy the hypotheses of Löwdin's Theorem/4/, modified to allow for essentially real wavefunctions /12/, i.e. those for which $\Psi = a\Psi^*$ where a is a constant.

One may reach the same conclusion regarding the inversion symmetry of $\rho(\vec{p})$ for bound states of real Hamiltonians, by noting that such systems obey the principle of microreversibility /4,13/. It follows that the probability density of momentum before and after time reversal /4,13/ are related by:

$$\Pi_{rev}(\vec{p}_1,\ldots\ldots\vec{p}_N,t) = \Pi(-\vec{p}_1,\ldots\ldots,-\vec{p}_N,-t)$$

and hence for stationary states, eqn. (5) holds.

Questions about momentum space properties such as the inversion symmetry of $\rho(\vec{p})$ and its asymptotic behaviour /10,11/ have arisen from the renewed interest in Compton profiles /10,14,15/. Since the various Compton profiles are obtained from the momentum density within the so-called impulse approximation, it is important that $\rho(\vec{p})$ calculated from approximate or model wavefunctions should resemble the exact momentum density as closely as possible. With regard to the inversion symmetry property, the theorem by Löwdin /4/ guarantees its validity for the diverse use of essentially real wavefunctions. However, there are a variety of ways to construct model wavefunctions with complex components /16-22/ or to use non-real basis orbitals in variational calculations. In addition the natural orbital analysis /5/ of variational wavefunctions may lead to complex orbitals as the pioneering work of Löwdin and Shull /23/ indicated.

In the next section, we present some conditions on the position space wavefunctions and natural orbitals in order for the inversion symmetry in momentum space to be valid. To our knowledge all molecular wavefunctions used in the literature so far for evaluating Compton profiles are covered in this section, and thus mathematically proven to fulfill the requirement of relation (5). That we do not have complete freedom in the choice of wavefunction is illustrated in Section III by an example related to Löwdin's concept of complex hybridization /24/:

II. SEVERAL SUFFICIENT CONDITIONS ON $\Psi(\vec{r})$ for $\rho(\vec{p}) = \rho(-\vec{p})$

A. Löwdin /4/ has shown, as mentioned above, that inversion symmetry in momentum space is guaranteed when the wavefunction is real. The one particle charge density matrix $\gamma(\vec{r},\vec{r}\,')$ from such a wavefunction is real. The reality of $\gamma(r',\vec{r}\,')$ is also true if all of its natural orbitals are real. Thus for a real $\gamma(\vec{r}, \vec{r}\,')$,

$$\begin{aligned}\rho(\vec{p}) &= \rho(\vec{p})^* = \{(2\pi)^{-3}\int e^{-i\vec{p}\cdot(\vec{r}-\vec{r}\,')}\gamma(\vec{r},\vec{r}\,')d^3r d^3r'\}^* \\ &= (2\pi)^{-3}\int e^{i\vec{p}\cdot(\vec{r}-\vec{r}\,')}\,\gamma(\vec{r},\vec{r}\,')d^3r d^3r' \qquad (6) \\ &= \rho(-\vec{p})\end{aligned}$$

We note that this case includes molecular orbital calculations where the molecular orbitals are expanded with real coefficients in a basis of Cartesian Gaussian type orbitals or Gaussian lobe functions.

As a corollary, we observe that for a single Slater determinant wavefunction D, the theorem is true if there exists a unitary transformation that brings the natural spin orbitals of D to real form. This is easily seen by considering the Fock-Dirac density matrix or fundamental invariant /25/ corresponding to D. This is related to a theorem of Brändas /26/ which states that D is essentially real, i.e. D = aD where a is a constant, if and only if such a unitary transformation exists.

B. If $\gamma(\vec{r},\vec{r}')$ is a function only of the scalar quantities $|\vec{r}'|$ and $|\vec{r}-\vec{r}'|$ or $|\vec{r}|$ and $|\vec{r}'|$, the corresponding $\rho(\vec{p})$ is spherically symmetric (and thus is inversion symmetric).

By hypothesis, $\gamma(\vec{r},\vec{r}') = f(|\vec{r}'|,|\vec{r}''|)$, where $\vec{r}'' = \vec{r}-\vec{r}'$. Then

$$\begin{aligned}\rho(\vec{p}) &= (2\pi)^{-3}\int e^{-i\vec{p}\cdot(\vec{r}-\vec{r}')}\,\gamma(\vec{r},\vec{r}')\,d^3r d^3r', \\ &= (2\pi)^{-3}\int e^{-i\vec{p}\cdot\vec{r}''}\,f(|\vec{r}'|,|\vec{r}''|)\,d^3r''d^3r', \\ &= \rho(|\vec{p}|).\end{aligned} \tag{7}$$

The other case is easily proven by considering the fact that $\pm\vec{r}$ have the same scalar magnitude.

C. Let $\underline{\Psi}(\vec{r})$ be a set of natural orbitals which all have the same occupation number n. If there exists a unitary matrix $\underline{U}$ that transforms them to a set of real orbitals $\underline{\chi}(\vec{r})$, then

$$\rho(\vec{p}) = \rho(-\vec{p})$$

By hypothesis,

$$\underline{\chi}(\vec{r}) = \underline{\Psi}(\vec{r})\underline{U} \tag{8a}$$

or

$$\underline{\Psi}(\vec{r}) = \underline{\chi}(\vec{r})\underline{U}^{\dagger}. \tag{8b}$$

Since the diagonal occupation number matrix $\underline{n}$ commutes with any other matrix of the same order, it follows that

$$\begin{aligned}\rho(-\vec{p}) &= \hat{\underline{\Psi}}(-\vec{p})\underline{n}\,\hat{\underline{\Psi}}^{\dagger}(-\vec{p}) \\ &= \hat{\underline{\chi}}(-\vec{p})\underline{U}^{\dagger}\,\underline{n}\,\underline{U}\,\hat{\underline{\chi}}^{\dagger}(-\vec{p}) \\ &= \hat{\underline{\chi}}^{*}(\vec{p})\,\underline{U}^{\dagger}\,\underline{U}\,\,\underline{n}\,\hat{\underline{\chi}}^{T}(\vec{p})\end{aligned}$$

$$= \{\underline{\hat{\chi}}(\vec{p})\ \underline{n}\ \underline{\hat{\chi}}^{\dagger}(\vec{p})\}^*$$

$$= \underline{\hat{\chi}}(\vec{p})\ \underline{n}\ \underline{\hat{\chi}}^{\dagger}(\vec{p})$$

$$= \underline{\hat{\Psi}}(\vec{p})\ \underline{n}\ \underline{\hat{\Psi}}^{\dagger}(\vec{p})$$

$$= \rho(\vec{p}) \tag{9}$$

If the hypotheses of this theorem are satisfied for the set of natural orbitals associated with each different occupation number n_i, then

$$\rho(\vec{p}) = \sum_i \rho_i(\vec{p}) = \sum_i \rho_i(-\vec{p}) = \rho(-\vec{p}). \tag{10}$$

D. Let $\underline{\Psi}(\vec{r})$ be a set of natural orbitals which are expressible as a linear combination of atomic orbitals, each of which in turn is a product of a real radial function and a spherical harmonic. If the coefficient matrix is denoted by $\underline{c}$, then

$$\underline{\Psi}(\vec{r}) = \underline{\phi}\ (\vec{r})\ \underline{c} \tag{11}$$

and

$$\begin{aligned} \rho(\vec{p}) &= \underline{\hat{\Psi}}\ (\vec{p})\ \underline{n}\ \underline{\hat{\Psi}}^{\dagger}(\vec{p}) \\ &= \underline{\hat{\phi}}\ (\vec{p})\ \underline{c}\ \underline{n}\ \underline{c}^{\dagger}\ \underline{\hat{\phi}}^{\dagger}(\vec{p}) \\ &= \underline{\hat{\phi}}\ (\vec{p})\ \underline{D}\ \underline{\hat{\phi}}^{\dagger}(\vec{p}) \end{aligned} \tag{12}$$

Consider a general two-orbital contribution to $\rho(\vec{p})$;

$$\hat{\phi}_i(\vec{p})D_{ij}\hat{\phi}_j^*(\vec{p}) + \hat{\phi}_j(\vec{p})D_{ji}\hat{\phi}_i^*(\vec{p}) =$$

$$\begin{aligned} &= g_i(p)g_j(p)\{D_{ij}e^{-i\vec{p}\cdot(\vec{R}_i-\vec{R}_j)}(-i)^{\ell_i-\ell_j}\ Y_{\ell_i m_i}(\hat{\theta},\hat{\phi})Y^*_{\ell_j m_j}(\hat{\theta},\hat{\phi}) \\ &\quad + D_{ij}{}^*e^{i\vec{p}\cdot(\vec{R}_i-\vec{R}_j)}(i)^{\ell_i-\ell_j}\ Y^*_{\ell_i m_i}(\hat{\theta},\hat{\phi})Y_{\ell_j m_j}(\hat{\theta},\hat{\phi})\} \end{aligned} \tag{13}$$

Here we have used the fact that $\underline{D}$ is hermitian and that the Fourier transform of an atomic orbital centered on $\vec{R}_k$ with the spherical harmonic $Y_{\ell_k m_k}$ and a real radial function is:

$$\phi_k(\vec{p}) = e^{-i\vec{p}\cdot\vec{R}_k}(-i)^{\ell_k} g_k(p) Y_{\ell_k m_k}(\hat{\theta},\hat{\phi}) \tag{14}$$

where $g_k(p)$ is a real function of the scalar p.

Application of the parity property of the spherical harmonics under inversion enables us to write:

$$\hat{\phi}_i(-\vec{p})D_{ij}\hat{\phi}_j^*(-\vec{p}) + \hat{\phi}_j(-\vec{p})D_{ji}\hat{\phi}_i^*(-\vec{p}) =$$

$$= g_i(p)g_j(p)\{D_{ij}e^{i\vec{p}\cdot(\vec{R}_i-\vec{R}_j)}(i)^{\ell_i-\ell_j} Y_{\ell_i-m_i}(\hat{\theta},\hat{\phi})Y^*_{\ell_j m_j}(\hat{\theta},\hat{\phi})$$

$$+ D_{ij}^* e^{-i\vec{p}\cdot(\vec{R}_i-\vec{R}_j)}(-i)^{\ell_i-\ell_j} Y^*_{\ell_i m_i}(\hat{\theta},\hat{\phi})\, Y_{\ell_j m_j}(\hat{\theta},\hat{\phi})\}. \tag{15}$$

Comparison of (13) with (15) shows that a sufficient condition for the inversion symmetry of $\rho(\vec{p})$ is that for each pair of indices (i,j) either or both of the following conditions holds:

1. $\ell_i - \ell_j$ is even and $\vec{R}_i = \vec{R}_j$

2. $D_{ij} = D_{ij}^* \delta_{m_i,m_j}$

Combinations of these conditions are allowed. As an example, consider the explicit case when atomic orbitals with spherical harmonics are coupled together to form a real orbital (combine p_+, p_- to p_x and p_y) such as

$$\Psi(\vec{r}) = af_1(r)Y_{\ell_1 m}(\theta,\phi) + (-1)^m\, a^*f_1(r)Y_{\ell_1 -m}(\theta,\phi)+bf_2(r)Y_{\ell_2 0}(\theta,\phi)$$

$$= af_1(r)Y_{\ell_1 m}(\theta,\phi)+\{af_1(r)Y_{\ell_1 m}(\theta,\phi)\}^*+bf_2(r)Y_{\ell_2 0}(\theta,\phi) \tag{16}$$

When the radial functions f and the coefficient b are real, the orbital $\Psi(\vec{r})$ is itself real. When this orbital is combined with other real orbitals, the requirements of case A are fulfilled. This example is not covered by our condition D1 when $\ell_1-\ell_2$ is odd such as occurs in sp hybridization. Although real combinations of the real hybrid atomic orbitals are covered by condition A, we must invoke other constraints when considering the possibility of complex hybridization. This is discussed in the next section.

III. COMPLEX HYBRID ATOMIC ORBITALS

The basic idea behind hybridization /27-32/ is the construction of atomic orbital combinations which are oriented in certain specified geometrical directions. One essentially makes a unitary transformation of a set of linearly independent (usually orthogonal) orbitals (such as the 2s, $2p_x$, $2p_y$ and $2p_z$ atomic orbitals) to another such set. This was usually restricted to real unitary transformations. Löwdin /24/ suggested the generalization to allow for complex mixing coefficients in order to increase the flexibility. Although the proposed advantage of constructing hybrids with directions less than 90° apart has been shown by Coulson and White /20/ to be an artifact of the definition of direction /19/, the method has been successfully applied in the literature to a number of problems of chemical interest /19,33,34/.

In this paper we choose a complex hybrid as an example of the possible construction of orbitals which violate the fundamental rest condition (4).

Consider the orbital

$$\Psi(\vec{r}) = c_i\phi_i(\vec{r}) + c_j\phi_j(\vec{r}) \tag{17}$$

where $\phi_i(\vec{r})$ and $\phi_j(\vec{r})$ are centered at the same atom which is chosen as the origin for the sake of convenience. We further specify that $c_j = ic_i = i/\sqrt{2}$ and that the orbitals have real radial functions with the respective spherical harmonics $Y_{\ell_i m_i} = Y_{00}$ and $Y_{\ell_j m_j} = Y_{10}$. If we assume that this is the only occupied orbital the total momentum distribution is:

$$\rho(\vec{p}) = \frac{1}{8\pi}\,(g_i^2(p) + 3g_j^2(p)\cos^2\hat{\theta}) + \frac{\sqrt{3}}{8\pi}\,g_i(p)g_j(p)\cos\hat{\theta} \tag{18}$$

where the notation of equation (14) is used. The second term does not contribute to the norm but the expectation value of the p_z component of $\vec{p}$,

$$\langle p_z\rangle = \int p_z\rho(\vec{p})d^3p = \frac{1}{2\sqrt{3}}\int_0^\infty p^3 g_i(p)g_j(p)dp \tag{19}$$

is normally different from zero.

This violation of relation (4) implies that in order to use these complex hybrids we must ensure that there exist other orbitals with similar properties so that the total electronic center-of-mass system is at rest.

IV. SUMMARY

Some sufficient conditions for inversion symmetry in the momentum density have been presented. Cases A through C are more general than cases D1 and D2 in that there are no constraints put on the actual form of the natural orbitals. The most commonly used molecular wavefunctions are constructed however in the LCAO-form with real coefficients. A basis of Cartesian Gaussian type functions or Gaussian lobe functions is then sufficient according to case A. On the other hand Gaussian and Slater type orbitals with spherical harmonics suffice as long as we do not mix different m quantum numbers in the NO's (case D2). Thus, nearly all molecular orbital schemes result in an inversion symmetric momentum density.

Although the exact $\rho(\vec{p})$ has inversion symmetry, this may not always be the case for approximate wavefunctions. It is hoped that the ideas presented here will be of assistance in the construction of wavefunctions fulfilling this fundamental requirement.

The authors wish to dedicate this article to Professor Per-Olov Löwdin on the occasion of his sixtieth birthday. We have both had the privilege to have been members of the Uppsala Quantum Chemistry Group and thus to have personally experienced the enthusiastic and dedicated scientific leadership and teaching of Per-Olov Löwdin.

Support of this research by the National Research Council of Canada is gratefully acknowledged.

REFERENCES

1. R. McWeeny and C.A. Coulson, Proc. Phys. Soc. 62, 509 (1949).
2. R. McWeeny, Proc. Phys. Soc. 62, 519 (1949).
3. P.A.M. Dirac, Quantum Mechanics (Oxford University Press, London, 1958) page 97.
4. P.-O. Löwdin, in Advances in Quantum Chemistry edited by P.O. Löwdin (Academic Press, New York, 1967), vol. III, 323.
5. P.-O. Löwdin, Phys. Rev. 97, 1474 (1955).
6. V.H. Smith, Jr. and J. Harriman, Report WIS-TCI-379, Theoretical Chemistry Institute of the University of Wisconsin (1970).

7. R. Benesch, S.R. Singh and V.H. Smith, Jr., Chem. Phys. Letters 10, 151 (1971).

8. R. Benesch and V.H. Smith, Jr., Chem. Phys. Letters 5, 601 (1970).

9. R. Benesch and V.H. Smith, Jr., Int. J. Quant. Chem. 4S, 131 (1971).

10. R. Benesch and V.H. Smith, Jr., in Wave Mechanics - the First Fifty Years edited by W.C. Price, S.S. Chissick and T. Ravensdale (Butterworths, London, 1973).

11. P. Kaijser and P. Lindner, Phil. Mag. 31, 871 (1975).

12. S.T. Epstein, Report WIS-TCI-431, Theoretical Chemistry Institute of the University of Wisconsin (1971).

13. A. Messiah, Quantum Mechanics (North Holland Publishing Company, Amsterdam, 1965) Vol. II, chapter XV.

14. M. Cooper, Adv. Phys. 20, 453 (1971).

15. I.R. Epstein, in MTP International Review of Science (Butterworths, London, 1975). Physical Chemistry Section, Theoretical Chemistry Volume, Second Series.

16. F.E. Harris and H.A. Pohl, J. Chem. Phys. 42, 3648 (1965).

17. F.E. Harris, J. Chem. Phys. 43, S17 (1965).

18. C.F. Bunge, Phys. Rev. 154, 70 (1967).

19. O. Mårtensson and Y. Öhrn, Theor. Chim. Acta 9, 133 (1967).

20. C.A. Coulson and R.J. White, Mol. Phys. 18, 577 (1970).

21. J. Felsteiner, R. Fox and S. Kahane, Phys. Rev. B6, 4689 (1972).

22. J. Hendekovic, Int. J. Quantum Chem. 8, 799 (1974).

23. P.-O. Löwdin and H. Shull, Phys. Rev. 101, 1730 (1956).

24. P.-O. Löwdin, Acta. Vålådalensia, Part I, 17 (1958).

25. P.-O. Löwdin, Phys. Rev. 97, 1490 (1955).

26. E. Brändas, J. Mol. Spectroscopy 27, 236 (1968).

27. L. Pauling, Proc. Nat. Acad. Sci. U.S. 14, 359 (1928).

28. L. Pauling, J. Am. Chem. Soc. 53, 1367 (1931).

29. J.C. Slater, Phys. Rev. 37, 481 (1931).

30. C.A. Coulson, Proc. Roy. Soc. (Edinburgh) 61, 115 (1941).

31. C.A. Coulson, Valence (University Press, Oxford, 1952).

32. P.-O. Löwdin, J. Chem. Phys. 21, 496 (1952).

33. E. Brändas and O. Mårtensson, Chem. Phys. Letters 3, 315 (1969).

34. P. Lindner and O. Mårtensson, Acta. Chem. Scand. 23, 429 (1969).

BONDING CHARACTER OF INNER-SHELL ORBITALS IN DIATOMIC MOLECULES

Osvaldo Goscinski

Department of Quantum Chemistry
Uppsala University
Box 518, S-751 20 UPPSALA, Sweden

INTRODUCTION

The concepts of bonding and antibonding orbitals were introduced by the pioneers Mulliken, Hund and Herzberg[1]. They became invaluable tools for the analysis of molecular spectra long before a quantitative basis could be established. To quote Mulliken: "Bonding and antibonding electrons are respectively electrons which tend to make the U(R) curves attractive or repulsive and nonbonding electrons are those which do not affect the U(R) curves"[2]. The K-shells in diatomic molecules like CO or N_2 were assigned non-bonding character ("although we have no direct experimental evidence of their presence..."[2]).

K-shells became observable through the advent of photoelectron spectroscopy, or ESCA[3]. Further recent development in experimental techniques made the bond length changes following inner-shell ionisation conspicuous[3]. Thus the concept of bonding in this connection has to be reexamined: The changes are observable and their magnitude and sign are site-dependent to an extent that a conventional non-bonding characterisation is not acceptable.

To strive for formal accuracy in order to assert the significance and eventual limitation of durable concepts is a path that Per-Olov Löwdin follows all the time, leaving clear prints and guidemarks. To follow them is stimulating and a challenge. I am not the only one, I am sure, who felt upon their sight like a contemporary Robinson... "how many wild ideas were found every moment in my fancy, and what strange unaccountable whimsies came into my thoughts by the way". The present paper is an attempt to illustrate the concept of bonding orbitals in the context of inner-shell ionisation. It is also a token of gratitude.

FORCE THEOREMS AND THE CONCEPT OF BONDING

It is clear that, insofar the ground state, the insight of Mulliken was confirmed by accurate calculations. The accurate *ab initio* work of the sixties[4] showed indeed that the lowest molecular orbitals did not experience very appreciable changes upon changes of bond-lengths, i.e., the intuitive description based on a MOLCAO picture with small distortion of the atomic orbitals is valid.

An important tool for the quantification of the concept is the generalized Hellman-Feynman theorem[5], discussed by Löwdin[6] and specifically for the present problem by Hurley[7]. The basic result is:

$$\frac{dE}{dR} = \langle\Psi_{el}(R)|\frac{\partial H(R)}{\partial R}|\Psi_{el}(R)\rangle \tag{1}$$

where it is assumed that H(R) is a hamiltonian depending parametrically on the internuclear distance R, E(R) and Ψ(R) are the electronic energy and wavefunction respectively, both dependent parametrically on R. Equation (1) is valid for exact solutions of the Schrödinger equation as well as for certain variational approximations[6,7], in particular for restricted Hartree-Fock (RHF) calculations. As a consequence equation (1) can be expressed in the form:

$$\frac{dE}{dR} = Z_B\{\int\rho(\vec{r},\vec{r})\frac{\cos\Theta_B}{|\vec{r}_B|^2}\,d\vec{r} - \frac{Z_A}{R^2}\} \tag{2}$$

where Z_A and Z_B are the nuclear charges, $\vec{r}_B = \vec{r} - \vec{R}_B$, Θ_B is the angle between $\vec{r}_B$ and the internuclear axis (the origin being at nucleus A). $\rho(\vec{r},\vec{r})$ is the spinless first-order reduced density matrix[8]. Equation (2) was written by Bader and Jones in the form[9]:

$$\frac{R^2}{Z_B}\cdot\frac{dE(R)}{dR} = -Z_A + \sum_{shells} f_i(R) \tag{3}$$

where

$$f_i(R) = n_i R^2\langle\phi_i(\vec{r};\vec{R})|\frac{\cos\Theta_B}{|\vec{r}_B|^2}|\phi_i(\vec{r};\vec{R})\rangle \tag{4}$$

is the contribution from the i-th shell to the force balance. n_i is the occupancy of the shell. At equilibrium one has

$$\sum_{\text{shells}} f_i(R) = Z_A \tag{5}$$

and since at infinity $f_i(R) \sim 1$ it was natural[10] to assign bonding (B), non-bonding (NB) or antibonding (AB) character to an orbital according to the value of f_i: $f_i > 1$, $f_i \sim 1$ and $f_i < 1$.

It should be emphasized that this characterisation refers to the contribution of each orbital to the electronic force counteracting the internuclear repulsion. It is not simply related to what happens during ionisation. Hence the characterisation of the $1\sigma_g$ orbitals in B_2 (NB), N_2 (B), O_2 (B), F_2 (B)[10] and the 1σ orbitals in C^*O (NB), CO^* (B), B^*F (NB), BF^* (B)[11] is not relevant for analysis of an ionisation process even though it is a pertinent one for our understanding of the ground state.

The Hellman-Feynman theorem is not valid for the ionic state unless one performs a separate variational optimisation, in what is known as ΔE_{SCF} calculation. This leads to rather accurate results for description of inner-shell ionisation[12] but to a set of f_i^* for the upper-state, thus to a loss of a simple orbital description.

KOOPMANS' THEOREM, GENERALISATIONS AND SLOPE OF UPPER STATE

If one abandons the conventional force theorems and uses instead Koopmans' theorem[13], thus deliberately ignoring relaxation effects one can write:

$$E_i^*(R) - E(R) = -\epsilon_i(R) \tag{6}$$

where $E_i(R)$ is the electronic energy of the ion calculated with orbitals for the ground state. $-\epsilon_i$ denotes the orbital energy associated to the ionisation. If $\bar{R}$ denotes the equilibrium distance of the ground state one can conclude that

$$\left.\frac{dE_i^*}{dR}\right|_{\bar{R}} = -\left.\frac{d\epsilon_i}{dR}\right|_{\bar{R}} \tag{7}$$

The slope of the upper curve at the equilibrium distance of the ground state is simply given: minus the slope of the orbital eigenvalue of the ground state, i.e., if the slope of $\epsilon_i(R)$ is positive (negative) the upper state will have a longer (shorter) equilibrium distance. One cannot ignore relaxation effects in a deep-shell ionisation. This implies that instead of equation (6) one has to consider instead:

$$\Delta E_{SCF}^{(i)}(R) = -\epsilon_i(R) - E_{Rel}^{(i)}(R) \qquad (8)$$

where $E_{Rel}^{(i)}(R)$, is the relaxation correction. It has been shown that (8) can be approximated to good accuracy by the eigenvalues ϵ_i^T of a transition operator[14]:

$$\Delta E_{SCF}^{(i)}(R) \sim -\epsilon_i^T(R) \qquad (9)$$

The relevant transition operator appeared first in Löwdin's 1955 paper[8] ($h_i^T = \frac{1}{2}(h^N + h_i^{N-1})$ where h^N and h_i^{N-1} are the Hartree-Fock operators of ground and i-th ionized state respectively). Whereas in Löwdin's treatment h_i^T was constructed with orbitals optimized for the ground state it has been convenient to construct it self-consistently from "democratic" trace optimisation[14]. As a consequence, instead of (7) one obtains for the slope of the ionized state:

$$\frac{dE_i}{dR}(SCF)\Big|_{\bar{R}} = -\frac{d\epsilon_i}{dR}\Big|_{\bar{R}} - \frac{dE_{Rel}^{(i)}}{dR}\Big|_{\bar{R}} \qquad (10a)$$

$$\sim -\frac{d\epsilon_i^T}{dR}\Big|_{\bar{R}} \qquad (10b)$$

It is concluded that the slope of the ionic curve is determined by (7) if one ignores relaxation and by (10) if one includes it. One can also attach significance to a specific orbital contribution. Comparison of (2), (3), (7) and (10) indicates that the ground state description based on $f_i(R)$, or rather on $Z_B f_i(R)/(n_i R^2)$ includes only the nuclear attraction part of $-d\epsilon_i/dR$.

In the present treatment it is assumed that correlation corrections are of less importance. Their inclusion lies beyond the simple orbital picture.

EVALUATION

The slopes of $\epsilon_i(R)$ for core orbitals is schematically depicted to be zero or slightly positive in the early papers [1-3]. It is nevertheless systematically negative[15] when one analyzes in detail the results obtained with accurate RHF calculations[4,16]. This would imply that when relaxation is disregarded, deep-hole ionisation leads inevitably to a shortening of the bond-length. The slope of $\epsilon_i^T(R)$ is not available yet. On the other hand, calculations by Clark and Muller[17], as well as a simple theoretical analysis[15], leads to negative values of $dE^{(i)}_{Rel}/dR$ thus counteracting the previous result.

In Table 1 preliminary values for N_2 and CO are given. It is seen that rather different behaviour is encountered in the two ionisation sites of CO. To ignore relaxation is not possible since it leads to a lengthening of the bond for an Oxygen hole in CO, confirming results of other authors[17].

In conclusion it might be said that the notion of bonding and/or antibonding orbitals can be given a precise formal and computational sense. The criterion which is relevant when analysing the ground state-orbital contribution to the force via the Hellman-Feynman theorem as in equations (2) - (5) has to be replaced by minus the derivative of the appropriate orbital energy, as in (7) and (10) when one "breaks" or removes an inner shell orbital. In this latter sense a "bonding orbital" has positive derivative $d\epsilon^T_i/dR$. The concept of orbital itself is process dependent, optimal ground state orbitals differ from those adequate for describing an ionisation except at the lowest order of approximation.

Interpretation of the present results, more accurate calculations of $d\epsilon^T_i/dR$ as well as applications to the line shapes of deep-hole ESCA peaks are in progress[15].

TABLE 1

Molecule	$-\frac{d\epsilon}{dR}\Big\|_{\bar{R}}$	$-\frac{dE_{Rel}}{dR}\Big\|_{\bar{R}}$	$-\frac{dE^*}{dR}\Big\|_{\bar{R}}$
C^*O	6.53[a] (AB)	-0.26[c]	6.27 (AB)
CO^*	0.68[a] (AB)	-2.40[c]	-1.72 (B)
NN^*	5.33[b] (AB)	-1.26[c]	4.07 (AB)

Slopes in eV/bohr (AB leads to shortening and B to lengthening of the bond upon ionisation).

a) Computed from data given in Ref. 4.
b) Computed from data given in Ref. 16.
c) Computed from data given in Ref. 17.

REFERENÇES

1. F. Hund, Z. Phys. 63, 731 (1930);
G. Herzberg, Z. Phys. 57, 616 (1929);
R.S. Mulliken, Rev. Modern Phys. 4, 1 (1932).

2. R.S. Mulliken, Chem. Reviews 9, 347 (1931).

3. K. Siegbahn, Conference Proceedings, Namur 1974, J. Electron Spectroscopy 5, 3 (1974) (for a review). See also U. Gelius, S. Svensson, H. Siegbahn, E. Basilier, A. Faxälv and K. Siegbahn, Chem. Phys. Letters 28, 1 (1974).

4. A.D. McLean and M. Yoshimine, IBM J. Res. Dev. 12, 206 (1968). References and Tables therein.

5. H. Hellman, "Einführung in der Quanten Chemie". Franz Deuticke, Leipzig and Vienna (1937);
R.P. Feynman, Phys. Rev. 56, 340 (1939).

6. P.O. Löwdin, J. Mol. Spectroscopy 3, 46 (1959).

7. A.C. Hurley in "Molecular Orbitals in Chemistry, Physics and Biology". Edited by P.O. Löwdin and B. Pullman, Academic Press, New York (1964).

8. P.O. Löwdin, Phys. Rev. 97, 1474; 1490 (1955).

9. R.F.W. Bader and G.A. Jones, Can. J. Chem. 39, 1253 (1961).

10. R.F.W. Bader, W.H. Henneker and P.E. Cade, J. Chem. Phys. 46, 3341 (1967).

11. R.F.W. Bader and A.D. Bandraux, J. Chem. Phys. 49, 1653 (1968).

12. P.S. Bagus, Phys. Rev. 139, A619 (1965);
W. Meyer, J. Chem. Phys. 58, 1017 (1973).

13. T.A. Koopmans, Physica 1, 104 (1933).

14. O. Goscinski, B.T. Pickup and G. Purvis, Chem. Phys. Letters 22, 164 (1973);
O. Goscinski, G. Howat and T. Åberg, J. Phys. B 8, 11 (1975);
O. Goscinski, M. Hehenberger, B. Roos and P. Siegbahn, Chem. Phys. Letters 33, 427 (1975);
O. Goscinski, Int. J. Quantum Chem. Symp. Vol. (1975).

15. O. Goscinski and A. Palma, to be published.

16. P.E. Cade, K.D. Sales and A.C. Wahl, J. Chem. Phys. 44, 1973 (1966).

17. D.T. Clark and J. Muller, private communication. See also D.T. Clark, I.W. Scaulau and J. Muller, Theor. Chim. Acta 35, 341 (1974).

A NEW FORMULATION OF THE CORRELATION PROBLEM

G. G. Hall

Department of Mathematics, University of Nottingham

University Park, Nottingham NG 7, 2RD, ENGLAND

1. INTRODUCTION

Ever since Löwdin /1/ in 1959 wrote his monumental survey of the correlation problem it has been recognized that this problem is one of the central issues in many-electron quantum mechanics. A clear insight into the problem is needed not only when wavefunctions of high accuracy are calculated but also when assessing the possibility of using a relatively poor wavefunction to predict experimental quantities. Although the form of the wavefunction changes greatly in its details from the He atom through molecules to solids and plasma there is a common problem of understanding how the electrons are correlated and how this correlation can be incorporated into the wavefunction.

The two most utilized approaches to the correlation probelm rely on the variation principle and on perturbation theory respectively. In the configuration interaction approach the trial wavefunction is expressed as a linear sum of determinants and the calculation becomes a tedious but understandable problem in optimizing this energy. It is well known that the marginal contribution of the last determinant to the wavefunction falls away rapidly so that the convergence of the expansion is slow. While the leading terms do provide some insight into the problem the marginal insight becomes zero very rapidly. On the other hand, the approach through perturbation theory (several forms are described in Sinanoğlu and Brueckner /2/) gives much more insight into the pair structures in the wavefunction and enables the problem to be divided into smaller sub-problems for computational purposes. While the lower orders of perturbation give tolerable equations there is a problem over higher orders since, at least in the form of unlinked clusters, these are important to the wavefunction but their calculation becomes difficult because of the dependence of the equations on accurate solutions to the lower order equations. There is, therefore, a clear need for another approach to the correlation problem which provides the interpretive insight of the pertubation approach without its infinite set of coupled equations.

Another approach to the correlation problem can be traced back to a calculation by Hylleraas /3/ on He. His wavefunction was a product of an orbital part and a correlation factor in which the interelectronic distance r_{12} appeared explicitly. The result was a compact but accurate wavefunction. The extension of this idea to systems with more than two electrons has often been proposed but abandoned because the application of the variation principle depends on the evaluation of many-electron integrals of great complexity. A significant advance in using this form of wavefunction was made in 1969 by Boys and Handy /4/ when they proposed their trans-correlated system of equations for the wavefunction. They discarded the variational approach in favour of a weaker but more flexible bi-variational approach /5/. Despite encouraging results the method has attracted little support possibly because of doubts about the validity of the equations.

In this paper a new formulation of the correlation factor approach is given. The wavefunction is factorized exactly into an orbital part and a correlation factor and the equations determining each are derived in §2. In §3 iterative methods of solving the equations are discussed and in §4 comparisons are made between this approach and earlier ones. The connections with the Sinanoğlu theory and that of Boys and Handy are sufficiently close to demonstrate the potential of the theory in practical calculations.

2. DERIVATION OF THE EQUATIONS

The Schrödinger equation for a many-electron system in the Born-Oppenheimer approximation is

$$H\psi = W\psi, \tag{1}$$

where ψ is the electronic wavefunction and H is the sum of kinetic energy operators for all the electrons together with the classical potential energy. The eigenfunction ψ is now decomposed into the product form

$$\psi = e^{L}\varphi, \tag{2}$$

where φ is a simple determinant of spin orbitals and L is a <u>function</u> to be found. It will be assumed in this paper that L is a real function though in magnetic problems a complex generalization is appropriate. In accordance with the Pauli principle, L will be symmetric in the electrons. Since two unknown functions L,φ cannot be determined uniquely from a single equation it is possible to impose additional conditions on them. This takes the form of the restriction on φ described below.

The equation (1) becomes

$$He^L\varphi = We^L\varphi \tag{3}$$

and, since e^L is non-singular, this takes the transcorrelated form

$$e^{-L}He^L\varphi \equiv (H+[HL]+\tfrac{1}{2}[[HL]L])\varphi = W\varphi, \tag{4}$$

where the commutator expansion terminates after two terms because the kinetic energy operator involves second derivatives. It should be noted that, since L is Hermitian, [HL] is anti-Hermitian and [[HL]L] Hermitian so that the division into Hermitian and anti-Hermitian parts /6/ is immediate.

The orbital factor φ is defined, as a determinant of spin orbitals

$$\varphi = \mathrm{Det}\{\omega_1(1),\ldots,\omega_n(n)\} \tag{5}$$

by minimizing the functional derived from (4)

$$\varepsilon(\varphi) = \langle\varphi(H+\tfrac{1}{2}[[HL]L])\varphi\rangle/\langle\varphi,\varphi\rangle. \tag{6}$$

In this variation L is treated as a fixed function. ε contains a term, additional to the usual functional, which can be written as

$$\tfrac{1}{2}[[HL]L] = -\tfrac{1}{2}\sum_i(\nabla_i L)^2 \tag{7}$$

so that it is clearly negative definite and consequently, if L is not correctly chosen ε may fall below the exact eigenvalue W. The optimum orbitals satisfy one-electron equations so that if H_o is defines as the sum of the effective one-electron Hamiltonians φ is its eigenfunction with an eigenvalue w_o, which is the sum of the one-electron eigenvalues,

$$H_o\varphi = w_o\varphi. \tag{8}$$

It is assumed in this paper that w_o is non-degenerate and this will be so for a closed shell system. An error potential V is then defined as

$$V = H + \tfrac{1}{2}[[HL]L] - H_o. \tag{9}$$

The kinetic energy terms in H and H_o do cancel so that V can legitimately be regarded as a potential though there are non-local exchange terms in H_o.

With this definition of V, equation (4) becomes

$$(H_o + V + [HL])\varphi = W\varphi \tag{10}$$

or, with $W = w_o + w_1$,

$$[HL]\varphi = (w_1 - V)\varphi, \qquad w_1 = \langle\varphi V\varphi\rangle / \langle\varphi,\varphi\rangle. \tag{11}$$

Since the commutator is written as

$$[HL] = -\sum_i (\tfrac{1}{2}\nabla_i^2 L + \nabla_i L\cdot\nabla_i) \tag{12}$$

this is an inhomogeneous partial differential equation for L. Since ψ must be normalizable there are constraints on L. In particular for large values of the variables it cannot increase more rapidly than linearly so that $\nabla_i L$ is bounded. This becomes a boundary condition on the solution. Equation (11) is identical in form to the first-order perturbation equation used in the Dalgarno-Lewis theory /7,8/, where V is the perturbation and $L\varphi$ the perturbed wavefunction. To fix L uniquely the additional condition

$$\langle\varphi L\varphi\rangle = 0 \tag{13}$$

is imposed.

It is important to note that equation (11) is <u>not</u> an operator equation but an equation identifying the images of two different operators acting on a single given function φ. Since [HL] is anti-Hermitian and $(w_1 - V)$ Hermitian this equation cannot be generalized into an operator equation without changing the nature of L. One interpretation of these equations is that they embody a form of infinite-order perturbation theory /9/. The initial function is φ with $L\varphi$ the first perturbation and $e^L\varphi$ is the compact form of the solution with the unlinked cluster terms arising from L included to all orders. The exact energy takes the form of the energy up to second order except that φ is not fixed but itself depends on L. The correspondence with Löwdin's operator formalism is not exact since ψ is not here decomposed into two orthogonal parts.

The basic equations of this theory are (6), whose optimization defines φ and hence V, and (11) which, with (13), difines L. For a solution, both equations have to be satisfied and the value of E is the exact eigenvalue W. Practical procedures for solving these equations are discussed in the following section.

From these basic equations a number of additional relations can be deduced which are of importance in choosing approximate forms for L. The set of determinants $\{\Phi'_r\}$ is defined as those determinants which can be obtained from φ by substituting for one of the occupied orbitals a virtual orbital satisfying the same Hartree-Fock equations. These single replacements are also eigenfunctions of H_0 but have eigenvalues different from w_0 so that

$$<\Phi'_r, H_0\varphi> = 0. \tag{14}$$

According to the Brillouin theorem, when ω_r are the optimum orbitals,

$$<\Phi'_r, (H+\tfrac{1}{2}[[HL]L])\varphi> = 0. \tag{15}$$

It follows that

$$<\Phi'_r, V\varphi> = 0 \tag{16}$$

and hence using (11) that

$$<\Phi'_r, [HL]\varphi> = 0. \tag{17}$$

Since (17) is satisfied for all members of the set this equation is equivalent to a strong-orthogonality condition of the form

$$<\varphi, [HL]\varphi>_{n-1} = 0 \tag{18}$$

where the integration extends over the coordinates of any (n-1) electrons.

3. METHODS OF SOLUTION

A. Variational Aspects

The orbital factor φ can be calculated by minimizing $E(\varphi)$ for fixed L. Because of the additional term E is not bounded below by the exact W. On the other hand for L to be acceptable in the correlation factor $\nabla_i L$ must be bounded so that this term is bounded above and below. There is thus a lower bound to E but it depends on L and could be below W. Some insight into these bounds can be gained by examing separately the two terms in W.

$$w = <\varphi H\varphi>/<\varphi,\varphi>; \quad w_2 = <\varphi\tfrac{1}{2}[[HL]L]\varphi>/<\varphi,\varphi>. \tag{19}$$

The lower bound to w will be the Hartree-Fock energy though this will not be attained even for the exact solution since φ is defines by optimizing E whereas φ_{HF} optimizes w. The major advantage of formulating the calculation of φ as an optimization of E is that the calculated energy will depend quadratically on the errors in φ so that stable and accurate energies can be calculated.

The calculation of L poses greater problems. It can be estimated by optimizing the functional

$$F(L) = \langle\varphi(L(V-w_1) + (V-w_1)L - \tfrac{1}{2}[[HL]L])\varphi\rangle, \qquad (20)$$

where L is varied and φ is fixed, since this is equivalent to solving (11). The maximum value of F is just the corresponding w_2. This functional is quadratic in the errors in L so that it gives a good estimate of w_2 which can be compared to the direct estimate given in (19).

The calculation of the solution thus proceeds by using the functionals E and F alternately to calculate and refine both φ and L till a consistent solution is found. The energy is $w + w_2$ and the progress of the iterative can be monitored by examining the behaviour of each term.

B. A Perturbation expansion

Some aspects of the solution of these equations become clearer when the process is examined by perturbation theory. Since φ becomes an exact solution for V = 0, the perturbation treatment starts by substituting εV for V and expanding in powers of ε. From (11) the leading term in L will be of order ε and the correction to the orbital Hamiltonian is of order ε^2. Thus the zero-order solution to the equation will be

$$\varphi_o = \varphi_{HF}: L_o = 0: W = W_{HF} \qquad (21)$$

The first order equation for L_1 now becomes

$$[H,L_1]\varphi_{HF} = (w_1-V_{HF})\varphi_{HF} \qquad (22)$$

where V_{HF} is the error potential of the Hartree-Fock solution and so consists of the interelectronic potentials less the Coulomb and exchange one-electron potentials. Since this is a linear inhomogeneous equation its solution can be expressed as the sum of terms, one for each of the terms in V_{HF}. Thus L_1 is a sum of one-electron and two-electron parts. The situation closely resembles that arising in the Sinanoglu perturbation theory /10/ except that (22) is replaced by the more complicated equation

$$[H_o,L_1]\varphi_{HF} = (w_1-V_{HF})\varphi_{HF}. \qquad (23)$$

Since the correction to the orbital equations has order ε^2 there is no first-roder correction to φ_{HF} and the total energy, correct to ε^2, will be

$$W = w_o + \varepsilon w_1 + \varepsilon^2(\tfrac{1}{2}\langle\varphi_{HF}[[HL_1]L_1]\varphi_{HF}\rangle/\langle\varphi_{HF}\varphi_{HF}\rangle), \qquad (24)$$

where $w_o + \varepsilon w_1 = W_{HF}$.

Since V depends on the orbital equations its first corrections will not be the next order in ε but in ε^2 i.e.

$$V = V_{HF} + \varepsilon^2 V_3 \qquad (25)$$

Similarly, the correction to L will be of the form

$$L = L_1 + \varepsilon^2 L_3, \qquad (26)$$

where

$$[HL_3]\varphi_{HF} = (w_{13} - V_3)\varphi_{HF} - [HL_1]\varphi_2 + (w_{11} - V_{HF})\varphi_2 \qquad (27)$$

$$w_1 = w_{11} + \varepsilon^2 w_{13} \qquad (28)$$

and φ_2 is the correction to φ_{HF} due to the additional term in the Hamiltonian. It is clear from this that φ, V, L become expansions in ε^2. In the energy the even powers of ε arise from w_o, the sum of the orbital energies, and the odd powers from w_1, the mean value of the error potential. The nature of these expansions suggests strongly that this process would converge rapidly.

Another feature of the equations which emerges from this analysis concerns the nature of L. Since V_{HF} has one and two-electron terms so has L_1. This means that the next orbital equation, which depends on $(\nabla_i L_1)^2$, will have three-electron terms and this will be true of V_3 and L_3. As the alternation between the equations continues so L acquires multi-electron terms. Thus, although L will be dominated by the leading term which is a sum of pairs, it does contain all higher order terms.

C. Expansion in a Finite Basis

The possibility of solving the equations for amolecule will depend on finding an adaptation of the equations to a finite basis set such that reasonable accuracy can be obtained from the computational investment.

The expansion of molecular orbitals as a linear combination of fixed basis functions gives rise to a set of equations which are cubic in the coefficients /11/ and have to be solved iteratively. When L is expressed with pair terms only, the matrix form of the orbital equations becomes identical with those discussed by Hall and Solomon /12/ in the context of the transcorrelated wavefunction. Thus the spin orbitals in φ are expanded in terms of a basic set $\{\eta_u : u = 1, \ldots, m \geq n\}$, where η_u are functions of space and spin,

$$\omega_r = \sum_u X^u{}_r \eta_u \tag{29}$$

and L is a sum of pairs

$$L = \tfrac{1}{2}\Sigma'_{ij}\, G_{ij} \tag{30}$$

so that the additional term in (6) is

$$\Sigma'_{ijk}\, F_{ijk}: \; F_{ijk} = -\tfrac{1}{2}\nabla_i G_{ij}\cdot\nabla_i G_{ik}. \tag{31}$$

The equations for $X^u{}_r$ take the form

$$\sum_v (f_{uv}-\mu S_{uv})X^v{}_r = 0: \; S_{uv} = \langle uv\rangle \equiv \langle \eta_u, \eta_v\rangle, \tag{32}$$

where μ is the orbital energy and the effective Hamiltonian is

$$f_{uv} = \langle uhv\rangle + \sum_{wx} P^{wx} \langle uw(\frac{1}{r_{12}} + 2F_{122})(ux-xv)\rangle +$$

$$\sum_{wxyz} 3\, P^{wx}P^{yz}\langle uwyF_{123}(vxz-vzx-xvz-zxv+zvx+xzv)\rangle \tag{33}$$

with h the sum of the one-electron terms. These equations are quintic in the coefficients but can still be solved iteratively. The bond order matrix is defined as

$$P^{uv} = \sum_r^n (X^u_r)^* X^v_r. \tag{34}$$

Several approaches are possible to the calculation of L. Since the slow convergence of the configuration interaction calculations is due to the large number of one-electron functions needed to build up a suitable function to describe the pair behaviour, it is natural to reject any attempt at representing L in terms of orbitals. A more useful approach is to use a set of two-electron functions and express

$$G_{ij} = \sum_\ell D_\ell G^\ell(\underline{r}_i, \underline{r}_j) \tag{35}$$

where D_ℓ are linear parameters. The optimal values D_ℓ are determined, using (20), by

$$\sum_m M^{\ell m} D_m + N^\ell = 0 \tag{36}$$

with

$$M^{\ell m} = \Sigma'_{ijk} \langle\varphi(\nabla_i G^\ell_{ij}\cdot\nabla_i G^m_{ik})\varphi\rangle + \Sigma'_{ij}\langle\varphi(\nabla_i G^\ell_{ij}\cdot\nabla_i G^m_{ij})\varphi\rangle \tag{37}$$

$$N^{\ell} = \sum_{ij}{}' <\varphi(G^{\ell}_{ij}(V-w_1) + (V-w_1)G^{\ell}_{ij})\varphi>. \qquad (38)$$

These are linear equations in D_m and easily solved once the integrals have been evaluated. One possible function for use in this context might be

$$G^1_{ij} = r_{ij} \qquad (39)$$

since this gives ψ the form expected from studies of the wavefunction at $r_{ij} = 0$. Boys and Handy /4/ use functions such as

$$G^2_{ij} = r_{ij}/(1+r_{ij}) \qquad (40)$$

since this produces a cusp but remains bounded everywhere. The advantage of supplementing these two-electron functions with sums of one-electron functions so that the strong orthogonality conditions (18) are satisfied has also been noted by Boys and Handy.

4. COMPARISON WITH OTHER THEORIES

As the discussion in the previous section has shown, the first stages in this solution as a perturbation in the potential are very similar to the first stages in the Sinanoğlu perturbation theory only that (23) replaces (11). The subsequent stages are quite different. In this theory there are just two equations and the solution for ψ and L proceeds till consistency is attained although L acquires multielectron character in the process. In the Sinanoğlu perturbation theory the first correction to the Hartree-Fock function always remains a sum of pair terms and the next correction would require the solution of the second order perturbation equation. The wavefunction, in the present notation, takes the form

$$\psi = (1+L_1+\tfrac{1}{2}L^2_1)\varphi_{HF} \qquad (41)$$

where the term in L^2_1 is an estimate of the second order term arising from the second order unlinked clusters. In the present theory the use of the exponential form of correlation factor ensures that these cluster terms are included to all orders.

The Boys-Handy theory is essentially an attempt to solve the transcorrelated equation (4). Their orbitals are determined using the converse of Brillouin's Theorem and adjusting the orbitals until

$$<\phi'_r, e^{-L}He^{L}\varphi> = 0. \qquad (42)$$

Since the full transcorrelated Hamiltonian is not Hermitian this is a more complicated process than in this theory. In practice the difference in orbital definitions appears to have little numerical significance. The redundance equations which they consider next are given in several forms but can all be regarded as equivalent to the orthogonality condition

$$\langle\phi_r'[HL]\varphi\rangle = 0 \tag{43}$$

for some restricted set of one-electron substituting functions. The correlation equations are sometimes written as

$$\langle\phi_r'', e^{-L}He^{L}\varphi\rangle = 0, \tag{44}$$

where ϕ_r'' is a double replacement of ψ, and sometimes as

$$\langle G_{ij}\varphi, e^{-L}He^{L}\varphi\rangle = 0, \tag{45}$$

which is also a form of double-replacement equation. If φ is now taken as defined by (6) rather than by their orbital equation it is readily seen, using the same argument as before, that (44) is equivalent to

$$\langle\phi_r'', ([HL]+V-w_1)\varphi\rangle = 0. \tag{46}$$

The correlation equation is thus the condition that (11) should be satisfied in the function space of double replacements. Since they do not envisage L as containing functions of more than two electrons this is the most important space. Their energy equation is identical to W above. The Boys-Handy theory can be regarded, therefore, as a form of this theory adapted for pratical calculations. It can now be improved by adopting the more accurate and simpler definition of the orbitals given here and by using the variation principle (20) to determine L instead of taking a least squares solution of the correlation equations (44).

A different mathematical approach to the transcorrelated equations has been given by Armour /6/. He has stressed the fact that the transcorrelated equation can be divided into Hermitian and anti-Hermitian parts and that it can be conveniently solved in a matrix form using a rapidly converging perturbation expansion. He also shows that the truncation errors caused by the finite basis set are balanced to some extent by the errors due to an incorrect L. Although there are many insights in his discussion which impinge directly on the theory above, the emphasis here has been on the use of (11) to eliminate the anti-Hermitian operator and so avoid the problems caused by the choice of a fixed L.

The fact that the Sinanoğlu theory has produced consistently good estimates of correlation energies and that the Boys-Handy theory has given some of the best calculated energies on record suggests very strongly that this theory to which they approximate in slightly different ways should be capable of accurate but compact results.

5. CONCLUSION

The theory presented here reduces the correlation problem for a closed shell system to the solution of two coupled equations. The first is an orbital equation and the second has the form of a first-order perturbation equation. The equations can be solved variationally and both have been well studied in other contexts. Both the Sinanoglu perturbation theory and the Boys-Handy transcorrelated theory can be seen as approximate solutions to the equations and both can be improved and simplified by bringing them closer to these equations.

Although many practial problems of solving these equations have been solved, a number of major difficulties remain. The greatest is that of calculating the multicentre integrals over appropriate functions. Even if L is restricted to a sum of pair functions, there will be threecentre integrals to evaluate and formulae for these are required. The number and variety of these integrals will also cause data-processing problems. While the choice of functions in L is not prescribed in detail it seems difficult to find functions for which the resulting integrals can be evaluated without resorting to numerical integration.

Part of the significance of this theory lies in the fact that by factorizing the wavefunction into an orbital part and a correlation part it prepares for a more physical interpretation of accurate wavefunctions. The orbital part provides orbitals which are close but not identical to the Hartree-Fock orbitals and can, by a unitary transformation, be localised into chemically significant entities /13/. (They do not seem to be identical to the Brueckner orbitals.) At the same time the correlation factor can be interpreted in terms of the behaviour of an assembly of correlated electrons /14/ and related to concepts used in the theory of solids and plasma. Such interpretations applied to individual molecules should prove very illuminating.

ACKNOWLEDGEMENTS

The author wishes to thank Professor P.O. Löwdin who has greatly stimulated his interest in the correlation problem and Dr. E.A.G. Armour whose comments on the theory have been invaluable.

BIBLIOGRAPHY

1. P.O. Löwdin, Adv. in Chem.Phys. 2, 207 (1959)
2. O. Sinanoğlu and K.A. Brueckner, Three Approaches to Electron Correlation in Atoms (Yale, New Haven, 1970)
3. E.A. Hylleraas, Z.Physik 54, 347 (1929)
4. S.F. Boys and N.C. Handy, Proc. Roy.Soc. A309, 209 (1969), A310, 43, 63 (1969), A311, 309 (1969)
5. S.F. Boys, Proc. Roy.Soc. A309, 195 (1969)
6. E.A.G. Armour, Mol.Phys. 24, 163, 181 (1972)
7. A. Dalgarno and J.T. Lewis, Proc. Roy. Soc. A233, 70 (1965)
8. C. Schwartz, Ann. of Phys. 6, 156 (1959)
9. P.O. Löwdin, J. Math. Phys. 3, 969 (1962)
10. O. Sinanoglu, Proc. Roy. Soc. A260, 379 (1961)
11. G.G. Hall, Proc. Roy. Soc. A205, 541 (1951)
12. G.G. Hall and C.E. Soloman, Chem. Phys. Letters, 4, 352 (1969)
13. J.E. Lennard-Jones, Proc. Roy. Soc. A198, 1, 14 (1949)
14. N.H. March, W.H. Young and S. Sampanthar, The Many-Body Problem in Quantum Mechanics (CUP, Cambridge, 1967)

THE CHEMICAL BOND AS A MANY-ELECTRON PROBLEM

R. Daudel

Centre de Mecanique Ondulatoire, Appliquée

23, rue du Maroc, Paris 19, France

THE NEED FOR A LANGUAGE ADAPTED FOR VERY ELABORATE WAVE FUNCTIONS:

The orbital model, in which a molecular state is represented by a single determinant built of molecular orbitals, has been proven to be very useful, and associating orbitals with chemical bonds have become customary.

That model has, however, severe limitations. One of the earliest extended basis SCF calculations /1/ showed the Hartree-Fock energy of F_2 to lie more than 1 ev <u>above</u> the energy of two F atoms, leading to the wrong conclusion that the fluorine molecule would be unstable. It is now well-known that very elaborate wave-functions are necessary to properly represent a chemical bond.

Monoelectronic reorganizations themselves are not always conveniently represented by orbitals. Rohmer /2/ has recently shown that there is no relation at all between orbital energies and ionization potentials in ferrocene.

As further very elaborate wave functions are generated by electronic computers, there is a need for a language adapted to those kinds of functions. The simplest procedure for generating such a language is to use only concepts which are defined for any kind of wave function. Such are all the wave-mechanical observables.

DENSITY DIFFERENCE FUNCTION:

The density difference function $\delta(M)$ at point M in a molecule is simply the difference between the actual density $\rho(M)$ and the virtual density $\rho^v(M)$ which would result from the addition of the densities in the free atoms:

$$\delta(M) = \rho(M) - \rho^v(M) \quad /3/$$

Therefore in a point where δ is positive, the bonding has led to an increase of the electronic density. In a point where δ (M) is negative the bonding has led to a decrease of the electronic density. For that reason the δ function is also called the bond density function: is shows the effect of the formation of the bonds on the electronic distribution.

Many authors have calculated that function for various molecules. /4/ Figure 1 /5/ shows typical results of such calculations. For a homonuclear molecule such as N_2 the effect of binding is:

a) an increase of the electronic density in a central region of the molecule.

b) an increase of the electronic density in the "lone-pair" region.

c) a decrease near the nuclei.

For a heteronuclear molecule like LiF we see:

a) an increase of the electronic density in all the fluorine region except for a small volume near the nucleus.

b) a decrease of the density in the lithium region except for a small volume not far from the nucleus. The overall effect amounts to a transfer of charge from the lithium atom region to the fluorine one.

The effect of bonding on the electronic density near the nuclei has been proven experimentally by measuring the radioactive decay of various compounds. The effect of the chemical bond on the rate of decay of some radioactive nuclei has been predicted by Daudel /6/ in the case of electron capture and in the case of isomeric transitions. That prediction was based upon the fact that these radioactive decays depend on the electronic distribution near the radioactive nuclei. The prediction for electron capture has also been made by Segré. /7/

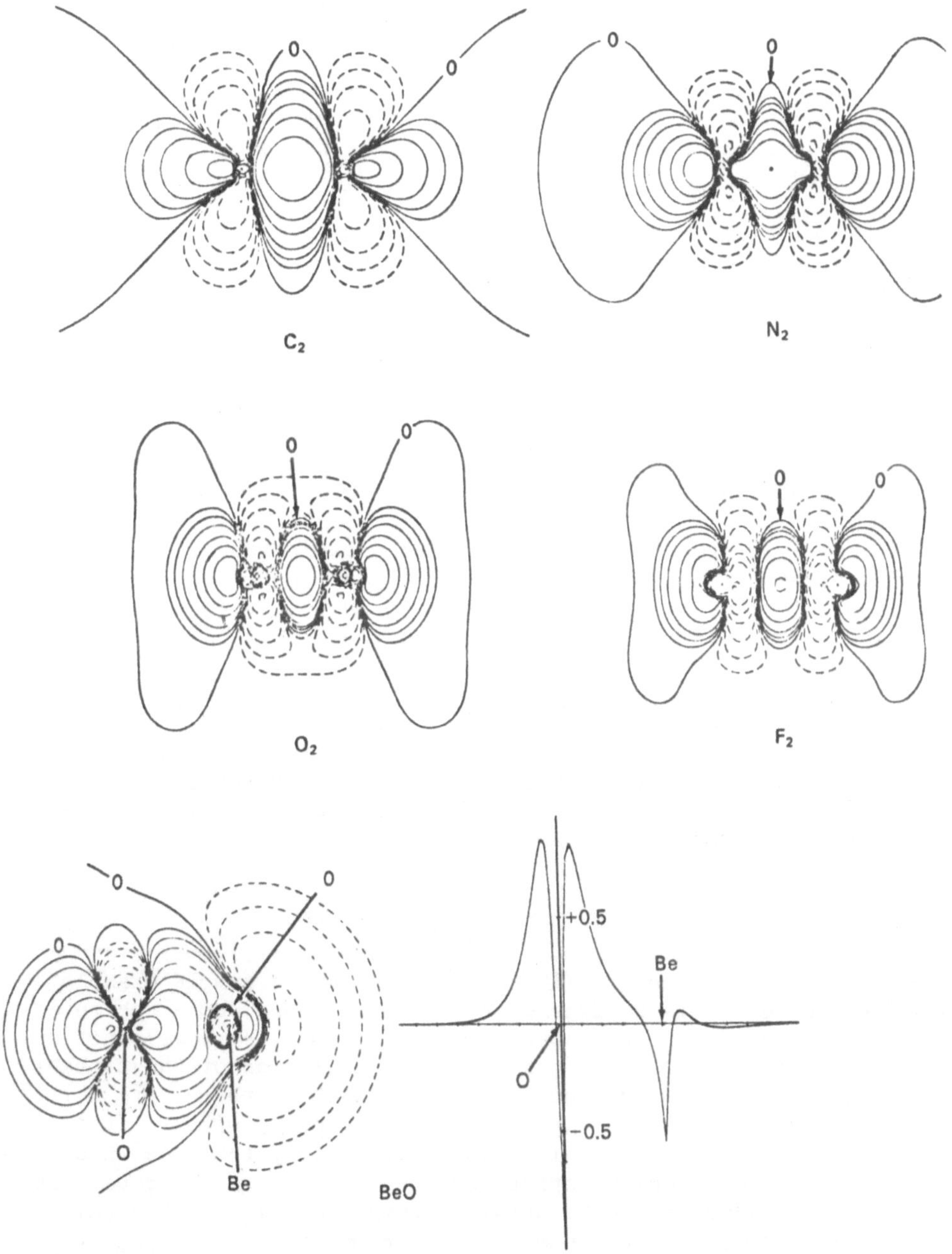

Figure 1. Bond density maps. Solid contours correspond to positive values of the δ function, dashed contours are associated with negative values of that function (reproduced by permission of R.F.W. Bader /5/).

Many experiments have confirmed that prediction. The first one has shown that the decay of ^{7}Be is larger in metallic Be than in Be and BeF_2. The relation between the rate of decay of a radioactive nucleus and the electronic structure of a molecule has been found to be so sensitive that the measurement of decay is now a powerful procedure for chemical analysis of very small amounts of radioactive compounds. /8/

The measurement of the δ function outside of the nucleus regions is difficult but possible nowadays. In molecular crystals it can be achieved by using both X-ray and neutron diffraction.

The comparison of calculated and measured δ functions has led to the conclusion that minimal basis set SCF calculations underestimate the effect of bonding on the electronic distribution in molecules. /9/ To calculate density difference functions one needs at least large basis set SCF functions. It is better to include a certain amount of configuration interaction.

A density difference function can also be introduced to represent the effect of ionization or excitation on the electronic density in a molecule. It is simply the difference between the density $\rho_f(M)$ in the ionized or excited state and the density $\rho_i(M)$ in the initial state:

$$\delta\ (M) = \rho_f(M) - \rho_i(M)$$

In the case of ionization the δ function represents a density hole which results from the departure of an electron and which cannot always be represented by an orbital, as already noted.

LOGE THEORY AND VIRIAL FRAGMENTS

To recall loge theory, /10/ let us consider a simple example: the BH molecule. It is a six-electron problem. Consider a sphere of arbitrary radius r centered at the boron nucleus. The wave function associated with the molecule permits the calculation of the probability P_n of finding n electrons and n only in that sphere. Fig 2 /11/ shows the variation of the various probabilities P_n with r. It is seen that from all possible electronic events only that one which places two electrons in the sphere can reach a very high probability (0.85). This maximum value is obtained when r = 0.7 a.u. For this value of r all other events have a small probability. Finding two electrons in the sphere becomes the <u>leading event</u>.

To obtain the maximum amount of information about the localizability of the electrons of the molecule we are led to minimize the missing information function:

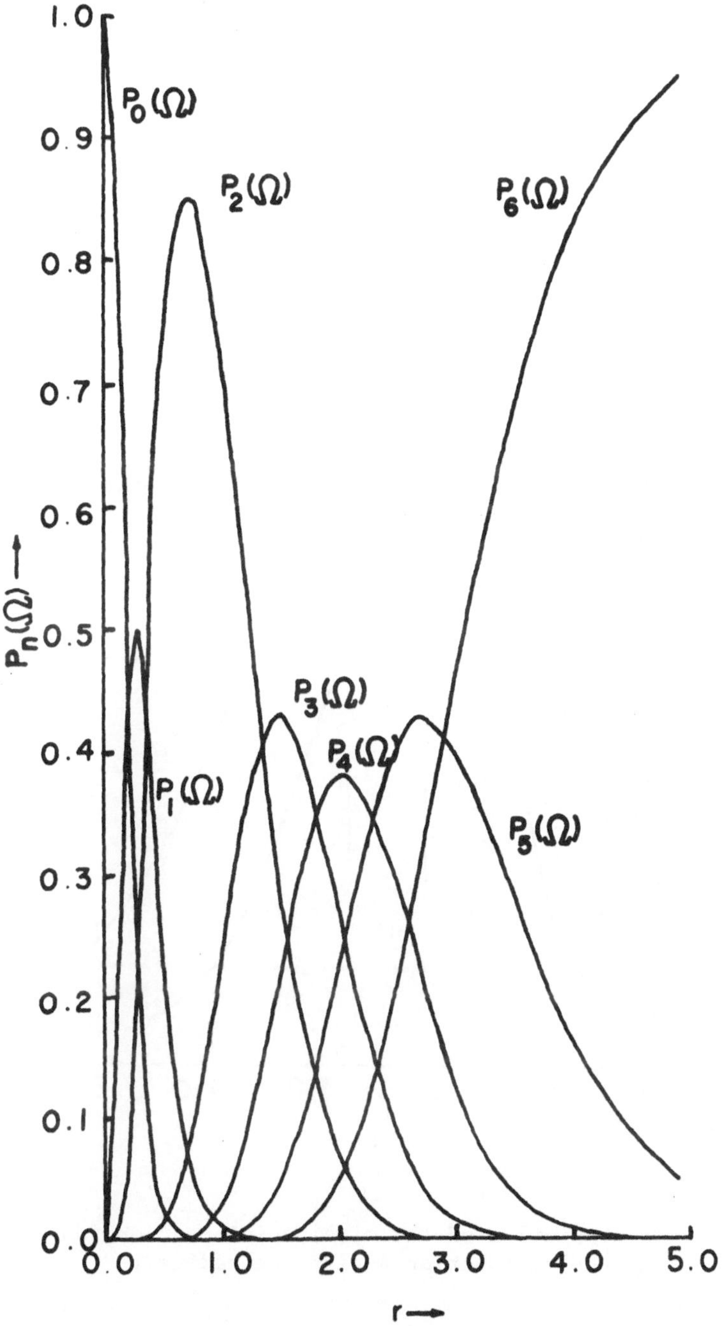

Figure 2a. The variation of the event probabilities P_i with the radius r.

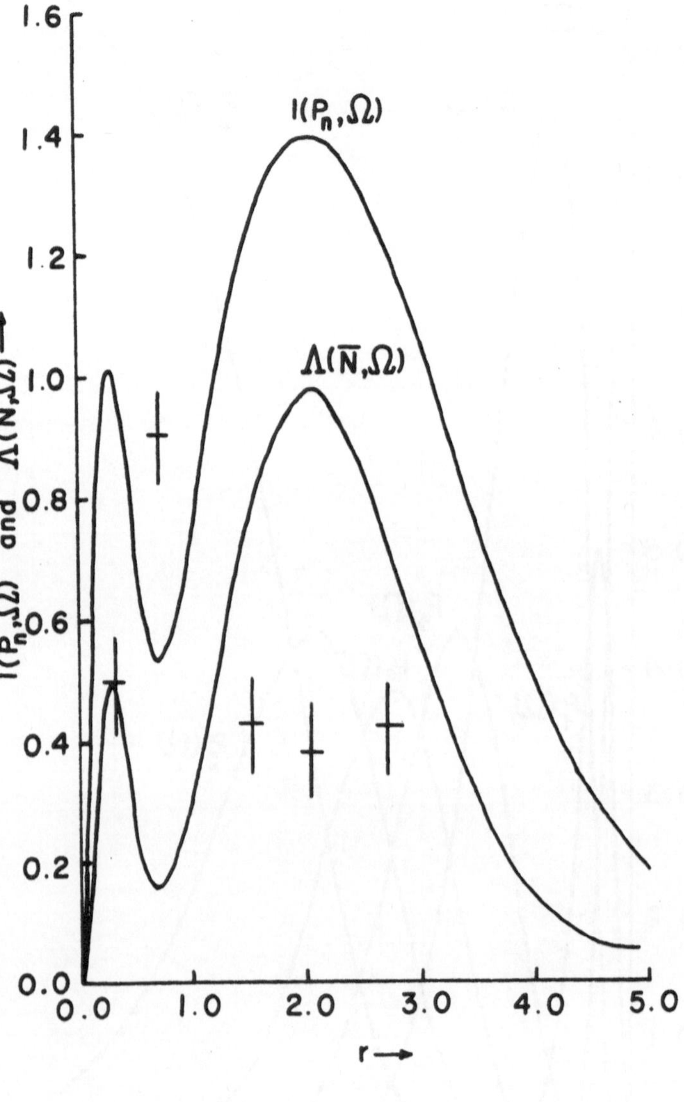

Figure 2b. The variation of the missing information function I and the fluctuation Λ with the radius r.

$$I = \sum_i P_i \log_2 P_i^{-1}$$

associated with the set of probabilities P_0, P_1, P_2.....P_6. The figure shows that the nontrivial minimum value of the function I is reached for r = 0.7 exactly when P_2 reaches its maximum value.

Then, <u>the corresponding sphere is said to generate the best partition in two loges of the molecular space</u>.

On figure 2 is also plotted the variation of Λ, the fluctuation of the number of electrons in the sphere:

$$\Lambda = \overline{N^2} - (\bar{N})^2$$

$\bar{N}$ is the average value of the number of electrons and $\overline{N^2}$ the average value of the square of that number. It is seen that the fluctuation and missing information functions reach their minimum values for the same value of z. <u>Therefore a good loge is a region of the space in which the number of electrons does not fluctuate much</u>.

It has also been demonstrated that for good loges, the correlation between the movement of electrons is maximized in the loges and minimized between the loges. /12/

For a molecule like BH we can also consider partitions in three loges by introducing a cone of angle ϕ (Fig. 3), the BH line being its axis and the boron nucleus its summit. The best partition into three such loges is obtaines when: /11/

$$r = 0.7 \text{ a.u.}$$

$$\phi = 73^{\circ}$$

The leading event places two electrons in each loge. Therefore, it can be said that the sphere corresponds to the atomic core, it is <u>a core loge</u>. It appears that the lone pair loge is much more bulky. More precisely, if an electronic density contour is taken as an arbitrary limit of a loge such that it contains 95 per cent of its electron charge, the ratio of nonbonded to bonded loge is equal to 2.11. <u>This result is in accord with one of the basic postulates of the very powerful Gillespie theory of molecular geometry</u>. /13/

In some cases loges are very analogous to virial fragments /11/; that is to say molecular fragments considered by Bader in which the virial theorem is valid. /14/ Finally, it has been observed /11/ that bond loges are transferable from molecule to molecule; for example the BeH loge in BeH_2 is very similar to the BeH loge in BeH. Therefore, in some cases bond loges have many interesting properties:

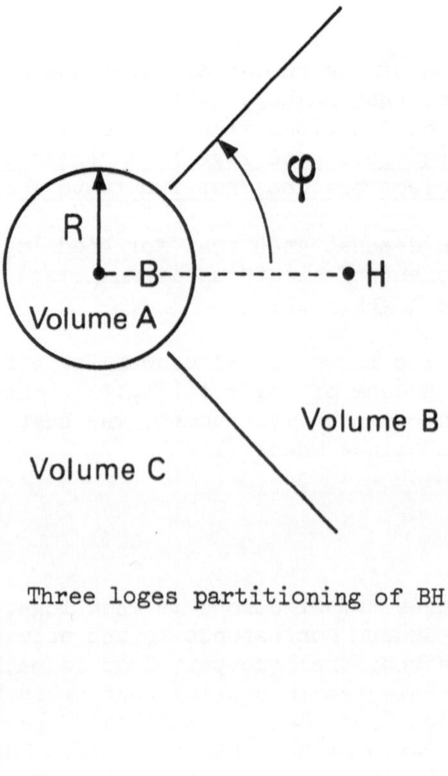

Figure 3. Three loges partitioning of BH molecule.

1) the virial theorem is satisfied

2) the missing information is a minimum

3) the fluctuation of the number of electrons is a minimum

4) properties of transferability appear

5) electronic correlation is maximized inside a loge and minimized between loges.

Associating the loge with the chemical bond is therefore a very attractive way of rationalization of that intuitive concept. Many fruitful applications are possible. Bader and Stephens /15/ have used the loge theory to study the localizability of the electrons in the bonds of hydride XH_p. They found that the localizability decreases when the atomic number of X increases and when the number of lone pairs increases. Furthermore, it has been demonstrated /10/ that any expectation value $\langle\psi|\Omega|\psi\rangle$ can be expressed as a sum of contributions associated with each loge and of contributions associated with each loge pair. This relation is a very good starting point for studying the additive and non-additive properties of molecules.

Finally, general relations of the type

$$p^{\gamma}v = \text{constant}$$

are found /16/ between the volume of a loge v and the mean value p of the electric potential which acts on the electron in that loge.

After this description of the main properties of the loges it is interesting to analyze more deeply their relations with the wave function ψ. /17/ To each electronic event λ it is possible to associate an event function ψ_λ which is identical to the wave function when there is in each loge a number of points equal to the number of electrons which is found in the loge when this event occurs. The function ψ_λ vanishes for all other positions of the points. It is readily seen that

$$\psi = \Sigma_\lambda \psi_\lambda$$

Now each event function can be expressed in terms of completely localized loge functions $L_{i\lambda}$ and a correlation function f_λ. A completely localized loge function for a given event λ is a function depending on a number of points identical to the number of electrons which is found in the loge for that event and which vanishes for other positions of the points. /18/ The event function ψ_λ can be written as

$$\psi_\lambda = a\,\Pi_i\, L_{i\lambda}\, f_\lambda\, \sigma$$

where a is an antisymmetrizer and σ the spin function.

Finally, any wave function can be rigorously written as

$$\psi = \sum_\lambda a\,\Pi_i L_{i\lambda}\, f_\lambda\, \sigma \ldots \tag{1}$$

Another useful but only approximate expression is obtained if the completely localized loge functions are replaced by loge (or group) functions $h_{i\lambda}$ only partially localized in the loges. Such functions allow a certain amount of correlation between the loges. Therefore it is possible to neglect (as an approximation) the correlation functions f_λ. The expression of ψ becomes

$$\psi = \sum_\lambda a\,\Pi_i\, h_{i\lambda}\, \sigma \ldots. \tag{2}$$

Both expressions for ψ suggest new methods of calculating wave functions. This problem has already been analyzed /17/ and we shall not here go into the details. We shall only discuss one new important aspect of that problem.

If there is a leading event ν expression (2) can be written as

$$\psi = a\,\Pi_i\, h_{i\nu}\, \sigma \; + \text{small terms} \ldots. \tag{3}$$

This is the expression introduced a long time ago by the author to represent atomic cores, lone pairs and bonds of a molecule. /19/ Two questions will be answered about this equation.

a) Given a wave-function ψ, is it possible to find the leading term of equation (3) by using a direct mathematical procedure? Constanciel /20/ has answered this question.

Consider the simple case in which

$$\psi^{(n)} = a\,\psi_A^{(p)}\,\psi_B^{(n-p)} \ldots \tag{4}$$

where $\psi^{(i)}$ means a function depending on i points. Let us introduce now $\hat{\rho}_n^{(p)}$ the p-particle reduced-density operator introduced by Löwdin. /21/ Constanciel has demonstrated that the idempotency of the p-particle reduced-density operator with respect to a Hilbert space containing $\psi_A^{(p)}$ is a necessary and sufficient condition for writing $\psi^{(n)}$ as in equation (4). In that case the reduced-density operator is simply a projector on $\psi_A^{(p)}$.

This theorem furnishes a powerful procedure to study the separability of a wave function into group functions. Usually this separability is not complete, we only have

$$\psi^{(n)} \approx a\, \psi_A^{(p)}\, \psi_B^{(n-p)} \ldots \qquad (5)$$

In that case we have to look for an operator as quasi-idempotent as possible.

When a wave-function is written as in equation (4) og (5), the relation with the loge theory is not obvious. The mathematics does not imply that the functions ψ_A and ψ_B are localized or quasi-localized in two non overlapping volumes A and B of the three-dimentional space. A Slater determinant of two molecular orbitals is like an expression (5) and both orbitals can be delocalized over the same molecular space. Therefore the functions ψ_A and ψ_B will correspond to loge functions, only if they are quasi-localized in two non overlapping volumes A and B.

When it is expected that a function given by the expression (4) could be a good representation of a state of the molecule we can use such an expression as a variational function. This amounts to the group-function method developed by Klessinger and McWeeny. /22/ A program for IBM electronic computers has been written which permits the calculation of wave functions of type (4) where the group functions are expanded on a basis set of gaussian functions. /23/ This program includes doing a local configuration interaction in each loge and a generalized self-consistent treatment between the various configuration interactions associated with the loges.

It has been observed /23/ that in a set of molecules like CH_4, C_2H_6, C_3H_8, C_4H_{10}, CH_3NH_2, CH_3OH, H_2O_2 and N_2H_4, group functions associated with a given kind of bond are well transferable from one molecule to another. Furthermore, it has been shown that it is easy to deduce the frontiers of the loges by plotting the density contour lines of the group functions.

Finally, let us add that P. Durand /24/ has used the potential created by atomic core loges to build a very efficient pseudo-potential in order to compute molecular wave functions.

It is obviously possible to use all the methods and concepts developed in the present paper to follow the evolution of the bonds of molecules along the path of a chemical reaction.

Figure 4 /5/ shows the evolution of the density difference function during the formation of a hydrogen molecule from two hydrogen atoms.

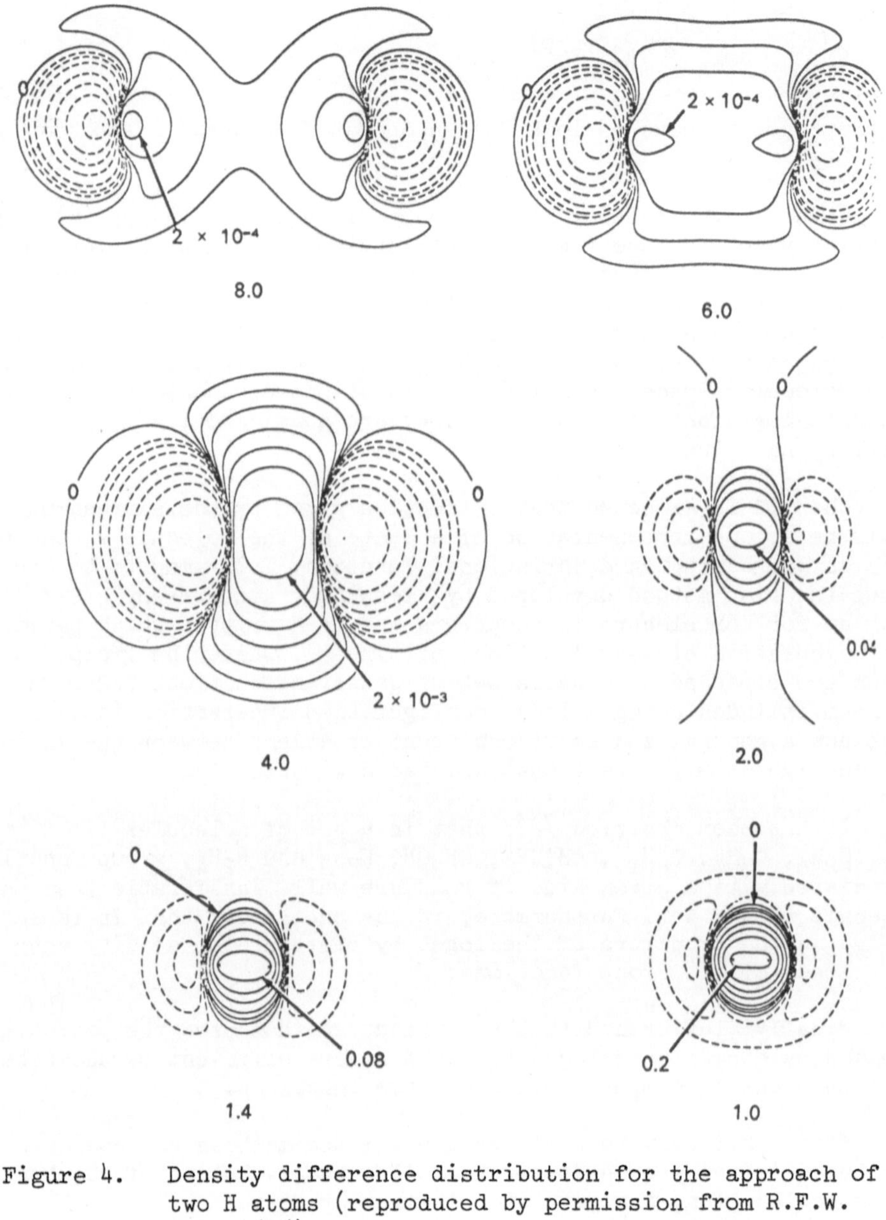

Figure 4. Density difference distribution for the approach of two H atoms (reproduced by permission from R.F.W. Bader /5/).

Burke et al. /25/ have followed the evolution of centroids of localized orbitals (which can be considered as centroids of loges) during the Diels-Alder reaction.

REFERENCES

1. A.C. Wahl, J.Chem.Phys. 41, 2600 (1964)
2. M.M. Rohmer, Thesis, Strasbourg, (1975)
3. R. Daudel, Compt.Rend. 235, 886 (1952)
   M. Roux and R. Daudel, Compt.Rend. 240, 90 (1955)
4. M. Roux, M. Cornille and L. Burnelle, J. Chem.Phys. 37, 1009 (1962)
   B.J. Ransil and J.J. Sinai, J. Chem.Phys. 46, 4050 (1967)
   M.H. Hazelrigg and P. Politzer, J.Am.Chem.Soc. 73, 10009 (1969)
   R.F.W. Bader, W.H. Henneker and P.E. Cade, J.Chem.Phys. 46, 3341 (1967)
   R.F.W. Bader, I. Keaveny and P.E. Cade, J.Chem. Phys. 46, 3381 (1967)
   P.E. Cade, Trans.Am.Crystallogr.Ass. 8, 1 (1972)
5. R.F.W. Bader, An introduction to the electronic structure of atoms and molecules, Clarke, Irwin and Co. Publisher (1970)
6. R. Daudel. Rev.Scientifique 87, 162 (1947)
7. E. Segré, Phys.Rev. 71, 274 (1947)
8. For a recent review see: J.I. Vargas M.T.P. Int.Rev.Sci. Serie 1, 8, 75 (1971)
9. See for example P. Coppens (in press)
10. For more details see:
    C. Aslangul, R. Constanciel, R. Daudel, P. Kottis Adv. Quantum Chem. 4, 93 (1972): Localization and Delocalization in Quantum Chemistry, Reidel (1975)
11. R. Daudel, R.F.W. Bader, M.E. Stephens and D.S. Borrett, Can.J.Chem. 52, 1310 (1972)
12. R.F.W. Bader in Localization and Delocalization in Quantum Chemistry, Reidel (1975) p. 15
13. R.J. Gillespie, Molecular Geometry, Van Nostrand Reinhold Co. (1972)
14. R.F.W. Bader and P.M. Beddall, J.Chem.Phys. 56, 3320 (1972)
    R.F.W. Bader, P.M. Beddall and J.Jr. Peslak, J.Chem. Phys. 58 557 (1973)
15. R.F.W. Bader and M.E. Stephens (in press)
16. Demonstrated by S. Odiot and R. Daudel for atoms; this relation

has been extended to molecules (Csizmadia, Daudel, Goddard, Kozmutza, Mezey, in press)

17. C. Aslangul, R. Constanciel, R. Daudel, L. Esnault and L.V. Ludena, Int. J. Quant. Chem. 8, 499 (1974)

18. E.V. Ludena and V. Amzel, J. Chem. Phys. 52, 5923 (1970)

19. R. Daudel, Les fondements de la Chimie Théorique, Gauthier Villars, Paris, 1956 p. 209

20. C. Constanciel, Phys. Rev. 11, 395 (1975)

21. P.O. Löwdin, Phys. Rev. 97, 1474 (1955)

22. M. Klessinger and McWeeny, J.Chem. Phys. 42, 3343 (1965)

23. M. Sanchez, P.D. Dacre, R. Daudel, S. Kwun, C. Valdemoro and McWeeny (in press)

24. P. Durand, Localization and Delocalization in Quantum Chemistry Volume II, Reidel (in press)

25. L.A. Burke, G. Leroy and M.. Sana (in press)

ORBITAL METHODS AND CORRELATION ERRORS IN EXPECTATION VALUES

S. Larsson and R.E. Brown

Department of Quantum Chemistry, University of Uppsala
Uppsala, Sweden
Instituto de Química, Universidade Estadual de Campinas
Campinas, Brazil

I. NATURAL SPIN ORBITALS

The concept of natural spin orbitals (NSO) is one of the most important contributions by Löwdin to quantum chemistry /1/. NSO are defined as eigenfunctions of the one-particle density matrix /1/ $\gamma(1, 1')$:

$$\gamma(1, 1') = N \int \psi(1, 2,..., N)\psi^*(1', 2,...,N)d2...dN \quad (1)$$

In the eigenfunction expansion

$$\gamma(1, 1') = \sum_i \nu_i \chi_i(1)\chi_i^*(1'), \quad (2)$$

ν_i is interpreted /1/ as the occupation number of the NSO χ_i. A pioneering work on the study of NSO for two-electron systems was carried out by Löwdin and Shull /2-4/. The subsequent theoretical work involved symmetry and convergence properties, whereas the practical work involved calculations on small atoms and molecules. For references we refer to a recent review article by Davidson /5/.

In this paper we will deal with states where a single Slater determinant is a fairly good approximation of the exact wave function. The occupation numbers of the "strongly occupied" orbitals are then depressed below unity but still larger than, say, 0.8. The deviations from unity are usually larger the more correlation terms are included in the wave function.

II. BRUECKNER SPIN ORBITALS

Brueckner spin orbitals (BSO) were first discussed by Nesbet /6/ and Löwdin /7/. The definition of BSO is at first sight straightforward. If the exact wave function ψ is at our disposal, they may simply be defined by

$$\int \psi \, \phi_o \, d1 \ldots dN = \max \tag{3}$$

where ϕ_o is a Slater determinant containing the BSO. They are thus, in this case, equivalent to best overlap orbitals. In calculations they may be obtained in iterative procedures based on eq. (2) /8/. For two-electron systems exact BSO can be made identical to exact NSO (BSO are defined up to a unitary transformation). For other systems this is not possible /9/. In calculations for the Li atom ground state the difference between BSO and corresponding NSO was found to be considerably smaller than the difference between NSO and unitarily transformed Hartree-Fock (HF) orbitals /8, 10/. Yet the latter difference was very small (see table 1). This situation, however, may be different in other cases.

Table 1

Overlap between unitarily transformed Hartree-Fock orbitals and NSO and BSO for the Li atom ground state /8, 10/.

	$1s\alpha$	$1s\beta$	$2s\alpha$
$\langle HF\|BSO\rangle$	0.9999997	0.9999995	0.9999564
$\langle HF\|NSO\rangle$	0.9999997	0.9999995	0.9999568

To obtain values close to unity in table 1 it is required that we have very accurate wave functions at our disposal. If this requirement is not met the NSO and BSO may depart considerably from the Hartree-Fock orbitals. The calculations on the Li atom ground state may be used as an example. We may calculate a contact term either directly from a certain wave function ψ, or using the Slater determinant of those BSO which are obtained from ψ via eq. (3). We compare three different ψ's accounting for 91, 96, and 99% of the correlation energy (see ref. 7). The respective expectation values of the contact term were 3.516, 2.812, and 3.155. The corresponding values if just the BSO are used are 3.534, 2.805, and 3.158. The unrestricted HF result is 2.824 and the experimental value 2.906.

Thus, in spite of the fact that we have obtained BSO from wave functions which account for most of the correlation energy, the contact term calculated from them will be erratic and far from the experimental value. This is, of course, not in accord with the idea that BSO should be improvements on the HF orbitals. One of us therefore formulated a "Brueckner-Hartree-Fock" method /11/, which requires information only of pair functions and does not use eq. (3). In this theory, an equation similar to the Hartree-Fock equation, but including operators accounting for correlation, is defined. If the pair functions are constructed from the wave function mentioned above, which accounted for 99% of the correlation energy, a contact term value of 2.908 was obtained. No erratic behaviour occurs when pair functions are obtained from different sources.

III. ONE-PARTICLE MODELS

In a one-particle model we use Schrödinger equations for single electrons. The total N-particle wave function is a single Slater determinant. The most well-known of such models are the Hartree and Hartree-Fock models. In the latter the exchange term may be interpreted as the influence of the Fermi hole on the electron in its motion. The Brueckner-Hartree-Fock method mentioned above is an extension where the electron is influenced also by the Coulomb hole. In recent years we have seen a number of orbital methods where the exchange and correlation terms are functionals of the charge density /12/. A well-known example is the $X\alpha$ method of Slater /13/, which uses a $\rho^{1/3}$ exchange potential. Other examples are the models suggested by Lundqvist and collaborators /14/, which in addition include correlation potentials. Whether correlation energy can be retrieved in one-particle models is a subject of its own and will not be discussed here. Instead we will concentrate on correlation effects in expectation values.

IV. CORRELATION EFFECTS IN ONE-PARTICLE EXPECTATION VALUES

The expectation value of a one-particle operator

$$\Omega = \sum_{i=1}^{N} \omega_i \qquad (4)$$

may be written

$$\langle\Omega\rangle = \mathrm{Tr}(\omega_1\gamma) = \sum_{i=1}^{\infty} \nu_i \langle\chi_i|\omega_1|\chi_i\rangle \qquad (5)$$

where γ is the one-particle reduced density matrix of the exact wave function (eq. (1)), and $\{\chi_i\}$ are the natural spin orbitals. In a one-particle model the expectation value of Ω will be

$$\langle\Omega\rangle_o = \sum_{i=1}^{N} \langle\phi_i|\omega_1|\phi_i\rangle \tag{6}$$

We may now consider the following identity /15/

$$\langle\Omega\rangle = \langle\Omega\rangle_o + \Delta_{orb} + \Delta_{occ} \tag{7}$$

where

$$\Delta_{orb} = \sum_{i=1}^{N} \langle\chi_i|\omega_1|\chi_i\rangle - \langle\phi_i|\omega_1|\phi_i\rangle \tag{8}$$

$$\Delta_{occ} = \sum_{i=1}^{N} (\nu_i-1)\langle\chi_i|\omega_1|\chi_i\rangle + \sum_{i=N+1}^{\infty} \nu_i\langle\chi_i|\omega_1|\chi_i\rangle \tag{9}$$

Δ_{orb} and Δ_{occ} may be considered correction terms to $\langle\Omega\rangle_o$. Δ_{orb} appears since our calculated orbitals are different from the natural spin orbitals. As can be understood from the discussion above, Δ_{orb} is completely negligible for BSO for the Li ground state. This is probably generally true if the BSO are obtained from the Brueckner-Hartree-Fock equations.

If the set $\{\phi_i\}$ are unrestricted Hartree-Fock (UHF) orbitals, Δ_{orb}^{UHF} may be fairly large. It represents a correlation error due to a correlation correction to the orbitals. Since NSO are almost identical to BSO, Δ_{orb}^{UHF} can be obtained if correlation operators are introduced in the UHF equations as in ref. 11. For the Li ground state contact term, Δ_{orb}^{UHF} is about 3% (but 30% for restricted Hartree-Fock).

For $\omega = r^2$, Δ_{orb}^{UHF} is of the order 1-2% /11/. This correlation corresponds to an overall contraction of the wave function. In the contact term case for the atoms B, C, N, O, and F, Δ_{orb}^{UHF} is of the same order of magnitude as $\langle\Omega\rangle_o$ but with reversed sign /16/. This is part of the explanation for the erratic results obtained in contact term calculations for these atoms. For other hyperfine structure expectation values Δ_{orb}^{UHF} is unimportant.

The term Δ_{occ} depends on the fact that the occupation numbers are different from unity and zero. This error cannot be avoided in one-particle models, where we calculate N spin orbitals for N electrons. Δ_{occ} is around 0.4% of $<\Omega>$ for the Li ground state contact term, of the same order magnitude and sign as $<\Omega>$ for the atoms B - F, and fairly unimportant for other hyperfine structure expectation values for these atoms except the B quadrupole term /16/. For $\omega = r^2$ for the Li ground state, Δ_{occ} is negligible, since the weakly occupied NSO are localized in the same region of space as the strongly occupied ones. Consequently there is no transfer of charge when correlation is "switched on".

The method of analysis presented here may be of great interest when discussing the accuracy of the UHF method for calculation of spin density distributions, a problem which has attracted much attention from Löwdin and his group. We have studied this problem for some atoms /17/ and have found that the UHF method gives a fairly accurate estimation of the spin density distribution at least in the valence shell. Δ_{orb} is unimportant, whereas Δ_{occ} may be fairly large in some cases and leading to a decrease of the spin polarization.

REFERENCES

1. P.-O. Löwdin, Phys. Rev. 97, 1474 (1955).
2. P.-O. Löwdin, and H. Shull, Phys. Rev. 101, 1730 (1956).
3. H. Shull and P.-O. Löwdin, J. Chem. Phys. 23, 1565 (1955).
4. H. Shull and P.-O. Löwdin, J. Chem. Phys. 30, 617 (1959).
5. E.R. Davidson, Revs. Mod. Phys. 44, 451 (1972).
6. R.K. Nesbet, Phys. Rev. 109, 1632 (1958).
7. P.-O. Löwdin, J. Math. Phys. 3, 1171 (1962).
8. S. Larsson and V.H. Smith, Jr., Phys. Rev. 178, 137 (1969).
9. W. Kutzelnigg and V.H. Smith, Jr., T.N. 130 from Uppsala Quantum Chemistry Group.
10. S. Larsson and V.H. Smith, Jr., Int. J. Quant. Chem. 6, 1019 (1972).
11. S. Larsson, J. Chem. Phys. 58, 5049 (1973).
12. R. Gáspár, Acta Phys. Hung. 3, 263 (1954); W. Kohn and L.J. Sham, Phys. Rev. 140, A1133 (1965).
13. J.C. Slater, Adv. Quant. Chem. 6, 1 (1972).
14. S. Lundqvist and C.W. Ufford, Phys. Rev. 139A, 1 (1965); L. Hedin and S. Lundqvist, J. Phys. C4, 2064 (1971).
15. S. Larsson, Acta Universitatis Upsaliensis 215 (1972).
16. S. Larsson, R.E. Brown, and V.H. Smith, Jr., Phys. Rev. A6, 1375 (1972).
17. R.E. Brown and S. Larsson, J. Chem. Phys. 59, 4192 (1973).

LONG-RANGE INTERACTION IN SOME TWO-ELECTRON SYSTEMS[+]

W. Kołos

Quantum Chemistry Laboratory, University of Warsaw, and

Max Planck Institute for Physics and Astrophysics, Munich

1. INTRODUCTION

There seems to be no area of quantum chemistry which has not been influenced by Löwdin's work. In the theory of inter-molecular interactions there is the accurate calculation by Hirschfelder and Löwdin /1/ of the dispersion interaction between two ground state hydrogen atoms. However, first of all, there is the pioneering work on the properties of ionic crystals /2/ which, although published almost 30 years ago, still represents the highest achievement in this field. The work clearly shows the enormous complexity of the problem and the necessity of more approximate treatments of the interaction between closed shell systems.

The present work approaches the problem of interaction from a different direction. It deals with very simple systems for which accurate results can be obtained. These results give some insight into the nature of the interactions and into various approximations commonly made in the theory of intermolecular interactions. They also show the reliability, or unreliability, of the approximate approaches.

In the next section we discuss the results obtained by employing the perturbation theory to study the H...H interaction in the X $^1\Sigma_g^+$ and B $^1\Sigma_u^+$ states, and the He^+...H interaction in the A,a $^{1,3}\Sigma^+$ states. It is shown that in all the above cases, with the exception of the X state of H_2, the second-order exchange energy plays a very important role.

[+] Dedicated to Professor Per-Olov Löwdin on his 60-th birthday.

In Section 3 the effect of intraatomic electron correlation in the helium atom on the He...H^+ interaction energy is discussed. It is shown that over a wide range of internuclear distances the interaction energy is in error by 5-7% when calculated without taking into account the electron correlation. The error seems to be smaller only in the vicinity of R = 5 a.u. where a change of its sign should occur.

In the last section we discuss very special nonadiabatic effects which appear in the excited states of the nonsymmetric hydrogen molecules, such as HD, and which affect the asymptotic energy and the asymptotic interaction in these systems.

2. INTERNAL STRUCTURE OF THE ENERGY OF INTERACTION BETWEEN SOME ONE-ELECTRON ATOMIC SYSTEMS

To introduce the notation let us write the hamiltonian for a two-electron molecule in the form

$$H = H_a(1) + H_b(2) + H_{ab}(1,2) \tag{1}$$

where a,b refer to the nuclei and 1,2 to the electrons. The interaction of the two atoms, represented by $H_{ab}(1,2)$ will be treated as a perturbation and the spin-independent wavefunction will be assumed in the form

$$\Psi = \psi^{(o)} + \sum_{i=1} \psi^{(i)} \tag{2}$$

where

$$\psi^{(o)} = \left[\chi_{ak}(1)\chi_{bl}(2)+\lambda\chi_{bk}(1)\chi_{al}(2)\right] + \sigma\left[1\leftrightarrow2\right] \tag{3a}$$

$$\psi^{(i)} = \sum_{n,m\neq k,l} c^i_{nm}\{\left[\chi_{an}(1)\chi_{bm}(2)+\lambda\chi_{bn}(1)\chi_{am}(2)\right] + \sigma\left[1\leftrightarrow2\right]\} \tag{3b}$$

χ_{an} and χ_{bm} are eigenfunctions of the one-electron hamiltonians H_a and H_b, respectively, $\lambda = 1$ (-1) for the Σ_g (Σ_u) and Π_u (Π_g) states, $\sigma = 1$ (-1) for the singlet (triplet) states, and the symbol $[1\leftrightarrow2]$ denotes the term which is obtained from the preceding one by interchanging 1 and 2. If the molecule dissociates into two atoms in identical states we have k=l and the second term in each square bracket in (3a) and (3b) disappears.

Using (3) and one of the symmetry adapted perturbation theories (see, e.g., /3/ and references therein) one can - at least in principle - calculate the correct interaction energy $E^{(i)}$ in 2-3 lowest orders of the perturbation theory. In practice, however, no completely reliable values of $E^{(2)}$ have been obtained even for the interaction of two ground state hydrogen atoms.

The long-range interaction energy can also be computed more easily by making use of the polarization approximation (see, e.g. /3/) in which the perturbational wavefunction (2) is assumed in the form

$$\psi^{(o)}_{pol} = \chi_{ak}(1)\chi_{bl}(2) + \lambda\sigma\left[a1 \leftrightarrow b2\right] \qquad (4a)$$

$$\psi^{(i)}_{pol} = \sum_{n,m=k,l}\{c^{i}_{nm}\ \chi_{an}(1)\chi_{bm}(2) + \lambda\sigma\left[a1 \leftrightarrow b2\right]\} \qquad (4b)$$

In each order of the perturbation theory we can now define the exchange energy as the difference

$$E^{(i)}_{exch} = E^{(i)} - E^{(i)}_{pol} \qquad (5)$$

between the correct interaction energy $E^{(i)}$ obtained in this order using wavefunctions (3) and the polarization energy $E^{(i)}_{pol}$ obtained in the same order using wavefunctions (4).

The total polarization energy

$$E_{pol} = \sum_{i=1} E^{(i)}_{pol} \qquad (6)$$

reliably approximates the interaction energy only in the region where the overlap of wavefunctions of the interacting systems is negligible. The importance of overlap in the calculation of the interaction energy has been recognized by Löwdin /2,4/ already in the early years of quantum chemistry. In our approach the overlap affects the interaction energies in two different ways: (1) It gives rise to the exchange energy; (2) It gives rise to the charge-overlap effects /5,6/ neglected when using the asymptotic expansion of the interaction hamiltonian in powers of R^{-1}, where R is the internuclear distance.

Various contributions to the interaction energy, and the reliability of some most common approximations have been studied in details for two hydrogen atoms in the X $^1\Sigma^+$ and B $^1\Sigma^+_u$ states /7,8/. The results listed to Table I show the charge overlap effect mentioned above. This effect is represented by the difference between the correct value of the polarization energy, $E^{(i)}_{pol}$, and the value of $E^{(i)}_{pol.as.}$ obtained using the well known asymptotic multipole expansion of the interaction hamiltonian. For the B state the values of $E^{(2)}_{pol.as.}$ given

Table I

Polarization energies calculated in various orders of the perturbation theory for the X and B states of H_2 using the correct form of the interaction hamiltonian ($E^{(i)}_{pol}$) and its asymptotic multipole expansion ($E^{(i)}_{pol.as.}$)[a]

	R	$E^{(1)}_{pol}$	$E^{(1)}_{pol.as.}$	$E^{(2)}_{pol}$	$E^{(2)}_{pol.as.}$	$E^{(3)}_{pol}$	$E^{(3)}_{pol.as.}$
	4			-560.0	div.	-109.9	
	6			-53.67	-52.8	1.40	
$X\ ^1\Sigma_g^+$	8			-8.03	-8.28	0.298	div.
	10			-1.81	-1.83	0.025	0.02
	12	-155.41	-140.96	-44.8	div.	-8.3	
	15	-73.37	-72.17	-7.4	-6.9	-0.3	
$B\ ^1\Sigma_u^+$	18	-41.86	-41.77	-1.9	-1.9	0.0	
	20	-30.46	-30.45	-0.88	-0.90	0.00	

[a] R in atomic units, energies in cm^{-1}

in Table I have been improved by including the term proportional to R^{-10} /9/ inadvertently not taken into account previously /8/. It is seen that the values of $E^{(1)}_{pol.as.}$ are fairly reliable, and the values of $E^{(2)}_{pol.as.}$ either diverge (denoted as "div." in Table I) or are also reliable. The results seem to indicate that in higher orders $E^{(i)}_{pol.as.}$ may become divergent even at relatively large internuclear distances and the calculated values, as illustrated by $E^{(3)}_{pol.as.}$ are likely to be either negligible small or divergent. Bukta and Meath /9/ have also calculated the lowest order (in R^{-1}) term in $E^{(3)}_{pol.as.}$. It represents the resonant energy and is proportional to R^{-9}. This term, however, constitutes only a very small fraction of $E^{(3)}_{pol}$ listed in Table I and has the opposite sign.

In Table II the total interaction energy of two hydrogen atoms

Table II

Polarization and exchange energies in the X and B state of H_2 in percent of the accurate variationally computed interaction energy (E_{var})

	R(a.u.)	E_{pol}	E_{exch}	$E_{var}(cm^{-1})$
	6	39.0	61.0	-183.31
$X\ ^1\Sigma_g^+$	8	67.8	32.2	-12.19
	10	93.8	6.2	-1.92
	12	22.7	77.3	-927.2
$B\ ^1\Sigma_u^+$	15	69.2	30.8	-117.3
	18	97.3	2.7	-45.2

in the X and B states is split into E_{pol} and E_{exch} /7,8/. The results show that in the regions under consideration in both states the relative contributions of E_{pol} (and of E_{exch}, respectively) are of comparable magnitude. However, it is seen in Table III that the internal structure of E_{pol} and E_{exch} is completely different in the two states /7,8/. In the X state the dominant contributions are due to $E^{(1)}_{exch}$ and $E^{(2+)}_{pol}$ (defined as $\sum_{i=2} E^{(i)}_{pol}$), whereas $E^{(1)}_{pol}$ and $E^{(2+)}_{exch}$ (defined as $\sum_{i=2} E^{(i)}_{exch}$) are almost negligible. Exactly the opposite holds for the B state, where $E^{(1)}_{pol}$ and $E^{(2+)}_{exch}$ play the dominant role.

The large value of $E^{(1)}_{pol}$ in the B state is understandable. In the X state $E^{(1)}_{pol}$ is entirely due to the charge-overlap effect which vanishes exponentially with increasing R whereas in the B state it represents the well known resonance interaction which vanishes only like R^{-3}. It is, however, difficult to find a physical reason for the large value of $E^{(2+)}_{exch}$.

The above mentioned resonance interaction emerges if the dissociation products are in different electronic states and an electric multipole transition between the two states is allowed. The sign of this interaction is determined by $\lambda\sigma$ in Eq. (4a). Hence in the C $^1\Pi_u$ state of H_2 the resonance interaction is repulsive and gives rise to the well known potential maximum at large internuclear distances

Table III

First- and higher-order contributions to the polarization and exchange energies in H_2 in percent of the accurate variational interaction energy

	R	$E^{(1)}_{pol}$	$E^{(1)}_{exch}$	$E^{(2+)}_{pol}$	$E^{(2+)}_{exch}$
	6	7.1	53.8	31.9	7.2
$X\ ^1\Sigma_g^+$	8	3.2	28.1	64.6	4.1
	10	0.6	5.3	93.2	0.9
	12	16.8	9.0	5.9	68.3
$B\ ^1\Sigma_u^+$	15	62.5	6.2	6.7	24.6
	18	92.7	1.3	4.6	1.4

/10-12/, whereas in the B $^1\Sigma_u^+$ state the interaction is attractive. In both cases it goes like R^{-3}. Nonadiabatic perturbation of the resonance interaction in HD will be discussed in Section 4.

A still different structure of the interaction energy has been found in the interaction between H and He^+, both in their ground states /13/. The results are shown in Table IV. The polarization energy is here decomposed into induction and dispersion energy

$$E_{pol} = E_{ind} + E_{disp}$$

and $E_{ind}(5)$ in Table IV denotes $\sum_{i=2}^{5} E^{(i)}_{ind}$, whereas E_{pol} includes $E_{ind}(5)$ and $E_{disp}(2)$. Ih the H...He^+ interaction the induction energy plays obviously the most improtant role. It is seen that for $R \geq 8$ the interaction energy is fairly well represented by $E^{(1)}_{exch} + E_{pol}$, i.e., neglecting $E^{(2+)}_{exch}$ in the interaction energy. It is, however, surprising to note that also in this region the almost negligible $E^{(2+)}_{exch}$ is essential for a correct prediction of the relative order of the singlet and triplet states.

Table IV

Interaction energies in the A $^1\Sigma^+$ and a $^3\Sigma^+$ states of HeH^+ which dissociate into H(1s) and $He^+(1s)^a$

	A	$E^{(1)}$	$E_{ind}^{(5)}$	$E_{disp}^{(2)}$	$E_{exch}^{(1)}+E_{pol}$	E_{var}
	6	-20.05	-529.94	-4.18	-554.17	-348.26
A $^1\Sigma^+$	8	-0.417	-137.27	-0.66	-138.35	-133.72
	10	-0.008	-52.84	-0.16	-53.01	-52.75
	6	21.28	-529.94	-4.18	-512.84	-441.85
a $^3\Sigma^+$	8	0.436	-137.27	-0.66	-137.50	-137.29
	10	0.008	-52.84	-0.16	-53.00	-53.10

^{a}R in atomic units, energies in cm^{-1}

3. THE EFFECT OF INTRAATOMIC ELECTRON CORRELATION ON THE INTERACTION ENERGY

Most theories of intermolecular interactions are based on SCF treatment of the isolated systems. Therefore an important question arises whether it is possible to obtain reliable interaction energies when neglection any changes of intraatomic correlation energies due to the interaction. An extensive study of this problem has been made for the He...He interaction /14-16/.

Recently a still simpler system has been studied, viz., the HeH^+ ion in its ground state which dissociates into $He+H^+$ /13,17/. In Table V we show the changes of the correlation energy (E_{corr}) of the helium atom caused by the interaction with the proton. For the region $1.2 \leq R \leq 4.5$ they were obtained from the accurate variational /17/ and SCF /18/ results. For $R \geq 6.0$ the numbers given in Table V result from the multipole expansion of the interaction energy in which the helium atom polarizabilities calculated with and without

Table V

Change of the correlation energy of the helium atom due to the interaction with a proton

R	ΔE in % of E_{corr} at R = ∞	ΔE in % of E_{int}
1.2	6.21	-4.5
1.4	8.02	-4.5
1.6	9.02	-5.2
1.7	9.18	-5.6
1.8	9.11	-6.0
2.0	8.47	-6.7
2.2	7.33	-7.2
2.8	3.57	-7.7
3.5	1.21	-6.8
4.0	0.57	-6.1
4.5	0.29	-5.3
...	....	....
6.0	-0.097	6.9
7.0	-0.052	7.2
8.0	-0.031	7.4
9.0	-0.019	7.5
∞	0.0	7.8

electron correlation /19,20/ are employed. It is clearly seen that in the region where the multipole expansion is applicable the correlation energy changes very little. However, in this region the interaction energy is small, and it is very interesting to note that the absolute value of the relative effect of intraatomic electron correlation has almost the same value for a very wide range of R including the very large internuclear distances and the vicinity of the equilibrium. Only in the region where the polarization approximation breaks down a change of sign of the correction to the correlation energy occurs and in the vicinity of this point the correction, in absolute value, must be very small.

The intraatomic electron correlation which, by making the helium atom less polarizable, decreases the binding at large R has the opposite, i.e., attractive effect in the region where the overlap is not negligible. If in larger systems the internal electron correlation had the same relative effect on the interaction energy it would change the energy of, e.g., a typical hydrogen bond by not more than 0.5 kcal/mole.

4. ASYMPTOTIC NONADIABATIC EFFECTS IN NONSYMMETRIC HYDROGEN MOLECULES

In the nonsymmetric hydrogen molecules, such as HD, the long-range interaction is of particular interest since in these systems two qualitatively new effects appear. If the hydrogen molecule dissociates in the B or C state one of the hydrogen atoms is in the ground state (n=1) and one is in an excited state (n=2). In the case of HD the energies of H*+D and H+D* differ by 22.4 cm^{-1}, and the B state, as well as the C state, dissociate into the lower energy products, i.e., into H +D /21/. The adiabatic approximation, however, does not distinguish between these two possibilities and gives an average energy which is thus 11.2 cm^{-1} higher than the true energy of the dissociation products.

In Table VI the experimental dissociation energies of HD for several vibrational levels in the B state /21/ are compared with the theoretical values obtained directly in the adiabatic approach /22/ and with those referred to the correct nonadiabatic energy at $R = \infty$. It is seen that the dissociation energies of the lowest vibrational levels are considerably improved by the correction. For intermediate vibrational levels the discrepancy between the theoretical (corrected) and experimental levels reaches its maximum. This discrepancy is mainly due to those nonadiabatic effects which are present also in H_2 and D_2. For still higher vibrational levels the discrepancy decreases. It is clearly seen that the appearance of the v=44 level in the adiabatic approximation is entirely due to the failure of this

Table VI

Dissociation energies (in cm^{-1}) of HD for various vibrational levels in the B state

v	D_v		
	exper.[a]	adiabatic	adiabatic corrected
0	28266.0	28275.7	28264.5
1	27118.8	27128.3	27117.1
2	25999.7	26008.8	25997.6
20	9959.8	9962.5	9951.3
30	3965.0	3965.3	3954.1
40	184.9	186.5	175.3
41	63.0	67.4	56.2
42	18.0	25.5	14.3
43	3.3	10.5	-0.7
44		4.0	-7.2

[a] Reference /21/.

approximation to describe correctly the dissociation of HD.

Although a potential energy curve in the nonadiabatic approach has no meaning it is tempting to try to take into account the effect under consideration by lowering the potential energy curve at large R by 11.2 cm^{-1}. Test calculations have shown, however, that the adiabatic potential energy curve must be used almost to the crossing with the new dissociation limit which occurs at $R \approx 28.25$ a.u. If the R^{-3} extrapolation to the new limit is made from a somewhat smaller R value the v=44 will not be eliminated.

The second effect which appears in HD at large internuclear separations is also related to the different energies of H^*+D and $H+D^*$. The asymptotic wavefunction of HD in a molecular state which dissociates into two differently excited hydrogen atoms should represent either H^*+D or $H+D^*$, and should not have the g or u symmetry. The wavefunction should thus be given either by the first or by the second term in Eq. (4a) but not by their linear combination. The resonance interaction present in the B and C state of the hydrogen molecule is due to the fact that the excitation can be on either atom, and the zero-order wavefunction has the form (4a). If, however, the excitation is localized on one of the atoms, like in HD, there should be no resonance interaction and at large R the potential curve should go like R^{-6} rether than R^{-3}. This point has been recently raised by Dabrowski and Herzberg /21/ who have measured the vibrational levels in the C state of HD and asked the question whether or not the well known potential maximum in the C state existed in HD.

Table VII

Some vibrational quanta (in cm^{-1}) in the C state of HD

v	$\Delta G(v+\frac{1}{2})$		
	theor.	exper.[a]	Δ
11	885.72	885.15	0.57
12	753.19	752.55	0.64
13	603.13	602.38	0.75
14	425.05	426.1	-1.0

[a] Reference /21/

In Table VII we list the theoritical /23/ and experimental /21/ differences between the highest vibrational levels in the C state og HD. It is seen that the discrepancy, Δ, for the last level is not consistent with those for lower levels. It is again tempting to try to remove this discrepancy by a "nonadiabatic" correction to the potential energy curve. This correction should eliminate the resonance interaction for sufficiently large internuclear distances. Test calculations have shown, however, that any modifications of the potential curve which decreased or eliminated the potential maximum worked in the wrong direction. Any modifications which decreased the width of the potential barrier, by assuming a more rapid decrease of the potential for large R, had only a small effect on the energy of the highest vibrational level, and shifted it also in the wrong direction. Thus we may conclude that the small discrepancy which exists between the experimental and theoretical energies for the highest vibrational level in the C state of HD is not due to the nonadiabatic effect which switches off the resonance interaction. The results reported above suggest that the potential maximum in the C state of HD does exist, and that the resonance interaction disappears at so large internuclear distances that it has little, if any, effect on the energies of the vibrational levels.

ACKNOWLEDGEMENT

The author is very grateful to Dr. G.H.F. Diercksen for his very kind hospitality at the Max Planck Institute for Physics and Astrophysics in Munich. He is also very much indebted to Dr. G. Herzberg for sending his results prior to publication.

BIBLIOGRAPHY

1. J.O. Hirschfelder and P.O. Löwdin, Mol. Phys. 2. 299 (1959), 9, 491 (1966).
2. P.O. Löwdin, Arkiv f. mat. astr. o. fysik 35A, No. 9,30 (1948) and "A Theoretical Investigation of Some Properties of Ionic Crystals" (Almquist & Wicksell, Uppsala, 1948).
3. D.M. Chipman, J.D. Bowman and J.O. Hirschfelder, J. Chem. Phys. 59, 2830 (1973).
4. P.O. Löwdin, J.Chem. Phys. 18, 365 (1950).
5. H. Kreek and W.J. Meath, J. Chem. Phys. 50, 2289 (1963).
6. G. Chałasinski and B. Jeziorski, Mol. Phys. 27, 649 (1974).
7. W. Kołos, Int. J. Quantum Chem. 8S, 241 (1974).
8. W. Kołos, Int. J. Quantum Chem. 9, 133 (1975).

9. J.F. Bukta and W.J. Meath, Mol. Phys. 25, 1203 (1973).

10. G.W. King and J.H. Van Vleck, Phys. Rev. 55, 1165 (1939).

11. R.S. Mulliken, Phys. Rev. 120, 1674 (1960).

12. W. Kołos and L. Wolniewicz, J.Chem.Phys. 43, 2429 (1965).

13. W. Kołos, Int.J.Quantum.Chem., in publication.

14. P. Bertoncini and A.C. Wahl, Phys. Rev. Letters 25, 991 (1970).

15. D.R. McLaughlin and H.F. Schaefer III, Chem. Phys. Letters 12, 224 (1971).

16. B. Liu and A.D. McLean, J. Chem. Phys. 59, 4557 (1973).

17. W. Kołos and J.M. Peek, to be published.

18. S. Peyerimhoff, J. Chem. Phys. 43, 998 (1965).

19. T.R. Singh, Chem. Phys. Letters 11, 598 (1971).

20. W.D. Davison, Proc. Phys. Soc. 87, 133 (1966).

21. I. Dabrowski and G. Herzberg, Can. J. Phys. in publication.

22. W. Kołos and L. Wolniewicz, Can. J. Phys. in publication.

23. W. Kołos, to be published.

THE GENERATOR COORDINATE METHOD ILLUSTRATED ON THE HYDROGEN MOLECULE

Bernard Laskowski†

Quantum Theory Project, Chemistry and Physics Departments

University of Florida, Gainesville, Florida 32611, U.S.A.

I. INTRODUCTION

The problem of correlation, as pointed out by P.O. Löwdin, /1/ is a central one in many body physics. It is realized that correlations in the particle motion are to be included in the mathematical description of many body systems in order to reach an agreement with experimental data. However, this is a nearly impossible task except for a few simple systems or for some simple kinds of correlation.

Usually one tries to see what can be done if the independent particle model is pushed to its limits. In that approximation one sticks to wave functions in which the particles are essentially assigned to definite orbitals, see e.g. Löwdin /2,3,4/. In the unrestricted Hartree Fock (UHF) approximation the orbitals are allowed to take their optimal shape within some chosen model space. This picture has however to be refined to account for fundamental symmetries (like the Pauli principle and angular momentum). This refinement is introduced with the help of projection operators; in technical language: Projected - UHF(PHF). If we consider correlation as any departure from a pure product wave function then the projections introduce a certain amount of correlation because projected products are sums of products. As long as we remain with the fundamental kinemetical symmetries like angular momentum or spin, these correlations are of a rather trivial kinematical nature. It is to be expected that these types of correlations will be insufficient to bridge the gap between experimental results and more primitive

†On leave of absence from Faculty of Science, Rijksuniversitair Centrum Antwerp, Belgium.

Restricted Hartree Fock (RHF) results.

The difficulties and limitations of the independent particle model, pushed to its limits, have become recognized. In the Generator Coordinate Method (GCM), one considers a class of trial functions more general than Slater determinants or projected Slater determinants and so, one hopes, better approximations to the exact solutions. Approximations to the physical states of the system are written as continuous linear superpositions of 'intrinsic' states $\Phi(x,\alpha)$ depending on a set of parameters α (generator coordinates). For the moment we restrict the notation to the case of one such generator coordinate

$$\Psi(x) = \int \Phi(x,\alpha) f(\alpha) d\alpha \tag{1}$$

where x, stands for all degrees of freedom (spin and space coordinates of N particles). The functions $\Psi(x)$ generated by $\Phi(x,\alpha)$ form a subspace H_Φ of the complete Hilbert space H. The accuracy of the GCM depends on how well the states in H_Φ can approximate or imitate the states in H. This depends on the choice of $\Phi(x,\alpha)$.

Whereas $\Phi(x,\alpha)$ in (1) is usually a determinant the GCM trial function corresponds to a sum of determinants. PHF is a special case of GCM. Angular momentum projection is obtained when the generator coordinates are chosen to be the Euler angles. This was pointed out by Peierls and Yoccoz /5/. However, for other generator coordinates GCM is more general and can go beyond the independent particle model. GCM can consider dynamical correlations whereas in PHF only a restricted class of correlations are built in. GCM reduces the correlation error and besides an improved ground state a set of excited states are obtained. In fact the GCM was specially designed to describe collective motion without explicitly introducing collective degrees of freedom /6/. No redundant variables are used.

II. GCM ILLUSTRATED ON H_2

In this section we explain the main aspects of the GCM on a simple model. For this purpose we choose the smallest neutral molecule, namely H_2. For this molecule GCM is essentially equivalent to configuration interaction (C.I.), it gives no new physical results but illustrates how the method works.

If we suppose on each hydrogen only one s type A.O., an LCAO version of RHF gives the two well known gerade and ungerade MO's ϕ_1 and $\phi_{\bar{1}}$. With two electrons there are three possible configurations namely a: (ϕ_1^2), b: $(\phi_{\bar{1}}^2)$ and c: $(\phi_1\phi_{\bar{1}})$. The configurations (a) and

(b) give rise to two gerade singlets ${}^1\Sigma_{ga}$, ${}^1\Sigma_{gb}$ distinguished by a and b and being Σ since we only use s-type A.O.'s. From the third configuration one can get two ungerade states, a triplet and a singlet: ${}^3\Sigma_u$, ${}^1\Sigma_u$.

From symmetry considerations, one sees at once that in a C.I. treatment only ${}^1\Sigma_{ga}$ and ${}^1\Sigma_{gb}$ will interact giving the improved states ${}^1\Sigma_{g1}$ and ${}^1\Sigma_{g2}$, whereas, ${}^3\Sigma_u$ and ${}^1\Sigma_u$ remain unchanged.

As intrinsic function in (1) we use the Alternant Molecular Orbital (AMO). AMO is a special kind of UHF (DODS) and was suggested by Löwdin /7,4/.

The correlation is simulated by spatial separation of electrons with antiparallel spin which in the RHF approximation occupy the same orbital. For electrons with parallel spin the RHF Fermi hole accounts partly for the Coulomb hole. For the hydrogen molecule we have

$$\Phi(\text{UAMO})(x_1,x_2,z) = \det|a(\vec{r})\alpha, b(\vec{r})\beta|\sqrt{\tfrac{1}{2}} \tag{2}$$

where we recall that $a(\vec{r})$ and $b(\vec{r})$ depend on a mixing parameter z and are given by

$$a(\vec{r}) = (\phi_1(\vec{r}) + z\phi_{\bar{1}}(\vec{r}))/\sqrt{(1 + z^2)} \tag{3}$$

$$b(\vec{r}) = (\phi_1(\vec{r}) - z\phi_{\bar{1}}(\vec{r}))/\sqrt{(1 + z^2)} \tag{4}$$

In (2) U stands for unprojected. The AMO function (2) is not an eigenfunction of the total spin but can be expanded in its singlet and triplet components

$$|\Phi(\text{UAMO}),z\rangle = |S = 0,z\rangle + |S = 1,z\rangle \tag{5}$$

with

$$|S = 0,z\rangle = (|{}^1\Sigma_{ga}\rangle - z^2|{}^1\Sigma_{gb}\rangle)/(1 + z^2) \tag{6}$$

$$|S = 1,z\rangle = z\sqrt{2}|{}^3\Sigma_u(M_S = 0)\rangle/(1 + z^2) \tag{7}$$

One sees that $|S = 0,z\rangle$ does not contain the $|{}^1\Sigma_u\rangle$ state. This is due to a general theorem in AMO stating that the even spin components of $|\Phi(\text{UAMO}),z\rangle$ only contain configurations with an even number of excited electrons (excited to the paired level).

Table 1

R Internuclear distance, RHF ground state, UAMO and optimal mixing coefficient, SC-GCM and variationally obtained g. (all energies in atomic units, shielding factor in Slater orbitals is taken equal to 1).

R	RHF	UAMO	z_0	SC-GCM	g_0
0.5	-0.2275	-0.2275	0.0	-0.2337	10.494
1.0	-0.9860	-0.9860	0.0	-0.9965	6.337
1.5	-1.0973	-1.0973	0.0	-1.1145	4.133
2.0	-1.0808	-1.0808	0.0	-1.1085	2.746
2.5	-1.0338	-1.0380	0.2907	-1.0768	1.890
3.0	-0.9828	-1.0159	0.5231	-1.0468	1.357
4.0	-0.8954	-1.0027	0.7632	-1.0128	0.834

In Table 1 we give the improved ground state obtained with the function (2) and note that the proper dissociation limit is obtained.[†] However, for small internuclear distances one notices

[†] The numerical values of the basic integrals can be looked up in e.g. Slater /8/; see also Weinbaum /9/.

that RHF is stable against deformations of the AMO type. Sometimes this is loosely denoted as a phase transition to an antiferromagnetic state. We shall show in section IV that this transition disappears if the correlation is introduced in a proper way.

The vector $|\Phi(\text{UAMO})\rangle$ contains the configurations $|^1\Sigma_{ga}\rangle$ and $|^1\Sigma_{gb}\rangle$. There are the only two configurations which interact in a C.I. The expansion coefficients of these configurations are in the CI function obtained by solving a secular equation whereas in the UAMO function, they are prescribed by the nature of the function. To get the same flexibility as in CI we can consider $|\Phi(\text{UAMO})\rangle$ not only for the optimal mixing parameter z_0 but also for values z_i centered around z_0 and add the $|\Phi(\text{UAMO}),z_i\rangle$

$$|\Psi(\text{GCM})\rangle = \sum(i)f(z_i)|\Phi(\text{UAMO}),a_i\rangle \tag{8}$$

where the $f(z_i)$ are coefficients to be determined in a variational way. Three values of z_i would be sufficient to reproduce the CI spectrum except ${}^1\Sigma_u$, which is not contained in $|\Phi(\text{UAMO}), z\rangle$.

However, for mathematical elegance and for other systems we replace the sum in (8) by an integral, so that we expand $|\Psi(\text{GCM})\rangle$ in a continuous set

$$|\Psi(\text{GCM})\rangle = \int f(z)|\Phi(\text{UAMO}),z\rangle dz \tag{9}$$

From the foregoing it must be obvious that $|\Psi(\text{GCM})\rangle$ is a better trial vector than $|\Phi(\text{UAMO}),z\rangle$ since we reproduced the CI result. For systems with more particles $|\Phi(\text{UAMO}),z\rangle$ will miss, contrary to (5), configurations which would interact with the RHF ground state in a CI treatment. Therefore GCM, with AMO as generating function, will give an improved ground state but will not reproduce the full CI result for a large system.

III. GCM IN GAUSSIAN OVERLAP AND HARMONIC APPROXIMATION

Minimization of the energy E calculated with (1) leads to an integral equation for the weights $f(\alpha)$, first derived by Hill and Wheeler /10/.

$$\int (H(\alpha,\beta) - EI(\alpha,\beta)f(\beta)d\beta = 0 \tag{10}$$

with eigenvalues E_i and eigenfunctions f_i. This Hill-Wheeler equation contains the energy and overlap kernels

$$\begin{bmatrix} H(\alpha,\beta) \\ \\ I(\alpha,\beta) \end{bmatrix} = \int \Phi^*(x,\alpha) \begin{bmatrix} H \\ \\ 1 \end{bmatrix} \Phi(x,\beta)dx \tag{11}$$

For convenience one assumes that the origin of the generator coordinate α has been chosen such that

$$E(\alpha) = H(\alpha,\alpha)/I(\alpha,\alpha) \tag{12}$$

has a minimum at $\alpha = 0$.

Remembering that the overlap of two determinants $|\Phi\rangle$ and $|\Psi\rangle$ equals the determinant of the overlap of their one particle functions ϕ_α and ψ_β it is straightforward to obtain

$$I(\alpha,\beta) = ((1 + zz')/\sqrt{(1 + z^2)(1 + z'^2)})^N \tag{13}$$

where $z = z_0 + \alpha$, $z' = z_0 + \beta$ and $N = 2$ for the hydrogen molecule. To calculate the energy kernel $H(\alpha,\beta)$ we need the formulae first derived by Löwdin /3/ (see also, Brink /11/) for the one and two body operators T and V

$$\langle\Phi|T|\Psi\rangle = \langle\Phi|\Psi\rangle \sum_{\alpha,\beta} \langle\phi_\alpha|t|\psi_\beta\rangle (B^{-1})_{\beta\alpha} \tag{14}$$

$$\langle\Phi|V|\Psi\rangle = \tfrac{1}{2}\langle\Phi|\Psi\rangle \sum_{\alpha,\beta,\gamma,\delta} \langle\phi_\alpha\phi_\beta|v|\psi_\gamma\psi_\delta\rangle \left((B^{-1})_{\gamma\alpha}(B^{-1})_{\delta\beta} - (B^{-1})_{\gamma\beta}(B^{-1})_{\delta\alpha}\right) \tag{15}$$

where B^{-1} is the inverse of the overlap matrix. We get for the hydrogen molecule (however, for a many particle system no higher order terms in z and z' will appear)

$$\frac{H(\alpha,\beta)}{I(\alpha,\beta)} = \frac{1}{(1+zz')^2}(H_{00}+H_{02}z'^2+H_{11}zz'+H_{20}z^2+H_{22}z^2z'^2) \tag{16}$$

with

$$H_{00} = 2\langle\phi_1|t|\phi_1\rangle+\langle\phi_1\phi_1|v|\phi_1\phi_1\rangle + \frac{1}{R} \tag{17a}$$

$$H_{20} = -\langle\phi_1\phi_{\bar{1}}|v|\phi_{\bar{1}}\phi_1\rangle = H_{02} \tag{17b}$$

$$H_{11} = 2(\langle\phi_1|t|\phi_1\rangle + \langle\phi_{\bar{1}}|t|\phi_{\bar{1}}\rangle + \langle\phi_1\phi_{\bar{1}}|v|\phi_1\phi_{\bar{1}}\rangle - \langle\phi_1\phi_{\bar{1}}|v|\phi_{\bar{1}}\phi_1\rangle + \frac{1}{R} \tag{17c}$$

$$H_{22} = 2\langle\phi_{\bar{1}}|t|\phi_{\bar{1}}\rangle + \langle\phi_{\bar{1}}\phi_{\bar{1}}|v|\phi_{\bar{1}}\phi_{\bar{1}}\rangle+ \frac{1}{R} \tag{17d}$$

where R is the internuclear distance.

It is obvious that for a general system it is practically impossible to obtain an analytical solution of the Hill-Wheeler equation (10) for the weights $f(\alpha)$. For large N one can use the approximation

$$(1 + x)^N \sim e^{Nx} \tag{18}$$

to write the overlap kernel in the Gaussian form

$$I(\alpha,\beta) = \exp(-\frac{s}{2}(\alpha-\beta)^2) \qquad ; \; s = N/(1+z_0^2)^2 \qquad (19)$$

This presupposes that $\alpha\sqrt{N}$ remains finite as N becomes large. The parameter α is indeed assumed to be small, since intuitively we can expect that only small vibrations around the optimal intrinsic state should be important. This assumption leads to the Harmonic approximation

$$\frac{H(\alpha,\beta)}{I(\alpha,\beta)} = E(0) + \tfrac{1}{2}(\alpha B\alpha + \beta B\beta + 2\alpha A\beta) + \ldots \qquad (20)$$

where E(0) is defined in (12) and A, B (matrices in the many generator case) are rather complicated expressions of the H_{ij} given in (17). Within the Gaussian Overlap (GO) (19) and the Harmonic approximation (HA) (20) the Hill-Wheeler equation (10) (with integration limits $\pm\infty$) is equivalent to a harmonic oscillator problem and yields the solutions

$$f_0(\alpha) = \exp(-\tfrac{1}{2}g(A,B)\alpha^2) \qquad (21a)$$

$$f_1(\alpha) = \alpha\exp(-\tfrac{1}{2}g(A,B)\alpha^2) \qquad (21b)$$

...

where g is obviously a function of A and B. In this approximation a harmonic spectrum is obtained

$$E_n = E(0) - (\frac{1}{2s}A - \tfrac{1}{2}\omega) + n\omega \qquad (22)$$

with

$$\omega = \frac{1}{s}\sqrt{(A^2 - B^2)} \qquad (23)$$

($|B|<|A|$ as long as the intrinsic state is stable).

As is suggested by the equidistant spectrum (22) the GCM in GO and HA (19,20) is equivalent to the Random Phase approximation (RPA) if we choose the generator coordinates to be the particle-hole amplitudes in

$$|\Phi(x,\alpha)\rangle = \exp\{\sum_{\substack{m>N \\ i\leq N}} \alpha_{mi}\, c_m^+ c_i\}|0\rangle \qquad (24)$$

and the matrices A and B become the well known RPA matrices. For details we refer to Jancovici and Schiff /12/ and to the proceedings of the first working seminar on GCM /13/.

IV. THE SELF CONSISTENT GENERATOR COORDINATE METHOD

In RPA the correlation is introduced in an inconsistent way obtained from the requirement that no quasi-boson phonons are present in the ground state, the variational upper bound property is lost and on top of this the excitation frequency of the spectrum can become imaginary; a most inconvenient situation.

The occurence of an imaginary frequency means that the system is stabilized only through higher order terms than the harmonic ones in the expansion of (20). We /14/ discussed a self-consistent generator coordinate method (SC-GCM) based on a variational principle and corresponding to an infinite summation of a $H(\alpha,\beta)/I(\alpha,\beta)$ expansion. The overlap $I(\alpha,\beta)$ is treated exactly. In SC-GCM one proposes to make in (1)

$$\Psi_i(x) = \int \Phi(x,\alpha) f_i(\alpha) d\alpha \tag{25}$$

the ansatz

$$f_0(\alpha) = e^{-g\alpha^2} \tag{26a}$$

$$f_1(\alpha) = \alpha e^{-g\alpha^2} \tag{26b}$$

$$\dots$$

in which g is obtained from the variational principle condition

$$\frac{\partial}{\partial g} E(g) = 0 \quad \text{where } E(g) = \langle\Psi|H|\Psi\rangle/\langle\Psi|\Psi\rangle \tag{27}$$

By its construction, one can always find at least one g which is positive. Like in RPA, the correlation of the ground state in SC-GCM is still obtained as a sum of two particle-two hole, four particle-four hole,... excitations. This follows since the ansatz (26a) resembles (21a). The correlation energy, however, is now calculated in a consistent way, i.e. a rigorous upper bound is obtained with $\Psi_i(x)$ being N representable.

In the ground state condition (27) gives rise to a "secular equation"

$$\iint e^{-g\beta^2}(\beta^2+\alpha^2)e^{-g\alpha^2}(H(\alpha,\beta) - E(g)I(\alpha,\beta))d\beta d\alpha = 0 \tag{28}$$

which is most conveniently solved by a self consistent procedure. For the hydrogen molecule, a two electron system, it is possible to get the optimal g_0 analytically from (28)

$$2g_0 = -\frac{H_{22} - H_{00}}{H_{02} + H_{20}} \pm\sqrt{\left(\frac{H_{22} - H_{00}}{H_{02} + H_{20}}\right)^2 + 1} \tag{29}$$

where the H_{ij}'s are defined in (17). Since $H_{22}>H_{00}$ and $H_{02}+H_{20}<0$ the first term in (29) is positive. The minus sign in (29) is spurious. The SG-GCM ground state energy $E_0(g_0)$ is obtained by putting g_0 in

$$(4g^2 + 1)\, E_0(g) = 4g^2H_{00}+2g(H_{02}+H_{20}) + H_{22} \tag{30}$$

We list $E_0(g_0)$ in table 1 and we notice that we obtained the exact C.I. result (see e.g. Slater /8/), a singlet, for all internuclear distances. It is realized that the occurence of an imaginary excitation frequency (23) (RHF-UHF phase transition) and related divergencies in the GCM are consequences of the Gaussian overlap and Harmonic approximation in section III. From table 1 we notice that g_0 decreases with increasing internuclear distance giving the Gaussian weight a larger width resulting in higher contribution of the excited (RHF) states to the correlation energy. The importance for the correlation energy of the low lying excited states (RHF) for large internuclear distances was pointed out by Löwdin /15/ and Nesbet /16/. Using the ansatz (26b) we find $E_1(g) = \frac{1}{2}H_{11}$ which is just the energy of the first excited state the ${}^3\Sigma_u$, $M_s = 0$ as one convinces oneself easily by calculating $\langle {}^3\Sigma_u|H|{}^3\Sigma_u\rangle$ and comparing with (17c). The secular equation in this case is identically zero. The situation is different for the second excited state.

V. DISCUSSION

In essence SC-GCM is a CI based on all the double excitations with expansion coefficient obtained in a variational way. There is an interesting parallelism between the SC-GCM and self-consistent Random Phase approximation (SC-RPA) /17/. The SC-RPA gives the exact result for the ground state of the two particle system. The reason is simply that in construction of the ground state, which satisfies the RPA annihilation condition, one makes use of the boson commutation rules for $a_i^+a_j$ and its adjoint which in the two particle case include all the terms in the actual expansion. Ratner /18/ used a method similar to SC-RPA, the self consistent polarization propagator approximation (SPPA) /19/ or geometric approximation. He also obtained the exact C.I. result but found convergence difficulties for large internuclear distances, i.e. the region where RHF is instable against UHF deformations. One does not have this difficulty within the SC-GCM. For N>2 we emphasize the advantage of using the SC-GCM because the properties of N representability and variational upperbound are conserved.

ACKNOWLEDGEMENTS

This article is dedicated to Professor P.O. Löwdin on his sixtieth birthday. The author is indebted to Professor P.O. Löwdin for showing him problems in many electron theory and giving many inspiring suggestions and comments. He is sincerely thankful for the experience of working in an international group and for the great hospitality shown to him in so many ways. Professors Y. Öhrn and E. Brändas are thanked for helpful comments and generous support. This work has been partly supported by a grant from National Science Foundation, GP 42477X.

1. P.O. Löwdin, Advan. Chem. Phys. 14, 283 (1969).
2. P.O. Löwdin, Phys. Rev. 97, 1474 (1955).
3. P.O. Löwdin, Phys. Rev. 97, 1490 (1955).
4. P.O. Löwdin, Phys. Rev. 97, 1509 (1955).
5. R. E. Peierls and J. Yoccoz, Proc. Phys. Soc. (London) A70, 381 (1957).
6. J.J. Griffin and J.A. Wheeler, Phys. Rev. 108, 311 (1957).
7. P.O. Löwdin, Symposium on Molecular Physics, Tokyo, Morusen Co., p. 13 (1953).
8. J.C. Slater, Quantum Theory of Molecules and Solids, Vol. 1, New York, McGraw-Hill Book Co. (1963).
9. S. Weinbaum, J. Chem. Phys. 1, 593 (1933).
10. D.L. Hill and J.A. Wheeler, Phys. Rev. 89, 1102 (1953).
11. D. Brink, Proc. of Int. School of Physics, "Enrico Fermi," Course 36, Acc. Press, Inc., London (1966).
12. B. Jancovici and D.H. Schiff, Nuclear Phys. 58, 678 (1964).

13. M.V. Mihailovic and M. Rosina (editors) "Generator Coordinate Method for nuclear bound states and reactions" Fizika 5 supplement (1973).

14. B. Laskowski and E. Brändas, Preliminary Research Report 407, Quantum Theory Project, Williamson Hall, University of Florida, Gainesville, Florida 32611.

15. P.O. Löwdin, Rev. Mod. Phys. 35, 496 (1963).

16. R.K. Nesbet, Rev. Mod Phys. 35, 498 (1963).

17. M. Ostlund and M. Karplus, Chem. Phys. Letters 11, 450 (1971).

18. M. Ratner, Int. J. Quantum Chem. 6, 1165 (1972).

19. J. Linderberg, P. Jørgensen, J. Oddershede and M. Ratner, J. Chem. Phys. 56, 6213 (1972).

PROJECTED HARTREE-FOCK CALCULATIONS ON THE GROUND AND FIRST EXCITED $^1\Sigma_g^+$ STATES OF THE HYDROGEN MOLECULE

G. Howat* and S. Lunell

Quantum Chemistry Group, Uppsala University, Box 518

S-751 20 Uppsala 1, Sweden

I. INTRODUCTION

The projected Hartree-Fock (PHF) method was formulated by Löwdin /1/ in order to introduce a consideration of the electron correlations which are absent in the ordinary Hartree-Fock (HF) method. To recap briefly, the wave function in the PHF method is written in the form

$$\Psi = O\Phi \tag{1}$$

where Φ is a single determinant:

$$\Phi = |\psi_1(\vec{x}_1)\psi_2(\vec{x}_2)........|. \tag{2}$$

O is a projection operator which, by construction, selects the desired symmetry of the electronic state. Although the spin orbitals in (2) can, in principle, be quite general,

$$\psi(x) = \phi^+(\vec{r})\alpha + \phi^-(\vec{r})\beta,$$

they have in most applications been chosen as pure α or β spin orbitals, which is often denoted as "different orbitals for different spin method (DODS). Calculations with general spin orbitals (GSO) have so far been reported only for two- and three-electron atoms /2,3/. In this paper, both types of spin orbitals are used in a study of the lowest $^1\Sigma_g^+$ states of H_2. The hydrogen molecule differs

*Present address: Chemistry Department, Edinburgh University, West Mains Road, Edinburgh EH9 3JJ, Scotland.

from the previously studied system primarily in its spatial symmetry, which introduces new problems compared to the atomic case. PHF calculations on H_2 using DODS have previously been reported by Davidson and Jones /4/ Goddard /5/ and Morrison and Gallup /6/.

II. METHOD

The singlet component of the determinant (2) is selected by a projection operator of the form /1/

$$O_s = \frac{1}{2}(2 - \hat{S}^2) \tag{3}$$

Since S = 0 in this case, pure S_z symmetry is obtained without further projection.

The Σ_g^+ symmetry can be achieved by different methods. In this investigation, Σ^+ symmetry was achieved by restricting the basis in which the spin orbitals were expanded to orbitals of σ-symmetry. However, it is possible to set up a projection operator also for this symmetry and relax this restriction on the orbitals /6/.

The g symmetry has in other work been obtained either by restrictions on the orbitals /4,5/ or by projection /6/. If the first method is chosen, there are two different alternatives which lead to a total wave function of pure g symmetry. The simplest one is to restrict both orbitals to g symmetry, which automatically ensures total g symmetry. This method was investigated by Davidson and Jones /4/ and describes what can be called the "in-out" correlation /4/. The second possibility, which was also studied by Davidson and Jones /4/ and later by Goddard /5/, is to force the spatial parts of the orbitals to satisfy the relation

$$\hat{i}\psi_1(\vec{x}) = \psi_2(\vec{x}) \tag{4}$$

where $\hat{i}$ is the inversion operator.This type of wavefunction is capable of describing the aptly named "left-right" correlation, which is found to be more important than the "in-out" correlation at equilibrium and larger distances /4/.

The other alternative is to let the orbitals vary freely without restrictions, and construct g symmetry by means of a projection operator /6/

$$O_g = \frac{1}{2}(\hat{i} + 1). \tag{5}$$

The total wave function obtained in this manner includes the two previous ones as special cases. Thus, it accounts for both "in-out" and "left-right" correlation, and necessarily gives a lower energy than either of all previous methods at all distances. In the present work all three types of wave-functions were investigated in the DODS case. In the general spin orbital case the fully projected function

$$\Psi = O_g O_s |\psi_1(\vec{x}_1)\psi_2(\vec{x}_2)| \quad (6)$$

was used; ψ_1 and ψ_2 were varied freely without imposing restrictions on their form.

The optimum spin orbitals were determined by a computational method described elsewhere /2,7/, which can be used for both ground and excited states of the same symmetry. The orbitals were expanded in a basis of Slater type orbitals (STO's) of s- and p_σ-symmetry centred at the hydrogen nuclei.

III. RESULTS AND DISCUSSION

A. Ground State

In the case of two-electron atoms it was previously found that a PHF function with general spin orbitals is equivalent to a multi-configuration self-consistent field (MC-SCF) function with three configurations /2/ rather than four, as suggested by Bunge /8/. The projected GSO function for the hydrogen molecule is, for similar reasons, expected to correspond to a configuration interaction (CI) calculation with three $(\sigma_g)^2$ and three $(\sigma_u)^2$ configurations. In a basis of not more than three σ-type basis functions at each centre, a PHF calculation with general spin orbitals should therefore give the same result as a full CI calculation.

Table I shows the ground state electronic energies obtained with different methods and different basis sets. The basis sets used by Goddard /5/, Morrison and Gallup /6/ and Fraga and Ransil /9/ consist of one 1s and one 2s and one $2p_\sigma$-function, while McLean and Ransil /9/ use an additional 1s-function at each centre. Davidson and Jones /4/ use an extensive set of functions expressed in elliptic coordinates, which is probably close to complete for the ground state, but could not be used in the present calculations.

Full CI results and projected GSO results can be directly compared in the cases where only six basis functions are used. One finds that the GSO energy -1.873594 a.u. in Morrison and Gallup's basis agrees with their full CI result -1.87358 a.u. /6/. The small discrepancy in the fifth decimal occurs also in the pure spin

Table I

Calculated electronic energy, in a.u., at R = 1.4 a.u., for the ground state of hydrogen with different methods and basis sets. Exact energy -1.88876 a.u. /17/.

Basis	Type of wavefunction (see text)				
	Hartree-Fock	In-Out	Left-right	DODS	GSO
Goddard	-1.847753	-1.855216	-1.865812[a)]	-1.872447	-1.873432
Morrison & Gallup	-1.847698	-1.855410	-1.865720	-1.872590[b)]	-1.873594[c)]
Fraga & Ransil[d)]	-1.846760[e)]	-1.854256	-1.864735	-1.871394	-1.872392[f)]
McLean et al.	-1.847594	-1.855529	-1.865538	-1.872727	-1.873652
Davidson & Jones[g)]	-1.847859	-1.856119	-1.866358	-	-

a) Ref. 5

b) Morrison and Gallup report -1.87257 a.u.

c) Morrison and Gallup report the full CI result -1.87358 a.u.

d) R = 1.402 a.u.

e) Ref. 9

f) Fraga and Ransil report the full CI result -1.87246 a.u.

g) From Ref. 4

orbital case, where we obtain -1.87259 a.u. using full projection, compared to Morrison and Gallup's -1.87257 a.u. which indicates that their integral evaluation may be slightly inaccurate. (We believe our own figures to be correct since we reproduce Goddard's /5/ result as well as Fraga and Ransil's /9/ HF results to six decimals). The agreement with Fraga and Ransil's full CI results, which are lower than the GSO results, is not as good, but we attribute this to a possible error in their CI program.

The improvement in energy when going from pure to general spin orbitals can be seen to be about 0.001 a.u., independent of basis set, which is approximately the value suggested by Bunge /8/.

In bypassing, the results in Table I can illustrate the importance of optimization of the basis set for the specific functional form which is used, if small basis sets are used. The basis sets in the first three rows of Table I all contain STO's of the same type, but optimized for different wave-functions. Goddard's basis was optimized for the left-right correlated function, and does give lower energy for this function than any of the other STO basis sets, including the larger basis used by McLean et al. The basis used by Morrison and Gallup, in contrast, was optimized for the CI calculation, as well as McLean's and thus gives a lower energy than Goddard's for the GSO function as well as the fully projected DODS function.

It should also be mentioned that we explored the possible existence of general spin orbital solutions of Overhauser /11/ type in the unprojected single determinant case, a question raised, among others, by Musher /12/. The result was negative, in the sense that the GSO solutions gave exactly the same energy as the ordinary unprojected DODS solution of Coulson-Fischer /13/ type. In this respect there was no difference between the molecular case and the atomic systems investigated before /14/.

B. Excited State

The first excited ${}^1\Sigma_g^+$ state of H_2 has been shown to have an unusual total energy versus internuclear distance curve, with two minima at 1.9 and 4.3 a.u. separated by a maximum at 3.3 a.u. /15/ Davidson attributed the existence of the double minimum to the effects of prohibited energy curve crossing. As the internuclear separation increases from 1.9 a.u. to 4.3 a.u., the nature of the wave function changes from a covalent structure, ($1s\sigma_g$, $2s\sigma_g$) to the more ionic state $(2p\sigma_u)^2$. At the maximum around 3.3 a.u. strong mixing occurs between these two configurations. The nature of the wave function is thus very different at the distances 1.9, 3.3 and 4.3 a.u. This is also reflected in the computational aspects (*vide infra*).

Early calculations /9/ failed to reveal the double minimum and led to erroneous assignments of the excited states, due to the use of a poorly constructed basis (optimized for the ground state). Later CI calculations have confirmed the double minimum structure /16/ demonstrating that the choice of basis for the excited state calculations is crucial. From physical considerations, it is clear that a ground state basis is inappropriate since the diffuse $2s\sigma_g$ and $2p\sigma_u$ orbitals are poorly described by such a basis. As an example, Goddard's ground state basis gives a GSO energy of

-0.18924 a.u. for the first excited $^1\Sigma_g^+$ state, which is far above the accurate value -0.70295 a.u. reported by Kolos and Wolniewicz /17/. A reduction of the values of the 2s and 2p orbital exponents, to 0.5 and 0.8, respectively, lowers the energy drastically to -0.67323 a.u., which is still somewhat high but physically reasonable. No further exponent optimization was however done at this internuclear distance, since it was not judged to be very interesting.

A further problem is presented by the observation that the nature of the wave function changes radically with internuclear separation (vide supra). The optimal values of the orbital exponents are therefore expected to be rather different at different distances, which should ideally necessitate a complete reoptimization of the orbital exponents at every distance. For reasons of economy, this was however done only for the two inflexion points at 1.9 and 3.3 a.u. (due to problems with convergence, complete optimisation was not carried out at the internuclear distance 4.3 a.u.). The PHF energies at these points are shown in Table II both for pure spin orbitals (DODS) and general spin orbitals (GSO), together with the results of Davidson /15/ and Gerhauser and Taylor /16/, as well as the experimental energy, where available.

The calculations at R = 1.9 a.u. and shorter distances offered no computational problems, and converged rapidly to the final solutions both in the DODS and GSO cases. The convergence was in fact much faster than in the ground state, where the GSO function converged rather slowly. The PHF energy is about 0.003 a.u. higher than the CI results of Davidson and Gerhauser and Taylor and 0.005 a.u. above experiment, compared to 0.016 a.u. for the ground state. It can be noted that the energy improvement with general spin orbitals is only 10^{-4} a.u., one order of magnitude smaller than in the ground state, which indicates that the natural spin orbital expansion /1/ for the excited state is more rapidly convergent than in the ground state.

For the internuclear distance 3.3 a.u. the PHF results were quite close to the previous CI calculations. The GSO energy is here 0.001 a.u. above Davidson's and 0.004 a.u. above Gerhauser and Taylor's results. Moreover, the difference between the DODS and GSO results is considerably larger, 0.001 a.u., and the GSO energy is actually lower than Gerhauser and Taylor's two-configuration result. This indicates that the natural spin orbital expansion is less rapidly convergent at this distance, which is to be expected from the strong configuration mixing between the states ($1s\sigma_g$ $2s\sigma_g$) and ($2p\sigma_g$ $2s\sigma_g$) and $(2p\sigma_u)^2$, as discussed above.

Finally at the distance 4.3 a.u. very serious convergence problems were found. The best solution for this distance was about 0.01 a.u. above the previous CI results. The reason for these difficulties is probably that the two first coefficients in the

Table II

Total energy, in a.u. for the first excited $^1\Sigma_g^+$ state of H_2 with different methods.

R (a.u.)	PHF DODS	PHF GSO	Davidson[15]	Gerhauser & Taylor[16]	Experiment
1.9	-0.71262[a]	-0.71274[a]	-0.71618	-0.71643	-0.71812
3.3	-0.68214[b]	-0.68327[b]	-0.68443[d]	-0.68709	-0.69235[f]
4.3	-0.69254[c]	-0.69697[c]	-0.70067[e]	-0.70522	-0.71438[f]

a) Basis: 1s (1.2063), 2s (0.3537), $2p_\sigma$(1.512), 3s (0.5047).

b) 1s (1.135), 2s (1.25), $2p_\sigma$ (1.119), 3s (0.579).

c) 1s (1.1), 2s (0.9), $2p_\sigma$ (0.9), 3s (0.6); not fully optimised.

d) At 3.25 a.u.

e) At 4.35 a.u.

f) Accurate theoretical values from ref. 17.

natural spin orbital expansion have the same sign /15/, which is not possible to reproduce by a PHF function unless the orbitals are complex. These problems have not yet been solved.

IV. CONCLUSION

The PHF method with general spin orbitals is seen to provide a good description of the lowest $^1\Sigma_g^+$ states of H_2, and reproduces the unusual double minimum character of the first excited state potential curve. For two-electron systems, the method is computationally simple. However, the extension to larger numbers of electrons is likely to lead to greater difficulties in interpretation and computation than would be presented by conventional CI calculations.

ACKNOWLEDGEMENT

One of us (GH) thanks the Science Research Council (U.K) for the award of a NATO Fellowship.

PERSONAL NOTE (GH)

For two years, I had the pleasure of participating as a member of the Uppsala group and owe a great deal to Per-Olov, not only scientifically, but also for the exceptional (and legendary) hospitality extended towards my family and myself. It is therefore an honour for me to dedicate this paper to him on the occasion of his 60th birthday.

REFERENCES

1. P.-O. Löwdin, Phys. Rev. 97, 1474, 1490, 1509 (1955).
2. S. Lunell, Phys. Rev. A1, 360 (1970).
3. N.H.F. Beebe and S. Lunell, J. Phys. B. (to be published).
4. E.R. Davidson and L.L. Jones, J. Chem. Phys., 37, 1918, (1962).
5. W.A. Goddard, III, Phys. Rev., 157, 81, (1967).
6. R. Morrison and G.A. Gallup, J. Chem. Phys., 50, 1214 (1969).
7. S. Lunell, Phys. Rev. A7, 1229 (1973).
8. C.F. Bunge, Phys. Rev. 154, 70 (1967).
9. S. Fraga and B.J. Ransil, J. Chem. Phys. 35, 1967 (1961).
10. A.D. McLean, A. Weiss, and M. Yoshimine, Revs. Mod. Phys., 32, 211 (1960).
11. A.W. Overhauser, Phys. Rev. Lett., 7, 415, 462 (1960).
12. J. Musher, Chem. Phys. Lett. 7, 397 (1970).
13. C.A. Coulson and I. Fischer, Phil. Mag. 40, 386 (1949).
14. S. Lunell, Chem. Phys. Lett., 13, 93 (1972).
15. E.R. Davidson, J. Chem. Phys., 33, 1577 (1960); 35, 1189 (1961).
16. J. Gerhauser and H.S. Taylor, J. Chem. Phys., 42, 3631 (1965).
17. W. Kolos and L. Wolniewicz, J. Chem. Phys., 50, 3228 (1969).

DEFORMED ATOMS AND THE PROJECTED HARTREE-FOCK METHOD

P. Van Leuven

Dienst Theoretische en Wiskundige Natuurkunde

Rijksuniversitair Centrum, Antwerp, Belgium

1. INTRODUCTION

Atoms are usually considered as spherical systems. The self-consistent field is commonly supposed to be a central field and the atomic orbitals are taken as eigenfunctions of the angular momentum ℓ. It is well known that in the case on non-closed shell configurations these assumptions are inconsistent. The valence electrons will deform the core and give rise to a non-central field. These effects can be described in two equivalent ways: either by introducing deformation parameters in the self-consistent potential and working with orbitals in a deformed field or by considering the orbitals as mixtures of eigenvalues of angular momentum with different ℓ-values. It is, of course to be expected that due to the preponderance of the nuclear field, this deformation and these admixtures will be small and hence will not influence the binding energies very much. Nevertheless certain operators which are particularly sensitive to special modes of deformation might show marked effects in their expectation values.

If one departs from the Restricted Hartree Fock (RHF) picture with symmetric orbitals, the many-electron wave functions for closed-shell or one-particle configurations are no longer eigenfunctions of total angular momentum L. In order to get wave functions which are physically admissible it was proposed by Löwdin /1/ to project onto the subspace of desired angular momentum and use this projected function for the physical state. Then immediately the question arises of how to determine the orbitals selfconsistently. One can either determine the orbitals by minimizing the energy before projection and use this optimized unprojected determinant or one can determine the orbitals after projection, i.e. by minimizing the energy calculated with the projected function. The former method we shall call "Unrestricted Hartree-Fock" (UHF) and the latter "Projected Hartree-Fock" (PHF). Another possibility is to use the projected wave function with the UHF-orbitals; this will be indicated by the notation HFP. From variational arguments it is obvious that PHF is the better of these methods, i.e. it leads to the lowest ground state energy. It is not clear, however, how good or bad the other approximations are as far as other physical quantities are concerned. It should be stressed that the calculation of the energy with projected wave functions is very complicated and hence it is improtant to ascertain what exactly is the improvement to be expected from such a procedure.

Entirely parallel to the above mentioned questions about angular momentum is the problem of spin polarization. If the equivalence restriction is lifted and different spatial orbitals are assigned to electrons of a spin pair, the total wave function is not an eigenfunction of total spin S anymore. Spin projection is needed and different methods, analogous to the ones described above, can be considered. These problems have been studied quite intensively in the past. Spatial deformation of atomic orbitals on the contrary has been investigated rather seldom, probably bacause the effects are expected to be small and possibly also because of the great complexity of angular momentum projection.

In this paper we want to study, by means of the simplest possible example, the relation between UHF and PHF as to the influence of spatial deformation and L-projection on the expectation values of physical operators. We shall only treat the orbital angular momentum aspect and not consider spin effects again. The purpose is to contribute some further insight in the Projected Hartree-Fock method, a field of research, the foundations of which have been laid by Per Olov Löwdin.

2. THE Li-ATOM

The simplest, non-trivial system suitable for our investigation is the Li-atom. In the RHF scheme the ground state would be represented by the configuration $(1s)^2 2s$ and the first excited state by $(1s)^2 2p$. For our purpose the excited state is the simpler because of the orthogonality of core and valence orbitals due to different angular momenta. Furthermore, this state has total angular momentum L=1 and hence expectation values of tensor operators with rank 0,1 and 2 can be studied. We know that due to spin polarization, the core orbitals will be different, but what we want to describe here is the fact that the core orbitals are no longer pure s-waves but are deformed by the valence electron. The origin of this deformation can be understood as follows: the mean field felt by the core electrons is of the form

$$(s|\frac{1}{r_{12}}|s) + (p|\frac{1}{r_{12}}|p)$$

(brackets indicate integration over one of the coordinates 1,2). If the Coulomb repulsion is developed in multipoles we see that this field contains monopole and quadrupole terms. Hence the core orbitals move in a field with quadrupolar deformation. The valence electron, on the other hand, feels a mean field of the form $(s|1/r_{12}|s)$. Its deformation is only a second order effect due, indirectly, to the deformation of the core orbitals. From these considerations it seems plausible to assume for the core orbitals a form s+d (small admixture of quadrupolar d-wave to the s-wave) and for the valence electron to keep its pure p character. This d-admixture can be considered as an extra freedom in the wave function that will be used to lower the energy of the system. We shall therefore try to introduce ideas analogous to Löwdin's Alternant Molecular Orbital theory /1/ in which two electrons of a spin pair are placed in orbitals localized in different regions of space. In terms of d-admixture this can be done by assigning to the α and β core electrons the orbitals $s+\varepsilon d$ and $s-\varepsilon d$ respectively. Those orbitals have prolate and oblate probability distributions (to first order in ε), consequently the model corresponds to partially concentrating one core orbital in the direction of the z-axis and the other in the xy-plane. This picture is the rotational analogue of the translational delocalization in the Alternant Molecular Orbital theory.

The proposed UHF wave function of the first excited 2^2P-state, is (unnormalized)

$$\phi = \det\,[(s+\varepsilon d)\alpha, (s-\varepsilon d)\beta, p\alpha] \qquad (1)$$

The orbitals will be taken as eigenfunctions of ℓ_z with eigenvalues 0 for d and 1 for p. We thus assume axial symmetry for the field, which is a consistent symmetry. In fact we could as well take $\ell_z = 0$ or -1 for the p orbital, this would merely change the total L_z value. With the present assumptions ϕ is an eigenfunction of L_z with value 1 and of S_z with value $\frac{1}{2}$, hence $J_z = \frac{3}{2}$. We shall further consider s, p and d to be normalized to unity. We can develop the wave function in a power series of ε:

$$\phi = \phi_0 + \varepsilon\phi_1 + \varepsilon^2\phi_2 \qquad (2)$$

The value of ε in the UHF approximation can be calculated by considering the energy in the state ϕ to second order in ε. It is straightforward to show the following relations:

$$(\phi\phi) = N_0 + \varepsilon^2 N_2$$

$$(\phi T\phi) = T_0 + \varepsilon^2 T_2$$

$$(\phi V\phi) = V_0 + \varepsilon V_1 + \varepsilon^2 V_2$$

Here T and V are the one-and two electron parts of the Hamiltonian and $N_0 = (\phi_0\phi_0)$, $T_0 = (\phi_0 T\phi_0)$ and $V_0 = (\phi_0 V\phi_0)$. The first and second order corrections can be written as $T_2 = (\phi_1 T\phi_1)$ and $V_1 = 2(\phi_0 V\phi_1)$ where $\phi_1 = \det(d\alpha, s\beta, p\alpha) - \det(s\alpha, d\beta, p\alpha)$. Finally, if v stands for the Coulomb repulsion between 2 electrons

$$V_2 = 2(sd|v|sd) - 2(sd|v|ds) + 2(sp|v|sp) - (sp|v|ps) + 2F_0(d,p) \quad (3)$$

To obtain the relation (3) we have introduced two approximations: 1) we have neglected the matrix element $(ss|v|dd)$ with respect to V_2 and 2) the multiplet structure of the configuration pd has been neglected, i.e. the Slater radial integrals F_2, G_1 and G_3 are neglected with respect to F_0. The former approximation (neglect of configuration interaction) is justified on account of the small differential overlap (sd) and the second assumption is reasonable from inspection of the numerical values of F_k and G_k for the configuration (2p, 3d).

It can easily be seen, by means of angular momentum algebra, that the effect of L-projection on the above matrix elements is to be described by the following relations

$$(\psi\psi) = N_0 + \varepsilon^2 C_1^2 N_2$$

$$(\psi T\psi) = T_0 + \varepsilon^2 C_1^2 T_2$$

$$(\psi V\psi) = V_0 + \varepsilon V_1 + \varepsilon^2 C_1^2 V_2$$

where C_1 is the Clebsch-Gordan coefficient $(2011|11) = 1/\sqrt{10}$. The same approximations concerning the second order potential energy V_2 have been made here. The only influence is to multiply all second order terms by C_1^2 and to leave zeroth and first order terms unaltered; hence, after minimizing the energy in the UHF resp. PHF approximations, we get the relations

$$\varepsilon^{UHF} = V_1/2\ [\ (T_0 + V_0)N_2/N_0 - (T_2 + V_2)\ ] \quad (4)$$

$$\varepsilon^{PHF} = \varepsilon^{UHF} / C_1^2 = 10\varepsilon^{UHF} \quad (5)$$

Let us now turn to the expectation value of an arbitrary one body operator Ω.
It is easily seen that the only one-electron operators Ω, which can have a first order correction with respect to their restricted Hartree-Fock value, are of the form

$$\Omega = \sum_i \omega(r_i) Y_{20}(\theta_i \phi_i) s_z(i) \tag{6}$$

One such operator would be the spin dipolar term in the magnetic hyperfine structure. For an operator of the form (4) we would get (to first order)

$$(\phi\Omega\phi) = \Omega_0 + \varepsilon^{UHF}\Omega_1$$

$$(\Psi\Omega\Psi) = \Omega_0 + \varepsilon^{PHF} C_1^2 \Omega_1$$

Taking into account (5) we conclude that, in the framework of our approximations and up to and including first order

$$(\phi\Omega\phi) = (\Psi\Omega\Psi) \tag{7}$$

3. DISCUSSION

We have presented a wave function for states of the Li-atom belonging to the configuration s^2p in which all inconsistent symmetry restrictions have been removed. The deformation of the self-consistent field, due to the unclosed shell structure, is measured by means of the d-wave amplitude in the core-orbitals. We have schematized the model in such a way that only one variational parameter ε remains. The basic orbitals s,p,d are assumed to be known, e.g. from a RHF calculation. If we optimize the wave function without taking care of the total angular momentum (UHF) we obtain an amplitude ε given by (4), but if we optimize the L=1 component of the wave function (PHF) we obtain an amplitude ε which is enhanced $C_1^{-2} = 10$ times.

We can distinguish between three energy values

$$E_{UHF} = H_0 - H_1^2/4H_2 \tag{8}$$

$$E_{PHF} = H_0 - H_1^2/4H_2C_1^2 \tag{9}$$

$$E_{HFP} = H_0 - (2 - C_1^2)H_1/4H_2 \tag{10}$$

where $H_0 = (T_0+V_0)/N_0$, $H_1 = V_1/N_0$ and $H_2 = (T_2+V_2-H_0N_2/N_0)/N_0$.

The energy gain with respect to RHF is of the order of H_1^2/H_2 and hence expected to be small compared to the total correlation energy. The effect of straight projection, i.e. $E_{UHF}-E_{HFP}$, is to increase the energy gain by a factor of about 2, whereas the effect of deformation and projection taken together, i.e. $E_{UHF}-E_{PHF}$, is to increase the energy gain by a factor $C_1^{-2} = 10$.

We have also seen that, in the simple model used here, expectation values of one-particle operators independent of spin are not changed with respect to their RHF value in first order. Operators of rank 2 and of the type $F = \sum_i f(i)s_z(i)$ do get a first order correction. This correction is decreased by a factor C_1^2 in HFP, but if deformation is taken into account consistently, i.e. in PHF theory, the same value as in UHF is found. We conclude that angular momentum projection has two antagonistic effects: it enhances the deformation in the wave function by a factor C_1^{-2} but it reduces the expectation values in first order by a factor C_1^2. As far as expectation values of one-body operators are concerned, PHF and UHF give the same answer to second order. Of cource, these results are based on the validity of a perturbation expansion, which would not be true in nuclear physics, where deformations are much larger.

The formulae of section 2 can be generalized immediately to the case of a configuration $s^2\ell$; the only modification is to replace C_1 by $C_\ell = (20\ell\ell|\ell\ell) = [\ell(2\ell-1)/(\ell+1)(2\ell+3)]^{1/2}$. A generalization to other configurations and an arbitrary number of particles has not been performed.

A rather similar analysis regarding spin projection has been carried out by Marshall /2/; essentially the same conclusions were reached. However, in the present case of orbital angular momentum, the effects are more important due to the higher values of the parameters occuring in the Clebsch-Gordan coefficients.

Recently a numerical PHF-calculation has been performed /3/ for the 2^2P-state of Li, using the complete expression for the energy, which contains terms up to an including fourth order. There, the s, p and d orbitals were expanded on a limited STO-basis and determined self-consistently after projection of both spin and angular momentum. No a priori relationship between the core orbitals was imposed. The results of that calculation confirm the analysis presented in this paper.

REFERENCES

1. P.O. Löwdin, Phys. Rev. 97, 1474, (1955)

2. W. Marshall, Proc. Phys. Soc. 78, 113, (1961)

3. S. Lunell et al., to be published

MCSCF STUDIES OF CHEMICAL REACTIONS:

NATURAL REACTION ORBITALS AND LOCALIZED REACTION ORBITALS

Klaus Ruedenberg and Kenneth R. Sundberg

Ames Laboratory-ERDA and Department of Chemistry

Iowa State University, Ames, Iowa 50010

INTRODUCTION

Since its inception by Hund and Mulliken /1/, molecular orbital theory has proved to be extremely fruitful. On the one hand, it has offered to molecular scientists a simple conceptual model, namely the pseudo-one-electron picture, for understanding molecular spectra and chemical structure. On the other hand, it has been amenable to rigorous mathematical formulation in terms of a single antisymmetrized product of molecular orbitals,

$$\Psi = A\{(\phi_1\alpha)(\phi_1\beta)(\phi_2\alpha)(\phi_2\beta)\ldots(\phi_{\frac{1}{2}N}\alpha)(\phi_{\frac{1}{2}N}\beta)\} , \qquad (1)$$

and optimal MO's for wavefunctions of this type can be straightforwardly determined by the Hartree-Fock self-consistent field procedure /2/. Numerous implementations on modern digital computers have made ab-initio as well as semi-empirical HF-SCF-MO calculations important research tools for experimental as well as theoretical chemists. For the interpretation of such calculations, two particular orbital representations have proved especially illuminating: The canonical MO's are fundamental to most problems concerning electronic spectra, and the localized MO's greatly simplify discussions concerning the structure of related molecules /3/.

However, in general a single configuration of this type cannot adequately represent large enough portions of a potential energy surface to describe a chemical reaction. For example, for most chemical bonds HF-SCF-MO theory predicts dissociation into ions rather than atoms or radicals, and in many reactions, where the products have spin multiplicities different from those of the reactants, a wavefunction such as that of Eq. (1) does not contain the necessary spin-recouplings. Wavefunctions that properly describe such reactions must be written as superpositions of several configurations, i.e.,

$$\Psi = c_1\Psi_1 + c_2\Psi_2 + C_3\Psi_3 + \ldots , \qquad (2)$$

where the various configurations involve various orbital products and various spin couplings, e.g.,

$$\Psi_K = A\{\phi_{K,1}\phi_{K,2}\cdots\phi_{K,N}\Theta_K\}. \qquad (3)$$

Here Θ_K is a spin eigenfunction of the appropriate multiplicity /4/. The determination of optimal orbitals for wavefunctions of this type is more difficult than the determination of HF-SCF orbitals. But several solutions of this "MC-SCF" problem have recently been developed, among them the methods of Das and Wahl /5/, of Hinze and Roothaan /6/, of Grein and Chang /7/, and of Cheung and Ruedenberg /8/. As these methods become available, MCSCF calculations will be reported more frequently.

It is apparent that the interpretation of such wavefunctions requires a wider perspective than the simple MO picture, and it is the objective of this note to develop interpretative concepts concerning the description of chemical reactions in terms of such a MCSCF approach.

ORBITAL REACTION SPACE AND CONFIGURATIONAL REACTION SPACE

Our first objective is to formulate a multiconfiguration wavefunction including only those configurations beyond the HF-SCF-MO approximation that are <u>necessary to describe a specific reaction</u> and, moreover, those additional configurations whose <u>energy lowering contributions will change during the reaction by comparable amounts</u>. Usually chemical reactions involve the cleavage, formation, or rearrangement of only a few bonds in a molecule, and the most affected orbitals are usually easily identified. Clearly the MCSCF treatment should be applied to this reactive part of the molecule, while the rest of the system can still be described after the fashion of the HF-SCF-MO approximation.

Suppose the reactants as well as the products of a reaction have the HF-SCF-MO wavefunctions

$$\Psi_{React} = A\{\phi'_1\alpha\phi'_1\beta\ldots\phi'_\pi\alpha\phi'_\pi\beta\psi'_1\alpha\psi'_1\beta\ldots\psi'_n\alpha\psi'_n\beta\} , \qquad (4)$$

$$\Psi_{Prod} = A\{\phi''_1\alpha\phi''_1\beta\ldots\psi''_\pi\alpha\sigma''_\pi\beta\psi''_1\alpha\psi''_1\beta\ldots\psi''_n\alpha\psi''_n\beta\} . \qquad (5)$$

The orbitals $\phi'_1,\ldots,\phi'_\pi,\phi''_1,\ldots,\phi''_\pi$ describe those chemical functions that are common to both systems, such as inner shells, lone pairs, and bonds remote from the reactive region. The orbitals ϕ'_i will differ only slightly from the orbitals ϕ''_i. On the other hand, the orbitals $\psi'_1,\ldots,\psi'_n,\psi''_1,\ldots,\psi''_n$ describe the reactive portions of the molecule, and the ψ'_k are very different from the ψ''_k. Nonetheless, some of the ψ'_k may go over into certain of the ψ''_k through continuous smooth transformations during the course of the reaction, and such a pair of orbitals ψ'_k, ψ''_k can be considered to be <u>one deformable orbital</u> which essentially retains constant occupancy during the course of the reaction. If this is true for all the orbital pairs, ψ'_k, ψ''_k, then an HF-SCF-MO description of the reaction is possible. For most molecoles, however, there are some orbitals ψ'_k that simply do not appear among the ψ''_k, and vice versa. In this case the very simplest wavefunction that can smoothly describe the reaction is

$$\Psi = C_{React}\,\Psi_{React} + C_{Prod}\,\Psi_{Prod}\,. \tag{6}$$

But such a simple superposition is usually inadequate. To see this, let us divide the orbitals into three identifiable groups as follows:

Orbital	Deformation	Occupancy
"Unaffected"	Essentially Undeformed	Essentially Unchanged
"Affected"	Essentially Deformed	Essentially Unchanged
"Replaced"	Severely Deformed	Essentially Changed

Along certain portions of the reaction's potential energy surface, those orbitals which are being replaced and those orbitals replacing them have approximately the same occupation numbers, and there usually exist other configurations, in addition to the two of Eq. (6), in which the orbitals have similar occupation. Some of these latter configurations will involve the "affected" orbitals in various occupations, and they will mix into the wavefunction to a significant extent.

For the purpose of the present analysis we take the point of view that the "reaction orbitals" ψ'_k, ψ''_k lie within a certain "orbital reaction space" spanned by a basis ψ_1, $\psi_2,\ldots\psi_m$, and that the wavefunction Ψ lies within the "configurational reaction space" spanned by <u>all possible configurations</u> whose space part is a product of the orbitals ψ_1, $\psi_2,\ldots\psi_m$ times a closed shell core made from the "unaffected" orbitals $\phi_1, \phi_2, \ldots, \phi_\pi$. That is, Ψ is a superposition of all configurations of the form

$$\Psi_{K,\kappa} = A\{\phi_1\alpha\phi_1\beta\ldots\phi_\pi\alpha\phi_\pi\beta\psi_{K,1}\ldots\psi_{K,n}\Theta_\kappa\} \quad , \tag{7}$$

where the $\psi_{K,\kappa}$ are chosen from the reaction orbital basis $\psi_1, \psi_2, \ldots \psi_m$, and the orbital products defined by Eq. (7) contain them in all singly and doubly occupied ways. Each orbital product must have the proper spatial symmetry and be multiplied by all possible spin eigenfunctions Θ_κ of the proper spin multiplicity to discribe the reaction. Moreover, the reaction orbital basis is presumed to obey the orthogonality conditions

$$\langle\phi_i|\phi_j\rangle = \langle\psi_i|\psi_j\rangle = \delta_{i,j}, \langle\phi_i|\psi_j\rangle = 0 \; . \tag{8}$$

We say that the $\Psi_{K,\kappa}$ span the "complete configurational reaction space (CCRS), because it cannot be enlarged without adding a new reaction orbital or opening one of the closed shells.

Just how large the CCRS is, i.e., how many configurations exist, depends on the number of reaction orbitals $\psi_1, \psi_2, \ldots \psi_m$ and the number of reactive electrons; naturally this varies from case to case. It is often practical to use the same number of reaction orbitals as there are components of the minimal AO basis* located in the reactive region. Thus, the dissociation of molecular H_2 to atomic H involves two MO's, as shown in Fig. 1a and, for singlet states, the CCRS configurations. The dissociation of a double bond, like ethylene, involves four reaction orbitals like those shown in Fig. 1b, which, assuming D_{2h} symmetry generate a singlet state CCRS with eight configurations. These systems have the same number of reaction orbitals as there are reactive electrons; for some systems, however, there are fewer MO's. For instance, the acid base reaction shown in Fig. 1c uses three orbitals to describe four electrons; the CCRS uses six configurations to treat the singlet state.

NATURAL REACTION ORBITALS

The completeness of the CCRS implies that it is invariant to orthogonal transformation among the reaction orbitals. Thus, if a new basis ψ_k is defined in the reaction orbital space by an orthogonal

---

* By this we mean the "conceptual" minimal basis, not the computational atomic orbital basis, which is usually much larger. The explicit form of this conceptual minimal basis is one of the results of the MCSCF calculation (see the localized reaction orbitals below).

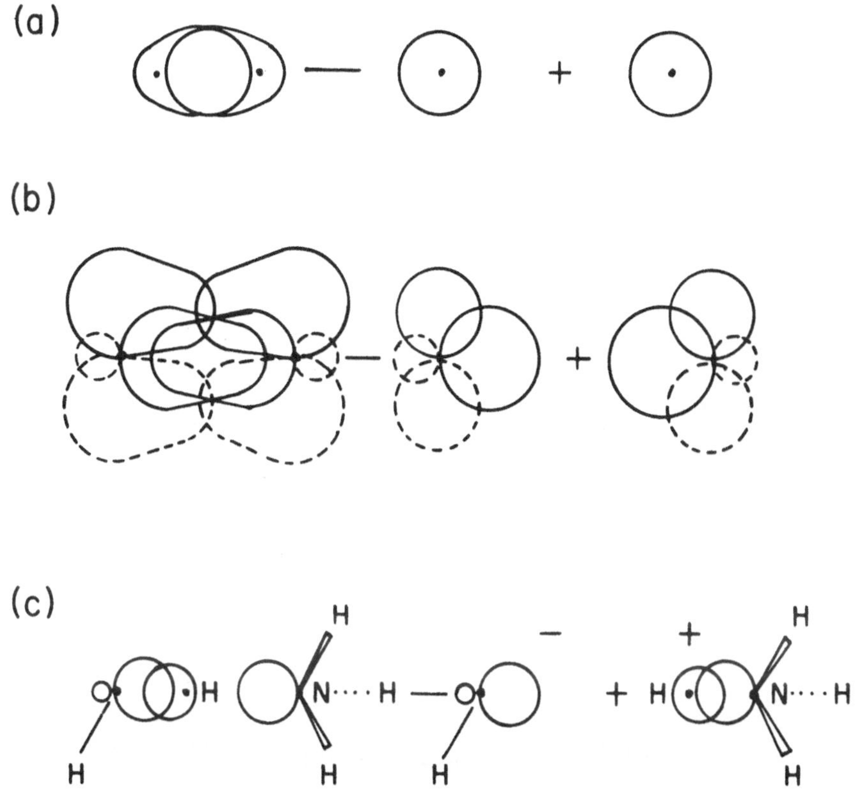

Figure 1

Schematic Diagrams of minimal reaction-orbital bases for (a) the dissociation of H_2; (b) the dissociation of ethylene; (c) the acid base reaction of H_2O and NH_3.

transformation,

$$\bar{\psi}_k = \sum_j \psi_j T_{jk} , \tag{9}$$

and if, now, a new CCRS basis is generated using these new orbitals, i.e.,

$$\bar{\Psi}_{K,\kappa} = A\{\phi_1\alpha\phi_1\beta\ldots\phi_\pi\alpha\phi_\pi\beta\bar{\psi}_{K,1}\bar{\psi}_{K,2}\cdots\bar{\psi}_{K,n}\Theta_\kappa\}, \tag{10}$$

then these new configurations are related to the configurations $\Psi_{K,\kappa}$ Eq. (7) by an orthogonal transformation,

$$\overline{\Psi}_{K,\kappa} = \sum_{J,\gamma} \Psi_{J,\gamma} T_{J,\gamma;K,\kappa} \tag{11}$$

These new configurations span the same configurational space, namely the CCRS, so the wavefunction of the system can be expressed in either basis:

$$\Psi = \sum_{K,\kappa} \overline{C}_{K,\lambda} \overline{\Psi}_{K,\kappa} = \sum_{J,\gamma} C_{J,\gamma} \Psi_{J,\gamma} \ . \tag{12}$$

The MCSCF procedure optimizes the coefficients $C_{K,}$ in Eq. (12) as well as the orbitals $\phi_1,\ldots,\phi_\pi$, $\psi_1,\ldots,\psi_m$ by optimizing their expansions in a much larger basis set of atomic orbitals. However, from the orbital invariance just discussed, it is apparent that neither the coefficients $C_{K,\kappa}$, nor the molecular orbitals ϕ_j, ψ_k are unique. The orbitals are only determined to within two orthogonal transformations; one transformation between the $\phi_1,\ldots,\phi_\pi$ and another between the $\psi_1,\ldots,\psi_m$; the coefficients $C_{K,\kappa}$, of course, change accordingly. In general, however, there is a distinction between these orbitals because the first-order density matrix of the MCSCF wavefunction Ψ,

$$\rho(1,1') = 2(n+\pi) \int d\tau_2 \ldots d\tau_{2\pi+n} \ \Psi(1,2,\ldots)\Psi(1',2,\ldots), \tag{13}$$

is diagonal in the orbital space $\phi_1,\ldots,\phi_\pi$, but it is not diagonal in the orbital space $\psi_1,\ldots,\psi_m$. As a result, the orbitals $\psi_1,\ldots,\psi_m$ are uniquely fixed by requiring that $\rho(1,1')$ be diagonal,

$$\rho = \sum_j 2\phi_i\phi_j + \sum_k n_k\psi_k\psi_k \ . \tag{14}$$

As a result, <u>the completeness of the CCRS makes it possible to express the wavefunctions in terms of configurations constructed from molecular orbitals that are also natural orbitals</u> /9/. We propose the name <u>"natural reaction orbitals"</u> (NRO's) for these orbitals.

We expect, that, in the MCSCF theory, the NRO's play a role similar to that taken by the canonical orbitals in the HF-SCF-MO approximation. In concrete applications /10/, we have found that each NRO is characterized by two distinctive features: the occupation number and the actual shape and directionality of the function. They reveal (1) what role the orbital plays in the energy balance, <u>i.e.</u>, whether it is bonding, antibonding, or correlating, (2) where it plays this role, and (3) how important the role is for the wavefunction. Both occupation number and orbital shape usually change along the reaction path because the orbital changes its role and its importance. Thus, an examination of the NRO's at various stages of the reaction provides insight into how the reactive electrons are redistributed over the bonding, antibonding, and correlating orbitals. The NRO's therefore yield a convenient analysis of the essential characteristic changes that occur during the reaction.

The NRO's can be obtained, according to Eq. (9), from another set of optimal MCSCF orbitals, when the transformation T_{jk} has been determined by diagonalizing the first order density matrix. On the other hand, it is possible to design a MCSCF optimization method to calculate the NRO's directly /8/.

LOCALIZED REACTION ORBITALS

An alternative criterion which uniquely fixed another set of reactions orbitals is that of greatest localization. Using one of well known localization procedures /3/, one can obtain an orthogonal transformation which, used in Eq. (9), yields a set of reaction orbitals that are maximally localized and separated from one another. If the dimension of the reaction orbital space is equal to the number of components of the minimal atomic orbital basis set in the reactive region of the molecule, then these "localized reaction orbitals" (LRO's) will be essentially located on individual atoms. Indeed they will nearly represent the minimal-basis-set hybrid atomic orbitals in the environment of the reactive region of the molecule. They are linear combinations of the much larger atomic orbital basis set actually used to make the numerical calculations, and they will be symmetrically orthogonal to each other, though it would be possible to deorthogonalize them to make even more atom-centered functions.

In view of this character, the LRO's will undergo only relatively small changes during the reaction, which means that the larger changes associated with the reaction are reflected in the configurational expansion coefficients of Eq. (12). The interpretation of the first-order density matrix in terms of the LRO's is facilitated by the fact that the LRO's are orthogonal. The matrix, in terms of the LRO's $\hat{\psi}_\alpha$, has the form

$$\rho = \sum_i 2\phi_i\phi_j + \sum_{\alpha\beta} q_{\alpha\beta}\hat{\psi}_\alpha\hat{\psi}_\beta \ , \qquad (14)$$

and the diagonal terms $q_{\alpha\alpha}$ are the atom-like populations /11/ associated with the LRO's. The cross terms ($q_{\alpha\beta}, \alpha\neq\beta$) are interference terms /12/ between these basically atomic functions.

The LRO's bring out the persistent characteristics of a reaction orbital space; they are really its (orthogonalized) "atomic" basis. For instance, the populations of the LRO's are more stable than the corresponding quantities for the NRO's, and the changes they do make are reflective of subtle charge transfers such as those associated with changes in the polarity of the bonds.

ILLUSTRATIONS OF THE NRO's AND LRO's IN 1,2-DIOXETANE

The two types of orbitals introduced, the NRO's and the LRO's, represent complementary aspects of the changes in electron distribution that accompanu a chemical reaction: the NRO's embody the changes in bonding and correlation created by the reaction; the LRO's describe the more permanent aspects of the system such as its atomic populations. Both types of orbitals originate in an MCSCF calculation, which, in principle can involve rather lengthy expansions. However, for the dissociation of 1,2-dioxetane to ground state formaldehyde, see Fig. 2a, the singlet state CCRS for the reaction orbital

(a)

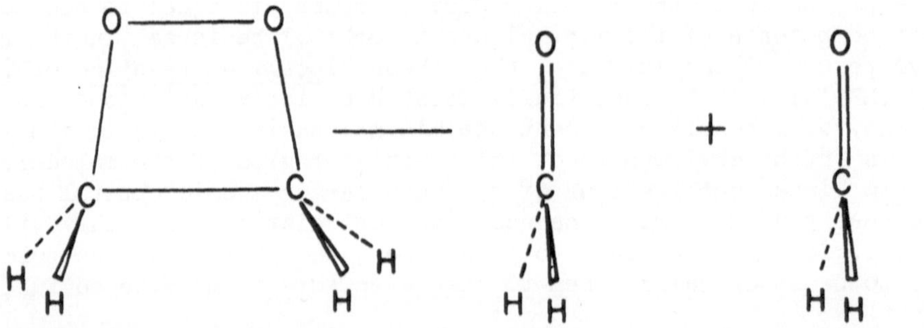

(b)

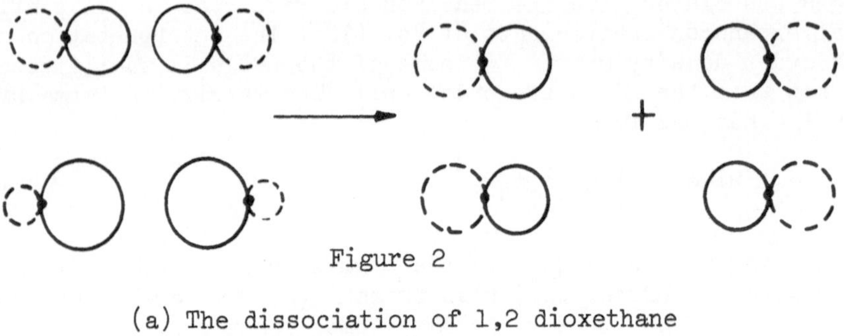

Figure 2

(a) The dissociation of 1,2 dioxethane into two formaldehydes;
(b) Schematic diagrams of the localized reaction orbitals for this reaction.

space contains only twelve configurations. This reaction and this wavefunction have been studied in our laboratory in the manner outlined here /10/. The four LRO's are indicated schematically in Fig. 2b, and the four NRO's are shown in Fig. 3 for molecular 1,2-dioxetane,

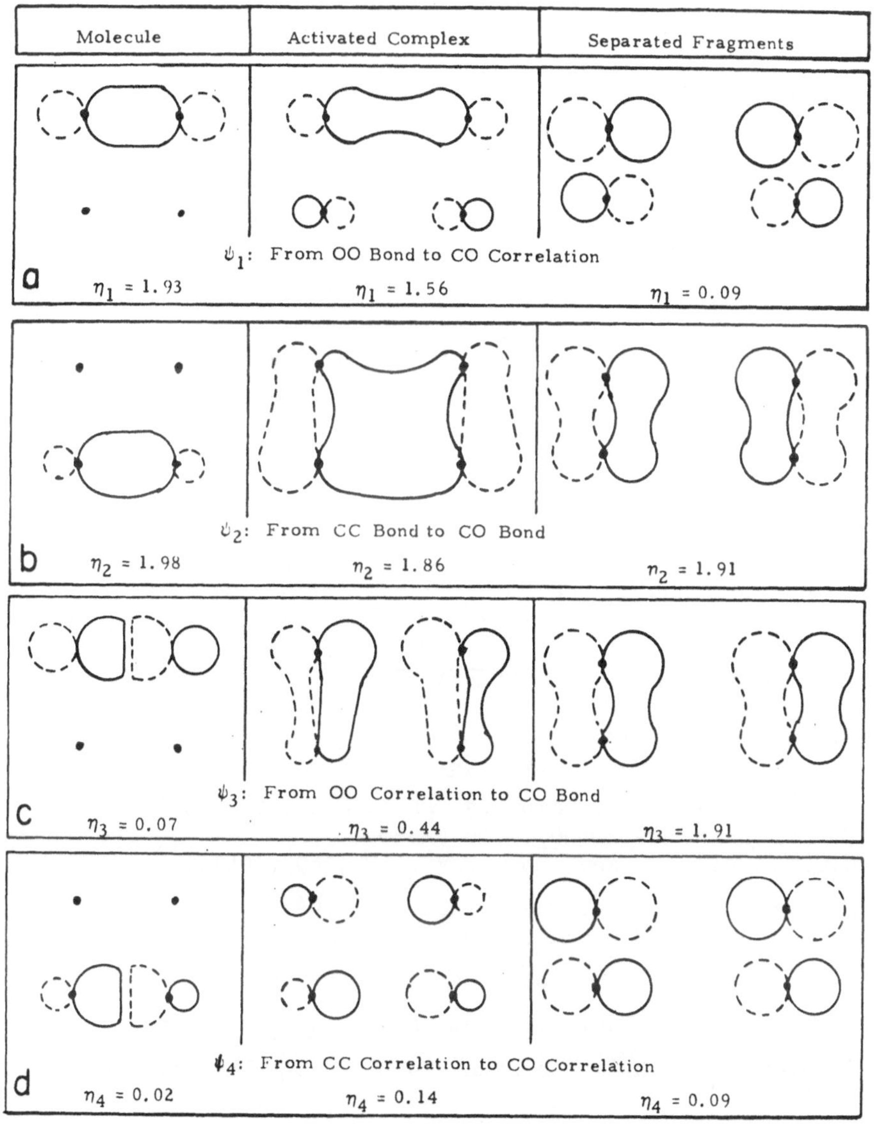

Figure 3

Schematic sketches of the actual natural reaction orbitals for the reaction shown in Figure 2 (a).

the activated complex, and the dissociated formaldehyde systems. It is apparent from the occupation numbers and orbital plots, that: (1) the molecule is joined by two sigma bonds (ψ_1, ψ_2) which are right-left correlated by ψ_3 and ψ_4 respectively along the direction of the bonds; (2) in the activated complex these sigma bonds (ψ_1, ψ_2) undergo delocalization but are not broken; and (3) the separated system consists of two carbon-oxygen pi bonds (ψ_2, ψ_3) that are right-left correlated by ψ_1 and ψ_4 along the direction of the CO bonds. Lastly, with reference to the three orbital types discussed previously, we see from their occupation and deformation that orbital ψ_1 (Fig. 3a) is replaced by orbital ψ_3 (Fig. 3c); that orbital ψ_2 (Fig. 3b) is an affected orbital; and that orbital ψ_4 (Fig. 3d), though slightly occupied nevertheless plays a correlation role throughout the reaction, undergoes a large deformation, and it is an affected orbital.

Finally, it is apparent that an expansion of the NRO's in terms of the LRO's will relate the bonding effects described by the NRO's to interactions between the LRO's. For this reason we expect that such an expansion will yield an interpretaiton of the interactions that occur between chemically reacting atoms.

DEDICATION

The authors wish to congratulate Professor Per-Olov Löwdin on his 60th birthday and on the 20th birthday of his natural orbitals.

REFERENCES

1a. F. Hund, A. Physik, 51, 759 (1928).
b. R.S. Mulliken, Phys. Rev., 32, (1928).

2a. D.R. Hartree, Proc. Cambridge Phil. Soc., 24, 89 (1928).
b. V. Fock, Z. Physik., 61, 126 (1930).
c. C.C.J. Roothaan, Rev. Mod. Phys., 23, 69 (1951).

3a. S.F. Boys, Rev. Mod. Phys., 32, 296 (1960).
b. C. Edminston and K. Ruedenberg, Rev. Mod. Phys., 35, 457 (1963).

4. W.I. Salmon and K. Ruedenberg, J. Chem. Phys., 2776 (1972).

5. G. Das and A.C. Wahl, J. Chem. Phys., 44, 87 (1966).

6. J. Hinze and C.C.J. Roothaan, Suppl. Prog. Theoret. Phys., 37 (1967).

7. F. Grein and T.C. Chang, Chem. Phys. Letters, 12, 44 (1971).

8. L.M. Cheung and K. Ruedenberg, to be published.

9. P.-O. Löwdin, Phys. Rev., 97, 1474 (1955).

10. K.R. Sundberg, L.M. Cheung and K. Ruedenberg, to be published.

11. R.S. Mulliken, J. Chem. Phys., 23, 1833 (1955).

12a. K. Ruedenberg, Rev. Mod. Phys., 34, 326 (1962).
b. K. Ruedenberg, Localization and Delocalization in Quantum Chemistry, Vol. I, (D. Reidel Publishing Company, Dordrecht-Holland, 1975).

THE PHOSPHATE GROUP IN QUANTUM BIOCHEMISTRY

Alberte Pullman and Bernard Pullman

Institut de Biologie Physico-Chimique, Laboratoire de Biochimie Théorique associé au C.N.R.S., 13, rue P. et M. Curie - Paris 75005 - France

1. INTRODUCTION

The phosphate group represents one of the essential components of a large variety of biological molecules and polymers: polynucleotides and nucleic acids, phospholipid constitutents of membranes, energy-rich (and poor) phosphates, a number of coenzymes, etc. Its importance in these structures resides both in its intrinsic electronic and conformational properties and in the fact that this group is frequently a center of interaction with external agents such as water, cations, etc. The perception of its importance in quantum biology and in particular in problems relating to the stereochemistry of nucleic acids and replication has been subjacent in Löwdin's work in this field. A more precise quantum mechanical knowledge of the properties of this group may be expected to stimulate further developments.

Different aspects of the role of the phosphate group have recently been investigated by refined quantum-mechanical methods and we wish to present here some essential results obtained. We shall limit, however, our study to the basic structure of the monophosphate anion, as exemplified in the phosphodiester linkage of Fig. 1 which shows also the numbering adopted. For easy recognition, the anionic oxygens carry the odd and the ester oxygens the even numbers. The case of the polyphosphates will be dealt with separately.

The principal problems which we wish to discuss here are:

1) The intrinsic conformation of the phosphate anion.

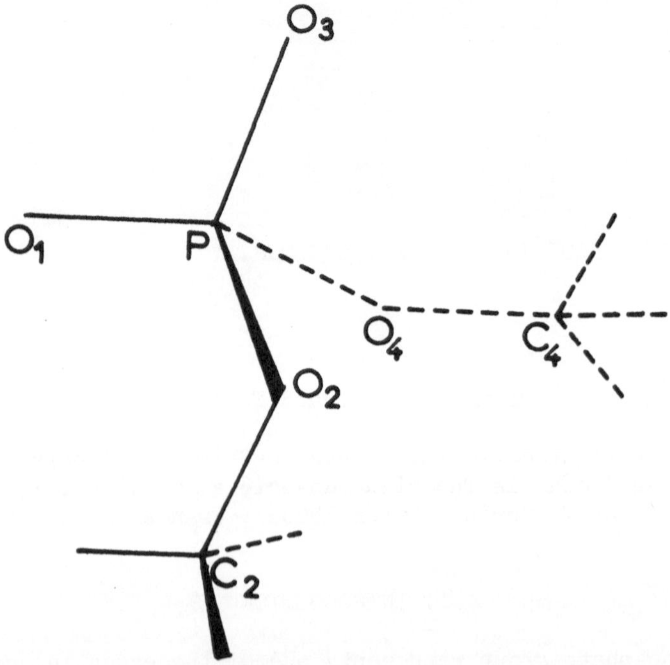

Figure 1. Numbering of the atoms in the phosphodiester linkage.

2) The molecular electrostatic potential of the phosphate group.

3) The hydration scheme of the phosphate group and the effect of hydration upon conformation.

4) The cation binding ability of the phosphate group and its effect upon conformation.

5) The effect of including or neglecting the d orbital of phosphorus upon the theoretical results.

2. THE CONFORMATIONAL PROPERTIES OF THE PHOSPHATE GROUP

The study of the conformational properties of the phosphate group is of particular importance in the field of nucleic acids and phospholipids. The significant torsion angles are those about the two $P\text{-}O_{ester}$ bonds which we shall denote $\Phi_{O_2-P} = \Phi(C_2 - O_2 - P - O_4)$ and $\Phi_{P-O_4} = \Phi(O_2 - P - O_4 - C_4)$. (We recall that the torsion angle Φ about the bond C-D is rotated relative to the near bond A-B. The cis-planar position of bonds A-B and C-D, represents $\Phi = 0^o$. The torsion angles are considered positive for a right-handed rotation: when looking along the bond B-C, the far bond C-D rotates clockwise relative to the near bond A-B. Alternatively, the positive angles are defined as 0 to 180°, measured for a clockwise rotation and negative angles as 0 to -180°, measured for a counterclockwise rotation). Following notations proposed by Sundaralingam, they are frequently denoted in the literature by the symbolds ω - ω' in the case of the nucleic acids and their constituents /1/ and by the symbols α_2 - α_3 in the case of phospholipids /2/.

Because of the obvious importance of the problem, it is certainly not astonishing that these properties have been amply investigated by many theoretical studies using both classical potential energy functions and various quantum mechanical methods and utilizing also different model compounds, anions of dimethylphosphate, disugar phosphate, dinucleoside phosphate, models of the polar head of phospholipids, etc. For a review of the computations carried out in relation with the nucleic acid and their components one may consult ref. 3, and for those performed explicitly in relation with the polar head of phospholipids ref. 4.

It is particularly interesting to observe that the essential results relevant to the intrinsic conformational properties of the phosphodiester linkage seem to be quite evident already in the simple model of dimethylphosphate (DMP^-) and are relatively only slightly modified in more complex models. On the other hand they depend strongly on the method of study used and from that point of view the results may be classified into two groups:

a) Those obtained by the classical potential energy functions (e.g. 7-8) and by the simple quantum-mechanical EHT method /9/. They are schematically represented by the conformational energy map of Fig. 2. Such computations generally indicate the existence of seven <u>equivalent</u> region of energy minima located around (Φ_{O_2-P}, Φ_{P-O_4}) = (60°, 60°), (300°, 300°), (60°, 180°), (180°, 60°), (180°, 300°), (300°, 180°) and (180°, 180°) which means that they essentially predict energy equivalence between the so-called gauche-gauche (<u>gg</u>) forms (the first two in the above enumeration), the gauche-trans or trans-gauche (<u>gt</u>, or <u>tg</u>) forms (the next four in the above enumeration), and the trans-trans (<u>tt</u>) form (the last in the above enumeration).

Attempts to refine some of the empirical computations carried out recently by different authors on more complex models, generally relevant to the structure of the nucleic acids, give results which have in common the property of disadvantaging the <u>gg</u> forms with respect to others. Thus, computations by Olson and Flory /6,7/ on

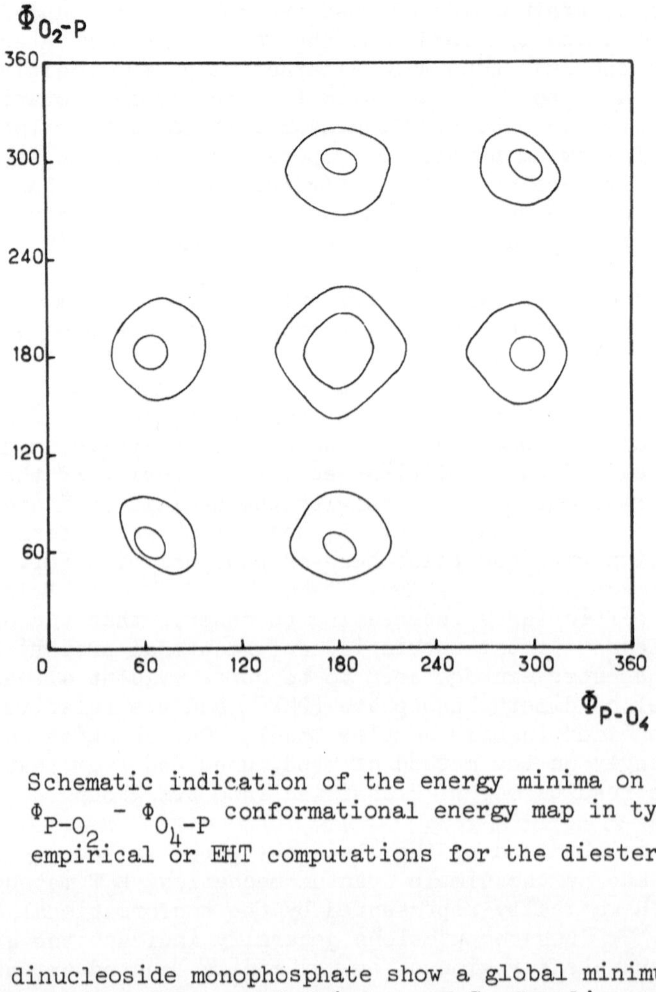

Figure 2. Schematic indication of the energy minima on a $\Phi_{P-O_2} - \Phi_{O_4-P}$ conformational energy map in typical empirical or EHT computations for the diester linkage.

a model of dinucleoside monophosphate show a global minimum for the tt conformation. Those of Yathindra and Sundaralingam /8/ on a similar model favour one of the gt conformations. On the other hand, in computations carried out by the EHT method, explicitly devoted to the α-chain of phospholipids, some of these minima merge into larger zones but the essential image remains as in Fig. 2 /10/.

b) Those obtained by the more refined quantum-mechanical methods: ab initio /11,12/, PCILO /4,12/ and CNDO /13/. Their main general features are illustrated on the PCILO map of dimethylphosphate presented in Fig. 3. This map clearly indicates a net preference for the gg form and predicts a relative instability for the tt form. Ab initio results are essentially similar to those of Fig. 3. More complex models give basically comparable indications and, when asymmetrical lead to a distinction between the two gg

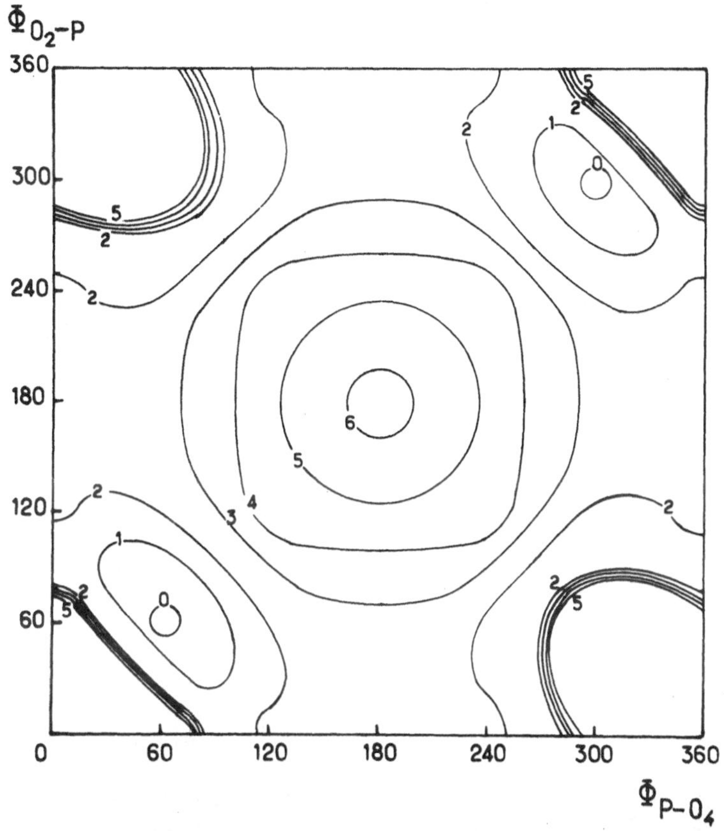

Figure 3. $\Phi_{P-O_2} - \Phi_{O_4-P}$ PCILO conformational energy map of dimethylphosphate. Isoenergy curves in kcal/mole with respect to the global energy minimum taken as energy zero.

forms, $\underline{g^+g^+}$ ($\Phi_{O_2-P} = \Phi_{P-O_4} = 60^\circ$) and $\underline{g^-g^-}$ ($\Phi_{O_2-P} = \Phi_{P-O_4} = 300^\circ$). Rules have been proposed which permit to predict the preponderance of one of these gg forms as a function of other structural characteristics of the group /12/ and of its environment /4/.

In view of such a disagreement between two groups of theoretical procedures, it seems natural to turn to experimental data for the estimation of their validity in this field. The most abundant and precise such data come from X-ray crystal structure studies. Those, whether performed for simple phosphate diesters e.g. diethyl phosphates or for different more complex oligo- and polynucleotides (for a review see 3) or on model compounds of the polar head of phospholipids and even for the one available crystal of a phospholipid (for a review see 4) constantly show a very strong preference of the phosphate group for a gg conformation and a total absence of a compound with with a tt conformation about the P-O bonds. Although the computations as described above have been carried out for the free molecule while the experimental results quoted refer to the solid state, the remarkable agreement between the predictions of the refined quantum mechanical methods and the situation in the crystals seems to indicate that the phosphate group has an intrinsic preference for a gg conformation about the P-O bonds and that this preference is conserved in the crystals. This appears a more appropriate viewpoint than that expressed by some defenders of the empirical computations who, finding by their calculations a preference of the phosphate group for other conformations, call upon subsidiary factors in order to account for the experimental predominance of the gg form. It would be simpler to recognize the insufficiency of the empirical (and EHT) computations in this type of compounds.

This viewpoint is still strengthened by the results of computations within the ab initio and PCILO methods, to be described in the following sections, which indicate that the preference for the gg form persists upon the hydration of/and cation binding to DMP^-.

3. MOLECULAR ELECTROSTATIC POTENTIAL

The knowledge of a molecular wave function allows the accurate calculation of the electrostatic potential $V(r_i)$ created in the neighbouring space by the nuclear charges and the electronic distribution. For a first order density function $\rho(1)$, the average value of the potential $V(r_i)$ at a given point i of space is

$$V(r_i) = - \int \frac{\rho(1)}{r_{1i}} d\tau_1 + \sum_\alpha \frac{Z_\alpha}{r_{\alpha i}}$$

where z_α is the nuclear charge of nucleus α /14/. The interaction energy between the molecular distribution (considered unperturbed) and, say, an external point charge q placed at point i is $qV(r_i)$, and is rigorously the first order perturbation energy of the molecule in the field of the charge q.

Its very definition makes this potential an appropriate index for studying chemical reactivity, at least in the early phase of approach of an external reagent. Taking q as a unit positive charge the interaction potential can in particular then be used for studying proton affinities (basicities) and hopefully electrophilic attacks. In fact, recent detailed investigations of a number of fundamental biological substrates have given extremely satisfactory results (for a review see 15).

Computations for the phosphate group have been carried out on the model of DMP^-, in its preferred gg conformation.

Isopotential maps have been drawn /16/ in the four planes: O_1PO_3 which contains the two equivalent anionic oxygens, O_1PO_2 and O_3PO_2 which both contain simultaneously one anionic and one esteric oxygen and $O_1O_2O_4$ which contains one anionic and two esteric oxygens. Due to the symmetry of the molecule the maps in the last three planes are identical to those in the planes O_3PO_4, O_1PO_4 and $O_3O_4O_2$, respectively.

The results are given in Fig. 4a, b, c, d in the order indicated. The study and the comparison of these maps indicate the existence in all the OPO planes of two main potential minima at a distance of about o.95 Å from the anionic oxygen in two directions situated at -120 and +120 degrees respectively from the PO_1 (or PO_3) axis. The depth of these minima is comprised between -221 and -228 kcal kcal/mole, the larger value being always situated inside the OPO angle.

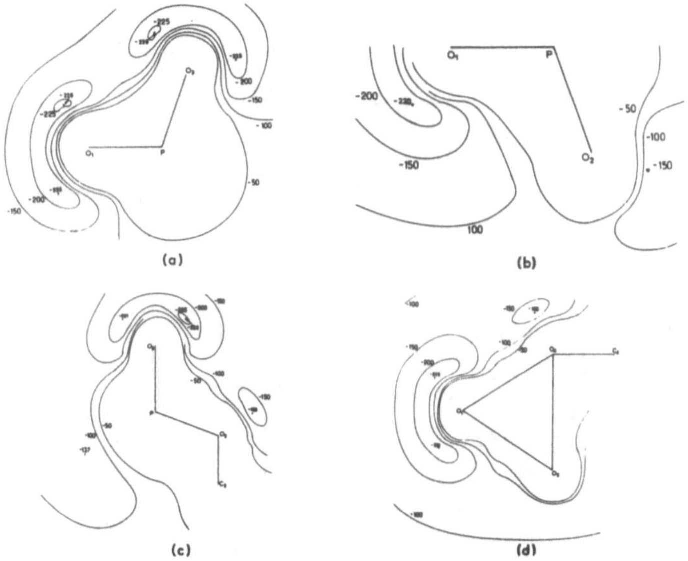

Fig. 4. Isopotential curves in the planes: a) O_1PO_3, b) O_3PO_2, c) O_1PO_2, d) $O_1O_2O_4$ (values in kcal/mole).

On the other hand a secondary potential minimum, appreciably less deep, is observed in the plane O_3PO_2 (and its equivalent O_1PO_4) (Fig. 4b). This minimum is situated at 0.95 Å from the esteric oxygen in a direction making an angle of 120 degrees with the PO_2 (or PO_4) axis, its value reaching only -168 kcal/mole. This secondary minimum is less deep in the other kind of anionic-ester plane namely O_1PO_2 (or O_3PO_4) because of the proximity of the methyl group. The same minimum appears in the plane $O_1O_2O_4$ (Fig. 4d) in the neighbourhood of O_4 approximately on the projection of the external bissectrix of the PO_4C_4 angle. The same situation exists on the side of O_2 in the equivalent $O_3O_4O_2$ -plane.

This analysis of the potential in the various OPO and OOO planes points to the existence of two main equivalent regions intrinsically strongly attractive for an electrophile and surrounding the anionic oxygens: in fact a finer exploration shows that the PO_1 (and PO_3) axis are entirely surrounded by a circular furrow where the values of the potential remain within 5% of the maximum of -229 kcal/mole. This result was obtained by computing first the potential around a circle described by rotating about the PO_1 axis a point M defined by O_1M = 0.95 Å and PO_1M = 120° (fig. 5), then searching in the plane of this circle for all the potential values between -218 and -229 kcal/mole with a grid-spacing of 1 kcal/mole. The circular zone obtained is slightly asymmetrical, with the two most negative regions situated on each side of the plane O_1PO_3, the deepest point being on the

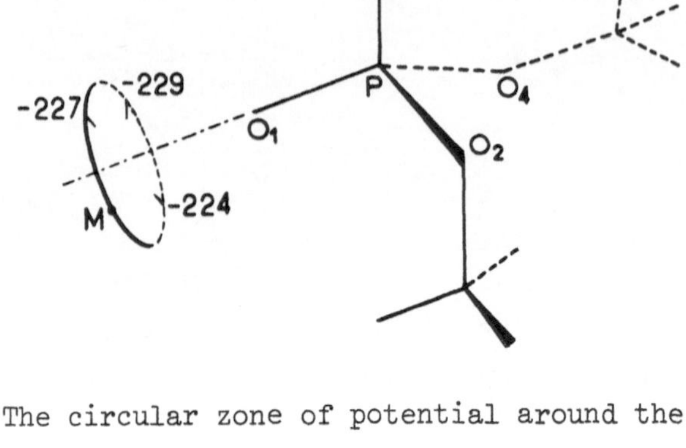

Figure 5. The circular zone of potential around the anionic oxygens.

the side of O_2).

The two main conclusions which may thus be drawn from this study point to: 1) the greater attractive character of the anionic oxygens in comparison to the ester oxygens towards electrophilic agents; 2) the absence of a strong directional preference in this respect about the anionic oxygens. This situation has important bearings also on the hydration scheme around the phosphate group and on the cation binding properties of this groups, indicating thereby the important large significance of this index of molecular electronic properties.

4. THE HYDRATION SCHEME AND ITS EFFECT ON CONFORMATION

The procedure utilized is based on the "supermolecule" approach in which the interaction energies are calculated by comparing the energies of the adduct (DMP^- + water)with the sum of those of the isolated components and which has been applied recently to the study of the hydration of a large number of fundamental biomolecules /17/. Interaction with one molecule of water was considered first in all possible details so as to determine both the most favorable site of attachment and the lability of the binding with respect to various degrees of freedom. Next, the addition of further water molecules was considered until completion of the first hydration shell. Then an exploration was performed on the possible existence and properties of a second (and further) hydration shell(s). The computations were carried out ab initio by the SCF LCAO procedure with an STO 3G Gaussian basis set.

A) Monohydration

The results concerning the possibilities of hydration by one water molecule are best visualized by examining first the situation in the various OPO planes of the phosphate. They are presented for the preferred gg conformation in Table I and Fig. 6a, b, c. The notations are explained in Table I /18, 19/.

The plane O_1PO_3 contains the site corresponding to the minimum minimorum (-28.6 kcal/mole) on the energy hypersurface: this occurs in two equivalent positions when water is bound to one of the anionic oxygens, O_1 or O_3, by one H-bond, at the exterior of the OPO angle (positions E_{13} and E_{31}). But there is also in that plane, another position E''_{13} (and its equivalent E''_{31}) only 1.2 kcal/mole less stable, where water is bound by one hydrogen to the other side of the anionic oxygen, with its second hydrogen turned towards the

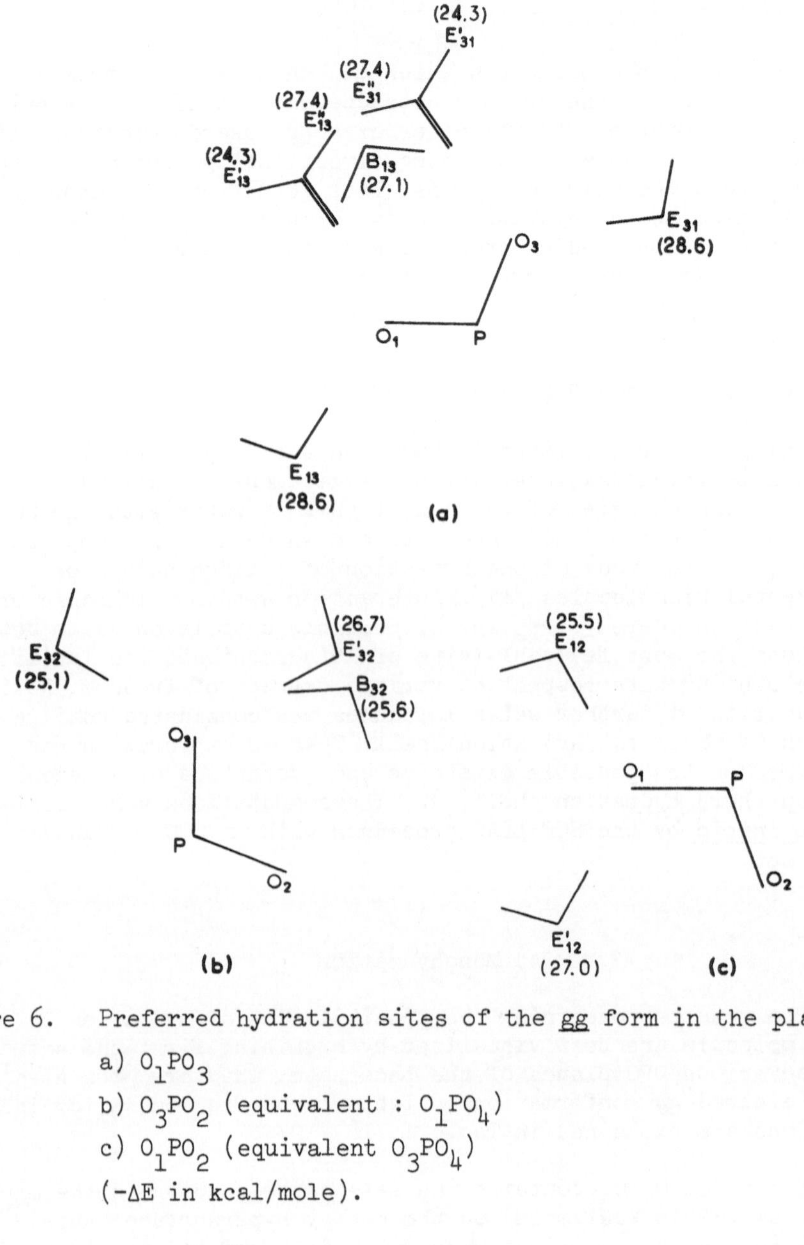

Figure 6. Preferred hydration sites of the gg form in the planes:
a) O_1PO_3
b) O_3PO_2 (equivalent : O_1PO_4)
c) O_1PO_2 (equivalent O_3PO_4)
($-\Delta E$ in kcal/mole).

interior of the OPO angle. The corresponding position of water rotated by 180° about the H bond axis (E'_{13}) lies 4 kcal/mole above (vide infra). Moreover, there is the possibility of a bridge adduct B_{13}, nearly as favorable in energy (see Fig. 6a). There are, however, also a number of other nearly as favorable binding sites

Table I

Monohydration of DMP$^-$ (STO 3G)

Plane	symbol$^{(b)}$	$d_{O_i \ldots H}$(Å)	θ	β	-ΔE(kcal/mole)
O_1PO_3	E_{13} E_{31}	1.45	120	180	28.6
	B_{13}	1.75	110	143	27.1
	E''_{13} E''_{31}	1.50	120	180	27.4
O_3PO_2	E'_{32}	1.50	120	180	26.7
	B_{32}	1.60	116	160	25.6
	E_{32}	1.50	120	180	25.1
O_1PO_4 -	equivalents: E'_{14} B_{14} E_{14}				
O_1PO_2	E'_{12}	1.45	120	180	27.0
	E_{12}	1.45	120	180	25.5
O_3PO_4 -	equivalents: E'_{34}, E_{34}				

(a) θ = angle between PO_i and $O_i \ldots H$ directions

β = angle between $O_i \ldots H$ and HO_{water} directions.

(b) The symbols E_{ij}, E'_{ij}, E''_{ij} correspond to water bound by one hydrogen bond to oxygen _i_ in the plane O_iPO_j. E stands for water external to the OPO angle, E' and E" for water internal to this angle. In the unprimed and primed E_{ij}-positions the second hydrogen of water is turned towards the PO_i axis. In the double-primed position this second hydrogen is turned away from the PO_i axis (see fig. 2a, b, c). The symbol B_{ij} corresponds to water making a bridge between oxygens i and j.

in the other OPO planes as shown in Table I and Figs. 6b and 6c: each OPO plane contains one external binding site E (to the anionic oxygen) similar to E_{13} and E_{31}, and another site E' obtained from E by a 180° rotation about the PO axis. Moreover, a favorable bridged position B appears also in the equivalent planes O_1PO_4 and O_3PO_2. The corresponding bridge positions in the planes $O_1PO_2^-$ and O_3PO_4 are less favorable.

On the whole, the distribution of the favorable hydration sites in the neighbourhood of the anionic oxygens points to the existence of a whole zone of attraction for water about these two atoms: this has been confirmed by studying the variations in the binding energy obtained by rotating the plane of the water molecule about the PO_1 axis, from its most stable position E_{13}, in such a way that O_1...OH describes a cone about this axis (Fig. 7). A circular zone of very large binding interaction with water was found, where the binding energy remains within 85% of its most favorable value.

B) Polyhydration in the First Shell

The monohydration scheme outlined above, indicating the existence of large attractive zones of similar energies and many possibilities for the fixation of water, should result in multiple possiblities of polyhydration.

The results obtained for simultaneous fixation of a number of water molecules to the phosphate ion (distance = 1.8 Å for sites B and 1.5 Å for sites E, based upon optimization of model cases) confirm this expectation as shown by the examples given in Table II. Up to six water molecules may be accommodated in the first hydration shell. (The addition of a 7th water molecule decreases the total binding energy). Intermediate hydration involving two to five water molecules may of course take place, a number of energetically nearly equivalent possibilities occurring for a given number of water molecules (3 for n = 2 to 4, 2 for n = 5). The average energy of water attachment is approximately -23 kcal/mole

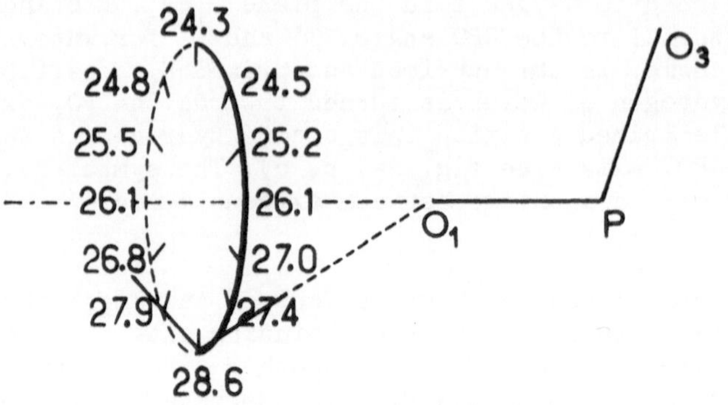

Figure 7. Variation of $-\Delta E$ upon rotation of water out of position E_{13} (O_1...HO sweeping a cone about PO_1 axis with the non hydrogenbonded hydrogen turned towards this axis).

Table II

Polyhydration of DMP^- with n water molecules in the first shell.

n	occupied positions	$-\Delta E_{tot}/n$ (kcal/mole)
2	E_{13} E_{31}	27.1
	B_{14} E'_{12}	23.6
3	B_{13} B_{32} B_{14}	23.0
	B_{13} E'_{32} E'_{14}	23.5
	B_{13} E_{31} E_{13}	24.7
4	E_{13} E_{31} E_{14} E_{32}	21.9
	E'_{12} E'_{34} E'_{14} E'_{32}	21.7
	E_{14} E_{32} E'_{14} E'_{32}	21.3
5	B_{13} B_{32} B_{14} E'_{12} E'_{34}	20.3
	B_{13} E'_{32} E'_{14} E'_{12} E'_{34}	20.3
6	E_{31} E_{32} E_{34} E_{12} E_{13} E_{14}	17.3

for n = 2 and 3 and decreases when n increases further, down to -17.3 kcal/mole for n = 6. The hydration scheme in each case may be easily visualized from the indications in Table II and the definitions of the symbols given in Table I.

It must be kept in mind that the STO 3G basis overestimates the binding energies and allows too close an approach of the water molecule to ionic species /20,21/. However the qualitative features of the binding are correctly reproduced (see references 17,21 and 22 for a discussion).

Experimental information about the hydration of the phosphate group comes mostly from studies of phospholipids. It is relatively abundant /23-31/ and although it does not lead to a unique scheme and does not fix precisely the preferred sites of hydration, it indicates a number of "bound" water molecules which altogether is comparable with that suggested by the theoretical studies. Depending upon the experimental conditions and techniques utilized the number of water molecules in the primary hydration shell (most strongly bound) varies at the polar head of phosphatidylcholine (which was more abundantly studied than phosphatidylethanolamine) from 2 to 6. As in these molecules no strong hydrogen bond is expected around the cationic head, this number may be considered as relevant essentially to the phosphate group. The preferential fixation of 4-6 water molecules on the phosphates of the nucleic acids has similarly been proposed as a result of infra-red studies /32,33/.

C) Effect on Conformation

The extension of the computations to the gt and tt forms of DMP^- has enabled the estimation of the influence of hydration on the relative stability of the three principal conformers. A schematic representation of the results is given in Table III. (For more details see ref. 19).

The principal conclusion of this comparison is that polyhydration is destabilizing the gg form relative to the gt and tt forms. The gg form still remains, however, the most stable one for all degrees of hydration with the exception of hydration with six water molecules when the gt form becomes energetically equivalent to it. The tt form always remains the least stable one.

D) Further Hydration Shells

An exploration of the possibility of existence of organized more remote layers of hydration around DMP^- indicates the probable presence of a second such layer but the improbability of finding organized "bound" water beyond the second shell, although residual organized such fragments may persist in the vicinity of some particularly favorable hydration sites. (For details see 19). This conclusion seems in agreement with the main experimental data of ref. 23-31.

5. BINDING OF CATIONS AND ITS EFFECT ON THE CONFORMATION OF THE PHOSPHODIESTERS LINKAGE

These studies also have been carried out /34/ with DMP^- as model compound. The computations have been performed within the SCF *ab initio* procedure using the STO 3G basis set. The cations studied are Na^+ and Mg^{++}. Studies are in progress on K^+ and Ca^{++} ions.

Table III

Hydration of DMP^- (ab initio STO 3G).
(Energies in Kcal/mole with respect to the gg form taken as energy zero).

H_2O		gg	gt	tt
	DMP^-	0	3.4	8.0
1			2.4-3.9	
2			2.9	
3			2.6	
4			1.2	
5			1.0_5	
6			0.0	1.7_5

The construction of complete conformational energy maps being highly expensive in an <u>ab initio</u> treatment the investigation was limited to the fixation of the metal cations in <u>a priori</u> most plausible sites and to the evaluation of the effect of such attachments upon the relative energies of the three fundamental conformations resulting from possible rotations about the $P\text{-}O_{ester}$ bonds: <u>gg</u>, <u>gt</u> and <u>tt</u>. In the selection of the most plausible sites of location of the cations we have been guided by the results obtained above for the preferred sites of fixation of protons and water molecules. Four directions of approach have been retained which were named by analogy to notations used for the water fixation scheme: B_{13} on the bissectrix of the O_1PO_3 angle, E_{13} and E'_{13} along directions which make an angle of 120° with the O_1P axis, respectively on the external and internal side of the O_1PO_3 angle; the fourth site is along the PO axis (Fig. 8). In each case the O... cation distance was optimized for each of the three conformations, <u>gg</u>, <u>gt</u> and <u>tt</u>.

It must be kept in mind that the cation binding energies computed with the STO 3G basis set are largely overestimated (scaling factor needed ≈ 0.6) and that the distances of approach

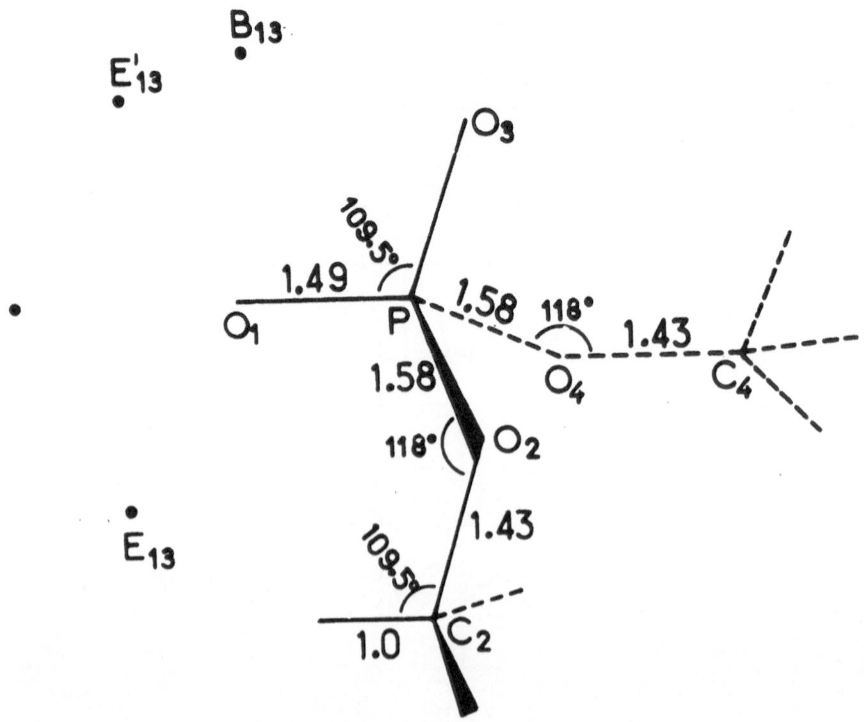

Figure 8. Dimethylphosphate anion and the principal sites of Na^{+} and Ng^{++} binding.

are too small /35/. However, a number of examples show that all the qualitative features of the binding are satisfactorily reproduced /36-38/ and allow the obtention of valuable information in an exploratory study.

The principal results are presented in Table IV. The essential conclusions which may be drawn from them are two-fold.

1) In the first place they indicate that the preferred site of attachment of the cation to DMP^{-} is along the bissectrix of the O_1PO_3 angle (site B_{13}), a position in which the cation is exposed strongly to the effect of both anionic oxygens. Among the three other binding sites explored in the case of Na^{+}, the next in the order of decreasing binding energy is the E'_{13} site already 32 kcal/mole less efficient than the B_{13} site. The E_{13} site and the binding site along O_1P are still weaker. In the case of Mg^{++} only the B_{13} and E'_{13} were taken into consideration. The latter is already

Table IV

Binding energies in DMP^-... cation adducts for different conformations of DMP^- (kcal/mole)

Site of cation binding	O...M^+ (Å) optimized distance	Energy of cation binding: ΔE with respect to DMP^- and cation at infinite separation			Difference in energy of cation binding		Difference in energy between conformers	
		gg	gt	tt	ΔE(gt-gg)	ΔE(tt-gg)	ΔE(gt-gg)	ΔE(tt-gg)
Free DMP^-							3.4	8.0
			$DMP^-...Na^+$					
E_{13}	1.9	-167.6	-167.6		0		3.4	
Along O_1P	1.9	-169.0						
E'_{13}	1.9	-182.5	-182.4		0.1		3.5	
B_{13}	2.0	-214.6	-214.0	-212.6	0.6	2.0	4.0	10.0
			$DMP^-...Mg^{+++}$					
E'_{13}	1.7	-436.3						
B_{13}	1.9	-497.8		-495.0	1.4	4.2	4.8	æ2-2

63 kcal/mole less stable than the former.

2) The second conclusion is that the cation binding leaves unperturbed the order of conformational preferences with respect to the torsion about the P-O ester bonds: the gg conformation remains the most stable one, followed by the gt one, followed in turn by the tt one, as in free DMP^-. In fact, the cation binding in its preferred B_{13} site seems even to increase the relative stability of the gg form with respect to the two others: the energy difference in *ab initio* computations on free DMP^- is 3.4 kcal/mole between the gg and gt form and 8.0 kcal/mole between the gg and tt forms /11/. The two values become, respectively, 4 and 10 kcal/mole in the Na^+ adduct and 4.9 and 12.2 kcal/mole in the Mg^{++} adduct.

The present results are in contradiction with results obtained recently for the same problem using the CNDO/2 method /13/, which predict that the tt conformation becomes the most favorable one in $DMP^-...Na^+$ and $DMP^-...Mg^{++}$ complexes. It seems probable that the indications of the CNDO method are artefacts of the procedure. The more so as the available although limited experimental indications seem rather in favor of the *ab initio* conclusions. Both crystallographic X-ray studies and different solution studies seem to indicate that the phosphodiester linkage conserves the gg conformation upon cation binding. Thus, e.g. the gg conformation is observed in the crystal of barium diethylphosphate /40/, of magnesium diethylphosphate /41/ and of glycerylphosphorylcholine $CdCl_2$ trihydrate /42/. It is also maintained in the aqueous solution of barium diethylphosphate /40/. A number of authors indicate also that the conformation of model compounds for the polar head of phospholipids /44,45/, which intrinsically prefer the gg orientation about the $P\text{-}O_{ester}$ bonds, are not influenced by the presence of cations. A similar lack of effect has been recorded recently for the conformation of nicotinamide mononucleotide in aqueous solution upon binding of the La^{3+} ion /46/.

6. THE EFFECT OF THE 3d ORBITAL OF PHOSPHORUS ON THE PROPERTIES OF THE PHOSPHATE GROUP

The studies described in the preceding sections of this paper have been carried out by neglecting the 3d orbitals of the phosphorus atom in the gaussian basis set. They have been refined recently by including these orbitals /47/. The results point to a very limited importance of this inclusion for the properties described above. The qualitative pictures obtained for the preferred conformational states with respect to the torsion about the $P\text{-}O_{ester}$ bonds, the general aspect of the molecular electrostatic potentials and the location of the principal minima, the general aspect of the hydration scheme remain unchanged or are only slightly perturbed. Even from the quantitative point of view the effects are small, with the exception of the values of the hydration energies which are reduced by about 20%. These properties may thus be satisfactorily explored without the explicit inclusion of d orbitals. On the other hand, the introduction of these orbitals has one strong incidence: a large decrease of the net electronic charges. (For details see ref. 47).

CONCLUSIONS

This wide theoretical exploration of the properties of the phosphate group provides a useful amount of information which enable a much more clearcut recognition of their significance for the structural role of this group in the fundamental biomolecules in which it is present. Thus e.g. manifestly the conformational behaviour of polynucleotides is due to a large extent to the intrinsic conformational preferences of this group. The detailed determination of its hydration scheme and of its modes of interaction with cations goes beyond the present day availability of experimental data and enables to put into evidence explicitly the role of these environmental effects in the behaviour of compounds containing this group. A particularly striking application of this procedure was given in a recent theoretical exploration of the effect of hydration of the phosphate upon the conformation of the polar head of phospholipids /48/.

REFERENCES

1. M. Sundaralingam, Biopolymers, 7, 821 (1969).
2. M. Sundaralingam, Ann. New York, Acad. Sci., 195, 324 (1972).
3. B. Pullman and A. Saran, Progress in Nucleic Acid Research and Molecular Biology, in press.
4. B. Pullman and A. Saran, Inter. J. Quant. Chem. Quant. Biol., Symp. 2, in press.
5. V. Sasisekharan and A.V. Lakshminarayanan, Biopolymers, 8, 505 (1969).
6. W.K. Olson and P.J. Flory, Biopolymers, 11, 25 (1972).
7. W.K. Olson and P.J. Flory, Biopolymers, 11, 57 (1972).
8. N. Yathindra and M. Sundaralingam, Proc. Natl. Acad. Sci. U.S., 71, 3325 (1974).
9. A. Saran and G. Govil, J. Theoret. Biol., 33, 407 (1971).

10. S.P. Gupta and G. Govil, FEBS Letters, 27, 68 (1972).

11. M.D. Newton, J. Am. Chem. Soc., 95, 256 (1973).

12. D. Perahia, B. Pullman and A. Saran, Biochim. Biophys. Acta, 340, 299 (1974).

13. R. Tewari, R.K. Nanda and G. Govil, J. Theoret. Biol., 46, 229 (1974).

14. R. Bonaccorsi, A. Pullman, E. Scrocco and J. Tomasi, Theoret. Chim. Acta, 24, 51 (1972).

15. A. Pullman, in Chemical and Biochemical Reactivity, Proceedings of the 6th Jerusalem Symposium in Quantum Chemistry and Biochemistry edited by E.D. Bergmann and B. Pullman (Reidel Publishing Company, Dordrecht, Holland, 1974) p.1.

16. H. Berthod and A. Pullman, Chem. Phys. Letters, 32, 233 (1975).

17. A. Pullman and B. Pullman, Quart. Rev. Biophys. 7, 505 (1975).

18. A. Pullman, H. Berthod and N. Gresh, Chem. Phys. Lett., 33, 11 (1975).

19. B. Pullman, A. Pullman, H. Berthod and N. Gresh, Theoret. Chim. Acta, in press.

20. M. Perricaudet and A. Pullman, FEBS Letters, 34, 222 (1973).

21. A. Pullman and A. Armbruster, Int. J. of Quant. Chem. S8, 169 (1974).

22. A. Pullman, Int. J. of Quant. Chem. Quantum Biology Symposium Nr. 1, 33 (1974).

23. G. Klose and F. Stelzner, Biochim. Biophys. Acta, 363, 1 (1974).

24. G.L. Jendrasiak and J.H. Hasty, Biochim. Biophys. Acta, 348 45 (1974).

25. G.L. Jendrasiak and J.H. Hasty, Biochim. Biophys. Acta, 337, 79 (1974).

26. W.V. Walter and R.G. Hayes, Biochim. Biophys. Acta, 249, 528 (1971).

27. D. Chapman, Annals New York Acad. Sci., 195, 179 (1972).

28. M.C. Phillips, E.G. Finer and H. Hauser, Biochim. Biophys. Acta, 290, 397 (1972).

29. Z. Veksli, N.J. Salsburg and D. Chapman, Biochim. Biophys. Acta, 183, 434 (1969).

30. R.L. Misiorowski and M.A. Wells, Biochemistry, 12, 967 (1973).

31. K. Henrikson, Biochim. Biophys. Acta, 203, 228 (1970).

32. M. Falk, K.A. Hartman Jr. and R.C. Lord, J. Amer. Chem. Soc., 85, 387 (1963).

33. K.A. Hartman, R.C. Lord, and G.J. Thomas, in Physical Chemical Properties of Nucleic Acids, edited by J. Duchesne (Academic Press, New York, vol. 2, 1973) p.l.

34. B. Pullman, N. Gresh and H. Berthod, Theoret. Chim. Acta, in press.

35. M. Perricaudet and A. Pullman, FEBS Lett., 34, 222 (1973).

36. A. Pullman, Int. J. Quant. Chem.; Quant. Biol. Symp. 1, 33 (1974).

37. A. Pullman and P. Brochen, Chem. Phys. Lett., 34, 7 (1975).

38. B.M. Rode, M. Breuss and P. Schuster, Chem. Phys. Lett., 32, 34 (1975).

39. R.K. Nanda and G. Govil, Theoret. Chim. Acta, 38, 71 (1975).

40. Y. Kyogoku and Y. Iitaka, Acta Cryst., 21, 49 (1966).

41. F.S. Ezra and R.L. Collin, Acta Cryst., B29, 1398 (1973).

42. M. Sundaralingam and L.H. Jensen, Science, 150, 1035 (1965).

43. G. Papakostidis and G. Zundel, Zeits. für Naturforschung, 28b 323 (1973).

44. R.L. Misiorowski and M.A. Wells, Biochemistry, 12, 967 (1973).

45. K.K. Yabusaki and M.A. Wells, Biochemistry, 14, 162 (1975).

46. B. Birdsall, N.J.M. Birdsall, J. Feeney and J. Thornton, J. Amer. Chem. Soc., 97, 2845 (1975).

47. D. Perahia, A. Pullman and H. Berthod, Theoret. Chim. Acta, in press.

48. B. Pullman, H. Berthod and N. Gresh, FEBS Letters, 53, 199 (1975).

TOWARDS THE THEORETICAL DETERMINATION OF THE CONFORMATION OF BIOLOGICAL MACROMOLECULES

János J. Ladik

Lehrstuhl für Theoretische Chemie der Friedrich-Alexander-Universität Erlangen-Nürnberg. BRD-852 Erlangen, BRD

INTRODUCTION

The theoretical determination of conformations of biological macromolecules like nucleic acids or proteins constitutes an important part of theoretical biology. To handle these rather large problems empirical and semiempirical schemes have been proposed ranging from empirical potential functions /1/ to various semi-empirical treatments of the relevant fragments of macromolecules. Among the semiempirical methods are Extended Hückel /2/, CNDO/2 /3/, MINDO/2 /4/, and PCILO in CNDO/2 parametrization /5/. In many cases these methods provide reasonable results, but in other cases the results have been unsatisfactory. Since the semiempirical methods in general lack a sound theoretical foundation it is difficult to predict their limits of applicability. Furthermore, in most of the conformational calculations entropy effects were neglected.

In principle, one could obtain the conformation of a macro-molecule by calculating the total energy of different conformations with the aid of the ab initio SCF LCAO MO method. In practice, how-ever, this would lead to astronomical computer times and will remain impractical for many years to come. Consequently, one has to look for other possibilities. In a macromolecule the number and the nature of covalent bonds is fixed. A fairly good approximation for the total energy of the macromolecule (measured relative to the sum of the total energies of its consitutent atoms) is therefore obtained as the sum of the energies of its covalent bonds and the sum of the interaction energies between its non-bonded parts. We can write this as

$$E = E_{cov.b} + \sum_{i<j}^{N} E_{i,j} \qquad (1)$$

where $E_{cov.b.}$ stands for the binding energies of all covalent bonds, $E_{i,j}$ for the interaction energy between the non-bonded groups i and j, and N for the number of groups.

In reality, the change of the interaction energies between the non-bonded parts of the macromolecule with the change of conformation will also influence the bond energies of the covalent bonds. On the other hand, if one is interested only in the change of total energy with different rotation angles - this is not most frequently the case - and does not consider steric hindrances, the lengths and the strengths of the covalent bonds will not change considerably.

Hydrogen bonds constitute an intermediate case between bonds and interaction between non-bonded groups. In the present scheme they may be considered as a special case of the latter.

In order to illustrate the energy partitioning of Eq. (1), let us consider the tripeptide part of a polypeptide chain shown in Fig. 1. In the first approximation, the bond energies are

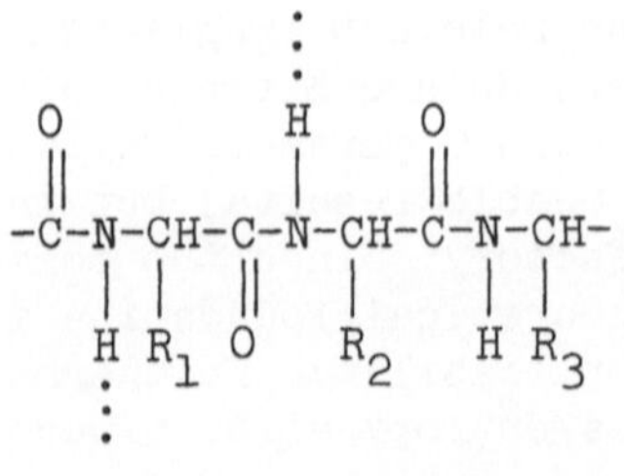

Fig. 1. The tripeptide part of a polypeptide chain. The dots below and above the hydrogens bonded to the nitrogen atoms denote hydrogen bonds (with the C = O groups of another polypeptide chain, or of another part of the same polypeptide chain in the case of an α helix or β sheet). R_1, R_2, R_3 stand for the different side chains.

considered to be independent of the conformation. The sum of interaction energies $E_{i,j}$ will contain the following parts:

1. the three hydrogen bond energies
2. the interaction energies between R_1 and R_2 and between

R_1 and R_3, respectively

3. the interaction energies between the side chains R_i (i = 1, 2, 3) and the (CO)NH peptide groups and with all the CH groups except the one to which they are covalently bound. In the latter case only the R_i-H interaction is taken into account.
4. the interaction energies between the peptide groups and the CH groups, respectively,
5. the interaction energies between non-neighbouring (CO)NH and CH groups.

Thus, even for a tripeptide there are interactions between 9 groups which have to be taken into account simultaneously, in addition to the hydrogen bond energies. Furthermore, in order to find the real conformation of a biological macromolecule, one also has to account for both the interaction with surrounding water molecules and entropy effects.

In most cases, perturbation-theoretical expressions have been used for non-bonded interactions. One of the most frequently used and more sophisticated schemes is that of Murrell et al. /6/, based upon the perturbation expansion with overlap included to second order. This scheme seems to work well for two interacting molecules at larger distances, but the results are questionable for those intermediate distances (3 - 4 Å) /7/ which occur between neighbouring side chains of a polypeptide chain and between the stacked nucleotide bases in nucleic acids. Furthermore, the perturbation-theoretical schemes are unable to account correctly for simultaneous interactions between many non-bonded groups because they work with fixed wavefunctions of the interacting units. In this way the interaction energy between units 1 and 2 is made independent of the presence of a third unit. This is far from correct, especially if the third unit is charged.

In order to help overcoming the formidable problems stated above, we will outline a newly proposed calculation scheme. The results for two and three interacting molecules obtained so far seem to be rather promising. Therefore, one might hope that, after further elaborations of the method, it will also be of value for the calculation of conformations of biological macromolecules. In this way problems like the stability and conformation of the replication plane of DNA consisting of four nucleotide bases suggested by Löwdin /8/ could be treated more accurately.

THE MUTUALLY-CONSISTENT-FIELD METHOD

Let us assume a system of N interacting molecules or non-bonded subgroups of a macromolecule. If there were no interactions between the subsystems we could write for each one an independent Hartree-Fock equation

$$F^{(\ell)}\,\phi_i^{(\ell)} = \varepsilon_i^{(\ell)}\,\phi_i^{(\ell)} \qquad (\ell=1,2,\dots,N) \tag{2}$$

with

$$F^{(\ell)} = H^{(\ell)} + G^{(\ell)} \tag{3}$$

where $H^{(\ell)}$ is the one-electron and $G^{(\ell)} = \sum_{j=1}^{n_\ell}(2J_j^{(\ell)} - K_j^{(\ell)})$ the two-electron part of the Fock operator of the ℓth subsystem.

If the units interact we can modify the one-electron operators

$$H^{(\ell)} = -\tfrac{1}{2}\Delta - \sum_{\alpha_i=1}^{M_\ell} \frac{Z_{\alpha_\ell}}{|\underline{r} - r_{\alpha_\ell}|} \tag{4}$$

($\underline{r}_{\alpha_\ell}$ is the position vector of nucleus α of unit ℓ and M_ℓ is the number of nuclei in the same unit) to

$$\tilde{H}^{(\ell)} = H^{(\ell)} + \sum_{s\neq\ell}^{N} (V_c^{(s)} + V_{x\alpha}^{(s)}) \tag{5}$$

Here

$$V_c^{(s)}(\underline{r}) = \int \frac{\rho^{(s)}(\underline{r})}{|\underline{r} - \underline{r}'|}\, d\underline{r}' - \sum_{\alpha_s=1}^{M_s} \frac{Z_{\alpha_s}}{|\underline{r} - \underline{r}_{\alpha_s}|} = V_{c,e\ell.}^{(s)} + V_{c,n.}^{(s)} \tag{6}$$

is the Coulomb potential of unit s, $\rho^{(s)}(\underline{r})$ is the charge density and $V_{x\alpha}^{(s)}$ is the local exchange potential of Slater /9/

$$V_{x\alpha}^{(s)}(\underline{r}) = -6\alpha\left(\frac{3}{8\pi}\rho^{(s)}(\underline{r})\right)^{1/3} \tag{7}$$

In the Hartree-Fock case $\rho^{(s)}$ is simply

$$\rho_{HF}^{(s)}(\underline{r}) = \sum_{i=1}^{n_s} 2\phi_i^{(s)*}\ \phi_i^{(s)} \tag{8}$$

where n_s stands for the number of doubly filled orbitals.

Substituting (5) into (2), one obtains the coupled Hartree-Fock equations of the interacting systems:

$$\tilde{F}^{(\ell)}\ \tilde{\phi}_i^{(\ell)} = \tilde{\varepsilon}_i^{(\ell)}\ \tilde{\phi}_i^{(\ell)} (\ell=1,2,\ldots,N) \tag{9}$$

where the modified Fock operators $\tilde{F}^{(\ell)}$ contain the modified one-electron operators $\tilde{H}^{(\ell)}$ defined through (5) - (8) instead of $H^{(\ell)}$. Furthermore, the Coulomb and exchange operators $J_j^{(\ell)}$ and $K_j^{(\ell)}$ have to be calculated with the aid of the modified wavefunctions $\tilde{\phi}_i^{(\ell)}$.

In order to solve the system of equations (9), one starts with the charge distributions of the non-interacting molecules or subgroups and iterates all the coupled equations simultaneously until a mutually consistent (MCF) solution is achieved. In this way, one obtains the charge distribution on the Hartree-Fock level of the interacting systems. For non-bonded subgroups of a macromolecule, one should modify the charge distribution of the free subgroups in the starting iteration by taking into account the covalent bonds to their neighbours. The results of ab initio super-molecule calculations of different segments could be used with two or three subgroups selected for this purpose.

After the MCF solution of the equation system (10), one can calculate the total energy $E^{(\ell)}$ of each interacting system. In this way one obtains the interaction energy between the p-th and q-th subsystem in the presence of all other units

$$E_{int} = \tilde{E}^{(p)} + \tilde{E}^{(q)} - E^{(p)} - E^{(q)} + \sum_{\alpha_p=1}^{M_p} \sum_{\beta_q=1}^{M_q} \frac{Z_{\alpha_p} Z_{\beta_q}}{|\underline{r}_{\alpha p} - \underline{r}_{\beta q}|}$$

$$-\sum_{\alpha_p=1}^{M_p} Z_{\alpha_p}\ \tilde{V}_c^{(q)}(\underline{r}_{\alpha_p}) - \sum_{\beta_q=1}^{M_q} Z_{\beta_q}\ \tilde{V}_c^{(p)}(\underline{r}_{\beta_q})$$

$$+ \int \frac{\tilde{\rho}^{(p)}(\underline{r}_1)\tilde{\rho}^{(q)}(r_2)}{|\underline{r}_1 - \underline{r}_2|} d\underline{r}_1 d\underline{r}_2$$

$$+ \tfrac{1}{2}\left(\int \rho^{(p)}(\underline{r})\tilde{V}^{(q)}_{x\alpha}(\underline{r})d\underline{r} + \int \rho^{(q)}(\underline{r})\tilde{V}^{(p)}_{x\alpha}(\underline{r})d\underline{r}\right) \tag{10}$$

The interaction energy automatically provides not only for the electrostatic energy but also for the polarization energy, and in a rather good approximation for the exchange terms of Murrell's perturbation scheme /6/. It is not clear, however, to what extent the charge transfer term of Murrell's scheme is incorporated this way.

CALCULATIONS

In the first calculations of two interacting HF and two interacting CH_2O molecules, respectively, the rather crude monopole approximation for the charge distributions $\rho^{(\ell)}$ was applied /10/ and the local exchange potential neglected. The results obtained show that the proposed method gives somewhat better results for the electrostatic and polarization terms than the perturbation-theoretical method using the monopole approximation in both cases.

In subsequent calculations, the monopole approximation was substituted by numerical integration of the charge densities $\rho^{(s)}(\underline{r})$ in (6) and (10), and the local exchange potentials $V^{(s)}_{x\alpha}(\underline{r})$ were incorporated into the operators $\tilde{H}^{(s)}$. Calculations have been performed for up to three interacting molecules. Using a 7/3 + 4 gaussian basis set we obtained the interaction energy of two H_2O molecules as -7.39 kcal/mole with the MCF method compared to -7.97 kcal/mole with the super-molecule (SM) method. The perturbation scheme gives a similar result of -7.36 kcal/mole, but requires much more work than the MCF calculation. For two HF molecules, the interaction energy difference between MCF and SM methods is also quite satisfactory: only -0.75 kcal/mole. For three interacting molecules (2 HF molecules and a H_3O^+ molecule), calculations have so far only been performed using the monopole approximation. In this case, the MCF interaction energies (in agreement with previous calculations /10/) are essentially smaller than the ones obtained from SM (-27.95 kcal/mole as compared to -37.47 kcal/mole). With the application of the numerical integration technique (which is under further improvement to decrease the necessary computer time /11/), one expects that the MCF results will be quite close to the SM interaction energies also in this case. The details of these calculations will be published elsewhere. /12/.

POSSIBILITIES FOR THE TREATMENT OF INTERMOLECULAR CORRELATIONS

The mutually-consistent-field method can be applied not only on the Hartree-Fock level, but also with electron correlation included. A CI wavefunction of the p-th subsystem can be written

$$\Psi^{(p)} = \sum_K C_K^{(p)} \Delta_K^{(p)} \tag{11}$$

where $\Delta^{(p)}$ is the Slater-determinant belonging to the K-th configuration of subsystem p and $C_K^{(p)}$ is the corresponding coefficient. The total energy of this subsystem can be expressed

$$E^{(p)} = \langle\Psi^{(p)}|H^{(p)}|\Psi^{(p)}\rangle = \sum_{K,L} C_K^{(p)*} C_L^{(p)} \langle\Delta_K^{(p)}|H^{(p)}|\Delta_L^{(p)}\rangle \tag{12}$$

where $H^{(p)}$ is its total hamiltonian and the normalization $\langle\Psi^{(p)}|\Psi^{(p)}\rangle = 1$ is assumed.

For N interacting units we may rewrite (12) as

$$\tilde{E}^{(p)} = \langle\tilde{\Psi}^{(p)}|H^{(p)} + \sum_{s\neq p}^{N} \tfrac{1}{2}(\tilde{V}_{c,e\ell}^{(s)} + \tilde{V}_{x\alpha}^{(s)}) + V_{c,n.}^{(s)}|\tilde{\Psi}^{(p)}\rangle \tag{13}$$

where $V_c^{(s)}$ and $V_{x\alpha}^{(s)}$ are the Coulomb and local exchange potentials defined by (6) and (7). The subsystem electron density $\tilde{\rho}^{(s)}$, however, can no longer be obtained with the aid of (8) but has to be calculated with

$$\frac{\tilde{\rho}^{(s)}}{2n_s} = \int\tilde{\Psi}^{(s)*}(\underline{r}_1,\underline{r}_2,\ldots,r_{2n_s})\tilde{\Psi}^{(s)}(\underline{r}_1\underline{r}_2,\ldots,\underline{r}_{2n_s})d\underline{r}_2\ldots dr_{2n_s}$$

$$= \sum_{K,L} \tilde{C}_K^{(s)*}\tilde{C}_L^{(s)}\int\tilde{\Delta}_K^{(s)*}(\underline{r}_1,r_2,\ldots r_{2n_s})\tilde{\Delta}_L^{(s)}(\underline{r}_1,r_2,\ldots,\underline{r}_{2n_s})d\underline{r}_2\ldots d\underline{r}_{2n_s} \tag{14}$$

where $2n_s$ is the number of electrons in subsystem s.

With expressions similar to (13) for all other subsystems $s \neq p$ one can again try to find the mutually consistent charge distributions $\tilde{\rho}^{(s)}$ of the subsystems. The best way to achieve this is probably to minimize the total energy

$$\tilde{E} = \sum_{s=1}^{N} \tilde{E}^{(s)} \tag{15}$$

of the system of N interacting units. For this purpose, one can vary $\tilde{E}$ to find its minimum

$$\delta\tilde{E} = \sum_{s=1}^{N} \delta\tilde{E}^{(s)} = 0 \tag{16}$$

This can be achieved either by the method of steepest descent or with the aid of some still more effective procedure /13/.

After the mutually consistent solutions for the CI wavefunctions have been obtained one can compute the sum of interaction energies of the interacting units as

$$E_{int} = \tilde{E} + N.R. - E = \sum_{s=1}^{N} (\tilde{E}^{(s)} - E^{(s)}) + N.R. \tag{17}$$

where the nuclear repulsion term N.R. represents the sum of repulsions between the nuclei of the interacting units.

In the procedure sketched here, the most difficult problem will be the choice of those configurations which are most important from the point of view of intermolecular interactions. In this respect, experience from super-molecule calculations on small interacting systems would be very useful.

Another possibility would be to avoid the general CI formalism, and to apply the idea of the mutually consistent field in a pair-correlation scheme for which easily applicable programs are available /14/.

CONCLUSION

The MCF formalism outlined above seems to be promising for the treatment of the electrostatic, polarization and exchange parts of intermolecular interactions as shown by preliminary results on the Hartree-Fock level. With correlated wavefunctions of the interacting systems one could expect to obtain an essential part of the dispersion term as well as correlation corrections to the terms mentioned.

After the perfection of the numerical integration procedures now in progress /11/ one is likely to obtain a sufficiently accurate and not very time-consuming *ab initio* scheme for the calculation of interaction energies between many simultaneously interacting molecules or subgroups. In this way, the problem of the theoretical determination of conformations of biological macromolecules could be solved.

When some conformational problems have been solved using this method one might hope that it will provide a basis for improved semiempirical methods for the rapid calculation of the large number of conformational problems in molecular biology.

ACKNOWLEDGEMENT

I should like to express my deep gratitude to Professor Per-Olov Löwdin who has inspired and supported my work in quantum biology for many years. I am further indebted to Dr. P. Otto for the very fruitful co-operation and for many useful discussions.

REFERENCES

1. See for instance: H.A. Scheraga, Adv. Phys. Org. Chem. 6, 103 (1968); G.N. Ramachandran and V. Sasisekharan, Adv. in Prot. Chem. 23, 283 (1968).
2. See for instance: S. Kang, C.L. Johnson, J.P. Green, J. Mol. Struct, 15, 453 (1973); L.B. Kier, J. Pharm. Sci. 57, 1188 (1968).
3. See for instance: Å. Støgard, Theoret. Chim. Acta, 33, 339 (1974); O. Gropen, H.M. Seip, Chem. Phys. Letters 11, 445 (1971).
4. M.J.S. Dewar, E. Haselbach, J. Am. Chem. Soc. 92, 590 (1970).
5. See for instance: B. Maigret, B. Pullman, Theoret. Chim. Acta, 35, 113 (1974); D. Perahia, A. Pullman, Chem. Phys. Letters, 19, 73 (1973); Ph. Courrière, J.L. Couheils, B. Pullman, Compt. Rend. Acad. Sci. Paris 272, 1697 (1971).
6. J.N. Murell, M. Randic and O.R. Williams, Proc. Roy. Soc. A 284, 566 (1965).
7. H. Lischka, Chem. Phys. 2, 191 (1973).
8. P.-O. Löwdin, Study Week on Molecular Forces, North Holland Publ. Co., Amsterdam 1967, p. 637.
9. J.C. Slater, Phys. Rev. 81, 385 (1951).
10. P. Otto and J. Ladik, Chem. Phys. 8, 192 (1975).

11. P. Otto and J. Ladik, to be published.

12. P. Otto and J. Ladik, submitted to Chem. Phys.

13. M. Rosenberg and F. Martino, J. Chem. Phys. (accepted).

14. See for instance: R. Ahlrichs, W. Kutzelnigg, J. Chem. Phys. 48, 1819 (1968); W. Kutzelnigg, Theoret. Chim. Acta, 1, 327 (1963).

ENERGETICS AND MECHANISM OF 2-AMINOPURINE INDUCED MUTATIONS

Robert Rein and Ramon Garduno

Department of Experimental Pathology, Roswell Park Memorial Institute, 666 Elm St., Buffalo, New York 14203
Department of Biophysics, State University of New York at Buffalo, 4234 Ridge Lea Rd., Buffalo, New York 14226

INTRODUCTION

Template specified base selection in nucleic acid synthesis underlies the expression of the genetic code. This comprises the storage replication and processing of information embedded in the chromosomal base sequences and is fundamental for the dynamics of the living state of matter. The fidelity of the gene copying is manifested in an error discrimination of the order of one error in 10^6-10^9 replicated bases /1/. What gives the fidelity to this selection specificity has intrigued physicists from the early days of genetics. To explain the stability of the genetic code, Delbruck /2/ and Schrödinger /3/ have considered the genome to be in a stationary quantum state. The discovery of the double helical structure of the DNA by Watson, Crick and Wilkins /4,5/ has provided the structural basis for a more explicit physical theory of the genetic code.

Implicit in the early ideas concerning the molecular basis of point mutations is the concept of mismatched base pairs /7/, that is, base pairs formed not according to the Watson-Crick complementarity principle. Pairing of non-complementary purine and pyrimidine bases, or two purines or two pyrimidines, would lead to transitional or transversional errors respectively. It has also been recognized that the purine and pyrimidine bases could exist in more than one tautomeric form /8, 9, 10/. Base pairing schemes involving bases in rare tautomeric forms have been suggested /11/ as mutational precursors.

Löwdin's proton tunneling theory of mutations /6/ appears

to be a synthesis of the Delbruck-Schrödinger quantum concepts on one hand and of the Watson-Crick structural model and base tautomerization on the other. Löwdin visualizes the genetic code as a proton-electron code determined by a probability distribution in the successive base pairs. With these ideas, quantum mechanical proton tunneling can be a mechanism leading to metastable states with different protonic distribution. These states lead to errors in copying during the replication steps, that is, a mutation.

Löwdin has succeeded in describing one of the most fundamental biological phenomena with a model which can not only be interpreted in terms of Quantum Mechanical concepts, but can also be described in quantitative form. Thus it is not surprising that his theory has inspired many follow-up studies in this field /12,13,14,15,16/.

One of the author's (R.R.) scientific interest has been greatly influenced by these ideas during his stay in Uppsala. The study of base analog induced mutations reported in this paper is a direct outgrowth of this interest.

BASE ANALOG INDUCED MUTATION

2-aminopurine (Ap) was studied theoretically first by the Pullmans /17/ and later by Danilov et al. /18/. We have chosen Ap for further investigation since it has several interesting simplifying features in relating mutational events to nucleotide interactions.

The mutagenic effect of Ap has been thoroughly investigated experimentally. It has been found that this base analog induces mutations with a frequency of many orders of magnitude higher than the frequency of spontaneous mutations /19/.

Due to the extent of the base analog incorporation into the helical strand of DNA, the template base to which Ap pairs has to be in its normal form. Thus, proton tunneling cannot be a possible mutational mechanism in this case. The effect of stereochemical factors of mismatched pairs on inducing errors has been discussed recently by Rein et al. /7/. The possible hydrogen bonding schemes between Ap and thymine and cytosine respectively are presented in figure 1. It should be recognized that these schemes do not imply a significant deviation from the double helical geometry. Hence, the stereochemical factor is not likely to have a significant influence on base selection either. This leaves the option for two major alternatives for the mechanism of erroneous coupling:

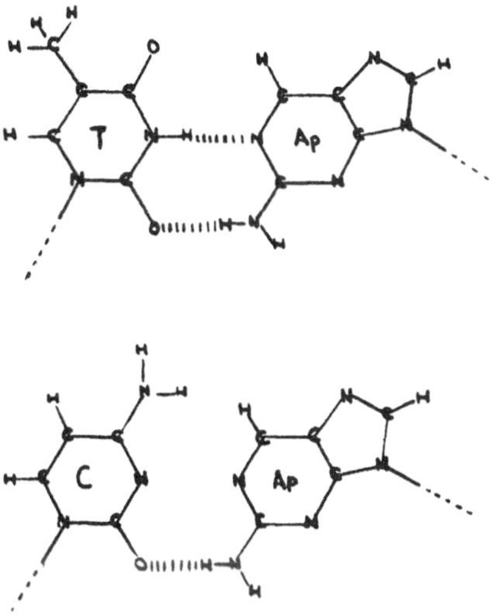

Fig. 1a - Base pairing of the normal tautomer of 2-aminopurine with thymine and cytosine.

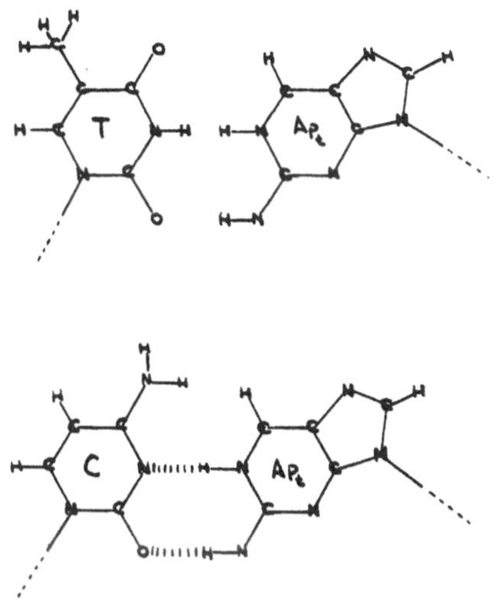

Fig. 1b - Base pairing of the rare tautomer of 2-aminopurine with thymine and cytosine.

i) 2-aminopurine in its normal tautomeric form can lead to mispairing with cytosine by a single hydrogen bond (figure 1a), and

ii) 2-aminopurine in its rare form can make a pair with cytosine (figure 1b) in which two hydrogen bonds are involved.

The rest of the paper is concerned with the calculation of the energetics of these two alternative pathways. The objective of this is to describe the pairing scheme responsible for mutations.

METHODS AND RESULTS

In order to distinguish which of the tautomeric forms of Ap is responsible for the mutational effects caused by this base analog, first we calculated the electronic energies for these tautomeric species by MO methods. This is followed by interaction energy calculations involving various Ap pairs.

The energy calculations give a quantitative estimate of the relative probabilities of the various tautomeric species of Ap. The interaction calculations show the stability of the various forms in pair formation. The combination of these two steps permits the comparison of the stability and relative probability of existence for the base pair complexes of Ap in the normal and tautomeric forms with thymine and cytosine respectively.

Quantum Mechanical calculations have always required a precise geometry for the compounds under study. This is particularly improtant if one is interested in comparing the energies of two similar structures, such as two tautomeric forms. Since the crystal structure of the rare tautomer of Ap is not known, our first aim was to predict the molecular geometry for this compound. This was accomplished by finding the structure which minimizes the total energy of the molecule. We used the OPTIMO/MINDO program /19/ for these purposes. We obtained optimized geometries for both the normal and rare tautomers. The starting geometry for the optimization procedure of the normal base was taken from the known crystal structure determined by X-ray diffraction methods /20/. To get the rare tautomer, the ring structure of the normal one was conserved but the proton of the N2 was shifted to the N1 position. The resulting geometry was the imput for the optimization program. The optimized bond lengths and bond angles are listed in tables I and II respectively. The total energy for the two tautomers of Ap obtained by MINDO/2 methods are listed in table III.

Calculations of the wave functions, multipole moments and interaction energies have been performed according to methods described in our earlier papers /21,22/. The results obtained for Ap-T, Ap-C, Ap_t-T and Ap_t-C are listed in table IV. Table V lists the interaction energies of these pairs. Relative probabilities of the various pairs are listed in table VI.

The data presented in table III clearly shows that the normal tautomeric species of Ap is the most stable. This is in agreement with earlier postulates by DeBusk /24/.

The interaction energy calculations for the base pairs involving Ap show that: the Ap-T pair is more stable than the Ap-C pair by 3kcal/mole; the Ap_t-C pair is more stable than the Ap-C pair by 1 kcal/mole. The later is understandable since the molecular structure for Ap_t presents the possibility for two hydrogen bonds with cytosine. However, the data in table III also give a tautomerization energy of 15 kcal/mole for the tautomeric shift Ap--Ap_t. This energy is undoubtedly higher than that obtained in the stabilization of the Ap_t-C pair.

The relative probabilities show that Ap-T pair has the highest probability among the base pairs involving this base analog. The probability for this pairing scheme is approximately 10^2 times less probable than A-T pair formation for any of the two normalized values. It is clear from this study that the normal tautomer of Ap must be the form involved in the transition of DNA bases. The tautomeric species of Ap has a lesser probability when coupled to thymine.

TABLE I - bond distances (Å)

OPTIMO/MINDO$_2$ OPTIMIZATION VALUES FOR 2-AMINOPURINE

Bond	Normal Taut.	Rare Taut.	Exp. /20/
N1-H13	-	1.123	-
N1-C2	1.315	1.407	1.37
N1-C6	1.338	1.387	1.32
C2-N3	1.322	1.343	1.34
C2-N12	1.344	1.257	1.39
N3-C4	1.321	1.292	1.33
C4-N9	1.378	1.379	1.34
C4-C5	1.395	1.422	1.41
C5-C6	1.355	1.329	1.39
C5-N7	1.397	1.399	1.35
C6-H11	1.188	1.194	-
N7-C8	1.287	1.285	1.30
C8-H10	1.181	1.184	-
C8-N9	1.394	1.394	1.37
N9-H15	1.114	1.114	-
N12-H13	1.111	-	-
N12-H14	1.113	1.106	-

TABLE II - bond angles (<°)

OPTIMO/MONDO$_2$ OPTIMIZATION VALUES FOR 2-AMINOPURINE (CONT.)

Angle Between	Normal Taut.	Rare Taut.	Exp. /20/
H13-N1-C6	-	115.57	-
C2-N1-H13	-	118.42	-
C2-N1-C6	122.98	126.01	117.8
N1-C2-N3	124.77	126.60	126.5
N1-C2-N12	115.49	116.84	116.0
N3-C2-N12	119.74	116.57	-
C4-N3-C2	113.78	118.88	113.1
N3-C4-N9	128.01	128.13	127.3
N3-C4-C5	123.94	124.77	125.8
N9-C4-C5	108.05	107.10	-
C6-C5-C4	119.35	120.00	116.1
C6-C5-N7	133.07	133.13	134.7
C4-C5-N7	107.58	106.87	-
H11-C6-N1	121.46	119.57	-
H11-C6-C5	123.37	126.65	-
N1-C6-C5	115.17	113.78	120.6
C8-N7-C5	106.56	107.65	104.6
H10-C8-N7	125.65	125.78	-
H10-C8-N9	120.76	121.29	-
N7-C8-N9	113.59	112.93	115.5
H15-N9-C4	128.21	126.71	-
H15-N9-C8	127.57	121.29	-
C4-N9-C8	104.22	105.44	104.0
H13-N12-H14	118.56	-	-
H13-N12-C2	122.91	-	-
H14-N12-C2	118.52	124.05	-

TABLE III - total molecular energy for tautomers of Ap.[a]

Ap (normal tautomer)	-40 258.41 kcal/mole
Ap_t (rare tautomer)	-40 243.42 kcal/mole
	14.99 kcal/mole = tautomerization energy.

[a] - Obtained by MINDO/2 /23/.

DISCUSSION OF RESULTS

The analysis of the data presented in Tables I and II show that the optimized geometries are somewhat distorted relative to the values reported from X-ray crystal structure determinations. The most significant deviations are: the bond lengths of N1-C2 and C4N9 for both tautomers, and N1-C6, C2-N12 and N3-C4 for the rare tautomer. Their values deviate in the range of 0.6 Å from the experimentally determined ones. The bond angles suffering slight deviations are the C2-N1-C6, C6-C5-C4 and N1-C6-C5 for both tautomers, and N1-C2-N3 and C4-N3-C2 for the normal and rare tautomers respectively. The deviations observed for these bond angles are between -7.0 to + 10.0 degrees. Considering the formidable difficulties in predicting the geometry of large polyatomic molecules, these deviations are not surprising.

TABLE IV - interaction energies for the base pairs of the normal and rare tautomers of 2-aminopurine.

	Ap	Ap_t
Thymine	-5.3738 kcal/mole	+2.9739 kcal/mole
Cytosine	-2.7579 kcal/mole	-3.3089 kcal/mole

TABLE V - relative probabilities for the base pairs of A_p

Base pairs	Boltzmann factor[a] $B_i = \exp(-E_i/RT)$	Relative probability ($P_i = B_i/\Sigma_i B_i$) Normalized to: A-T	G-C
G-C	1.8166×10^{17}		9.999×10^{-1}
A-T	7.8766×10^{5}	9.99×10^{-1}	4.3359×10^{-12}
Ap-T	8.7299×10^{3}	1.10×10^{-2}	4.8056×10^{-14}
Ap-C	1.0536×10^{2}	1.33×10^{-3}	5.7998×10^{-16}
Ap_t-C	2.6710×10^{2}	3.41×10^{-3}	1.4703×10^{-15}
Ap_t-T	7.8890×10^{-219}[b]	9.92×10^{-224}	4.3427×10^{-236}

a - T = 25°C. E_i = interaction energy.

b - This extremely low value is the result of having two interacting protons in a region close to their Van der Waal's radii.

TABLE VI - calculated energies of interaction for some normal and abnormal base pairs of DNA[a]

Base pair	Energy	Base pair	Energy	Sourse
A-T	-8.04	G-C	-23.534	b
Ap-T	-5.37	Ap-C	-2.758	This work

a Values given in Kcal/mole.

b Egan et al., 1974.

However, what is important from the point of view of this study, is that our energy calculations are made for two equally optimized structures, whatever their distortions are, thereby providing the best way for comparison of the relative energies.

Very often other studies in the field have been limited to frozen structures of the compounds /8, 9, 10/. The latter have the disadvantage of comparing energies of arbitrarily chosen structures. In view of the sensitivity of the energy to the structure, these comparisons seem to be of poor reliability.

In view of the fact that energy calculations and relative probabilities have shown the normal form of Ap to be the mutagenic species, the mechanism for the transition induced by this base analog can be discussed as follows:

i) Recent studies on the interaction energies for the normal base pairs G-C and A-T (22) show that the energy of interaction for these pairs corresponds to -23.5 and -8.04 kcal/mole respectively. These values can be compared with those obtained for the corresponding pairs involving Ap (see table VI). One should observe the energy difference of 3 kcal/mole and 20 kcal/mole for the (A-T) - (Ap-T) and (G-C) - (Ap-C) base pairs respectively. From this it is clear that the A-T pair is the most likely to be used in the miscoupling with Ap.

ii) Freese has considered errors of incorporation and errors of replication for the induced mutations by base analogs /25/. The distinction between these two steps for the Ap case is that, in the first pathway Ap competes with normal bases for the template site, while in the second, the normal bases compete for the Ap as template.

iii) These postulations and the examination of table IV show that, in the incorporation step, Ap could compete with adenine in a ratio of approximately 1:100 since the energies of pair formation are separated only by 2.67 kcal/mole. However, the competition of Ap with guanine is energetically a process completely unfavorable, thereby excluding an incorporation error.

iv) In the replication step the competition between thymine and cytosine for the Ap template site is separated by an energy difference of 2.612 kcal/mole. That is, the corresponding probabilities for the miscoupling of C for T are in the order of 1:100.

The mutation steps discussed in paragraphs (iii) and (iv), together with the alternative (the energetically forbidden pathway leading to transitions involving incorporations into the G-C template) are summarized in figure 2.

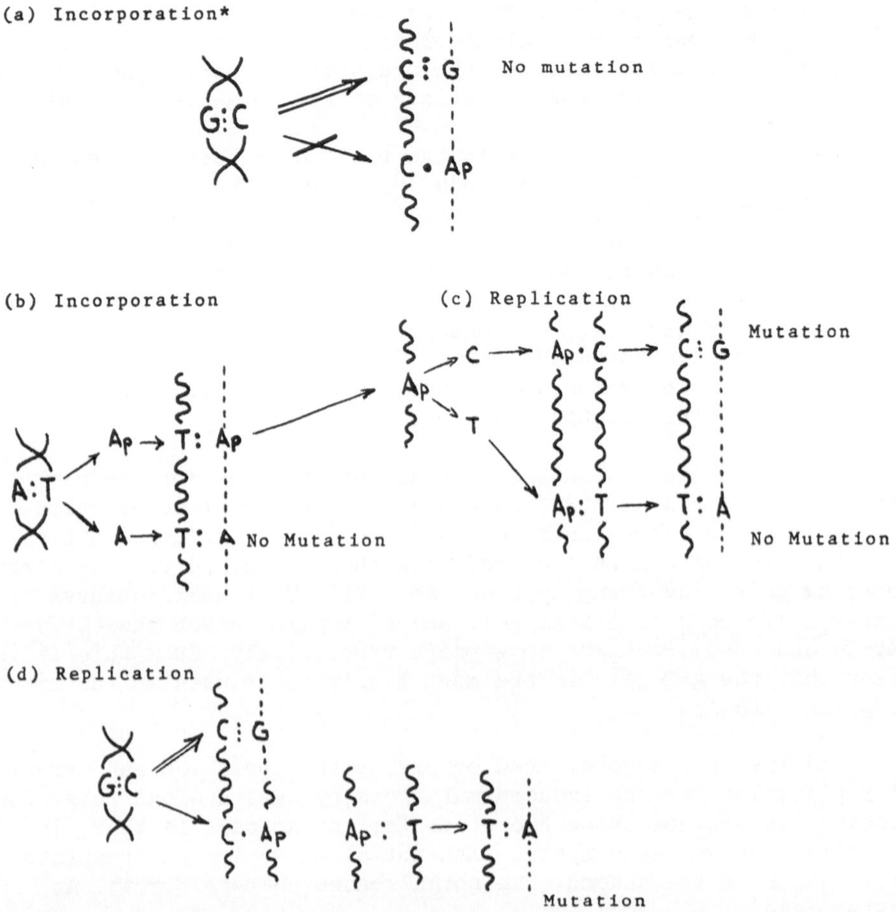

Figure 2. - 2-aminopurine mutagenesis by means of transitional erors.

*The closed loops indicate that the bases belong to a normal DNA double helix. Wiggled lines represent an open DNA, but the bases in this strand act as templates. Broken lines represent a DNA chain in construction.

From this study we can conclude that:

a) The mechanism in which Ap leads to transitional errors is undoubtedly represented by two major steps; errors of incorporation as discussed in (iii), and errors of replication as discussed in (iv).

b) The overall probabilities for a transitional error according to these mechanisms will appear as the product of the probabilities of replacing an A by an Ap in the incorporation step and the probability for the replication step when cytosine replaces thymine. The relative value is $1:10^4$.

c) The other interesting conclusion following this study is that A-T to G-C transitions should be prevalent and that G-C to A-T transitions could be allowed, according to our calculations, with a probability by nine orders of magnitude lower, assuming equal concentrations of G and Ap. This is probably an overestimate because solvent effects are not included. It has been shown in a recent study /7/ that, when solvent effects are included, the energy differences for the base pair formation are decreased. However, the trend going from A-T to G-C transitions is still predicted to be at least several orders of magnitude more probable than the G-C to A-T transitions. In fact this prediction is in agreement with recent experimental findings /26/. From this model we predict also that a guanine starvation should change the course for the probability of G-C to A-T transitions. This is a prediction which can be tested experimentally.

d) To answer the question regarding mutational mechanisms, as raised in the introduction, it is interesting to note that this study demonstrates, at least for the base analog 2-aminopurine, that the transitional errors occur via normal tautomeric forms.

ACKNOWLEDGEMENTS

This research was supported in part by a NASA Grant NGR-33-015-002. We would like to acknowledge the Computer Center of Buffalo's generous allotment of computer time. We would like to thank Mrs. Ruth Harvey for her fine typing of the manuscript, and Mr. Stephen Scott for valuable assistance. One of the authors, R.G., whishes to thank the C.O.F.A.A. of the Instituto Politecnico Nacional of Mexico and the Banco de Mexico S.A. for a fellowship during the development of this study.

REFERENCES

1. Drake J.W., J. Mol. Biol., 6,268-283 (1973).
2. Delbruck M., Proc. Nat. Acad. Sci. U.S.; 40, 783 (1954).
3. Schrödinger E., What is Life? The physical aspect of the living cell. (Cambridge University Press, 1945).
4. Watson D.J., Crick F.M., Nature 171, 737, 964 (1952).
5. Watson D.J., Crick F.M., Proc. Roy. Soc. (London), A233, 80 (1954).

6. Löwdin P., Biopolymers Symposia 1, 161-168 (1964).

7. Rein R., Coeckelenberg E., Egan T.J.; Int. J. Quantum Chem. (1976) In press.

8. Lee G.C.Y., Prestegard J.H., Chang I.S., J. Am Chem. Soc., 94 (3), 951 (1972).

9. Ibid., J. Am. Chem. Soc., 94(9), 3218 (1972).

10. Bertran J., Chalvet O., Daudel R., Anales de Fisica (Spain), Tomo LXVI, 247-257 (1970).

11. Watson D.J., Crick F.M., Cold Spring Harbor Symp. Quant. Biol. 18, 123-131 (1953).

12. Rein R., Ladik J., J. Chem. Phy. 40, 2466 (1964).

13. Rein R., Harris F.E., J. Chem. Phy. 41(11), 3393 (1964).

14. Rein R., Harris F.E., J. Chem. Phy. 42(6), 2177 (1965).

15. Rein R., Harris F.E., J. Chem. Phy. 43(12), 4415 (1965).

16. Rein R., Harris F.E., J. Chem. Phy. 45(5), 1797 (1966).

17. Pullman B., Pullman A., Biochim. Biophys. Acta 64, 703 (1962).

18. Danilov et al., Biofizika 12(4), 726-729 (1967).

19. McIver J.W., Komornick A., Chem. Phy. Lett. 10(3), 303-306 (1971).

20. Mazza F., Sobell H.M., Kartha G., J. Mol. Biol. 43, 407-422 (1969).

21. Rein R., Clarke G.A., Harris F.E., The Jerusalem Symposia on Quantum Chemistry & Biochemistry (Israel Acad. Sci. Humanities, 1970) p. 86.

22. Egan J., Swissler T., Rein R., Int. J. Quantum Chem. Q.B. Symp. 1, 71-79 (1974).

23. Dewar M.J.S., Hasselback E., J. Am. Chem. Soc. 92(3), 590-598 (1970).

24. DeBusk G.A., Molecular Genetics (The MacMillan Co., 1968) p. 59.

25. Fresee E.B., Brookhaven Symp. Biol. 12, 63-73 (1959).

26. Rogan G.E., Bessman J.M.; J. Bacteriology 103(3), 622-633 (1970).

EXTERNAL ELECTRICAL FIELD AND PROTON TRANSFER*

H. Chojnacki

Institute of Organic and Physical Chemistry, 50-370

Wroclaw, Wyb. Wyspianskiego 27, Poland

The role of the external electrical field has been found to be important for the proton transfer of a hydrogen bond. The contributions of the classical and tunneling effects were estimated. The relevant backward proton transfer seems to be essential in the process. Biological implications of the effects under examination have been pointed out.

According to the Löwdin's theory /1,2,3/ a hydrogen bond may be characterized as a proton shared between two electron lone pairs. The interaction of the proton with the lone pairs for some hydrogen bonded systems can be represented by a double-well potential involving two minima separated by a barrier. In such a case the proton transfer may occur and this effect seems to be of fundamental importance for some biochemical processes occurring in living systems like enzymatic activity or replication and transcription of DNA.

In some circumstances the barrier shape determining the proton transfer process may be modified under influence of any environment effects. This can lead to the induced proton transfer resulting in some biological systems in the irreversible mutations /2,3,4/.As suggested by Löwdin /2,3/ the double-well potential may have been disturbed through an outer electronic potential from electrical field. Therefore, the possible role of an external electrical field poses some interest either for chemical or for biological processes occurring with the relevant proton transfer.

In our case the influence of the electrical field on the proton

* A tribute to Professor P.O. Löwdin's 60 years birthday.

transfer has been examined as a function of the barrier dimensions assuming that its parabolic shape is adequate to represent the potential energy of a hydrogen bond itself. When the outer electrical field is applied to the one-dimensional barrier along the x axis /Fig. 1/ the potential energy of a tunnelling particle can be expressed by

$$V(x) = V_0(x) \mp eFx$$

where $V_0(x)$ is the respective potential energy without the field and the F is the field strength. The resulting probability of the transfer effect is the difference of passing forward P^+ and backward P^- against the field

$$P = P^+ - P^-$$

In a thermal equilibrium the total integral probability of the proton transfer should be expressed as the sum of the classical passage over the barrier and the tunnelling effect. Thus, the total probability of the forward or backward transfer against the field could

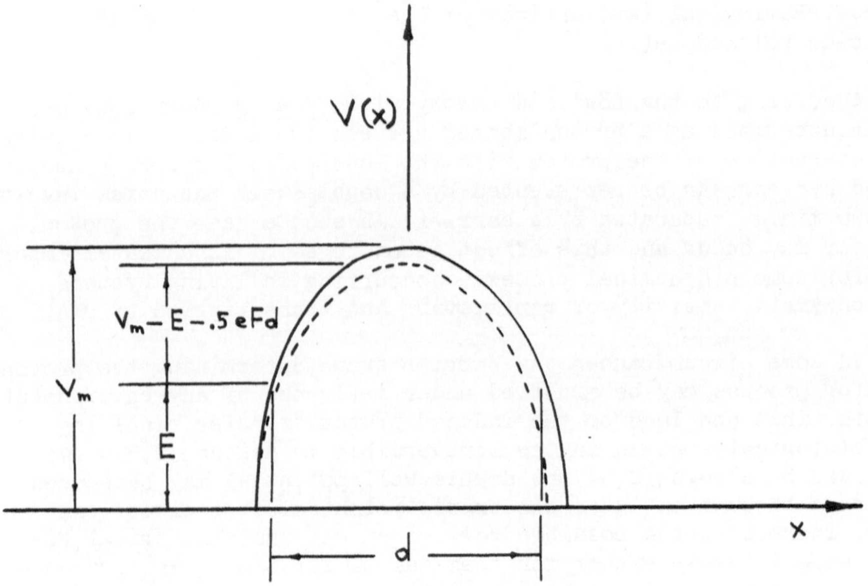

Fig. 1. Modification of the parabolic barrier in the presence of electrical field with the relevant notation used in the calculation.

be given by the two relevant terms

$$P^{\pm} = 1/kT \int_{A}^{\infty} \exp(-W/kT)dW + 1/kT \int_{0}^{A} D^{\pm}(W)\exp(-W/kT)dW$$

where $A = V_m - E \mp 0.5eFd$. E denotes here the kinetic energy of a tunnelling particle, V_m and d are, respectively, the barrier height and the barrier width without the field /Fig. 1/. Furthermore, $D^{\pm}(W)$ is the respective transmission coefficient for the forward or backward tunnelling process. With the above potential energy of a tunnelling particle in the presence of the electrical field $V^{\pm}(x)$, after integration

$$D^{\pm}(W) = \exp\{-2/\hbar \int_{x_1}^{x_2} \left(2m(V^{\pm}(x) - E)\right)^{\frac{1}{2}} dx\}$$

we obtain

$$D^{\pm}(W) = \exp\{-\pi/4\hbar(2m)^{\frac{1}{2}}d(V_m - E\mp 0.5eFd)^{\frac{1}{2}}$$

$$x\left(1 + \left(\frac{V_m-E\mp 0.5eFd}{V_m-E\pm 0.5eFd}\right)^{\frac{1}{2}}\right)\}$$

For F = 0 this result falls into the usual equation for the transmission coefficient of the parabolic barrier known from the semiclassical WKB approach /5/.

As it can be seen /Table 1/ in this model of the proton transfer both classical and tunnelling effects are important for the barrier heights of the order of 0.6 eV. When the potential barrier exceeds 1.0 eV the tunnelling effect is dominant and the classical passage over the barrier seems to be negligible. The classical proton transfer as well as the quantum mechanical effect are explicitly influenced by the field applied to the barrier enhancing the respective probabilities by several orders. The tunnelling effect is of much less importance for deuterated hydrogen bond. It should be noted that in general the backward proton motion is of essential importance for the total transfer effect.

It appears that the proton transfer effect is also temperature dependent but for the barriers as high as 1.2 and 1.8 eV the constant probability of the total transfer is predicted /Fig. 2/. For the intermediate barrier height of 0.8 eV deviation from the linearity in the Arrhenius plot /log P vs. 1/T/ was found. In general, however, the temperature dependence is a function of the barrier dimensions as well as the external field applied.

Table 1

CLASSICAL AND QUANTUM PROTON TRANSFER PROBABILITIES CALCULATED FOR THE PARABOLIC BARRIER OF THE WIDTH OF 1.0 Å AT ROOM TEMPERATURE

Barrier height (eV)	Field strength (V/cm)	$P_{class.}$	$P_{tunn.}$
0.6	0	0	0
	10^2	$1.58.10^{-15}$	$1.12.10^{-16}$
	10^5	$1.58.10^{-12}$	$1.10.10^{-13}$
	10^7	$4.76.10^{-10}$	$6.66.10^{-12}$
1.2	0	0	0
	10^2	$6.26.10^{-26}$	$1.24.10^{-21}$
	10^5	$6.26.10^{-23}$	$1.23.10^{-18}$
	10^7	$1.13.10^{-20}$	$7.87.10^{-17}$
1.8	0	0	0
	10^2	$2.48.10^{-36}$	$2.08.10^{-25}$
	10^5	$2.48.10^{-33}$	$2.06.10^{-22}$
	10^7	$4.48.10^{-31}$	$1.38.10^{-20}$

The results obtained here throw some light on the participation both of classical and quantum mechanical effects in the proton transfer processes. On the other hand the distinct role of the external electrical field seems to be important not only for chemical reactions but also for the respective biochemical induced phenomenas /4/. This conclusion is in accordance with a supposition that in some cases the local electrical fields in living systems are estimated to be as high as 10^6 volts per centimeter /6/. Besides, some evidence corroborating equally Löwdin's theory and our results on the influence of the outer electrical field on the proton transfer could be inferred indirectly from the experiment on the effect of the potential difference on cell mitosis /7/.

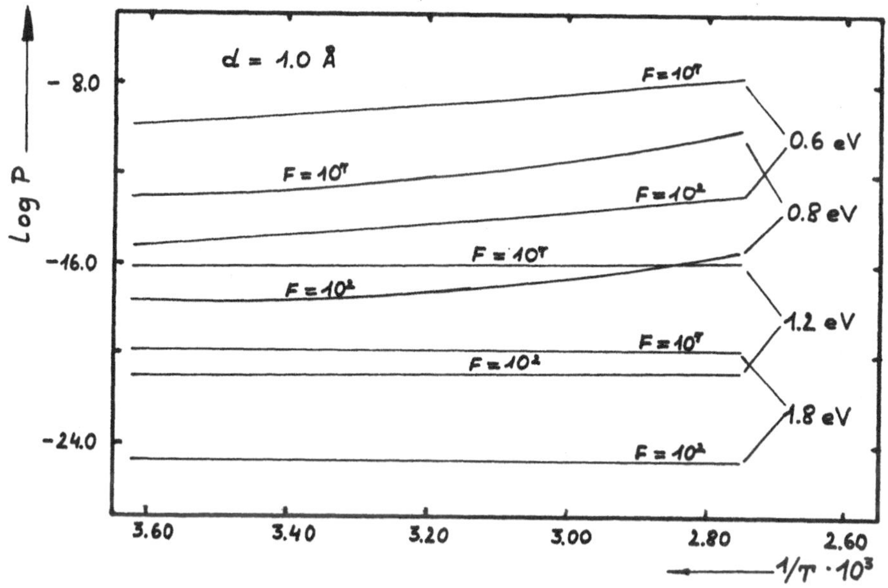

Fig. 2. The temperature dependence of the total transfer probability of the proton /log P vs. 1/T/ for different barrier heights and field strengths and for the barrier width of 1.0 Å. F is the field strength in volts per centimeter.

This work has been sponsored by the Polish Academy of Sciences.

REFERENCES

1. P.O. Löwdin, Rev. Mod. Phys., 35, 724 (1963).

2. P.O. Löwdin, Biopolymers Symposia, No. 1, 161 (1964).

3. P.O. Löwdin, Advances in Quantum Chemistry, Academic Press,

4. P.O. Löwdin, Electronic Aspects of Biochemistry, Academic Press New York, 1964.

5. N.F. Mott, I.N. Sneddon, Wave Mechanics and Its Applications, Oxford University Press, 1948.

6. R.B. Setlow, E.C. Pollard, Molecular Biophysics, Addison-Wesley, Massachussets, 1962.

7. L. Bozóky, Gy. Kiszely, T.A. Hoffmann, J. Ladik, Nature, 199, 1906 (1963).

REFERENCES

PROTON TUNNELLING IN DNA BASE PAIRS AND MUTAGENESIS

Suheil F. Abdulnur

Department of Chemistry, University of New Orleans*

New Orleans, Louisiana 70122, U.S.A.

The motivation for this work developed while the author was at the Quantum Theory Project, Gainesville, Florida, 1969-1973. During that period I had the privilege to attend the inspiring lectures of Professor Per-Olov Löwdin, to get exposed to his deep insight and undaunted scientific spirit, and to know him as a friend. It is with gratitude and affection that I wish him a very happy 60th birthday and many more years of happiness and scientific productivity.

INTRODUCTION

In 1953 Watson and Crick /1/ pointed out that the possible appearance of DNA bases in their unusual tautomeric forms can cause mispairings of the purines and pyrimidines and hence may lead to mutations. In 1963 Löwdin /2,3/ proposed that the transformation to the tautomeric forms could be affected by proton tunnelling between two energy wells in the hydrogen-bonds (H-bs) of the normal base pairs in the DNA helix, namely: Guanine-Cytosine (G...C) and Adenine-Thymine (A...T) (see figures 1 and 2).

Following this a number of researchers /4-6/ confirmed Löwdin's hypothesis through numerical calculations carried out on the base pairs. Due to the complexity of these systems the calculations were of necessity of a semi-empirical nature. In 1970 Clementi et. al. /7/ carried out an *ab initio* SCF calculation for the tunnelling of one proton in the GN_1-H---N_3C H-b of the G---C pair (see fig. 1). No double minimum was found in this case, although these authors

*Formerly Louisiana State University in New Orleans.

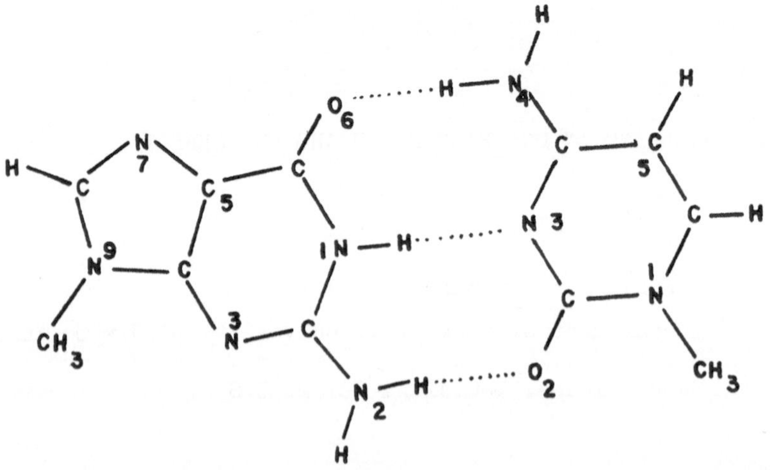

Figure 1. The Watson-Crick type base pair mG---mC. The H-b distances used are: $(G)N_1$---$N_3(C)$ = 2.92Å and (G) O_6---$N_4(C)$ = $(G)N_2$---$O_2(C)$ = 2.87Å.

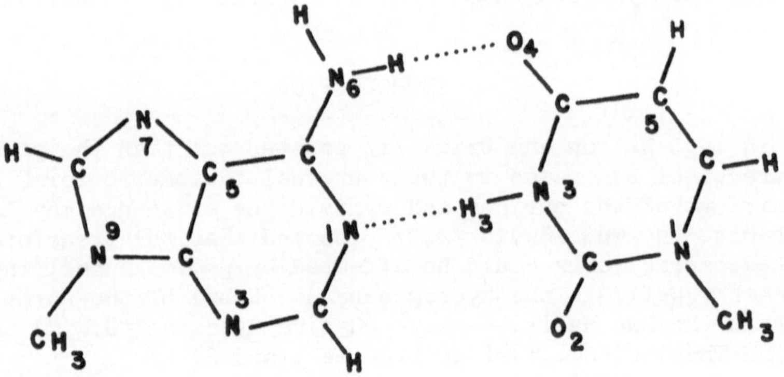

Figure 2. The Watson-Crick type base pair mA---mU. The H-b distances used are: $(A)N_1$---$N_3(U)$ = 2.91Å and $(A)N_6$---$O_4(U)$ = 2.82Å.

pointed out the dependence of such a result on the H-b distance and the particular geometry chosen. Their similar calculations on the dimeric form of formic acid did indicate however the presence of a double well potential when one considers the simultaneous coupled motion of two protons in two adjacent H-bs.

In this work we present some potential energy curves for single and double proton tunnelling in the DNA base pairs, using the CNDO/2 method /8/. The effect of methylation of the N_7 and O_6 positions of G, as a result of the action of certain alkylating agents, /9-13/ on such potentials is also considered. Although the results of the CNDO/2 method are not as reliable as those obtained from SCF _ab initio_ calculations, this method is one of the most sophisticated semi-empirical techniques currently available for handling such large systems at unprohibitive cost.

COMPUTATIONAL DETAILS

In order to obtain results pertaining to the behavior of the base pairs _in vivo_, the present calculations are carried on the Watson-Crick base pairs of (9-methyl) A and G and (1-methyl) Uracil and C (figures 1 and 2). These are referred to henceforth as mA, mG, mU and mC. The computations utilize a CNDO/2 program written in this laboratory by Howland and Flurry /14/.

The coordinates of the atoms in mA, mU and mG are those used in our previous study on the mutagenic nature of 5-Flourouracil /15/, taken from most recent X-ray crystallographic studies. Those for the heavy atoms in mC are taken from O'Brien's crystal data /16/. In all these cases standard bond lengths and angles /8/ are used to place the hydrogens as well as the methyl groups on the N_7 and O_6 positions of mG.

The H-b lengths in mG---mC and mA---mU are shown in Figures 1 and 2 and are in close agreement with those recently reported by Arnott et. al. /17/ for the A---T and G---C base pairs in DNA double helix. These same bond lengths were assumed to hold for $(N_7m)mG^{\pm}$--mC and $(O_6m)G^{\pm}$--mC as well. All bond lengths and angles were held fixed during the displacement of the hydrogens in the H-Bonds which are assumed to be linear. The curves in Figures 3 and 4 around the minima and maxima are drawn by parabolic interpolation from the calculated points in these regions.

RESULTS

Figure 3 shows the potential energy profiles resulting from the simultaneous displacement of the two protons in the bonds

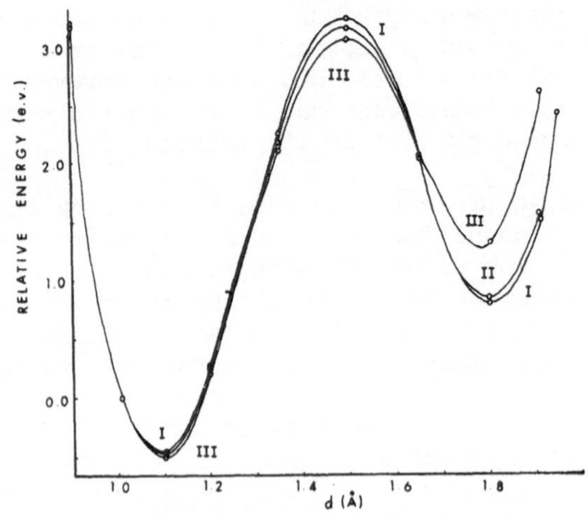

Figure 3. Potential energy profiles for the simultaneous displacement of two protons in mG---mC (curve I), $(N_7m)mG^+$---mC (curve II) and mA---mU (curve III). d is the distance of H-N_1 (in G) from N_1 and H-N_4 (in C) from N_4.

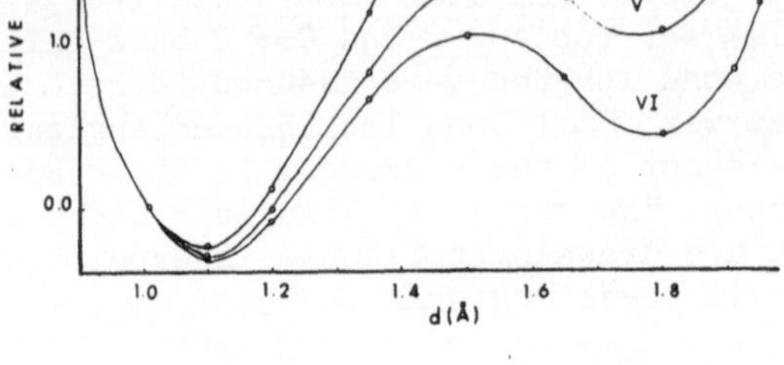

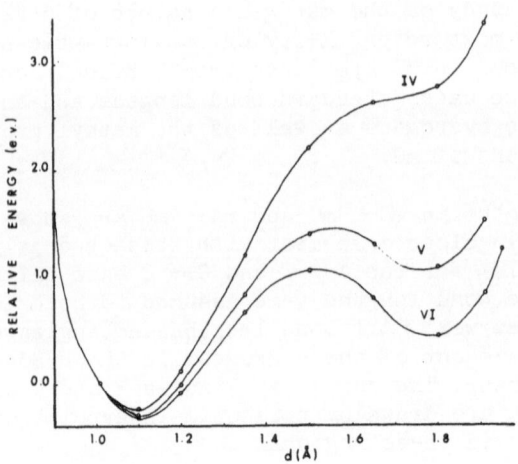

Figure 4. Potential energy profiles for the displacement of the proton in the H-b GN_1-H---N_3C for the base pairs mG---mC (curve IV), $(N_7m)mG^+$---mC (curve V), and $(O_6m)mG^+$---mC (curve VI). d is the distance of H-N_1 (in G) from N_1.

GN_1-H---N_3C and GO_6---H-N_4C of mG---mC (I) and $(N_7m)mG^+$---mC (II), and in AN_1---H-N_3U and AN_6-H---O_4U of mA---mU (III). A double well potential is observed in all three curves.

Figure 4 shows the potential energy curves resulting from the displacement of one hydrogen in the bond GN_1-H---N_3C of mG---mC (curve IV), $(N_7m)mG^+$---mC (curve V) and $(O_6m)mG^+$---mC (curve VI). A double minimum is observed in V and VI but not in IV.

Table 1 presents the potential energies of the maxima and right minima relative to the minima on the left for each of the cases in Figures 3 and 4.

Table 1

Relative Energies (e.v.) of Maximum and Right Minimum In the Energy Porfiles of Figures 3 and 4.

	Base Pair	Relative energy of maximum	Relative energy of right minimum
Double Proton Tunnelling	mG---mC	3.70	1.25
	mA---mU	3.54	1.77
	$(N_7m)mG^+$--mC	3.61	1.31
Single Proton Tunnelling	mG---mC	2.91	**
	$(N_7m)mG^+$--mC	1.74	1.35
	$(O_6m)mG^+$--mC	1.38	0.77

* Relative to the energy of the left minimum in each case.

** No visible minimum.

DISCUSSION

The results in Figure 3 support Löwdin's hypothesis /2,3/ that the tautomeric forms of the bases could result from double proton tunnelling in the Watson-Crick base pairs in DNA double helix. They also indicate that the equilibrium concentration of the tautomers will be greater in the case of mG---mC than in the case of mA---mU. This may offer a possible explanation for the larger A---T to G---C content in higher organisms as pointed out by Löwdin /3/ based on the earlier semi-empirical studies /4-6/. The introduction of a - CH_3^+ at the N_7 position of mG slightly reduces this probability of tautomer formation.

Figure 4 indicates that single proton tunnelling is not possible in the bond GN_1-H---N_3C in the case of the mG---mC base pair, as also found by Clementi et. al. /7/. However, the introduction of a - CH_3^+ at the N_7 or O_6 positions of mG makes such tunnelling possible.

On comparing the charge densities of the various atoms of the purine and pyrimidine bases in the vicinity of the left minimum to those of the right minimum in curves V and VI, one finds that the positive charge migrates almost completely from the purine to the pyrimidine after tunnelling. If strand separation occurs at this stage, the resulting cation H-N_3C^+, being unable to pair to any of the normal purines, may lead to irreversible mutations, as pointed out by Löwdin /3/. From the energy differences between the minima in these curves (see also Table I) one expects a greater tendence to form such a cation if the methylation occurs at the O_6 rather than at the N_7 position of G.

In 1969 Loveless /9/ proposed that the mutagenic effect of a number of alkylating agents results from alkylation at the O_6 rather than the N_7 position of G. This hypothesis has drawn considerable support from many investigators since then /10-13/. The above results are in agreement with it.

The question of how the proton's motion affects the electronic charge distribution in the base pair has been raised by Löwdin /3/. In Table 2, the charge densities on the atoms involved in hydrogen-bonding in the base pair (N_7m)mG^+---mC are presented at three protonic displacements corresponding to the vicinities of the minima and maximum for the cases of double and single proton tunnelling. In the former case most of the redistribution of charge takes place in the atoms that are involved in the H-bs where tunnelling occurs, namely: N_1,O_6,H-N_1 of G and N_4, N_3, H-N_4 of C, the direction of change being as expected from the motion of a proton towards or away from an isolated atom. In the latter case one sees a significant increase in the electron charge density on O_6 of G and a corresponding decrease in it on O_2 of C associated with the formation of H-N_3C^+, in addition to the expected redistribution in the atoms of the central H-b.

It is interesting to note that the proton drags with it a significant electron cloud as it moves from one end of the H-b to the other. In the coupled motion of two protons, the repulsion between these clouds as the protons pass each other results in a larger positive charge on them (compare the values for H-N_1 of G

Table 2

The Variation of the Net Charges on the Atoms Involved in Hydrogen Bonding in mG---mC and $(N_7m)mG^{\pm}$---mC base pairs with proton displacement

	Dis-* tance (Å)	Guanine Atoms O_6	C_6	N_1	C_2	N_2	$H\text{-}N_1$	$H\text{-}N_2$	Cytosine Atoms N_4	C_4	N_3	C_2	$+_2$	$H\text{-}N_4$
Double Proton Tunnelling in mG---mC	1.10	-0.424	0.365	-0.244	0.393	-0.268	0.156	0.174	-0.253	0.336	-0.373	0.427	-0.438	0.188
	1.50	-0.370	0.381	-0.377	0.386	-0.270	0.262	0.157	-0.377	0.340	-0.341	0.437	-0.425	0.273
	1.80	-0.250	0.346	-0.406	0.370	-0.268	0.199	0.147	-0.368	0.292	-0.230	0.439	-0.398	0.207
Single Proton Tunnelling in $(N_7m)mG^{\pm}$---mC	1.10	-0.377	0.374	-0.242	0.421	-0.256	0.200	0.200	-0.243	0.340	-0.390	0.431	-0.442	0.157
	1.50	-0.436	0.367	-0.371	0.406	-0.261	0.270	0.165	-0.238	0.366	-0.321	0.439	-0.396	0.189
	1.80	-0.486	0.354	-0.459	0.385	-0.262	0.256	0.134	-0.228	0.384	-0.217	0.442	-0.342	0.222

*In the double proton tunnelling case this is the distance of $H\text{-}N_1$ (in G) from N_1 and $H\text{-}N_4$ (in C) from N_4. In the single proton tunnelling case it is the distance of $H\text{-}N_1$ (in G) from N_1.

and H-N_4 of $^+$C in Table 2). The net transfer of one protonic charge from(N_7m)mG^+ to mC in single proton tunnelling results from an electronic density redistribution in all the atoms of the base pair, the atoms shown in Table 2 contributing only ~40% of it.

We are currently investigating the effect of the tunnelling time through the barriers in the various cases above on the rate of production of the tautomers during the DNA replication cycle time, and will report on them at a later date.

ACKNOWLEDGEMENTS

The author would like to thank Professor Robert L. Flurry, Jr. for his kind hospitality during the course of this investigation, for helpful discussions and for the use of his computer programs. Financial support by N.S.F. grant no. GP-38740 and the Computer Research Center of the University of New Orleans is greatly appreciated. Last but not least, I would like to thank my wife Karen for her patience and encouragement, who also wishes Professor Löwdin a very happy 60th Birthday.

REFERENCES

1. J.D. Watson and F.H.C. Crick, Nature 171, 964 (1953); Cold Spring Harbor Symp. Quant. Biol. 18, 123 (1953).
2. P.-O. Löwdin, Rev. Mod. Phys. 35, 724 (1963); Biopolymers Symp. 1, 161, 293 (1964); Electronic Aspects of Biochemistry, B. Pullman, Ed. (Academic Press Inc., New York, 1964), pp. 167-201.
3. P.-O. Löwdin, Adv. Quantum Chem. 2, 213 (1965).
4. J. Ladik, Preprint QB 8, Quantum Chemistry Group, University of Uppsala, Sweden (1963); R. Rein and J. Ladik, J. Chem. Phys. 40, 2446 (1964).
5. R. Rein and F.E. Harris, J. Chem. Phys. 45, 1797 (1966); 43, 4 4415 (1965); 42, 2177 (1965); 41, 3393 (1964); Science 146, 649 (1965).
6. S. Lunell and G. Sperber, J. Chem. Phys. 46,2119 (1967).
7. E. Clementi, J. Mehl, and W. von Niessen, J. Chem. Phys. 54, 508 (1971).
8. J.A. Pople and D.L. Beveridge, Approximate Molecular Orbital Theory, McGraw-Hill Book Co., New York (1970).
9. A. Loveless, Nature 223, 206 (1969).
10. P.D. Lawley, Prog. Nucl. Acid Res. Mol. Biol. 5, 89 (1966).

11. P.N. Magee and J.M. Barnes, Adv. Cancer Res. 10, 163 (1967).

12. L.L. Gerchman and D.B. Ludlum, Biochim. Biophys. Acta 308, 310 (1973).

13. P.N. Magee, J.W. Nicoll, A.E. Pegg and P.F. Swan, Biochem. Soc. Trans. 3, 62 (1975).

14. J.C. Howland and R.L. Flurry, Jr., Theoret. Chim. Acta (Berl.) 26, 157 (1972).

15. S. Abdulnur, J. Theoret. Biol. (in press).

16. E.J. O'Brien, Acta Cryst. 23, 92 (1967).

17. S. Arnott, S.D. Dover and A.J. Wonacott, Acta Cryst. B 25, 2192 (1969).

ON PROTON MOBILITIES IN INDIVIDUAL HYDROGEN BONDS

Mark A. Ratner[+] and J.R. Sabin

Department of Chemistry, Northwestern University, Evanston, Illinois, U.S.A.
Quantum Theory Project, Department of Physics, University of Florida, Gainesville, Florida, U.S.A.

I. Introduction

The hydrogen bond has been under intense scrutiny in physical and chemical systems since its definition by Latimer and Rodebush in 1920 /1,2/. Investigations in biological systems, however, are of much more recent vintage. Indeed, the pioneering investigations of Löwdin /3/ were among the first attempts to characterize the role of protonic motion in hydrogen bonds in the properties of biological macromolecules. In particular, Löwdin /3/ stressed the importance of proton tunneling, and showed that this provides a mechanism for transfer and for loss of information stored in the form of proton position in a double well. Recent experimental advances, particularly in the area of picosecond spectroscopy /4/, have made it possible to experimentally observe the motion of molecular subunits, and the promise of being able to delineate the tunneling process has reemphasized the need for an understanding of the dynamics of protons in hydrogen bonds.

We would like here to point out several processes which can seriously modify conclusions based only on the motion of uncoupled protons in one dimensional potentials. In hydrogen bonded ferroelectrics in which the double well model for individual proton motion has long been accepted /5/, the role of coupling between the protons was recognized fairly early /6/. The coupling of the protons to lattice phonons was discussed somewhat later /7/, but is necessary for understanding the directional polarization and large isotope

[+] Alfred P. Sloan Foundation Fellow

effect in systems such as KH_2PO_4, in which the critical temperature increases by 100K upon deuteration. For individual hydrogen bonded systems, however, the importance of coupling of the proton motion to the other degrees of freedom of the molecule, as well as to the phonons of the bath, has been invoked principally in connection with infrared spectra /8/, although discussions of the modification of proton tunneling due to this coupling have been given /9/. In fact, the coupling problem here is formally extremely similar to the case of polaron motion /10/, in which it is well known that the temperature dependence, as well as the rate of transition, can be completely dominated by coupling effects. After some comments anent the problem of proton tunneling in general, we will outline several features of the coupling of proton tunneling to other degrees of freedom and the effects which may be manifested in measurements of the proton motion due to this coupling.

II. TUNNELING OF PROTONS

It is useful to define precisely what is meant by this potential in which protons move. We will assume that the Born-Oppenheimer separation is valid, so that all nuclear motion occurs on a single potential surface, $V(q,Q)$, where q is the coordinate for proton motion and Q represents schematically all other nuclear coordinates. Two different types of one dimensional potential can then be defined. The rigid-ion potential, $V_R(q,Q_0)$ is defined as the potential for motion of the proton when the other, heavier, nuclei are frozen at their equilibrium positions. The diabatic potential, $V_D(q)$ will be defined as the potential felt by the proton at position q if all other nuclei are allowed to relax to their quilibrium positions with the proton at q. Clearly, these two potentials will be relevant to different timescales of protonic motion: $V_R(q,Q_o)$ describes experiments in which protonic motion is so rapid that the other nuclei have no opportunity to move ($\sim 10^{-14}$ sec.), whereas $V_D(q)$ describes the limiting case of slow proton motion (slower than 10^{-11} sec.). Figure 1 schematically illustrates the difference between these two potentials for the case of $(H_2O)_2H^+$ and Cl_2H^-. The difference between the potentials is quite marked. Neither is entirely correct, except in the relevant time limit. The general problem must ineluctably consider the full potential $V(q,Q)$. The cases V_R and V_D will emerge in the appropriate limit from a proper dynamical treatment. This coupling behaviour is outlined in section 3 below. Löwdin's original work was based on calculated potentials /11/ of the V_R type, and must therefore be considered only as a model, as Löwdin does.

Two additional complications occur, even in the one dimensional potential $V_R(q,Q_0)$. The first of these is that when the potential

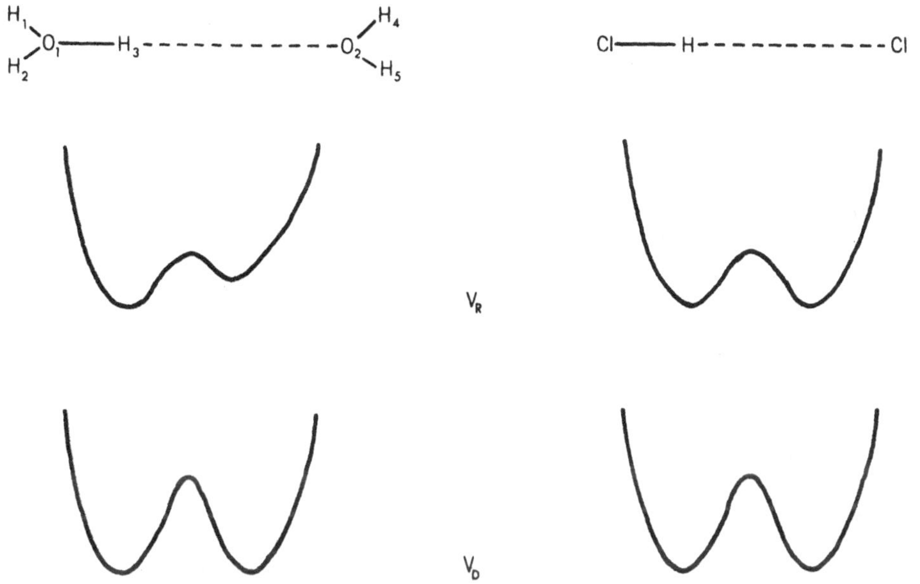

Figure 1. Schematic rigid ion and diabatic potentials for $(H_2O)_2H^+$ and $C\ell HC\ell^{-1}$.

V_R is exactly or very nearly symmetric, semi-classical theory (JWKB method) may be seriously in error in estimation of tunneling times. Thus for barrier heights of order 10000 cm^{-1}, the semi-classical estimate of the proton lingering time for small (<20 cm^{-1}) asymmetry is roughly three orders of magnitude too large /12/. This is due to resonance /12,13/ behavior: the wavefunction is inordinately sensitive to the introduction of asymmetry, becoming completely localized for even very small asymmetries. Thus the first order expansion about the classical limit will be insufficient for small asymmetries, due to the nearly discontinuous change in the wavefunction. For larger asymmetries, on the other hand, the semi-classical procedure works extremely well /3, 12, 13/.

The other point concerning one dimensional proton tunneling which deserves reemphasis involves "back tunneling". Consider a simple two level system, described by localized basis states ϕ_r and ϕ_ℓ localized in the right and left wells respectively. If we neglect overlap of ϕ_ℓ with ϕ_r, we can write the Hamiltonian for one dimensional motion in two equivalent ways:

$$H_0 = \varepsilon_\ell a_\ell^+ a_\ell + \varepsilon_r a_r^+ a_r + \Gamma(a_r^+ a_\ell + a_\ell^+ a_r) \tag{1}$$

$$H_0 = E_0 a_0^+ a_0 + E_1 a_1^+ a_1 \tag{2}$$

Here, a_r^+ and a_ℓ^+ create a proton in states ϕ_r and ϕ_ℓ with energies ε_r and ε_ℓ respectively. The tunneling energy Γ drops off very quickly with increasing barrier height: its size for typical hydrogen bonds varies from 100 cm^{-1} to essentially zero ($<10^{-5}$ cm^{-1}). The delocalized eigenstates ϕ_0 and ϕ_1 diagonalize the Hamiltonian; they are created by a_0^+ and a_1^+ with energies E_0 and E_1. The diagonalization is given by:

$$E_{0,1} = \tfrac{1}{2}\left(\varepsilon_r+\varepsilon_\ell \pm \sqrt{(\varepsilon_r-\varepsilon_\ell)^2+4\Gamma^2}\right) \tag{3}$$

$$a_0^+ = (a_r^+ + c a_\ell^+)/\sqrt{(1+c^2)} \tag{4a}$$

$$a_1^+ = (-c a_r^+ + a_\ell^+)/\sqrt{(1+c^2)} \tag{4b}$$

$$c = \frac{\varepsilon_r-\varepsilon_\ell \pm \sqrt{(\varepsilon_r-\varepsilon_\ell)^2+4\Gamma^2}}{2\Gamma} \tag{4c}$$

If, then, the proton is known at time t = 0 to be localized in state ϕ_ℓ, the probability of its being located in ϕ_r at time t is:

$$P_r(t) = \left(\frac{c}{1+c^2}\right)^2 \left(2-2\cos (E_1-E_0)t/\hbar\right) \tag{5}$$

The period of oscillation of the probability is then given by:

$$T = (E_0-E_1)^{-1}\hbar \equiv \Delta^{-1}\hbar \tag{6}$$

Thus, both the amplitude and period are precipitously reduced when small asymmetry is introduced. The lingering time, however, is increased as the potential becomes asymmetric; this is because resonant tunneling is no longer possible between ϕ_ℓ and ϕ_r. To see this, consider a measurement on an ensemble of systems. If N_r systems are localized in ϕ_r out of a total of N systems, and if first order decay is observed for the two level system, the rate constant, k, will be given by:

$$\frac{d}{dt} N_r = kN \tag{7}$$

$$k = \frac{dPr}{dt} = \frac{1}{\hbar}\left(\frac{c}{1+c^2}\right)^2 (2\Delta \sin\Delta t/\hbar) \tag{8}$$

$$\equiv k(t)$$

Thus the observed rate constant should be a harmonic function of time. This harmonic behavior is not observed, due to initial state preparation and to coupling of the protonic system with its surroundings and with the measuring apparatus. The initial rate is zero; the lingering time τ can be defined by:

$$\tau = \langle k^{-1} \rangle_{av}$$

$$= \hbar \left\langle \frac{1}{\sin\Delta t/\hbar} \right\rangle_{av} \left(\Delta \overline{W}_r\right)^{-1} \qquad (9)$$

$$\overline{W}_r \equiv 2 \left(\frac{c}{1+c^2}\right)^2 = \langle Pr \rangle_{av} \qquad (10)$$

The experimentally observed lingering time thus increases with asymmetry, and displays a resonance dip near $\varepsilon_r = \varepsilon_\ell$. The role of "back tunneling" is thus simply a manifestation of the delocalized character of the eigenstates in a near-symmetric potential.

III. COUPLING EFFECTS ON PROTON MOTION

Mechanical unharmonicity has been discussed in connection with hydrogen bonding for many years /14/. The potential $V(q,Q)$, when expanded into normal modes, contains third order terms which are not negligible, and, since the one dimensional protonic potentials are highly anharmonic, the protonic motion is subject to especially strong coupling effects. It is expected that the dominant coupling will be with the modes which contain a large component of the A----B stretch and A-H----B bend, for an A-H----B hydrogen bond. The most general case of proton dynamics, then, will involve complicated collective motions on the surface $V(q,Q)$ rather than simple one dimensional motion along either V_D or V_R. The problem can be simplified by noting that the coupling to one or a few normal modes will far exceed that to the other modes of the molecule, or to the bath /9/. As a result, the dynamical interaction with the weakly coupled modes can be replaced by a stochiastic term which characterizes the flow of energy into or out of these modes. The weakly coupled modes, for any condensed phase system, are sufficiently dense that a temperature can always be defined for them. Therefore, the general problem of motion on $V(q,Q)$ can be well approximated by motion on $V_c(q,Q_s)$ in a heat bath. Here V_c denotes the coupled potential, and the set of coordinates Q_s comprises only those which are strongly coupled. For the ground state, the coupling with the bend is probably not strong /9/, although this mode is, due to Fermi resonance /14/, extremely important in the spectroscopy. We will therefore content ourselves with considering two strongly coupled modes Q_s. The symmetric stretch will couple strongly, but it is expected that its coupling is approximately the same on the right and on the left, so that for the proton motion case (as opposed to the spectroscopic case /8, 15/ the coupling to the symmetric stretch adds no new features and can be omitted. The important modes, then, are those which couple differently when the proton is on the right or on the left. Such a coordinate, for instance, is the O_1---(H_1,H_2) distance in $H_5O_2^+$, as illustrated in Figure 1. Another coupled coordinate will be the O_2---(H_4, H_5) distance. These will clearly couple equally strongly, and are the two modes schematically designated by Q_s /16/.

The Hamiltonian can then be written as the sum of harmonic and anharmonic terms:

$$H = H_{HAR} + H_{ANH} \tag{11}$$

$$H_{HAR} = H_0 + \sum_s (b_s^+ b_s + \tfrac{1}{2})\hbar\,\omega_s \tag{12}$$

$$H_{ANH} = \sum_s (b_s^+ + b_s)\hbar\,\omega_s \left(g_r^s a_r^+ a_r + g_\ell^s a_\ell^+ a_\ell + L(a_r^+ a_\ell + a_\ell^+ a_r)\right) \tag{13}$$

Here the boson operator b_s^+ creates one quantum of excitation in the strongly coupled mode of frequency ω_s. The coupling constants g_r^s and g_ℓ^s refer to the oscillator displacement when the proton is localized in the left or right position of the double well. The coupling constant L describes the transfer of the proton accompanied by emission or absorption of one quantum. The quantities g_r^s, g_ℓ^s, a, b and L are all dimensionless; g_r and g_ℓ are of order unity /9/. In addition to the terms in eq. 11, the full Hamiltonian will include interactions with the bath. This has been studied exhaustively both for general two level systems /17, 18/ and for the particular case of hydrogen bonds /15/. For our illustrative case, we will assume that no important dynamical effects occur due to the bath. The Hamiltonian (11) is now subjected to a canonical transformation which allows the diagonal interaction term to be taken into account exactly; that is, it allows the strongly coupled vibration to follow the proton. The transformed Hamiltonian is given by:

$$\tilde{H} = e^u\, H\, e^{-u} \tag{14}$$

$$u = \sum_s (a_r^+ a_r g_r^s + a_\ell^+ a_\ell g_\ell^s)(b_s^+ - b_s) \tag{15}$$

$$\tilde{H} = \sum_s \hbar\omega_s (b_s^+ b_s + \tfrac{1}{2}) + a_r^+ a_r \tilde{\varepsilon}_r + a_\ell^+ a_\ell \tilde{\varepsilon}_\ell + \\ + \left(a_r^+ \left(\hat{\Omega}\Gamma + \sum_s L\hbar\omega_s \hat{\Phi}\right) a_\ell + \text{h.c.}\right) \tag{16}$$

$$\hat{\Omega} = \exp\left\{\sum_s \left((b_s^+ - b_s)(g_r^s - g_\ell^s)\right)\right\} \tag{17}$$

$$\tilde{\varepsilon}_r = \varepsilon_r - \sum_s g_r^{s^2} \hbar\omega_s \tag{18a}$$

$$\tilde{\varepsilon}_\ell = \varepsilon_\ell - \sum_s g_\ell^{s^2} \hbar\omega_s \tag{18b}$$

$$\hat{\Phi} = \exp \left(i\sum_s g_r^s (b_s^+ - b_s)\right)(b_s^+ + b_s)\left(-i\sum_s g_\ell^s (b_s^+ - b_s)\right) \tag{19}$$

The Hamiltonian (16) is of polaron type, and our discussion henceforth follows closely the case of a polaron hopping between two sites /18/. The operators $\hat{\Omega}$ and Φ indicate that the one dimensional (bare) proton tunneling and hopping, respectively, are modified (dressed) by the strongly coupled vibrations Q_s.

In general, response functions for (16) are given by Fourier transforms of two-time correlation functions. Then the time dependence will bring in forms like $\langle\hat{\Phi}(t)\hat{\Phi}(0)\rangle$ or $\langle\hat{\Omega}(t)\hat{\Omega}(0)\rangle$. These will result in alteration of the response property of the bare proton because of its coupling to the vibrational modes Q_s. For hydrogen bonded systems, this has been dealt with elsewhere /7,8,9,15/. Our point can be most simply illustrated, however, in the simple two site case. The proton motion can then be described most simply assuming that the double well is symmetric, defining $\varepsilon_r \equiv 0$, and omitting the oscillator term $\sum_s \hbar\omega_s (b_s^+ b_s + \frac{1}{2})$. The effective Hamiltonian for motion is then:

$$H_{eff} = (a_r^+ a_\ell + a_\ell^+ a_r)\left(\hat{\Omega}\Gamma + \sum_s L\hbar\omega_s \hat{\Phi}\right) \tag{20}$$

The bare tunneling energy, Γ, is now multiplied by a term $\hat{\Omega}$ which characterizes the modulation of the proton tunneling process by the rearrangement of the strongly coupled modes Q_s. The hopping process, described by the second term in H_{eff}, is also modulated by the vibrations; this is described by $\hat{\Phi}$. Thus the motion of the proton can occur by two mechanisms, one corresponding to tunneling reduced by a Franck-Condon factor, the other to activated hopping or diffusion, depending on the temperature range. The details can be quite complicated /9, 15, 17, 18, 19/. We can, however, differentiate two limits.

If the tunneling parameter Γ is large enough, the motion of the proton will be modulated by tunneling as described by the first term in eq. 20 above. In this case, the strong coupling results in decreased mobility, and as temperature is increased the scattering-like nature of $\hat{\Omega}$ means that the tunneling probability will decrease with temperature. At higher temperatures, and at smaller values of the tunneling integral Γ (corresponding to longer hydrogen bonds and higher barriers, or to deuteration), the tunneling term of eq. 20 will be negligible in comparison to the second, activated, term. The conditions for hopping to dominate tunneling are that the temperature be high, the decay of correlations in Q_s be short, and the tunneling integral be small. Thus, if τ_s is the correlation time for the coupled vibrations, the hopping limit will occur when:

$$\Gamma << kT/h,\ \hbar\tau_s^{-1} \tag{21}$$

The motion of the proton for temperatures in this range is diffusive, or activated. Thus it increases with temperature. The time scale, however, is now slower than that for the ordinary tunneling process, but is instead characteristic for diffusion. Arithmetically, when the vibrations are averaged, the tunneling modulation is:

$$<\hat{\Omega}> = \exp\left(-\sum_s (g_r^s - g_\ell^s)^2 \coth\beta\omega_s/2\right), \tag{22}$$

whereas the activation modulation is given by:

$$<\hat{\Phi}> = \left(1+\sum_s 2(N_s - g_r^s - g_\ell^s)\right) \exp\left(-\sum_s (g_r^s - g_\ell^s)^2 \coth\beta\omega_s/2\right) \tag{23}$$

where N_s is the equilibrium number of excitations in normal mode Q_s, and, as usual, $\beta = (kT)^{-1}$. We can see clarly that as the temperature increases, N_s increases exponentially and the second, diffusion or hopping term will come to dominate the band term.

These conclusions have some important applications for understanding of proton mobility in hydrogen bonds. For fairly low temperature and fairly small barriers, previous conclusions describing one dimensional tunneling in a rigid ion potential, V_R, are essentially correct and need only be multiplied by the Franck-Condon-like factor $<\hat{\Omega}>$. For higher temperatures and larger barriers, however, the thermally activated term will dominate. In ordinary chemical reactions, of course, this is precisely what happens: the role for quantum mechanical tunneling is essentially negligible

/20/, and the reaction rate is thermally activated, resulting in an Arrhenius type rate law. This should also be true for fairly long hydrogen bonds, where the proton is localized on one side for a time very long compared to inverse vibrational frequencies, and long compared to vibrational correlation times. Under these condition, the rate of proton motion will be controlled by the parameter L, and independent of the tunneling Γ. The relevant potential for this limit will be the diabatic potential V_D, which describes the situation in which the lattice and the modes Q_s can readjust to the proton's position, and are therefore always in thermal equilibrium. There is insufficient experimental data at present for determining which limit obtains in single hydrogen bonds, but one clearly expects that very long hydrogen bonds should not be controlled by the tunneling process. A calculation in which the tunneling is compared with activation due to dipole interaction with fluctuating solvent fields has been reported recently /21/. Coupling effects on the tunneling were not considered, so that the tunneling contribution is artificially large. Nevertheless, above room temperature, and for reasonable parameter choice, the proton mobility was indeed found to be dominated by the activated process, and to exhibit the expected exponential temperature increase (cf. figure 2).

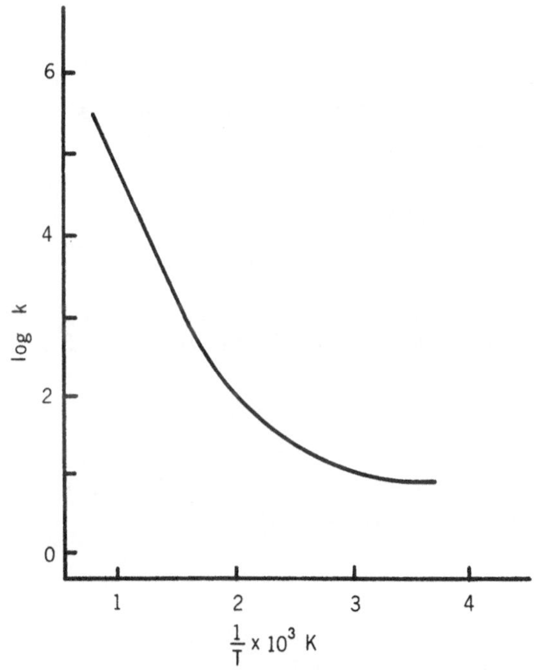

Figure 2. Schematic plot of rate constant for a proton in a double well potential undergoing an exothermic reaction vs. reciprocal temperature; from ref. 21.

Löwdin's early contributions stress the importance of tunneling as a mechanism for proton motion in hydrogen bonded systems, and in many real systems, in particular the ferroelectrics, the tunneling process, suitably modified by vibrational interaction, is in fact the dominant mechanism for low temperature mobility. In the case of very long hydrogen bonds, however, or of high barriers in general, the proton will not tunnel through (per) the mountain, but instead climb over it. For any real system, either or both of these processes may occur.

ACKNOWLEDGEMENTS

The authors are grateful to Abraham Nitzan, Ari Aviram, and G. Ludwig Hofacker for interesting and instructive discussion. The NSF is acknowledged for partial support of this work. M.R. is grateful to the donors of the Petroleum Research Fund administered by the American Chemical Society, for partial support of this work.

REFERENCES

1. W. Latimer and W.H. Rodebush, J. Am. Chem. Soc. 42, 1419 (1920).
2. M.L. Huggins, unpublished (cited by G.N. Lewis in "Valence and the Structure of Atoms and Molecules", Chem. Catalog Co., New York, 1923, p. 109).
3. P.O. Löwdin, Adv. Quantum Chem. 2, 213 (1965); Mutation Res. 2, 218 (1965); Pont. Acad. Scient. 31, 1 (1967); Ann. N.Y. Acad. Scient. 31, 1 (1967); Ann. N.Y. Acad. Sci. 158, 86 (1969).
4. cf. eg. P.M. Rentzepis, R.P. Jones and J. Jortner, J. Chem. Phys. 59, 766 (1973); W.S. Struve, P.M. Renzepis and J. Jortner, ibid, 59, 5014 (1973).
5. J.C. Slater, J. Chem. Phys. 9, 16 (1941).
6. P.G. deGennes, Solid St. Comm. 1, 132 (1963); Y. Takagi, J. Phys. Soc. (Japan) 3, 271 (1948); R. Blinc and M. Ribaric, Phys. Rev. 130, 1816 (1963).

7. K.K. Kobayashi, J. Phys. Soc. (Japan) 24, 497 (1968); R. Blinc and B. Zeks, Adv. Phys. 21, 693 (1972); K. Godzik and A. Blumen, Phys. Stat. Sol. B 66, 569 (1974).

8. B.I. Stepanov, Zh. Fiz. Khim. 19, 507 (1945); Y. Marechal and A. Witkowski, J. Chem. Phys. 48, 3697 (1968); S.F. Fischer, G.L. Hofacker and M.A. Ratner, J. Chem. Phys. 52, 1932 (1969); for reviews see M.A. Ratner and J.R. Sabin, in "Wave Mechanics, the First Fifty Years", eds. W.C. Price et. al., Butterworth, London, 1973; Y. Marechal, G.L. Hofacker, and M.A. Ratner in "Hydrogen Bonding", eds. P. Schuster et. al., North Holland, Amsterdam, in press.

9. L. Onsager and M. Dupuis, in "Electrolytes", ed. B. Pesce, Pergamon Press, New York, 1962; P. Gosar, Nuovo Cim. 30, 931 (1963); S.F. Fischer and G.L. Hofacker in "Physics of Ice", eds. N. Riehl et.al., Plenum, New York, 1969; P. Gosar, ibid; S.F. Fischer, G.L. Hofacker, and J.R. Sabin, Phys. kondens. Mat. 8, 268 (1969).

10. "Polarons and Excitons", eds. C.G. Kuper and G. Whitfield, Oliver and Boyd, Edinburgh, 1963.

11. cf. eg. R. Rein and F.E. Harris, J. Chem. Phys. 41, 3393 (1964); 42, 2177 (1965).

12. J. Brickmann and H. Zimmerman, J. Chem. Phys. 50 1608 (1969).

13. M.D. Harmony, Chem. Soc. Revs. 2, 211 (1973). It must be remembered that the considerations of Harmony and of Brickmann /12/ apply only to rigourously one-dimensional potentials. For real hydrogenbonded systems, the coupling results in a finite width for the localized double well states, and the tremendous reduction of tunneling rate for small asymmetrics predicted by (5) does not occur. Indeed, the "downhill" tunneling in these systems should be faster than the symmetric tunnelings essentially because the argument of the negative exponential in the JWKB expression is reduced by the asymmetry (A. Aviram, P.E. Seiden and M.A. Ratner, to be published).

14. N. Sheppard, in "Hydrogen Bonding", ed. D. Hadzi, Pergemon, Oxford, 1959.

15. N. Rösch, Thesis, T.U., München, 1971; Chem. Phys. 1, 220 (1973); N. Rösch and M. Ratner, J. Chem. Phys. 61, 3344 (1974).

16. M.D. Newton and S. Ehrenson, J. Amer. Chem. Soc. 93, 4971 (1971); R. Janoschek et.al. ibid 94, 2387 (1972).

17. A. Nitzan and R.J. Silbey, J. Chem. Phys. 60, 4070 (1974).

18. G. Sewell, in "Polarons and Excitations", eds. C.G. Kuper and G. Whitfield,Oliver and Boyd, Edinburgh, 1963.

19. T. Holstein, Ann. Phys. 8, 325, 389 (1959).

20. S. Glasstone,K.J. Laidler and H. Eyring, "Theory of Rate Processes", McGraw-Hill, New York, 1941.

21. R.G. Carbonell and M.D. Kostin, J. Chem. Phys. 60, 2047 (1974).

SUBJECT INDEX

Alternant hydrocarbon 48

Antibonding 34, 427

Antisymmetry 35

Aromatic bond 37

Bent bonds 45

Biorthonormal 77

Bond density 446

Bond energies 41

Branching diagram 357

Bracketing function 9, 315, 333, 381

Bracketing theorem 9

Cauchy relations 3

Cellular method 180

Charge density 219

Chemical shift 99

Chemisorption 36

Cohesion 3

- of metals 202

Cohesive energy 4, 123, 171

Cohesive properties of solids 4, 5

Complex hybrid 423

Compressed atoms 179

Compressibilities 179

Compton profile 141, 150, 419

Configuration interaction 7, 70, 106

Correlation 466

- dynamical 207
- effects 5, 461
- energy 4, 8, 43
- factor 434
- "in-out", "left-right" 492
- problem 4, 8, 201, 393, 433, 479

Crystal field theory 37
Coulson integral 96
Covalent 34, 35, 43, 44, 46
Defects 172
Density difference function 446
functional scheme 205, 206
matrix 5, 7, 38, 202, 459, 461
operator 97
Diamond crystal 216, 219
Dipole moments 37, 50
Directed valency 35
Dispersion interaction 465
Dispersion term 546
Dissociation energy 41
DNA 10
Double coset decomposition 128
Double minimum 495
Electron capture 446
Electronegativity 34, 44, 48
Electronic specific heat 190
Electron momentum distribution 141
Exchange
correlation 202, 206, 216
hole 204
indirect 124, 128, 132, 137
integrals 109
interactions 123, 124
parameter 105, 109, 120
phenomena 5
Exclusion principle 35
Fermi hole 203
Field gradients 173
Fock space 94

Formal charges 46
Franck-Condon 34
Free electron theory 48
Free radicals 37
Frontier electron 66
g tensor 99
Gaussian functions 40
Generator coordinate method 480
Genetic code 549
Golden rule 381, 389, 390
Greenian operator 38
Green's function 280
Hartree-Fock
    extended 8
    numerical 5
    projected 8, 491, 499, 500
    restricted 8
Hellman-Feynman 428
High pressure properties 3
Hybridization 39
Hydrogen
    atom 249
    bond 51, 528, 540, 561, 567, 577
    molecule 37, 491
Impulse approximation 141, 419
Impurity 174
Intermediate hamiltonian 321
Ionic character 41
    crystals 2, 3, 4, 43, 141, 143, 143, 171,201
    radii 34, 46
Isomeric transitions 446
Jost solution 296
Laguerre polynomials 305

Lattice dynamics 3
Ligand field theory 48, 43
Linked-cluster expansion 394, 395
Localizability 449
Localization 67
Loge theory 447
Lower bound 250, 315
Macromolecule 539
Magnetic ordering 123
Many body problem 7
Maximum localization criterion 69
Missing information 447
Molecular collisions 367
Molecular integrals 3
Moment expansion 94
Momentum space 417
Multiconfiguration 8, 506
Mutations 561, 567
Nonadiabatic 473
Non-bonding 34
Non-orthogonality 3, 5, 26, 28, 33, 36, 39, 42, 71, 135
N-representability 7
Nucleic acids 517, 541
Nucleotide 541
Occupation number 38, 40, 45
Operator inequality 313, 322
Orbital
    alternant molecular 5, 481
    atomic 3
    bond 34
    Brueckner spin 460
    energy 64
    equivalent 48

localized molecular 505
modified atomic 64
natural 7, 26, 27, 40, 510
natural spin 459, 462, 496
population 96
reaction 507
scanning molecular 411
Slater 40
tetrahedral 44, 49
Orthogonalization 5, 77
symmetric 38, 39, 42, 66, 72
Overlap 3, 33, 67, 69, 125, 153, 467
differential 5, 35
integrals 410
matrix 43, 64, 66
zero differential 70
Pade approximant 9
Pariser-Parr-Pople 5
Partitioning 9, 307, 393
Perturbation
localized 327
theory 9, 28
Phlogiston theory 49
Permutation symmetry 123
Phosphate 517
Phospholipids 517
Poisson's equation 218
Polarizabilities 50
Polarization 43
energy 468
Polypeptide 540
Populations 71
Pressure-density curves 187

Projection
inner 9, 321, 323, 349
operator 4, 8, 27, 48, 49, 357, 479, 492
Propagator 9, 93
Proton
tunneling 10, 549, 567, 577, 578
transfer 563
Pseudoquadrupole 409
Quadruple bond 42
Quadrupole coupling 99
Quantization principle 277, 279
Quantum Chemistry Group 1, 2, 3, 6
Quantum electrodynamics 2
Quantum Theory Project 2, 3, 6
Rare-earth elements 181
Rate constant 580
Reaction 46, 505, 506
channel 371
operator 397, 395
Reactivity 46, 48, 367, 522
Reduced-density operator 450
Relaxation 429
shift 207
Repulsive forces 37
Resolvent 333
reduced 9
Resonances 296
Rotational barriers 137
Scaling 250
Scattering theory 295
s-d mixing model 125
Second quantization 37
Self-energy 208